Chemistry of the Unpolluted and Polluted Troposphere

NATO ADVANCED STUDY INSTITUTES SERIES

Proceedings of the Advanced Study Institute Programme, which aims at the dissemination of advanced knowledge and the formation of contacts among scientists from different countries

The series is published by an international board of publishers in conjunction with NATO Scientific Affairs Division

A	Life Sciences	Plenum Publishing Corporation
B	Physics	London and New York
C	Mathematical and Physical Sciences	D. Reidel Publishing Company Dordrecht, Boston and London
D	Behavioural and Social Sciences	Martinus Nijhoff Publishers
E	Engineering and Materials Sciences	The Hague, London and Boston
F	Computer and Systems Sciences	Springer Verlag
G	Ecological Sciences	Heidelberg

Series C – Mathematical and Physical Sciences

Volume 96 – Chemistry of the Unpolluted and Polluted Troposphere

Chemistry of the Unpolluted and Polluted Troposphere

Proceedings of the NATO Advanced Study Institute held on the Island of Corfu, Greece, September 28 - October 10, 1981

edited by

H. W. GEORGII

Dept. of Meteorology and Geophysics, University of Frankfurt, F.R.G.

and

W. JAESCHKE

Center of Environmental Protection, University of Frankfurt, F.R.G.

D. Reidel Publishing Company

Dordrecht : Holland / Boston : U.S.A. / London : England

Published in cooperation with NATO Scientific Affairs Division

Library of Congress Cataloging in Publication Data

CIP

NATO Advanced Study Institute (1981: Corfu, Greece)

Chemistry of the unpolluted and polluted troposphere.

(NATO advanced study institutes series. Series C, Mathematical and physical sciences; v. 96)
Includes indexes.
1. Troposphere–Congresses. 2. Atmospheric chemistry–Congresses.
3. Air–Pollution–Congresses. I. Georgii, H. W. II. Jaeschke, W.
(Wolfgang) III. Title. IV. Series.
QC881.2.T75N37 1981 551.5'13 82-13339
ISBN 90-277-1487-8

Published by D. Reidel Publishing Company
P.O. Box 17, 3300 AA Dordrecht, Holland

Sold and distributed in the U.S.A. and Canada
by Kluwer Boston Inc.,
190 Old Derby Street, Hingham, MA 02043, U.S.A.

In all other countries, sold and distributed
by Kluwer Academic Publishers Group,
P.O. Box 322, 3300 AH Dordrecht, Holland

D. Reidel Publishing Company is a member of the Kluwer Group

Printed in The Netherlands

TABLE OF CONTENTS

PREFACE

The content of this book are lectures and research papers presented at the NATO Advanced Study Institute on "Chemistry of the Unpolluted and Polluted Troposphere" held from September 28 to October 10, 1981, on the Island of Corfu, Greece.

The realization of the scientific event was made possible by the sponsorship of the NATO Scientific and Environmental Affairs Division. We must express our gratitude first to this Institution for the important assistance and cooperation we received. For additional assistance we are much obliged to our distinguished co sponsors: Carl Zeiss Werk, Oberkochen, FRG; Biotronik GmbH, Frankfurt, FRG; Bodenseewerk Perkin Elmer u. Co GmbH, Ueberlingen, FRG; and TSI Deutschland Inc., Aachen, FRG.

To Mr. Dimitris Bouas, the Director of the Grand Hotel Glyfada, we would like to express our thanks and appreciation for the excellent services.

The Center of Environmental Protection of the University of Frankfurt, FRG, conducted the NATO Advanced Study Institute because of the rapidly increased interest in the problems of air chemistry and air pollution. This development has contributed in recent years to a significant expansion of knowledge in the field of atmospheric chemistry, where several of the classic disciplines like meteorology, physics and chemistry have joined in combined research. The Advanced Study Institute offered an ideal opportunity to give a comprehensive survey of the present knowledge in this relatively new field of atmospheric science.

Regarding the development of atmospheric chemistry, praise must go first and foremost to the meteorologists. They were the first to pay interest to the chemical composition of the troposphere: that layer of the atmosphere in which weather activity occurs. Based on the assumption that weather is determined by various circulatory systems, meteorologists attempting quantitative descriptions of atmospheric chemistry consequently introduced the observation of circulatory systems. Regarding geochemical atmospheric cycles, sources and sinks of trace compounds came into consideration, which influence the content of trace compounds in the atmosphere from the Earth's surface.In order to quantify the contribution of the various biological and geological sources and sinks, meteorologists soon became dependent upon the help of geologists and biologists. Physicists and chemists became involved with the development of sophisticated measuring techniques and with the investigation of chemical conversion processes. The experimental data concerning concentrations of trace compounds, kinetics of chemical reactions, and diffusion processes in the atmosphere, are then used by theoreticians for model calculations and global budget estimations. The scientific aspects briefly mentioned demonstrates the

H. W. Georgii and W. Jaeschke (eds.), Chemistry of the Unpolluted and Polluted Troposphere, vii–viii.

interdisciplinary character of atmospheric chemistry. Therefore, an intensive exchange of knowledge and information should take place between theoreticians and experimentalists working on the various aspects of atmospheric research.

The Advanced Study Institute was offered as a platform for communication and for the exchange of scientific information about the chemistry of the troposphere. According to this intention the proceedings which appear in this book are presented in such a way that a logical and coherent review is developed. This should represent the interaction of the various disciplines involved in the area of atmospheric ecology. The introductory part presents problems and methods of measuring trace gases and aerosols. Then the thermodynamic structure of the atmosphere, its influence on transport and distribution of trace compounds and the interactions between trace compounds and climate are treated. The next topic deals with atmospheric cycles of some of the more important trace elements and compounds thereof, in the order of their occurrence in the periodic system.

In this part, the chapters dealing with the cycle of carbon monoxide and with the cycle of halogen compounds are missing. We regret that we were not able to obtain the manuscripts from the respective lecturers.

The subsequent part deals with the fact that the troposphere is not a homogeneous gas phase, but much rather a region in which a great number of heterogeneous chemical and physical processes do occur in cloud and precipitation droplets or at the surface of particles. The knowledge and understanding of the naturally occurring cycles of trace compounds in homogeneous and heterogeneous systems of the troposphere is the basis for discerning man made changes, and for determining properly the extent to which these changes are harmful. Consequently, the final topic explicitly deals with pollution problems such as acid rain, air pollution monitoring and data handling.

The final topic shows that air chemistry is not only of scientific concern but also offers a solid basis for political decisions concerning air pollution and environmental protection. Therefore, in a final discussion of the Advanced Study Institute, all participants agreed, that 'air chemistry' should grow into the dimension of an independent discipline. As a symbol of this intention the "Tree of Air Chemistry" was planted by the participants of the ASI. It should symbolise that there was an organic interaction of the various root disciplines which finally formed a well shaped plant.

The organizers wish to thank all contributors for preparing their lectures and manuscripts as one of the branches of the tree. We do hope that the readers of this book will also gain the impression that air chemistry has become a coherent 'metabolism' and is no longer an inert gathering of independent knowledge.

Frankfurt am Main, May 1982 H. W. Georgii and W. Jaeschke

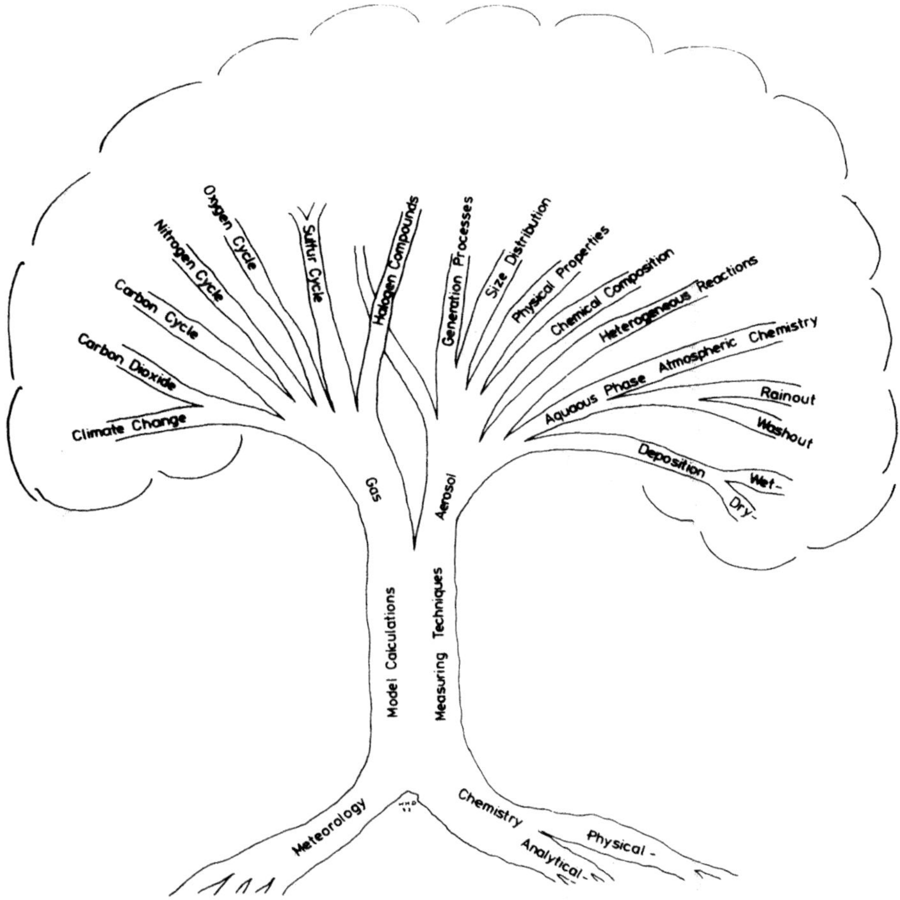
Oxygen Cycle
Nitrogen Cycle
Sulfur Cycle
Halogen Compounds
Generation Processes
Size Distribution
Physical Properties
Chemical Composition
Heterogeneous Reactions
Carbon Cycle
Carbon Dioxide
Climate Change
Aquaous Phase Atmospheric Chemistry
Rainout
Washout
Deposition
Wet-
Dry-
Gas
Aerosol
Model Calculations
Measuring Techniques
Meteorology
Chemistry
Physical -
Analytical -

PART I

MEASURING TECHNIQUES

PROBLEMS AND METHODS OF TRACE GAS MEASUREMENTS IN THE TROPOSPHERE

Wolfgang Jaeschke

Center of Environmental Protection
University of Frankfurt/Main
F.R.G.

ABSTRACT

The large number of measuring techniques for tropospheric trace gases may be divided into two categories: Continuously working methods where atmospheric air is led through an observation chamber and traces of the gas under investigation are detected by a physical effect which is proportional to the concentration; and Discontinuously working trapping methods, including a sampling procedure which traps and enriches the desired trace gas with a chemical reaction, when atmospheric air is led through an absorption medium. Afterwards the amount of collected substance is determined by a conventional analytical method. Advantages and disadvantages of both categories are discussed in reference to examples of their application.

I INTRODUCTION.

The analysis of trace components in the atmosphere is a highly sophisticated field of analytical chemistry. The complexity of the problem is characterized by extremely low mixing ratios of the trace components, requiring sensitive analytical methods and effective and reliable sampling techniques. Only few trace gases can be monitored continously. Most all of the trace gases have to be accumulated for analysis. Therefore the sampling techniques are of major importance. For the characterization a sufficient amount of a trace compound has to be collected from an abundance of diluting air. The composition of the diluting air is well known (see table I).

H. W. Georgii and W. Jaeschke (eds.), Chemistry of the Unpolluted and Polluted Troposphere, 3–38.

Table I Chemical composition of dry air [1]

Gas	Mixing ratio [%] (per volume)
N_2	78.09
O_2	20.94
Ar	0.93
CO_2	0.0325
Ne	0.0018
He	0.00052
CH_4	0.00015
Kr	0.00013

	99.99500
all other traces	0.005

	100.00000 [%]

It stays uniform to an altitude of about 80 km. Altogether 99.995% of the total volume of the air may be considered as a diluent for the trace compounds, wich are amounting to only as little as 0.005% (equivalent to 50 ppm). This relation offers a preliminary impression of the miniscule fraction of trace components in the atmosphere. The terrestrial atmosphere has a total air mass of 5.3×10^{15} t. Its density is decreasing exponentially with altitude. At the reference temperature of 0° C, the density of air at ground level has a value of 1.293 kg/m^3, whereas at an altitude of 5000 m the value has dropped to 0.73 kg/m^3. The density at very high altitudes is gradually approaching zero, indicating that the atmosphere has vanished and does not form a defined border towards space. Due to changes of the temperature gradient, the atmosphere has a layered structure. This book is concerned with the lowest layer, which is called ′troposphere′ and mainly dominates our environment. With an average magnitude of 11 km the troposphere embodies 8/10 of the total atmospheric air mass, i.e. 4×10^{15} t [2], including the fraction of the trace components. This fraction, with the relative amount of 0.005 % (see table I) is accounting for a total mass of 2.5×10^{11} t on a global scale. This is the result of a previous estimation, if we are assuming a mean density value of 1kg/m^3 for the air in the troposphere. Considering one particular trace gas, out of several hundreds, its total mass is in the order of 10^8 t. Since the geometrical volume of the troposphere is 5×10^{18} m^3, the mean concentration of a particular trace gas is in the order of 10^{-11} t/m^3 equivalent to 10 ug/m^3. This will demonstrate, that an analytical chemist has to cope with either mixing ratios in the part per million range or, expressed in mass per volume, with concentrations in the order of ug/m^3. Since the volume is dependent on temperature and pressure, one has to keep in mind that the concentrations of the analyzed sample will depend on these two variables. Thus the temperature and the pressure have to be determined accurately to get reliable concentration values by finally converting the data to conditions of Standard Temperature and Standard Pressure (STP), namely at 0° C and at 760 Torr. In table II a compilation of simple conversion formula between mixing ratios and concentrations are given.

Table II:

$[ug/m^3] = [ppm] \frac{M}{Vm} 10^{3}$	$[ppm] = [ug/m^3] \frac{Vm}{M} 10^{-3}$
$[Molec/ccm] = [ug/m^3] \frac{A}{M} 10^{-12}$	$[ug/m^3] = [Molec/ccm] \frac{M}{A} 10^{12}$
$[Molec/ccm] = [ppm] \frac{A}{Vm} 10^{-9}$	$[ppm] = [Molec/ccm] \frac{Vm}{A} 10^{9}$
A = Avogadro´s Number	$6.02\ 10^{23}$
Vm = Molecular Volume of Gases	22.4 [l] (at 0°C, 1 atm)
M = Molecular Weight	[g]

From this short consideration of the dimensions of the global atmosphere the analytical chemist will learn, that once the concentration of the trace compounds are determined, from air samples of only several cubicmeters the data and all experimental errors are likely to be blown up by a factor of 10^{18}, as soon as global dimensions are used in model calculations and budget estimations.
As there is no practical way to a complete and total analysis of air samples, the endevour to determine the concentrations of trace gases has to start with a consideration of the possible composition and the chemical properties of one particular sample. This can be achieved since the trace gases mainly originate from the elements C, N, O, S, and the halogens. All other elements and their compounds mainly occur in the atmospheric aerosol. The analysis of aerosols is discussed in the following paper.
In this paper a comprehensive survey of measuring techniques for atmospheric trace gases is compiled. The trace gases are presented in the sequence of the elements in the periodic system. Naturally a comprehensive survey cannot include all methods which are applicable for the analysis of trace gases. However, the author will present principles of the art, quoting methods and examples of his own research. Where other methods are quoted in this survey, references are made to specialized reports.

II CONTINUOUSLY WORKING METHODS.

The principle of a standard experimental setup for the continuous

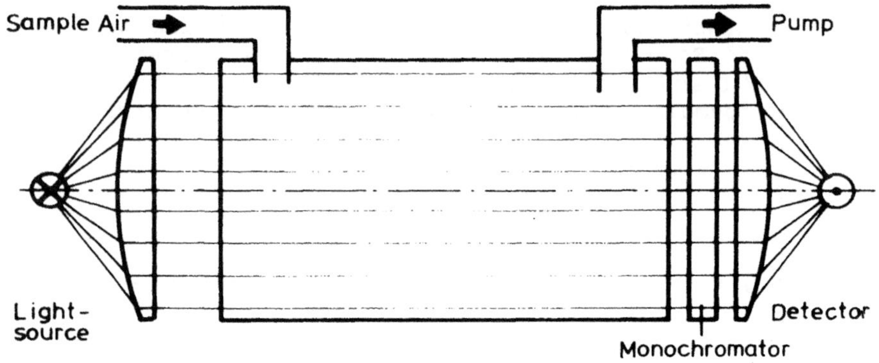

Fig. 1 Generalized flow cell for the continuous monitoring of atmospheric trace gases by the aid of light absorption.

monitoring of atmospheric trace gases is shown in figure 1. Atmospheric air, entering a flow cell continuously, should cause a physical effect which must be specific for a certain trace gas. The quantity should be determined in the volume of the cell without altering the sample. An approach for such a standard setup is the placement of a flow cell, as an optical cuvette, into the lightpath of a spectrometer. When radiation passes through a flow cell which contains an atmospheric mixture of trace gases, radiaton is absorbed at certain wavelengths. The type of compound trapped in the cell can be determined by the wavelength of absorption, and the quantity of the compound can be calculated from the amount of radiation absorbed, using the Lambert-Beer law.

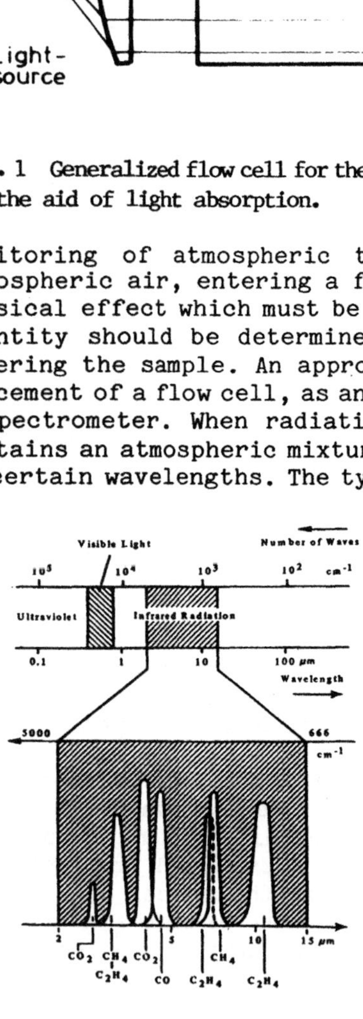

Fig. 2 Schematic of electromagnetic radiation spectrum with absorption bands for CH_4, C_2H_4, CO and CO_2 in the infrared range.

Especially carbon containing trace gases are detected with infrared light of a wavelength between 2 um and 15 um. Figure 2 schematically shows a number of compounds which, at usual urban atmospheric concentrations, exhibit an infrared absorption sufficiently intensive to be detected with reasonable path length. In the case of CO_2 the infrared absorption can even be applied for measuring the trend of background concentrations.

An instrument designed for this purpose is shown schematically in figure 3. It was developed by

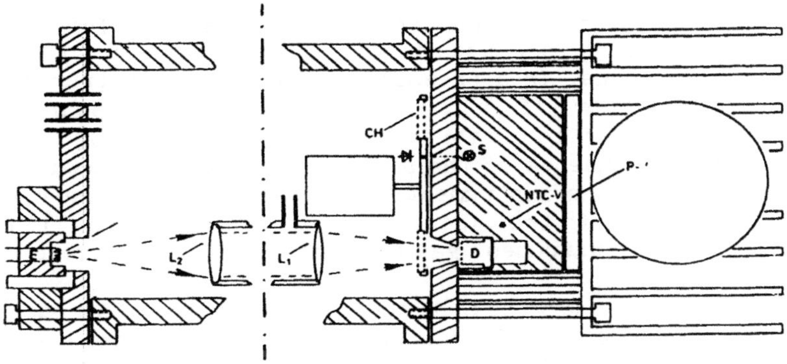

Fig. 3 Infrared absorption instrument for the detection of CO_2.
E: IR Lightsource, L: Lenses, CH: Chopper, D: Detector, NTC and P: Cooling elements.

Schmitt [3] at the Inst. of Meteorology of the University of Frankfurt in 1976. The infrared absorption of CO_2 at a wavelength of 4.3 um is suitable for detection. IR light of this wavelength is emitted by a resistance wire which is heated to 1000 K. The flow cuvette is a stainless steel tube which is allowing a pathlength of 135 mm. To both ends of the tube LiF_2 lenses are fitted as optical windows. The light source and the detector are placed in the focus of the lenses. Between detector and flow cell an optical interference filter, with a transmission at 4.2 um, is mounted. The detector is a semiconducting HgCdTe foto-resistor which is kept at - 32°C with the aid of a Peltier cooling element. In order to achieve a better signal to noise ratio, the method applies a split beam detection. Together with an interference filter, with a transmission at 4.2 um, a second filter, with a transmission at 3.6 um, is mounted in a chopper wheel. By rotation of the chopper the detector is alternately irradiated with light of the absorption wavelength of CO_2 (4.2 um) and with light in the 'window' range (3.6 um). The comparison of both signals allows the elimination of errors due to the drift of the irradiation intensity and of the detector's sensitivity. The instrument was designed for continuously moni-

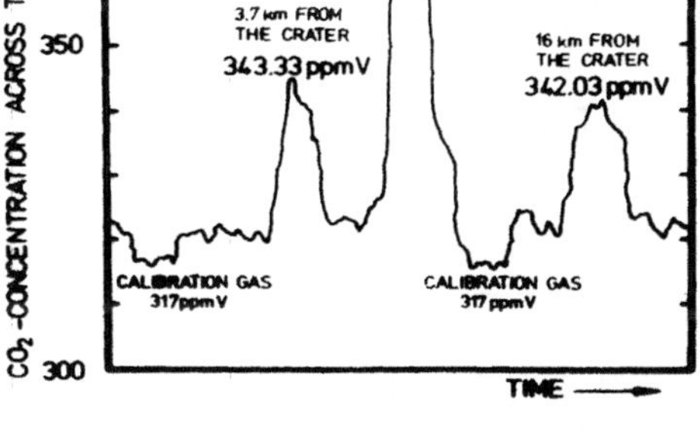

Fig. 4 Chart of the CO_2 instrument, recorded during a measuring flight. Three signals caused by calibration gases and two signals caused by the plume of Mt. Etna vulcano.

toring CO_2 concentrations during aircaft ascents.This kind of detector was chosen because it is insensitive to mechanical vibrations in a plane. In a pressurized cabin the instrument has a resolution of 0.02 ppm. If it is installed in a non- pressurized air-craft, it is necessary to calibrate the instrument during the flight because of the dependence of the signal upon the air pressure. As an example a chart of the instrument is presented in figure 4. It was recorded on a flight in the vicinity of Mt. Etna volcano [4]. For the calibration of the instrument gas mixtures of CO_2 in N_2 were used as standards.

The exploration of the global trend of atmospheric CO_2 background levels calls for a comparable calibration gas. It became available by the distribution of manometrically calibrated reference gas mixtures which had been used at Scripps Institute of Oceanography since 1959 by Keeling et al [5]. The complications in CO_2 data comparison will be discussed later.

Other instruments for the continuous monitoring of infrared absorbing compounds are commercially available. The detection is based on the principle of the non-dispersive infrared method (NDIR) [6][7]. The key of the NDIR method (Fig.5),is the detector. It is a cell filled with the gas in question, for instance CO. Two light sources emit infrared light which first traverses two filter cells, where radiation of unwanted wavelengths is removed by the IR absorption of filter gases. Afterwards the light is passing through the sample and through the reference cell, respectively. While a continuous stream of atmospheric air passes through the sample cell, the reference cell generally should be filled with a static volume of air containing all components except the gas in question. In reality this mixture is not available and purified dry nitrogen is used. In the end the light beam reaches the sensor cell which in our case is filled with CO. This sensor cell is divided into two chambers by an electrically conducting membrane. In the absence of any traces of CO

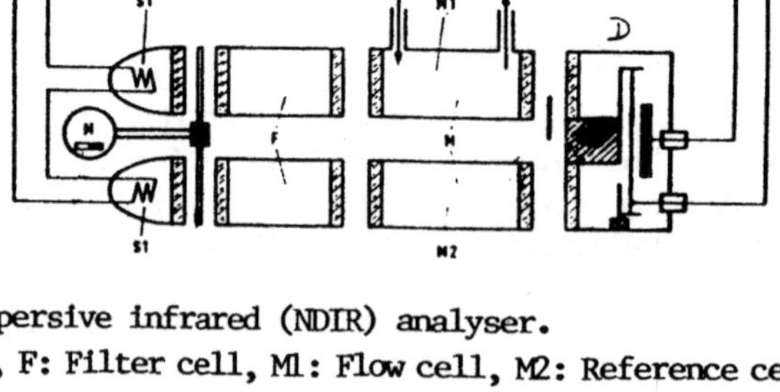

Fig. 5 Non-dispersive infrared (NDIR) analyser.
S1: Lightsource, F: Filter cell, M1: Flow cell, M2: Reference cell, D: Detector cell

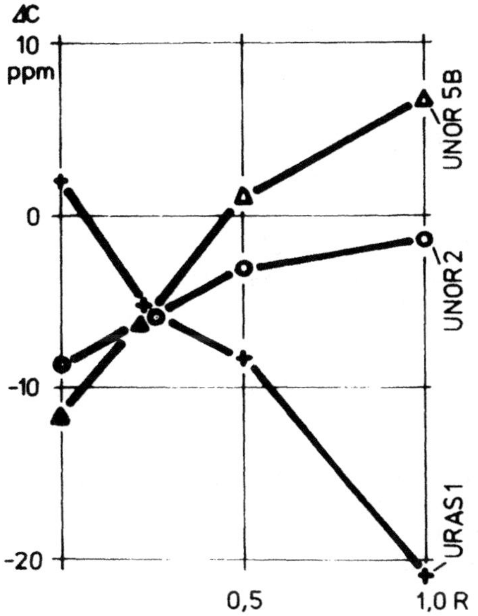

Fig. 6 The carrier gas effect shown for the different analyzers.
R is the ratio O_2/N_2+O_2.

in the sample cell, both chambers will be irradiated with the same intensity of IR light, and due to the equivalent irradiation the pressure in both chambers will be the same. In this case the membrane will not be displaced. If CO appears in the sample cell the upper light beam will be weakened and a displacement of the membrane will occur towards the cooler chamber of the detector cell. This movement of the membrane is sensed by a capacitance detector. In order to convert the movement into an electrical output, both light beams are alternately switched on and off by a chopper, with a typical frequency of 10 Hz. With the sample cell blank, the detector is exposed to the chopper frequency and both chambers of the detector are heated evenly. This exposure is resulting in a dark current. The presence of CO in the sample cell will bias the formerly even vibration of the membrane and will cause a signal current.

The CO measurements have an uncertainty of about 1 ppm. Some CO instruments without filter cells can have errors up to ± 5 ppm due to the interference of other IR absorbing gases, especially in the presence of water. When the same method is applied for CO_2, the precision is accurate to about ± 0.1 ppm. The accuracy depends of course on the kind of standard gases used for calibration. In the case of CO_2 Bischof [8] pointed out that an error occurs when standard gas compositions are used, which have different carrier gas (N_2) in comparison to the analyzed medium (air). Figure 6 shows that nearly identical results were obtained from three commercial analyzers when CO_2/air mixtures were used for calibration. CO_2/N_2 mixtures however yielded in an error of -11, -8 and +3 ppm re- spectively [9].

Determining background concentrations for CO, the NDIR method is not sensitive enough, since in remote regions the CO level is often below 0.1 ppm. A continuously working method which is applicable for studying global background levels of atmospheric CO, is the hot mercuric oxide method, first mentioned by Robbins et al, 1968 [10]. It was further developed and intensively applied by Seiler [11]. The key of this method is the reaction of CO with mercuric oxide

$$HgO + CO \xrightarrow{210^{o}C} Hg + CO_2 \qquad (1)$$

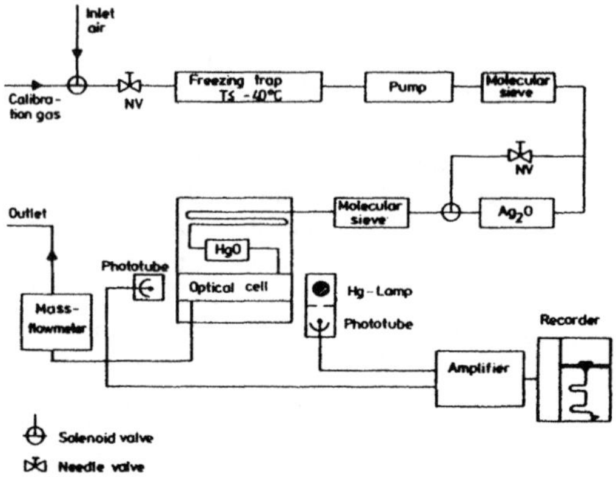

Fig. 7 Scheme of the indirect CO detection by the reduction of HgO and flame less atomic absorption of mercury vapor.

The mercury vapour, generated during this reaction, is determined quantitatively by flameless atomic absorption at 2537 A. Figure 7 shows a scheme of such a CO instrument. With a flow rate of 30 l/h ambient air is passed through a freezing trap (-40°C) and a column filled with molecular sieve for the removal of H_2O and other gaseous constituents such as SO_2 and higher hydrocarbons which interfere with the CO measuring effect. Since hydrogen, which also reacts with mercuric oxide, is not removed during this purification, the effect of this atmospheric trace gas has to be taken into consideration. It can be determined by passing the air for short periods of time through a bypass tube, filled with silver oxide. The bypass tube quantitatively removes CO but does not effect the H_2 content in the atmospheric air. The resulting signal is the zero value of the CO instrument. In a modification of the instrument, the H_2-effect is used to determine H_2 mixing ratios in the atmosphere [12]. The CO

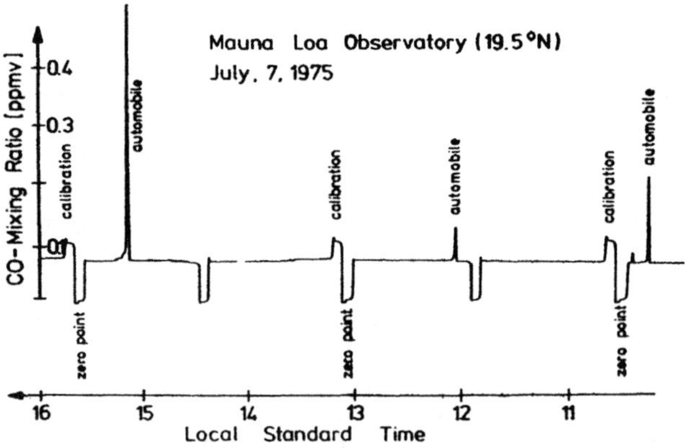

Fig. 8 Section of the continuous CO registration with the 'mercury' method.

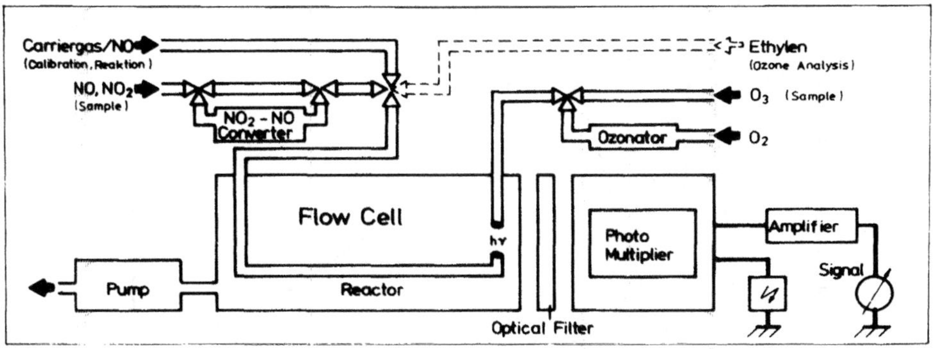

Fig. 9 Schematical gas flow configuration of multifunctional NO, NO_2, O_3 analyzer with chemiluminescence reactions.

instrument provides a continuous re-gistration of CO (figure 8). The lower detection limit is 0.1 ppb. The standard deviation is ±3 %, at typical background CO values of 50 - 100 ppb [13].
Apart from certain organic compounds, the absorption of other trace gases in an irradiated flow cell is too weak to be useful for analytical purposes . Therefore other techniques must be applied and the flow cell has to be modified.
A technique which has been developed in recent years, is the measurement of the chemiluminescence which occurs when certain trace gases undergo chemical reactions in a flow cell. The wave length, of the light emitted depends on the nature of the substances involved. Since the intensity is dependent on the concentration of the compounds, the chemiluminescence effect is the basis of a specific and highly sensitive detector, commonly used for monitoring oxides of nitrogen or ozone with a photomultiplier tube (figure 9).
Atmospheric NO concentrations are monitored by measuring the light intensity of wave lengths greater than 600 nm (figure 10) resulting from the chemiluminescence reaction (2) [14][15].

$$NO + O_3 \rightarrow NO_2^* + O_2 \qquad (2)$$

$$NO_2^* \rightarrow NO_2 + h\nu \qquad (2A)$$

Nitric oxide commonly exists in ambient air together with NO_2. The latter is quantitatively reduced to NO by means of a thermal converter. Since NO passes through the converter unchanged, the resulting signal is due to the total NO_x concentration equal to NO + NO_2. A fraction of the air taken up by the system is measured in certain intervals without having passed trough the converter. This NO signal is subtracted from the NO_x signal to yield the final NO_2 measurement. The NO and NO+NO_x measurements can be performed by either using a dual system or by working with one system in alternating cycles, with measuring periods of less than 1 minute.

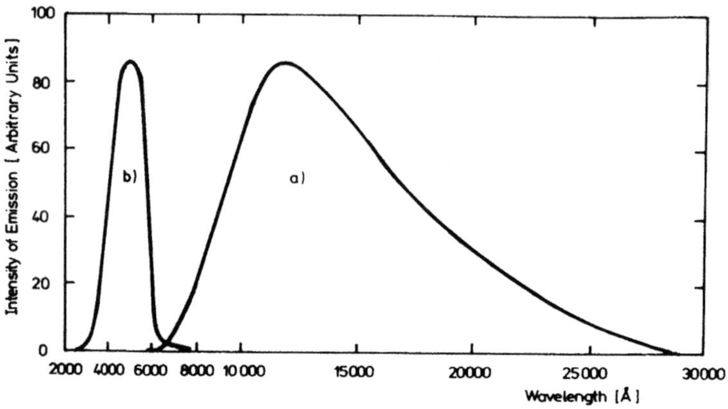

Fig. 10 Spectral distribution of the chemiluminescence emission of: a) the reaction NO + O_3 and b) the reaction O_3 + C_2H_4

For more details on the detection of nitrogen-containing trace gases we refer to the chapter of G. Helas presented in this book. The NO/O_3 reaction can also be used to monitor O_3. In this case the ozonator flow line (figure 9) is replaced by a flow line generating a mixing ratio of 0.5 % NO in N_2 as a carrier gas.
The basis of another detection method for ozone is reaction (3). Unsaturated compounds, mainly ethylene reacting with ozone generate a chemiluminescence in the blue region [16][17].

$$O_3 + C_2H_4 \rightarrow CH_2O^* + PRODUCTS \quad (3)$$

$$CH_2O^* \rightarrow CH_2O + h\nu \quad (3A)$$

The exact spectral distribution of the emitted light is also shown in

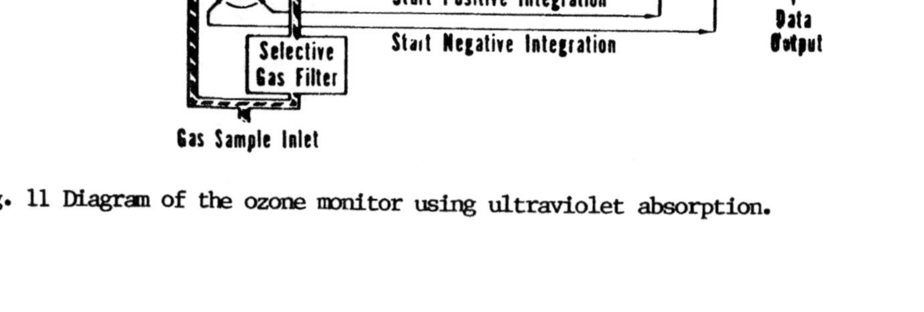

Fig. 11 Diagram of the ozone monitor using ultraviolet absorption.

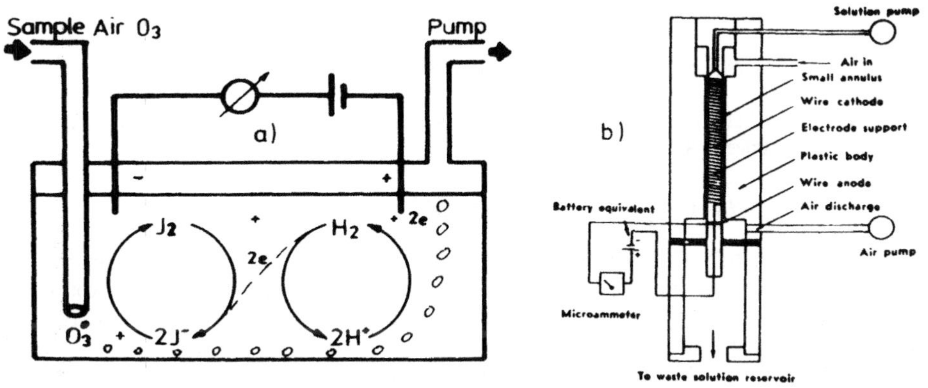

Fig. 12 A flow cell, housing a microcolomp sensor for O_3 monitoring. a) the principle b) how it really works.

figure 10. By means of an optical filter, the interfering chemiluminescence of the NO/O_3 reaction is discriminated. The gas flow system for monitoring ozone with this effect is in principle the same as shown schematically in figure 9. Instead of NO detection, ozone is the gas under investigation. It is detected by its reaction with an excess of ethylene.

Ultraviolet absorption has also been used for monitoring ozone. A flow cell (figure 11) is irradiated with a mercury light source and the absorption of ozone over the narrow band at 250 - 260 nm is monitored. The interference from sulfur dioxide or other UV absorbing compounds is eliminated by cycling the air sample through a selective gas filter, which quantitatively removes ozone, but does not effect the concentration of other UV absorbing gases. Now a certain amount of UV light is absorbed by the treated sample, and the resulting signal is stored as reference in the electronics of the instrument. In another cycle the air enters the cell without passing through the scrubber. The difference between the resulting signal for the absorption, and the one stored as a reference, is proportional to the concentration of ozone.

Finally a coulometric method for the continuous monitoring of atmospheric ozone should be mentioned [18]. It is a wet chemistry method, which applies a flow cell containing a water solution of potassium iodide (figure 12a). The detection of ozone in the air bubbled through the solution is accomplished by the oxidation-reduction cycle of potassium iodide. At a cathode a thin layer of hydrogen gas is produced by a polarization current, during the reduction cycle.

$$2\ e + 2\ H^+ \rightarrow H_2 \qquad (4)$$

The oxidation cycle is driven when free iodine is produced by the reaction with ozone.

$$O_3 + 2\ KJ + H_2O \rightarrow J_2 + O_2 + 2KOH \tag{5}$$

The free iodine in the solution will react immediately with the hydrogen formed at the cathode.

$$H_2 + J_2 \rightarrow 2\ HJ \tag{6}$$

The removal of the hydrogen from the cathode will result in a depolarization current since two electrons flow in the external circuit, reestablishing the hydrogen layer. Hence, with the rate of electrons exchanged , a current will result, which is directly proportional to the mass per unit time of ozone entering the flow cell. For example 1 ppmV of ozone will cause a current of 2 microamperes flowing in the external circuit.
Figure 12b demonstrates how the microcoulomb sensor really works. The solution flows in a thin film over the cathode. The air sample is pumped trough the sensor via a small anulus where it comes into intimate contact with the solution on the support of the electrode. A bias voltage is maintained across the cathode and anode and the output is a microampere current proportional to the oxidant present. The method has the disadvantage that a number of oxidizing agents, including NO_2 will also oxidize iodide-ions causing a positive error. Because of the variety of the methods it is necessary to adopt a measuring system as a reference. Ideal calibration can be carried out with calibration gases generated by an ozonator proposed by Regener et al [19] (figure 13). Oxygen flowing through a quartz-tube is exposed to the irradiation of a small, low pressure mercury lamp. An adjustable sleeve blocks off variable lengths of the lamp. The resulting ozone concentrations are directly proportional to the length of the exposed lamp and to the flow rate of the air. The generator itself is calibrated by titration with potassium iodide.

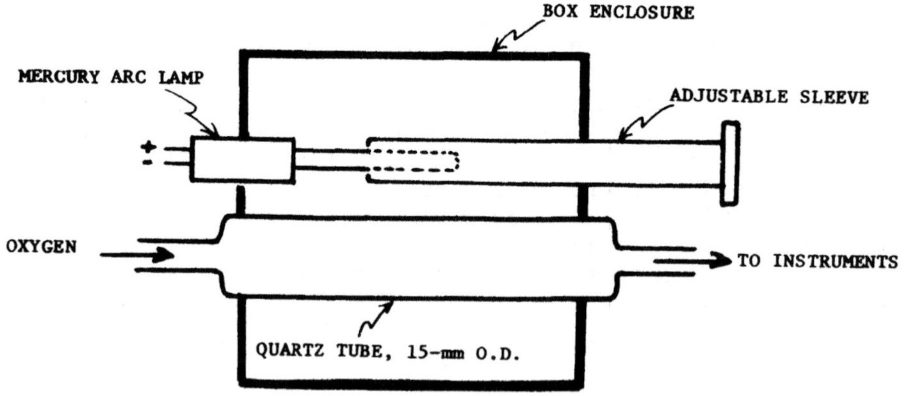

Fig. 13 Ozonator for the generation of O_3-containing calibration gas.

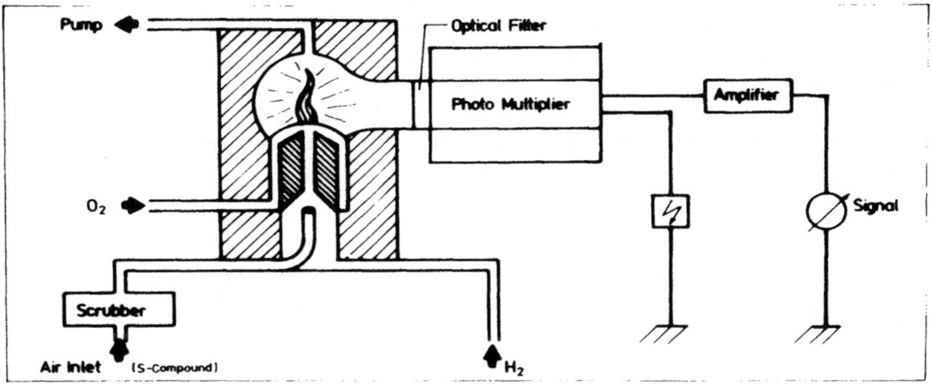

Fig. 14 Scheme of a flame photometric detector (FPD) for SO_2 monitoring.

For the detection of sulfur containing trace gases another modification of the flow cell has been developed [20][21] which is known as flame photometric detector (FPD). Stevens et al [22] were the first who reported the use of an FPD for the direct analysis of total volatile sulfur compounds in the atmosphere. It operates on the measurement of light emitted when sulfur compounds are burned in a reducing hydrogen-oxygen flame. Excited S_2 molecules emit light at a wavelength of 394 nm with an intensity approximately proportional to the square of the concentration of sulfur in the air sample. A scheme of a FPD is shown in figure 14. The air sample is mixed with hydrogen first and afterwards burned as hydrogen-oxygen flame. The intensity of emitted blue light is measured by a photo-multiplier tube. An optical filter is usually included in the optical path to discriminate light of other wavelengths.

The precision and reproducibility of the method depends upon the exact control of the flow rates for the sample and the hydrogen. A change in pressure changes the response of commercial detectors significantly. In figure 15 the dependence of the signal from the pressure is shown. It was measured during calibration of two commercial instruments by means of SO_2 sample concentrations in the ppb range in a pressure chamber,

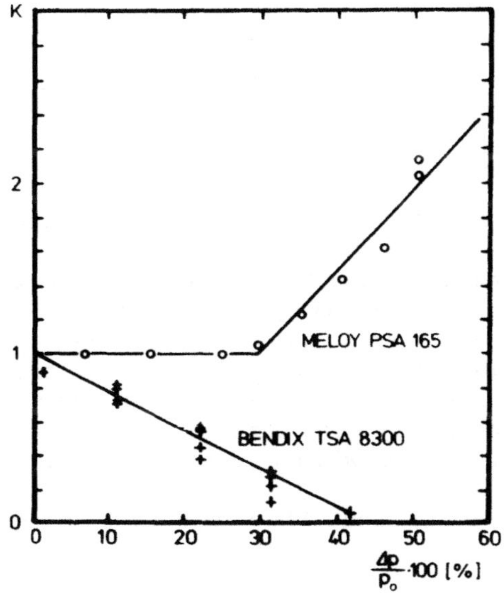

Fig. 15 The pressure dependance of two commercial FPD SO_2 monitors.

The detector is applicable for the direct analysis of total atmospheric sulfur compounds. In order to attain specific SO_2 determination, sulfate and the reduced sulfur gases must be removed with a scrubber.
A compilation of the continuously working methods for trace gase monitoring, which have been mentioned in this paragraph is given in table III.

TABLE III

CONTINUOUSLY WORKING METHODS FOR THE DETECTION OF ATMOSPHERIC TRACE GASES.

TRACE GAS	MEASURING EFFECT	CONCENTRATION RANGE
CO_2	JR ABSORPTION SPECTROSCOPY	RESOLUTION 0.2 PPM
CO	JR ABSORPTION SPECTROSCOPY	>1 PPM
$CO(H_2)$	ATOMIC ABSORPTION SPECTROSC.	>0.1 PPB
$NO(NO_2,NH_3)$	CHEMILUMINESCENCE EMISSION	>0.1 PPB
O_3	CHEMILUMINESCENCE EMISSION	>0.1 PPB
O_3	COULOMETRY	>1 PPB
SO_2	FLAME PHOTOMETRY	>10 PPB

II DISCONTINUOUSLY WORKING METHODS.

In view of the low concentration levels which are to be detected especially in remote areas, continuously working methods are often of insufficient sensitivity. Therefore it is necessary to concentrate the atmospheric trace compounds in the proceeding step of sampling. A suitable sampling train is shown schematically in figure 16. Atmospheric air is led through an absorption medium in order to trap and accumulate the gas under investigation. Besides wet scrubbing by the aid of a so-called ´impinger´, commonly used sampling devices are impregnated filters or containers filled with material for solid adsorption.
When the accumulation techniques are applied, five important topics are to be considered:

1. The precise measurement of the gas flow rate and the air sample volume.

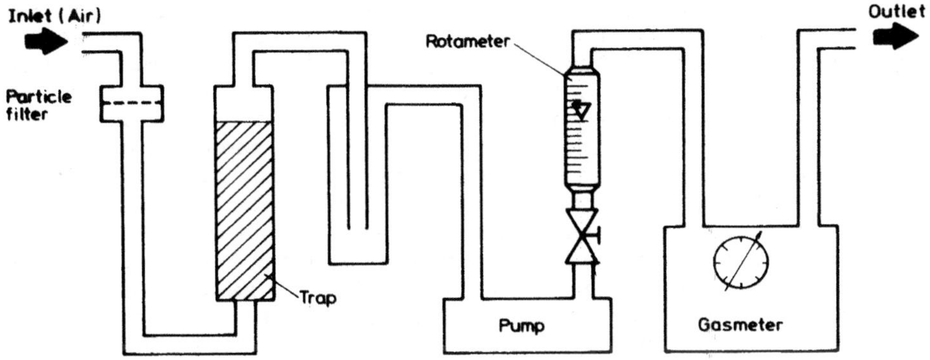

Fig. 16 Generalized sampling train for the accumulation of atmospheric trace gases.

2. The choice of a problem-oriented sampling device.

3. The transfer of the sampled compound to the subsequent analysis without contamination.

4. The calibration of the entire detection system i.e. sampling and analysis.

5. The choice of a sensitive method for analysis.

II_1 Measurement of the Gas Flow Rate and the Air Sample Volume.

A technique without accumulating the compound during the act of sampling is the so-called grab technique. This procedure is suitable to aquire a volume of ambient air rapidly by the aid of a container which is then transferred to the analytical laboratory. There the content is flushed through a preconcentration column and afterwards analyzed for specific compounds, frequently achieved by gas chromatography. The containers are usually made from flexible plastic, steel, and glass. Hydrodermic syringes can be used as well. A way of determining the volume of rigid containers is to weigh them empty and then fill them completely with a liquid of known density. The container is weighed again and the volume is calculated from the weight of liquid and its density. Another technique is to evacuate the container to a very low pressure. The evacuated container can then be attached to a volumetric device and the volume of air drawn in, is measured by a meter. There are several methods used as volumetric devices. Some of them will be discussed in the following part. Because of the expense and mass of large and rigid gas containers, large plastic bags have also been used to collect grab samples. The properties of a variety of materials, from which bags may be constructed were sumarized by Altshuller et al [24].
If preconcentration of a compound is desired to obtain for the

analysis a sufficient amount of the gas in question, large volumes of air must be led through the sampling device. The simplest approach for measuring these large volumes is the placement of a flow meter in the sampling train (figure 16) at the outlet of the pump [25]. The volume of the air sample can be calculated, when the air is flowing at a definite rate for a definite time through the sampling device. To obtain a proper integrated sample volume, the flow rate must be kept constant by the adjustment of a valve. This procedure requires the constant observation of the flow rate and errors may be caused by both, the observer and the instruments. One of the most commonly used flow rate measuring devices is the 'rotameter'. A gas, passing upward through a vertical tube, causes a V-shaped float to rise until its downward force is equal to the force of the gas flow. The float is rising to a height which is proportional to the flow rate. Thermal mass flow meters can also be applied for flow rate measurements. The key of all these instruments is the measurement of the amount of heat transferred by a gas passing over a heated surface. Depending on its heat capacity the cooling effect of the gas, versus a reference, is indicative of the gas flow rate. A review of the various thermal mass flow meters has been given by Benson et al.[26]. Since these instruments have an electrical output, for the determination of the volume it is possible to integrate the flow rate electronically. The disadvantage of this kind of instrument is its relatively high price. Wet or dry gas meters are cheaper instruments for the direct volume measurement. These meters contain two bellows of known volume and a sliding valve. Each of the bellows is alternately filled and emptied. The motion is converted by a piston and rod into a rotation of a dial indicating the total volume. A complete discription of dry and wet gas meters is given by Nelson [27]. Finally a more recent development by Haunold et al [28][29] should be mentioned. The measurement of gas flow and sample volume is effected by the sound of a whistle placed in a stream of gas (figure 17a). This sound is picked up by a microphone. Thus the instrument is called 'SONIC' gas flow meter. The intensity of the tone is proportional to the flow rate. Figure 17b shows the output current of the microphone as a linear function of the gas flow in the range from 6 l/min up to 16 l/min. Beside the flow rate, the

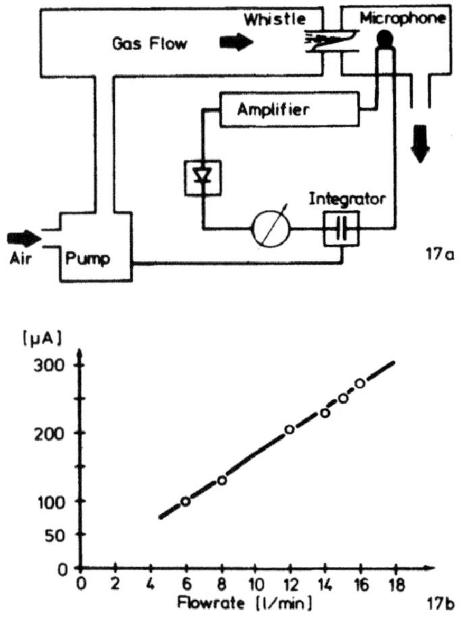

Fig. 17 a) Scheme of the 'sonic' flow meter. b) Current of the microphone as a function of the flow rate.

Fig. 18 Sonic gas flowmeter and personal sampler.

volume of the gas sample is measured by electronical integration of the output current of the microphone. When a chosen volume has been sucked through the sampling device, the pump is automatically switched off. The flow meter, together with the integrator, and a pump powered by a 12V motor cycle battery, is placed in a handy personal sampler, shown in figure 18. In the picture the air inlet is connected with a filter holder housing an impregnated filter as a sampling device. But, according to the actual measuring problem, other sampling devices may be connected to the sampler and will be discussed in the following section. They are shown schematically in figure 19.

II_2 The Choice of a Suitable Sampling Device.

<u>a) Impinger.</u> If wet scrubbing is the choice, atmospheric air is bubbled through a solvent which will dissolve and retain the components which are to be measured. The trapping is carried out in a

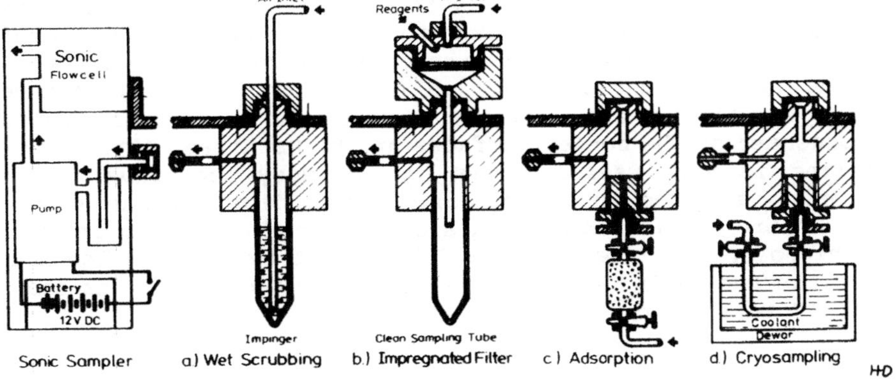

Fig. 19 Sonic sampler with four possibilities of connected sampling devices.

device called 'impinger' (figure 19a). In order to retain and accumulate the gas, dissolution or reaction rate with the solvent must be faster than the time needed for the bubbles to leave the liquid. The dispersion tube of the impinger can be open-ended or covered with a frit varying in shape and size. It is preferable to use solutions containing specific reagents which are reacting with the substance to be detected. For example, SO_2 can be absorbed by aspirating the sample air through a solution of potassium tetrachloro-mercurate, forming the dichloro-sulfito-mercurate complex. It is stable and not oxidized by the air which passes through the solution. A serious limitation in the use of impinger-samplers for gases, is the evapouration of the solvent [30].

b) Impregnated Filter. The loss of solvents in a scrubber is avoided if chemically impregnated filters are used (figure 19b). These filters offer several advantages. They can sample large quantities of air at high flow rates and they are non-fragile and inexpensive. Uncontrolled contamination during the preparation and the transport of the impregnated filters can be avoided by a new way of handling, which was developed in our laboratory [31]. The method will be described in the next paragraph.

c) Adsorption. Another sampling and accumulation technique is the adsorption of atmospheric trace gases on solid surfaces (figure 19c). The technique has not yet been developed to the point where specific gases are adsorbed selectively. Therefore only mixtures of gases are generally retained. Their subsequent separation and analysis is frequently performed by gas chromatography. Some of the commonly used adsorbents are: activated aluminas, silica gel, molecular sieves, charcoal, celite, and goldwool. The sample air is drawn through a tube with the adsorbent maintained at ambient or subambient temperatures. After sampling the tube is sealed and transferred to the laboratory for analysis.

d) Cryogenic sampling. In order to improve the sampling efficiency the sampling trap can be immersed in a coolant of subambient temperature. In general the trap consists out of a U-shaped glass tube (figure 19d). It may be empty or filled with a solid adsorbent. Drawing air through a cold trap, the vapour under investigation will be condensed at the cooled surfaces and be retained. Although cryo-sampling is very useful it is not an easy technique. Coolants frequently used are ice water, solid carbon dioxide, liquid oxygen, liquid nitrogen, liquid helium, and slushes of organic solvents such as chloroform or aceton with liquid nitrogen. A selection of solutions and their respective equilibrium temperature is shown in table IV taken from Altshuller [24].

Table IV: Summary of Cold Bath Solutions.

Coolant	Temperature (°C)
Ice and water	0
Ice and NaCl	-21
Carbon tetrachloride slush	-22.9
Chlorobenzene slush	-45.2
Chloroform slush	-63.5
Dry ice and acetone	-78.5
Dry ice and cellosolve	-78.5
Dry ice and isoprobanol	-78.5
Ethyl acetate slush	-83.6
Toluene slush	-95
Carbon disulfide slush	-111.6
Methyl cyclohexane slush	-126.3
N-Pentane slush	-130
Liquid air	-147
Isopentane slush	-160.5
Liquid oxygen	-183
Liquid nitrogen	-196

Some disadvantages of the cryo-sampling technique should be mentioned. Since all sub-0°C coolants tend to condense water, the trapping system can be plugged with ice or the condensing water may eventually react with other condensibles. This problem can be minimized by using a desiccant filter at the entrance of the trap. Liquid nitrogen as a coolant tends to condense oxygen which can react with other condensibles forming hazardous mixtures. This can be avoided by the use of liquid oxygen as coolant, however this requires non-combustible equipment and safety precautions.

II_3 Transfer of the Sampled Compound to the Subsequent Analysis.

When impingers are used for wet scrubbing the phase transfer from gas-to-liquid-phase is performed during the sampling and accumulation procedure, and the solvent of the impinger is suitable for direct wet chemistry analysis. The disadvantage of this phase transfer during the sampling procedure is the uncertainty whether a loss of the absorption liquid is caused by evapouration of the pure solvent, or by entrainment of solution droplets in the air stream. Another critical point is the determination of the impinger's blank, because the sampling-solution and the interior of the impinger may be contaminated during transportation and during the filling or emptiing of the device.

Faults caused by these critical steps are eliminated using impregnated filters, especially when impregnation and rinsing are performed under field conditions in ambient air. To achieve this goal a special kind of filter holder was constructed (figure 20). Its

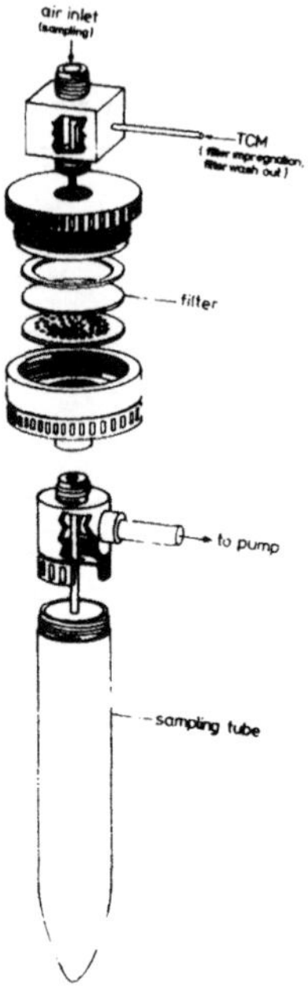

Fig. 20 Filter holder for the impregnation and rinsing of the filter under field conditions.

function is explained describing the sampling of atmospheric ammonia. The filter material (Delbag microsorban) is placed on a sieve of a filter holder, made of Teflon. As an impregnation liquid 5 ml of 0.1 n sulfuric acid are sprayed on the surface of the filter, in a way that it is completely covered with a thin layer of solution. By shortly operating the pump in the sampler (figure 19), the liquid is sucked through the filter and captured in a clean sampling tube. This procedure is repeated three times to ensure that the filter is totally washed and soaked with acid.
As a washing solution 15 ml are obtained which are subsequently analyzed by a photometric method (Indophenol reaction). The filter is clean and ready for sampling when the analysis of the washing liquid is equal to the blank value of the analysis. During the subsequent sampling procedure 50 - 1000 l of air are passed through the impregnated filter forming ammonium sulfate with the atmospheric NH_3. It is rinsed from the filter with three 5 ml equivalents of a fresh solution of sulfuric acid while the filter remains in its holder. Experience proves that the analysis of an additional washing procedure will result in a blank value. This indicates that all collected ammonia is washed out with the first fraction of the washing solution. Consequently this clean filter is used for the next sampling cycle. Since the filter material is being repeatedly used, it is possible to examine the blank value of the sampling device between each measurement. If these values correspond to the blank of the very sensitive analytical method, its detection limit can be applied for the entire detection method consisting of sampling procedure and analysis. Applying other solutions this filter method is suitable for the sampling of atmospheric H_2S and SO_2 [33][34].
A compound adsorbed on a solid surface can be desorbed by heating the adsorbent. The material is volatilized and should be directly flushed by an inert gas into the column of a gas chromatograph. The evapouration of cryo-sampled material may seem to be simple, but is accomplished with some difficulties. There is always the danger of condensed species reacting with each other, preferably when the sample is allowed to reach room temperature before analysis. Since

constituents of the air are frequently trapped along with the gases under investigation, the trap is often flushed with helium in order to remove the air before analysis [35].
This discussion was meant to demonstrate, the intricacies of the phase transfer between the sampling procedure and the analysis. The phase transfer is one of the most critical steps when trace gases are detected with discontinuously working methods. In order to test the sampling efficiency of the different sampling devices and to check the proper sample handling during the phase transfer it is necessary to calibrate the entire sampling train with carefully generated test gases.

II_4 Calibration of the Entire Detection System.

For calibration purposes standard gas mixtures of known composition are necessary. The motion of the gas mixtures determines whether a system is working in a static or dynamic mode. In a static dilution system, a known amount of the gas under investigation is added to a known volume of dilution gas. Air is used as a diluent most frequently, argon or nitrogen less so. The amount of diluted substance may be determined from pressure-, volume-, or mass measurements. The gas mixture may be stored in a container with flexible or rigid walls or in a cylinder under pressure. The biggest problem of these systems is the loss of standard gas during storage. The diluted gas can be adsorbed on the walls, can permeate through the walls, or may react with the walls or other contaminants present, or may decompose. Preconditioning of the containers is an important factor for maintaining the integrity of the standard. In a routine procedure for preconditioning the container is charged several times with the calibration gas of the desired concentration [36]. Despite these precautions, the concentrations of standard gases prepared under identical conditions and charged into similarly treated containers, may vary as much as 25 % [37]. Static systems are frequently used to calibrate continuously working methods because they require relatively low volumes of standard gas for a short period of time.
For calibration of discontinuously working methods large volumes of highly diluted standard gases must be introduced into the sampling device over relatively long periods of time. For these systems dynamic gas dilution devices have been proven to be a suitable technique. The gas dilution device requires a constant source of the gas under investigation which is steadily mixed with a metered flow of clean air. The most important method of preparation of such a constant source is the permeation tube, first developed by O'Keeffe et al [38][39]. It is based on the effect, that plastic materials are permeable for gases under pressure. Reliable permeation tubes can be home made and charged in a simple way. In our own model the pure standard gas is condensed into a cylinder of inert material by cooling its outside with liquid nitrogen. When the cylinder is partly filled

it is sealed with a cap of permeable material, such as Teflon. The sealed tube is then placed in a thermostate at 40°C. The temperature is kept as accurate as ±0.1°C. The vapour above the liquified gas diffuses through the cap at a constant rate, depending on temperature, the thickness, and the surface of the cap. The permeating compound is bleeding into a stream of dilution gas which is passed over the cap. The mixing ratio can be calculated from the diffusion rate and the diluting gas flow. The diffusion rate can be calibrated by measuring the weight loss of the tube as a function of time . Due to the heavy weight of commercial permeation tubes laboratory balances must be used, and repeated weighings over a period of several weeks are necessary to measure noticeable differences in weight. This handicap was the incentive for Beltz [40] to construct a permeation tube with a total weight of less than 2.5 g. This shortened the calibration time significantly since the weight loss in the range of ng could now be registered by the use of a micro balance (figure 21). The permeation tube is housed in one cylinder of the microbalance which can be kept at constant temperatures between 20°C and 40°C with an accuracy of ±0.1°C. The application of the permeation tube will be illustrated by the generation of a very low NH_3 mixing ratio. The gas dilution system is shown schematically in figure 21 also. Using a high volume compressor, outside air is pressed through a water vapour trap, several clean air filters (silica gel, activated charcoal) and aerosol filters (cotton wool) in order to obtain a carrier gas, which is free of ammonia, water vapour and aerosol. The purged stream of carrier gas is splitted, and the smaller part of it is passed through the microbalance acting as a diluent for the gas which is emanating from the permeation tube. This gas mixture is further diluted to an extent of three orders of magnitude when mixed with the main carrier gas stream supported by the

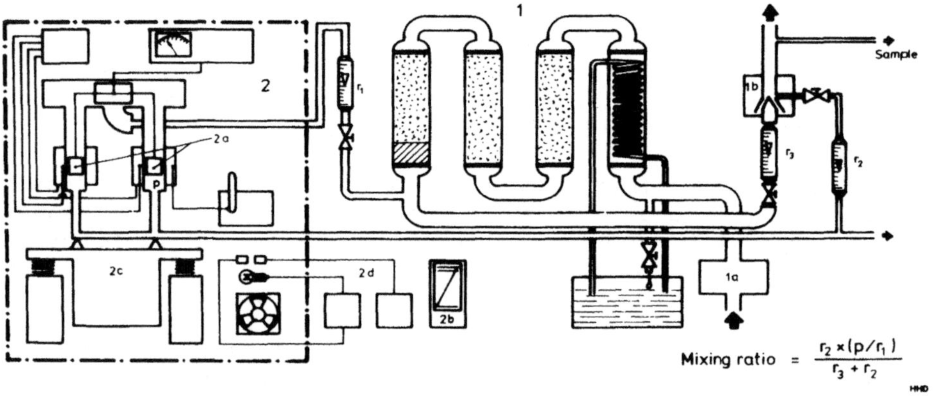

Fig. 21 Dynamic gas dilution system.
1: Gas cleaning train, 1a: Hig volume pump, 1b: Mixing orifice, 2: Micro balance, 2a: Permeation tube with conter weight, 2c: Table for micro balance, 2d: Thermostate, r: Rotameter.

venturi effect of a mixing orifice. Checking the permeation output (p) with the microbalance, and monitoring the total gas flow (r_1, r_2, r_3) with three flow meters, the NH_3 mixing ratio at the outlet of the system can be calculated with the formula given in figure 21. With this dynamic gas dilution system, mixing ratios between 0.05 and 30 ppbV NH_3 can be generated.

II_5 Analysis of the Sampled Compounds.

a) Wet chemistry. After transferring the gaseous atmospheric compound into the liquid phase a wet chemistry analytical method can be applied. Frequently colorimetric methods are used. The compounds are made detectable at low concentrations by their reaction with reagents producing colored products. The intensity of the color is measured by light absorption spectroscopy at a certain wavelenght in a colorimeter. The measurement is usually based upon the comparison of the light-transmittance of a solutions in a sample cell to that in a reference cell.

Formaldehyde, the most abundant aldehyde found in the atmosphere, is determined by such a colorimetric method. It is collected in aquous bisulfite solution. After sampling the absortion liquid is heated to remove nitrates and volatile organics [41]. Then it is treated with chromotrophic acid (4,5 dihydroxy-2,7-naphthalene disulfonic acid) in a sulfuric acid solution. In the presence of formaldehyde a purple chromogen will be formed with an absorption maximum at 580 nm [42].

Colorimetry also is the backbone of the Saltzman method [43] for the detection of NO_2. This trace gas is collected from the air by the aid of an impinger filled with an aqueous solution of N-naphtyl-ethylene-diamine hydrochloride, acetic acid and sulfanilic acid [44]. The NO_2 forms nitrous acid which diazotizes the sulfanilic acid and the resulting diazonium compound is coupled with N-naphtyl-ethylene diamine to form a colored product.

Another well known colorimetric method is the West and Gaeke method [30] for the determination of SO_2. Traces of this gas are removed from the air with an aqueous solution of sodium tetra chloro mercurate, either by wet scrubbing, or with impregnated filters. The resulting disulfito mercurate complex is then treated with formaldehyd and pararosaniline to form a colored product. A more sensitive method for the determination of the sulfite complex is the measurement of a chemiluminescence radiation. This effect occurs when the sulfite is oxidized in an acidic solution of potassium permanganate [45].

Another effect to determine colored products is their fluorescence. A sensitive method for the determination of hydrogen sulfide, first mentioned by Natush et al [46], is based on such a fluorescence measurement. Hydrogen sulfide is collected from the air by its reaction with silver nitrate on impregnated filters. The extend to which an extract of the filter quenches the fluorescence of a reagent containing fluorescein mercuric acetate is then determined [47].

A compilation of the most prominent wet chemistry analytical methods

is given in table V (at the end of this paragraph). They all have in common that the wet chemistry part of the detection method can be checked by introducing standard solutions of the colored reaction products to the analyzer. However, it is desirable to check the overall efficiency of the sampling and the analyzing procedure with standard gas samples. As an example of this procedure the determination of atmospheric ammonia by the aid of the indophenol reaction [48] will be discribed in more detail [49].
As already mentioned the gaseous ammonia is collected in a sulfuric acid solution to form ammonium-sulfate, either in an impinger or on impregnated filters. During the sampling act particles containing ammonium salts are removed by a prefilter. The ammonium ions present in the extract of the filter are reacted with anaqueous solution of penta cyano-nitrosylferrate, a sodium hydroxyde solution of phenol and an aqueous solution of sodium hypochlorite to form indophenol as a blue product.

$$NH_4^+ + OH^- \longrightarrow NH_3 + H_2O \qquad (7)$$

$$NH_3 + NaClO \xrightarrow{F^{++}} NH_2Cl + NaOH \qquad (8)$$

$$NH_2Cl + 2\,C_6H_5\text{-}OH \xrightarrow{60°C} HO\text{-}C_6H_4\text{-}N{=}C_6H_4{=}O + HCl \qquad (9)$$

The reactions are carried out automatically in an 'autoanalyzer (figure 22). It's heart is the proportioning pump which uses pump tubes that proportion the flow through the system in accordance with the requirements of the analysis. The sample probe of the sampler automatically dips into the sample cup on command of the pumps manifold. The volume of ammonium solution aspirated is determined by the pumping rate and by a timing-cam on the sampler. Between each sample a segment of air and washing solution is pumped through the

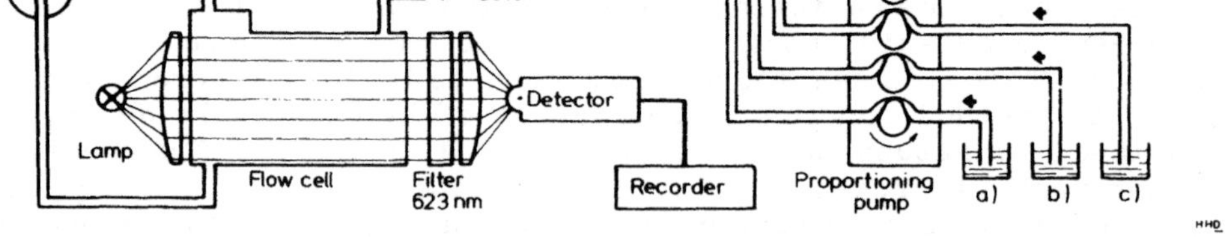

Fig. 22 Detection of ammonium by the indophenolblue-reaction, carried out in an autoanalyzer.
Reagents: a) Sodiumhypochlorite, b) Phenolat, c) Disodiumpentacyanonitrosylferrate.

system.This will clean the pump tubes and the mixing coils and will separate each sample. The reagent solutions are consecutively mixed with the sample solution in the mixing coil by the aid of the proportioning pump. The stream of liquid is divided into short segments by air bubbles. The different liquids of each segment are effectively mixed by the peristaltic effect of the traveling air bubbles. The well mixed solution is pumped through a thermostate with a controlled temperature of 60°C yielding the disired indophenol reaction. The heated section must be long enough so that each segment has acquired exactly the same temperature. After cooling the solution enters the flow cell of a colorimeter. In order to accomplish an unsegmented stream of solution an aspiration tube is connected to the flow cell to carry away air bubbles. An optical filter is used in the light path such that only a narrow band at 623 nm is measured. Following Lambert-Beer law the concentration value is obtained by comparing the light absorption of a solution which has undergone the indophenol reaction, with the absorption of a blank solution containing all reagents except ammonium. According to Kaiser et al [50], the mean of these blank values $\bar{J}$ plus 3 times the standard deviation σ will define the detection limit B of the method:

$$B = \bar{J} + 3\sigma \tag{10}$$

A calibration curve obtained from liquid standards is shown in figure 23 (upper curve). It is compared with the results of a calibration procedure with standard gases. The gas samples introduced to the sampling device have been produced by the previous described dynamic gas dilution device. The comparison made to examine the entire detection system showed a discrepancy. The absorption effects measured in solutions of a known ammonium content were always higher than those observed in washing solutions of impregnated filters

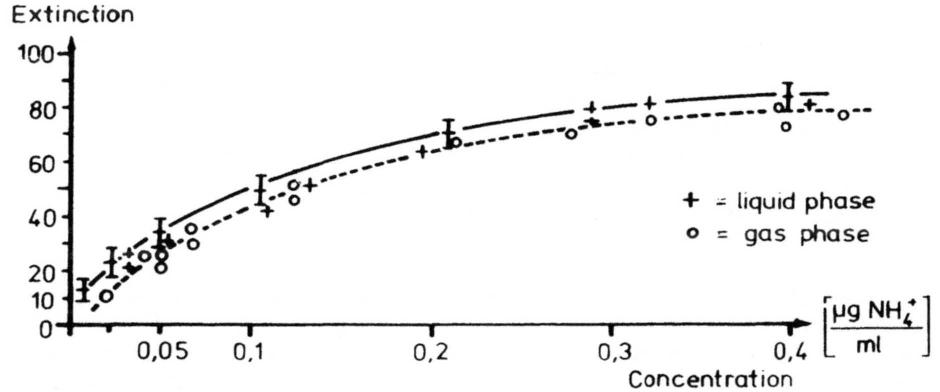

Fig. 23 Comparison of colibration curves. The detection method for ammonium with the indophenol reaction was calibrated in the liquid phase (+) and from the gas phase (o).

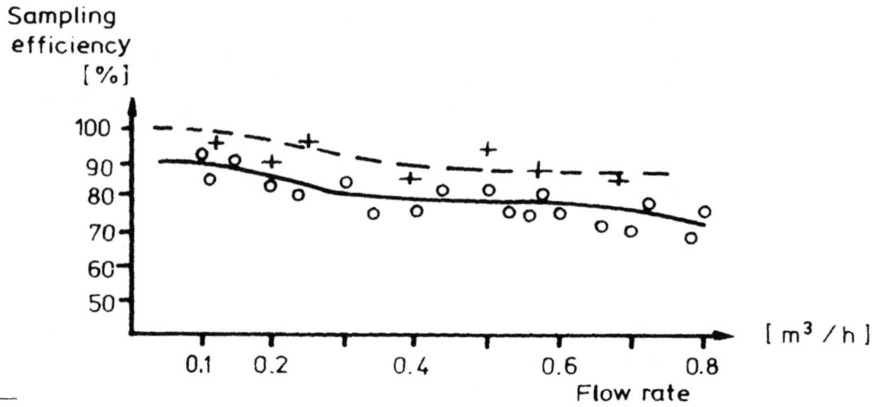

Fig. 24 Sampling efficiency of impregnated filters as a function of the flow rate and two different ammonia concentrations. (+): 50 ug/m^3, (o): 5 ug/m^3.

treated with the stochiometric amount of ammonia in the gas phase. Figure 24 demonstrates that the deviation is due to an uncomplete sampling efficiency. At flow rates higher than 0.3 m^3/h a breakthrough occurs yielding a sampling efficiency of about 80%. This efficiency remains constant up to flow rates of about 0.6 m^3/h. In view of the rather constant sampling efficiency at usual flow rates a correction term may be introduced. With equation (11) the atmospheric concentration C_{NH_3} can be calculated from the available data.

$$C_{NH_3} = \frac{A \times L}{V} \times \frac{P}{P_o} \times \frac{T}{T_o} \times \frac{1}{0.8} \qquad (11)$$

with	A	=	Analyzed concentration in the liquid phase		[ng/ml]
	L	=	Volume of washing solution		[ml]
	V	=	Volume of air sample		[m^3]
	P	=	Ambient pressure		[atm]
	P_o	=	Standard pressure	1	[atm]
	T	=	Ambient temperature		[K]
	T_o	=	Standard temperature	273	[K]
	$\frac{1}{0.8}$	=	Correction term (sampl.eff.)		

Equation (10) and (11) may be used to calculate the detection limit of the method. The optical absorption of a blank solution is uncertain within ±2%. This means that an absorption above 3σ = 6% is caused by traces of ammonia with a certainty of 99%. This absorption corresponds to a standard concentration of 23 ng NH_4^+/ml. When 15 ml of washing solution are used and 0.5 m^3 of air are drawn through the filter, according to equation (11) a detection limit of 0.85 ug NH_3/m^3

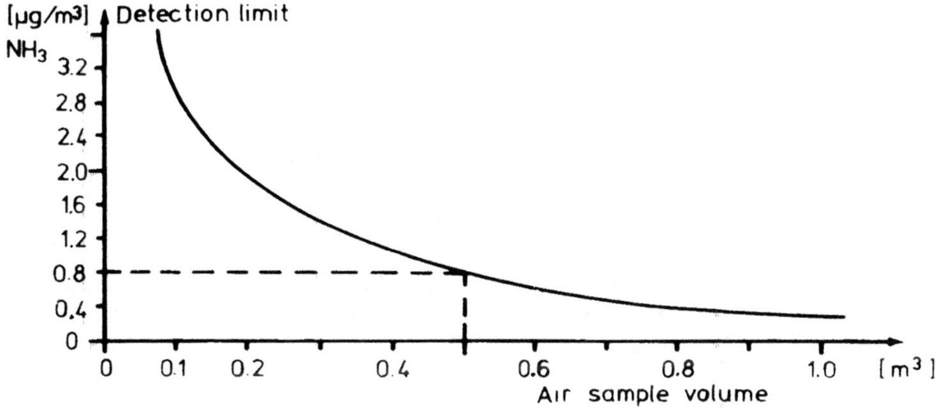

Fig. 25 The detection limit of the entire method for ammonia determination as a function of air sample volume.

is obtained. Of course the detection limit depends on the volume of sample air drawn through the filter. This dependence is demonstrated in figure 25.

b) Gaschromatography. Gaschromatography is basically the method for separating mixtures of trace gases which are present in gaseous grab samples or in solid adsorption and cryosamples. Grab samples are often enriched by cryogenic condensation from relatively modest volumes (100 - 1000 ml). This will allow to introduce the sample as a 'plug' into the head of the analytical column. Frequently the evapouration or desorption of the accumulated compound is combined with the sample introduction. In this case the compounds are flushed into the column with a carrier gas. The essential features of a chromatography column are schematically shown in figure 26.
The stationary liquid phase is confined in a long tube of approximately 2 m in length and some milimeter in diameter. In this so called column the stationary phase forms a thin film which is either distributed over an inert granular support (packed columnes) or applied as a coating on the inner surface of the column (wall coated open tubular column). At the top the column is connected to the inlet of the gas chromatograph and at the end it terminates in a detector cell. The stationary phase is continuously swept with a carrier gas as mobile phase. It is a gas which is inert with respect to the sample. Most commonly nitrogen or helium is used. The integrated system is adjusted to any suitable temperature by the aid of a thermostate. Assuming a mixture of two compounds is injected into the column both substances will be distributed between the stationary liquid phase and the mobile gas phase. Their equilibrium distribution between the two phases is governed by their distribution constants K_D. It is defined as the number of sample molecules per unit length in the stationary phase N_S, in relation to the number of sample molecules per

Distribution constant $K_D = \frac{N_s}{N_m} = \frac{\text{molecules per unit length in stationary phase}}{\text{molecules per unit length in mobile phase}}$

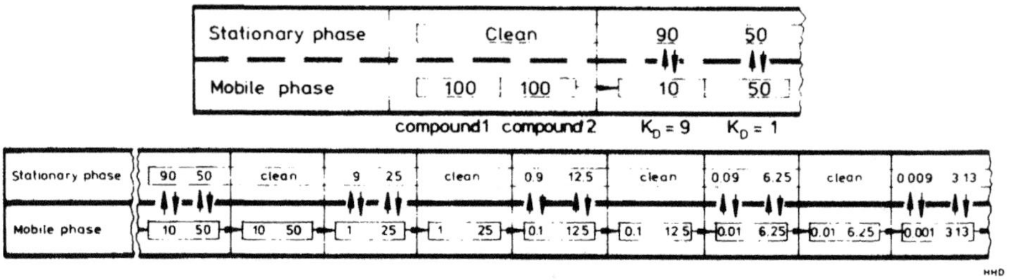

Fig. 26 Essential features of a chromatography columne.

unit length in the mobile gas phase N_m. The key for the separation of several compounds is the difference in the magnitude of their distribution constants. This is illustrated by an example in figure 26. It is assumed that a compound #1 with a large distribution constant and a compound #2 with a small constant are injected into a separating column. In a certain unit of length, hundred parts of the two compounds will be distributed between the stationary phase and the mobile phase according to their distribution constants $K_{D1}=9$ respectively $K_{D2}=1$. The parts remaining in the mobile phase will be flushed to a 'clean' surface in a adjunct unit of lenth. New equilibrium will be formed and the excess of compound #2 in the mobile phase increases. During the considered four examples of unit length in figure 26 the ratio of compound #1 and compound #2 in the mobile phase changes from five over one to approximately threethousand over one. This illustrates the process of separation. Compound #1 with a large magnitude of K_D is mainly attached to the stationary phase and moves through the column slowly. On the other hand most parts of compound #2 remain in the mobile phase. Thus it will reach much faster the end of the column and will be detected earlier.

The detector is the main part of a gas chromatograph. In general it consists of a flow cell in which the compounds are detected by physical effects. The type of compound is identified by its retention time and the quantity can be calculated from the area under the signal chart of the detector's output.

Four types of detectors which are commonly applied will be shortly discribed to give an idea of their function, their range and their limitations. The first detector which was used in gas chromatography is the thermal conductivity cell (figure 27). It detects the presence of a compound in the carrier gas by measuring the change of thermal conductivity. For a given gas molecule it is a function of molecular weight and collision cross-section. Therefore, the larger the difference between the molecular weight of the compound and molecular weight of the mobile phase, the greater will be the sensitivity. For this reason helium is often used as carrier gas. The technique is non-

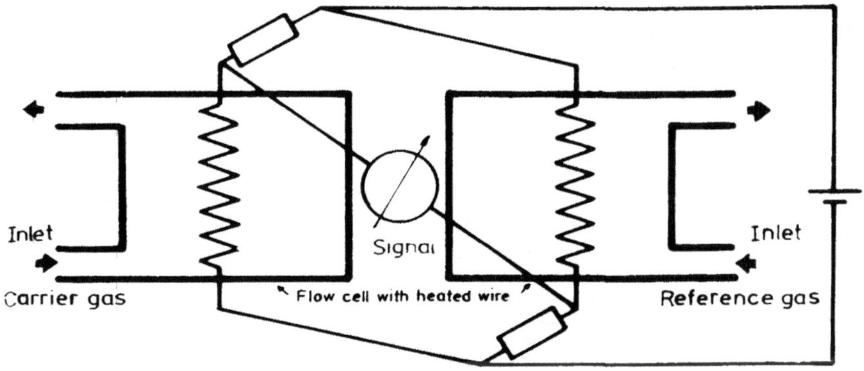

Fig. 27 Temperature conductivity detector (TCD).

selective and is effected by normal atmospheric components such as water vapour and carbon dioxide. It is useful for the detecting of high concentration levels but has limited application as a measuring technique for atmospheric trace gases in the ppb-range.
A development which improved the sensitivity is the flame ionisation detector (FID) (figure 28)[58]. In a flow cell hydrogen gas is mixed with oxygen to maintain a hydrogen-oxygen flame. The ionization of the effluent gas in the hot zone of the flame causes a charge transfer between two electrodes. The resulting increase of the electrical conductivity of the gas stream is the measured effect. The detector is mainly used for the determination of hydrocarbons and organic compounds containing a large number of CH-groups, since these compounds significantly increase the conductivity of the flame. It is insensitive to many atmospheric constituents including the rare gases, nitrogen, the oxides of nitrogen, carbondioxide, water, ammonia, and sulfur containing gases. A similar technique is the measurement of photo-ionization (PID). It is caused by the irradiation of the effluent gas with ultra-violet light. If the

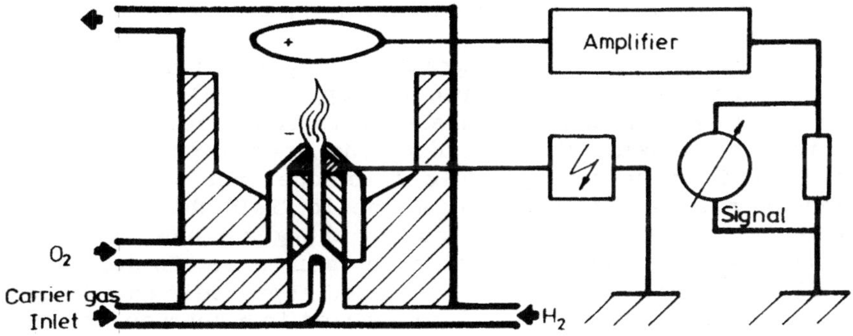

Fig. 28 Flame ionisation detector (FID).

energy of light is greater than the ionization potential of the irradiated substance it will be ionized and the ions are measured in a manner similar to that used in the flame ionization technique. By selecting an ultra-violet source which emits photons of an appropriate energy, it is possible to ionize and detect selected organic substances in the atmosphere. The device therefore shows some more selectivity and in this respect differs from flame ionozation. A third kind of detector is shown schematically in figure 29. It is called 'electron capture detector'(ECD)[59]. Between two electrodes

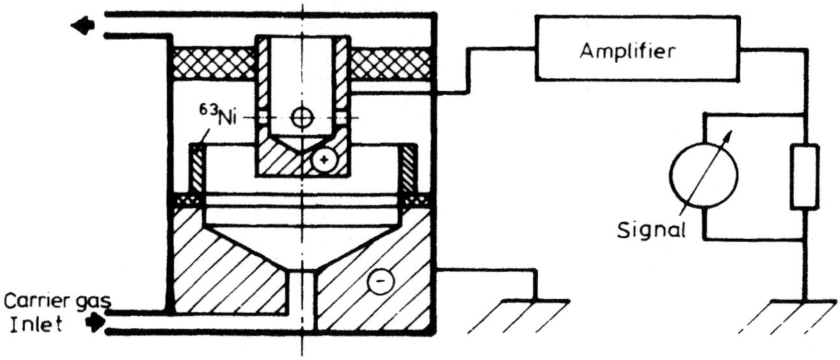

Fig. 29 Electron capture detector (ECD).

a low energy beta emitter such as ^{63}Ni or ^{3}H is located. The radiation ionizes the carrier gas between the electrodes and an ionization current can be produced by maintaining a potential difference between the electrodes. When an electronegative compound is entering the detector cell it captures electrons and reduces the standing current. The extend of this decrease depends on both, the number of electron capturing compounds present and their electronegativity. It is selective for highly electronegative species, such as compounds containing halogens. Furthermore oxygen containing compounds like peroxiacetylnitrate (PAN)[60] and compounds with unsaturated groups like polynuclear aromatic hydrocarbons [61] can be determined with this detector.

In recent years, gas chromatography using capillary columns was frequently combined with mass spectroscopy [62][63]. The gas sample enters a high vacuum chamber, where it is bombarded under standard conditions by a beam of electrons. The ions formed are given a resonably uniform energy by an electric acceleration field and they consequently are passing through a magnetic field which is located at right angle to their motion. The moving charges are deflected in the field by an angle inversely related to their mass. In dependence of their different masses they are collected separately on a target. In many cases a sample molecule will be cracked in a reproducable manner during the ionization process. The resulting unique pattern of fragments of different masses depends on the molecular structure of the substance and may be used for its identification. The

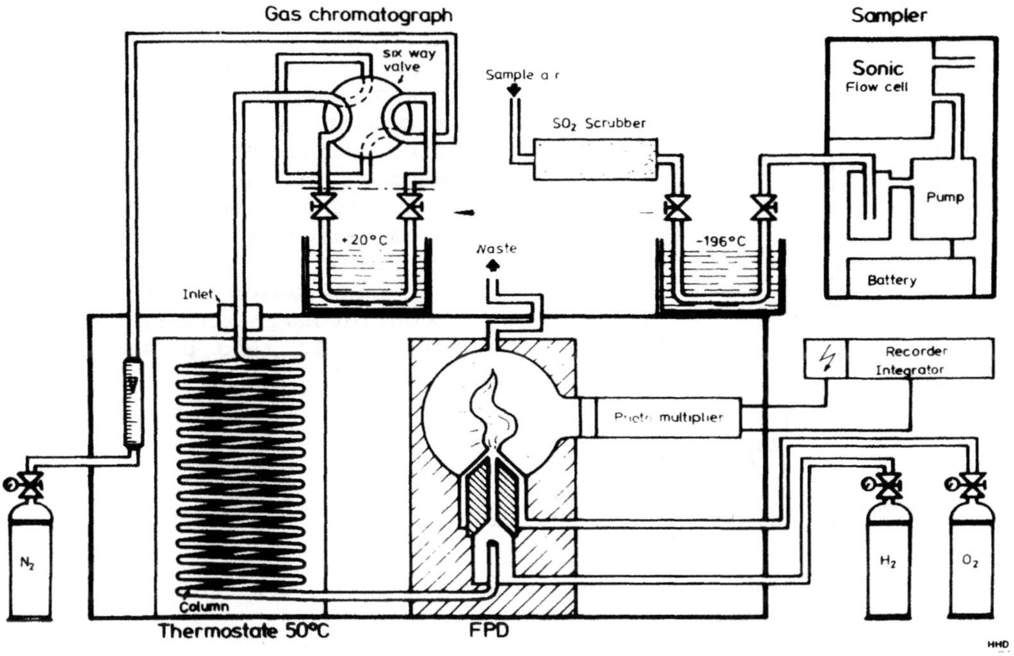

Fig. 30 Gas chromatographic system with flame photometric detector (GC/FPD).

concentration of the sample molecules is determined by the intensity of the mass peaks.

Finally an entire gas chromatographic detection system will be discribed by an example shown schematically in figure 30. In the system a flame photometric detector (FPD) is applied for the determination of reduced sulfur containing gases. The detector has already been discribed in section I. The reduced sulfur gases are pre-concentrated during a cryosampling procedure using i.g. the 'sonic' sampler. Ambient air is drawn through an U-shaped quartz tube filled with molecular sieve. The flow rate is in the range of 0.5 - 2 l/min. Sulfur dioxide is removed by a scrubber tube, which is located downstream to the sampling device. After collection the cold trap is tranferred to the laboratory, and there it is attached to a six way sample injection valve of the GC/FPD system. By rotation of the valve the nitrogen carrier gas flow is diverted through the sampling tube. Simultaneously the coolant is removed and the trapped compounds are suddenly evapourated by the aid of a water bath. With the help of this procedure the compounds are flushed as a plug into the separating column by the nitrogen carrier gas flow. The columns have a length of six feet and an inner diameter of 2 mm. They are packed with Carbopack B, a support which contains 10 % phosphoric acid as liquid stationary phase. The separated sulfur compounds are determined by the FPD at the end of the column and they can be identified by their different retention times. An example of the occuring signal is shown in figure

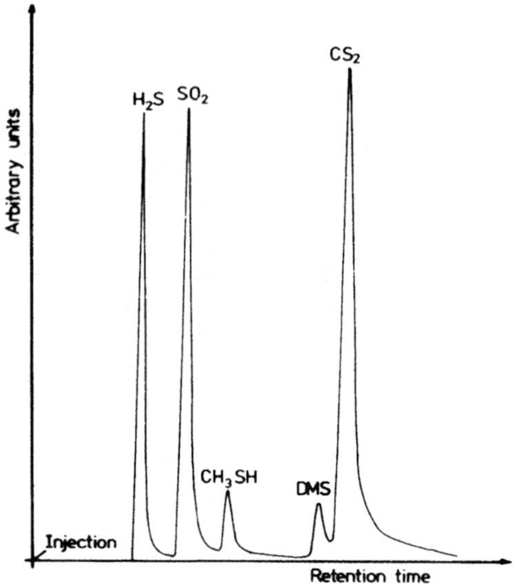

Fig. 31 Gas chromatrogram of sulfur containing gases. Peaks measured with FPD.

31. The method is calibrated qualitatively (re-tention time) and quantitatively (peak area) by checking samples of known sulfur gas content going through the entire procedure of sampling and analysis. The calibration gases are provided by the gas dilution system, which has been discribed in an earlier section.

A compilation of all discontinuously working methods mentioned in this paragraph is given in table V.

TABLE V

DISCONTINUOUSLY WORKING METHODS FOR THE DETECTION OF ATMOSPHERIC TRACE GASES.

TRACE GAS	ABSORPTION	SAMPLING FORM	IDENTIFICATION	DETECTION
CO_2	$Ba(OH)_2$	$BaCO_3$	$BaCO_3$	NEPHELOMETRY
CO	Pd^{++}	Pd	Pd	NEPHELOMETRY
CH_4	CRYOSAMPLING	PRECONCENTRATION	GASCHROMATOGRAPHIC	FID
HC	CRYOSAMPLING	PRECONCENTRATION	GASCHROMATOGRAPHIC	FID
FHC	CRYOSAMPLING	PRECONCENTRATION	GASCHROMATOGRAPHIC	ECD
N_2O	MOLEC. SIEVE	PRECONCENTRATION	GASCHROMATOGRAPHIC	ECD
NO_2	SULFANIL. ACID	DIAZOBENZENE DERIV.	AZO DYE	PHOTOMETRY
NH_3	SULFURIC ACID	AMON. SULFATE	INDOPHENOLBLUE	PHOTOMETRY
O_3	JODINE		J_2 BLEACHING	PHOTOMETRY
O_3	ELECTROLYTE		CATHODIC REDUC. OF O^3	COULOMETRY
H_2S	$AgNO_3$	Ag_2S	FLUORESCEIN MERC. AC.	FLUORESC. QUENCH.
H_2S,RSH	GOLDWOLL	PRECONCENTRATION	GASCHROMATOGRAPHIC	FPD
CS_2,COS	CRYOSAMPLING	PRECONCENTRATION	GASCHROMATOGRAPHIC	FPD
SO_2	H_2O_2	H_2SO_4	H_2SO_4	CONDUCTOMETRY
SO_2	TCM	BISULFITOMERCURATE	ROSANILIN	PHOTOMETRY
SO_2	TCM	BISULFITOMERCURATE	$KMnO_4$ OXIDATION	CHEMILUMINESCENCE
HCl	$HgCl_2$	$HgCl_2$	DIPHENYLCARBAZON	PHOTOMETRY
HCl	SILICAGEL	CYCLOHEXANOXID DERIV	GASCHROMATOGRAPHIC	HECD

SUMMARY.

The purpose of this chapter was to provide information about selected methods of analysis in a tutorial manner. I hope the reader will gain an idea of the problems, limitations and advantages of this kind of analytical chemistry in relation to its particular application. The chapter does not pretend to provide all the details actually required to go ahead and use a method. This survey is only meant to introduce the most prominent measuring methods, and it is addressed to the inquiring reader, who may whish to gain a better understanding of the results of atmospheric chemistry, which are discussed in the following chapters.
More recent developments, like Fourier transformation - infrared spectroscopy (FT-IR) or long path laser induced fluorescence or laser light absorption are omitted deliberately yet the methods will be discribed in the following chapters by experts using these methods.

REFERENCES.

1 Smithsonian Meteorological Tables, Sixth Ed. Washington (1951).

2 E. Neef (Ed.): Taschenbuch der physischen Geographie, Verlag Harry Deutsch, Thun und Frankfurt/Main (1977).

3 Schmidt, R.: Dissertation, Inst. f. Meteorologie der Universitaet Frankfurt (1976).

4 Berresheim, H.,Jaeschke, W., : this book.

5 Keeling, C.D.: Geochim. et Cosmochim., Acta 13, 322-334, (1958).

6 Luft, K.E.: Angew. Chemie, 19, 2, (1947).

7 Komhyr, W.D., Harries, T.B.: WMO Spec. Env. Rep., 10, 9-19, (1976).

8 Bischof, W.: Tellus, 27, 59-61, (1974).

9 Bischof, W.: WMO Spec. Env. Rep., 10, (1976).

10 Robbins, R.C., Borg, K.M., and Robinson, E.: J. Air Pollut. Contr. Ass., 18, 106, (1968).

11 Seiler, W. and Junge, C.: J. Geophys. Res., 75, (1970).

12 Schmidt, U.: Tellus, 26, 78-90, (1974).

13 Seiler, W.,Giel, H., and Ellis, H.: WMO Spec. Env. Rep., 10, 31-39, (1976).

14 Clyne, M.A.A., Thrush, B.A., and Wayne, R.P.: Trans Faraday Soc., 60, 359-370, (1964).

15 Fontijn, A., Sabadell, A.J., and Ronco, R.J.: Anal. Chem., 42, 575-579, (1970).

16 Pitts, J.N., Kummer, W.A., Steer, R.P., and Finlayson, B.J.: Adv. in. chem. Series, Nr. 113, 246-254, (1972).

17 Warren, G.J., and Babcock, G.: Rev. Sci. Instr., 41, 280- 282, (1970).

18 Mast, G.M. and Saunders, H.E.: Instr. Soc. Amer. Trans. 1, 325, (1962).

19 Regener, V.H.: J. Geophys. Res. 69, 3795-3800, (1964).

20 Draeger, B., Heinrich, Draegerwerk: West German Patent 1,113,918, (1962).

21 Crider, W.L.: Anal. Chem. 37, 1770, (1965).

22 Stevens, R.K., O'Keeffe, A.E., and Ortmann, G.C.: Env. Sci. Tech. 3, 652, (1969).

23 Berresheim, H.: Diplomarbeit, Inst.f.Meteorologie, Univ. Frankfurt, (1980).

24 Altshuller, A.P., Kopczynski, S.L., Lonneman, W.A., and Sutterfield, F.D.: Env. Sci. Techn. 4, 503, (1970).

25 Veillon, C., and Park, J.Y.: Anal. Chem. 42, 684, (1970).

26 Benson, J.M., Baker, W.C., and Easten, E.: Instr. Contr. Syst. 43, 85, (1970).

27 Nelson, G.O.: Ann Arbor Sci. Pub. 'Controlled Test Atmospheres', Ann Arbor, Michigan, (1971).

28 Haunold, W.: West. German Patent, SONIC P 2849122, (1980).

29 Jaeschke,W., Haunold, W.: Spec. Env. Rep. No. 10, 193, (1977).

30 West, P.W. and Gaeke, G.C.: Analyt. Chem. 28, 1816, (1956).

31 Jaeschke,W.: Atm. Env. 12, 715, (1978).

32 Hendrickson, E.R.: 'Airpollution (A.C. Stern Ed.) 2nd ed. Vol.II, 3-53, Ac. Press New York, (1968).

33 Jaeschke, W. and Herrmann, J.: Int. J. Env. Anal. Chem. 10, 107, (1981).

34 Meixner, F. and Jaeschke, W.: Int. J. Env. Anal. Chem 10, 51, (1981).

35 Feldstein, M. and Balestrieri, S.: J. Air Poll. Contr. Ass. 15, 177, (1965).

36 Conner, W.D. and Nader, J.S.: Amer. Ind. Hyg. Ass. J. 25, 291, (1964).

37 Jutze, G.A. and Lewis, R.J.: J. Air Poll. Contr. Ass. 15, 323, (1965).

38 O'Keeffe, A.E. and Ortmann, G.C.: Anal. Chem. 38, 760, (1966).

39 O'Keeffe, A.E. and Ortmann, G.C.: Anal. Chem. 41, 1598, (1969).

40 Beltz, N.: Diplomarbeit Inst. f. Meteorologie, Univ. Frankfurt, (1980).

41 Altshuller, A.P., Levy, L.J., and Wartburg, A.F.: Int. J. Air Water Pollut. 6, 381, (1962).

42 Altshuller, A.P., Miller, D.L., and Sleva, S.F.: Anal. Chem. 33, 62, (1961).

43 Saltzman, B.E.: Anal. Chem. 26, 1949, (1964).

44 Huygen, J.C.: Anal. Chem. 42, 42, (1970).

45 Stauff, J. and Jaeschke, W.: Atm. Env. 9, 1038, (1975).

46 Natush, D., Klonis, H., Axelrod, H., Teck, R., and Lodge, J.: Anal. Chem. 44, 2067, (1972).

47 Jaeschke, W. and Herrmann, J.: J. Geophys. Res., 85, 5639-5644,(1980).

48 Harwood, J.E. and Kuhn, A.L.: Water Res. 4, 805, (1970).

49 Goethel, M.: Diplomarbeit, Inst. f. Meteorologie und Geophysik, Univ. Frankfurt, (1980).

50 Kaiser, H. and Specker, H.: Z. anal. Chem. 149, 46, (1956).

51 Horn, O., Schwenk, U., Hachenberg, H.: Brennstoff Chem. 38, 116, (1957).

52 Thompson, B.: Fundamental of Gasanalysis by Gas Chromatography, Varian, Palo Alto, (1977).

53 Mindrup, R.: J. Chem. Sci. 16, 380, (1978).

54 Struppe, H.G. (Ed.): Handbuch der Gaschromatographie, Verlag Chemie, Weinheim, (1971).

55 Jennings, W.G. (Ed.): Application of Glases Capillary Gas Chromatography, Marcel Dekker, New York, Basel ,(1981).

56 Jennings, W.G.: Gaschromatography with Glass Capillary Columnes, Acad. Press, New York, (1980).

57 Proske, M.G., Bender, M., Schirrmeister, H., Boettcher, G.: Chromatographia 11, 715, (1978).

58 David, J.: Gas Chromatographic Detectors, Wiley, New York, (1973).

59 Lovelock, J.E., Maggs, R.J., and Adlard, E.R.: Anal. Chem. 43, 1962, (1971).

60 Darly, E.F., Kettner, K.A., Stephens, E.R.: Anal. Chem 35, 589, (1963).

61 Bjoerseth, A. and Eklund, G.: HRC and CC, 2, 22, (1979).

62 Henneberg, D., Henrichs, K., and Schomburg, G.: J. Chromatogr. 122, 343, (1975).

63 Stau, J. and Abraham, B.: Anal. Chem. 50, 2161, (1978).

TECHNIQUES FOR THE MEASUREMENT OF TROPOSPHERIC NITROGEN COMPOUNDS : NH_3, N_2O, NO, HNO_2, NO_2, HNO_3, NO_3, AND PAN

Guenter Helas

Max-Planck-Institut fuer Chemie, Otto-Hahn-Institut, Mainz, FRG

ABSTRACT

Measurement techniques for the title compounds are compiled. A short description of the methods is given and the areas of application are discussed.

1. INTRODUCTION

Nitrogen containing compounds are of importance in atmospheric chemistry because of their reactivity and their involvement in the cycles of many other trace gases. For this reason a compilation of measurement techniques for NH_3, N_2O, NO, HNO_2, NO_2, HNO_3, NO_3, and PAN will be useful. A short description of the methods employed is here given and the areas of application are discussed.

The nitrogen containing compounds play an important role in atmospheric chemistry as they are highly reactive and thus influence not only the mixing ratios of their own group, but they also affect other atmospheric constituents such as OH, O_3, SO_2. Some of the nitrogen containing compounds may have an impact on health, the visibility, and precipitation chemistry.

2. MEASUREMENT TECHNIQUES

In principle every measurement consists of two steps, one of acquisition of the sample and one of quantifying the amount of the trace component under investigation. This measurement can be done either in the aqueous phase or in the gas phase. In this sense the analytical procedures cited here are separated between liquid-phase

H. W. Georgii and W. Jaeschke (eds.), Chemistry of the Unpolluted and Polluted Troposphere, 39–55.

and gas-phase.

2.1. Measurement methods for ammonia

Detection methods are summarized in table 1.

Table 1

Detection methods for ammonia	~detection limit [ppbv]	interference
1.) liquid-phase		
indophenolblue	0.5 ($1 m^3$)	SO_2, H_2S, HCHO
ion-chromatography	0.5 ($1 m^3$)	amines
ion-selective electrodes	10^{-2} ($3 l\ min^{-1}$; 24h)	
2.) gas-phase		
FTIR (967.5 cm^{-1})	2 (1km)	?
optical absorption spectroscopy (208, 212nm)	0.2 (1km)	
chemiluminescence of NO after conversion and subtraction	10^{-2} ?	amines

liquid phase

Indophenolblue
This method, often called Berthelot's method, is based on the formation of an azo-dye, which in turn can be determined photometrically at 623 nm. Details of this reaction are described in another chapter of this issue (Survey of measuring techniques for atmospheric trace gases) and will not be repeated here.
In order to separate contributions from aerosol-borne ammonium and the gas-phase ammonia two filters are used, the first one to retain the aerosol, the second is an impregnated filter to retain the ammonia. After the sampling the ammonium ion is extracted with water and the solution is treated with the dye reagents. The lower detection limit of ammonia in solution with the indophenolblue method is in the range of 5 ng/ml. Thus, by sampling 1 m^3 of air a lower detection limit of about 0.5 ppbv can be achieved.
This often used method may suffer from interferences due to aerosol leading to larger values. This may happen because of chemical reactions on the aerosol filter. Minor interferences by SO_2, H_2S, formaldehyde, and amines are reported also [1].

Ion-chromatography
As sampling and measuring in the above procedure are two separate steps, the measurement can be exchanged with an ion- chromatographic determination. Here the conductivity of the solution is utilized to quantify the amount of ammonia after a liquid chromatographic separation.

Approximately 5 ppb by mass of ammonium in solution can be detected [2], which gives combined with the filter sampling procedure a similar lower detection limit of 0.5 ppbv of ammonia in 1 m^3 of sampled air.

Ion selective electrodes

Ferm [3] describes a technique for sampling ammonia which makes use of the different diffusion velocities of gas molecules and of aerosol particles. A gas flow is led through a tube, whose inner surface is coated with acidic material, e. g. oxalic acid. The diffusivity of the ammonia molecule is greater than that of aerosol particles. With appropriately chosen dimensions of air flow velocity and tube inner diameter gas phase ammonia will be deposited in the front part of the tube, while the aerosol borne ammonium will be carried further and be deposited in the rear part of the tube.
After washing out the different depositions Ferm used gas selective electrodes to determine the amount of ammonia. He stated a lower detection limit of 10 pptv after a sampling time of 24 hours. This sampling technique will be less subjected to aerosol interferences. Its limitation is given by the small throughput making long sampling times in clean air necessary.

gas-phase

Fourier-Transform Infrared Spectroscopy

The FTIR has been successfully used to detect ammonia in the troposphere. The experimental set-up consists of Michelson-type interferometers and a data acquisition system. An instructive description is given by Pitts et al.[4]. For ammonia the absorption bands at 931 and 967.5 cm^{-1} are used. Assuming a light path of 1 km length, a lower detection limit of 2 ppbv can be achieved [4]. The detection system is bound to the laboratory, so that mainly air in the environment of the laboratory can be measured, or comparisons with other instruments can be made.

Laser absorption spectroscopy

Absorption spectroscopy with a laser is also possible. Reid et al. [5] describe a technique consisting of tunable diode lasers and a White cell or a portable Wilks cell. They used bands near 1140 cm^{-1} and obtained with a pathlength of 40 m a lower detection limit of 0.2 ppbv. As the pathlength of 40 m is folded in a cell of 0.75 m, this system can be used as a field monitor. The detection limit of 0.2 ppbv ammonia should allow background air measurements.

Optical absorption spectroscopy

Absorption measurements in the UV range also were utilized to quantify the ammonia content of the troposphere [6]. The technique will be discussed in detail later in the NO_2- chapter. The absorption is measured at 208 and 212 nm. At a light path length of 1 km the lower detection limit is about 0.2 ppbv, and thus the method should be suitable for background air.

Chemiluminescence of NO after conversion
As it is possible to oxidize ammonia to NO, NO detectors can be applied for an indirect measurement of NH_3. Several groups [7,8,9], worked with commercially available or self-built NO_x-detectors. In principle, detection limits of about 10 pptv of ammonia should be possible with improved NO_x- analyzers. Interferences by amines are to be expected.

2.2. Measurement methods for nitrous oxide

The detection methods for nitrous oxide are summarized in table 2. Detection limits are usually not cited in experimental descriptions of N_2O measurement techniques, as they are not the limiting factor of the measurements.

Table 2

Detection methods for N_2O	~detection limit [ppbv]
GC with ECD	2
GC with UPS	–
LAS (1140.6,1140.7 cm^{-1})	3
FTIR (853 cm^{-1})	4

Gas-Chromatography with Electron-Capture-Detector
Gas-chromatographic methods have been used extensively for the analysis of N_2O in the troposphere. Whereas initially thermal conductivity detectors have been applied [10], the electron capture detectors are currently in use [11], as they are much more sensitive. The lower limit of detection of about 2 ppbv [12] is by far sufficient to measure ambient levels (about 300 ppbv) of nitrous oxide.

Gas-Chromatography with Ultrasonic-Phase-Shift-Detector
Other detection methods have also been put to use. Weiss et al. [13] describe the application of an ultrasonic phase shift (UPS) detector. This technique bases on the gas chromatographic separation of N_2O and CO_2 that is cryosampled from about 2 l of air. The ratio of the two signals is determined and calibration is obtained from an independent CO_2 determination by infrared technique. While the method discussed above requires reference to standard mixtures, the UPS- method makes reference to a stable atmospheric constituent, namely CO_2.

Laser absorption spectroscopy

An optical method for measuring nitrous oxide is described by Reid et al.[5] and Connel et al.[14]. They employ a tunable diode laser and use bands at 1140.6 and 1140.7 cm^{-1}. Interferences by ammonia can be filtered out. Using a portable Wilks cell of 75 cm and a pathlength of 40 m Reid and coworkers obtained a lower limit of detection of ca. 3 ppbv N_2O.

Fourier-Transform Infrared Spectroscopy

Similarly the Fourier-Transform technique can be used for the measurement of N_2O. Pitts et al. [4] use 1000 m pathlength and obtain with the band at 853 cm^{-1} a lower detection limit of about 4 ppbv.

2.3. Measurement methods for nitric oxide

Detection methods for NO are summarized in table 3. The techniques are:

Table 3

Detection methods for NO	~detection limit [ppbv]	interference
chemiluminescence ($NO + O_3$)	10^{-3}	?
optical absorption spectroscopy (215, 226 nm)	0.4 (1 km)	dust, fog
fluorescence (γ-bands)	1	SO_2
FTIR	1	H_2O
GC	10^4	H_2O

Chemiluminescence

Nowaday the most important method to measure the mixing ratios of NO is the chemiluminescence technique developed by Ridley and others[15,16,17]. It makes use of the chemiluminescence arising from the electronically excited $NO2^*$, which is formed by the reaction of NO with ozone. The air to be measured is continuously drawn into a reaction chamber and mixed with ozone. The luminescence is monitored by infrared sensitive photomultipliers. In the usual set-ups the luminescence is directly proportional to the mixing ratio of NO. If a 600 nm cut-off filter is used to discriminate from other luminescent reactions beside the $NO + O_3$ reaction, no atmospheric interferences

should occur. It should be mentioned, that the lower detection limit for these monitors is not given by the thermal noise band of the photomultipliers, but by a luminescence of unknown origin, presumably arising from ozone.
The lower detection limit of commercial available instruments is in the 1 - 10 ppbv region. Specially designed NO detectors, as are described by McFarland [18], Schiff [19], Galbally [20], and Helas [21] have limits of detection in the range of 1 to 10 pptv. These instruments may be suitable for measurements in background air, whereas the commercial instruments are applicable to monitor urban air, but usually not rural air.

Optical absorption spectroscopy
The NO molecule has absorption band in the near UV at 215 and 226 nm [6], which can be used for determining the mixing ratio of this trace component in the atmosphere. Wavelength- dispersive light intensity measurements of a given source result in a spectrum where the amplitude of the cited bands can be used to calculate the mixing ratios. The method is similar to that one used for NO_2 (further references to the method cited there).
The detection limit for this method using a light path of 1 km is about 0.4 ppbv [6]. As the near UV light is subjected to scattering by dust and fog, this measurement method is limited to urban and rural air and to good visibility.

Fluorescence
The fluorescence of the NO molecule from the gamma-bands has been used to detect NO in the troposphere. Hoeinghaus et al. [22] describe how to use a microwave discharge of a mixture of helium and air to excite the NO molecule in the continuously sampled air. The fluorescence is monitored by a photomultiplier , which measures the light of the (0,1)-, (1,0)-, and (0,2)- bands through a suitable filter.
This measurement technique has a lower limit of detection of about 1 ppbv. Positive interferences are said to exist for SO_2 [23]. But Hoeinghaus et al.[22] showed, that the SO_2 molecule in air can give only a minor contribution (100 ppmv SO_2 give a similar fluorescence intensity as 1.4 ppmv NO). So this method should be useful for measuring urban air.

Fourier-Transform-Infrared-Spectroscopy
The FTIR measurement technique seems to be not suitable for NO measurements in tropospheric air due to the interference of water. Under laboratory conditions, however, this technique is a powerful tool for studying reactions of NO [24] and has been applied to mixtures containing NO down to about 1 ppbv [25]. This qualifies the FTIR technique for calibration comparison purposes of other field measurement devices.

Gas-chromatography
GC methods have been described for detection of NO [26], but the

detection limit of about 10 ppmv is unfavorably high and prevents its use for atmospheric measurements.

2.4. Measurement methods for nitrous acid

Detection methods for nitrous acid are given in table 4. Like shown at the discussion of ammonia, the methods are separated between liquid-phase and gas phase, reflecting the difference in handling the sample.

Table 4

Detection methods for HNO_2	~detection limit [ppbv]
1.) liquid - phase	
colorimetry	1
ion - chromatography	5 (in sol.)
2.) gas - phase	
optical absorption spectroscopy (354, 368 nm)	0.02 (5 km)
FTIR (791, 853 cm^{-1})	5 (1 km)

liquid-phase

Colorimetry

The colorimetric methods for NO_2 potentially are able to measure HNO_2 also, as these methods depend on the coupling of the nitrite ion for the dye development. Nash describes a physical separation of NO_2 and HNO_2 by utilizing their different solubilities in water [27]. He took in sequence 0.1 N alkali, which is said to absorb all the nitrous acid and a small portion of the nitrogen dioxide, and secondly 0.1 N alkali with 0.004 M guaiacol (o-methoxy-phenol), which will collect nearly all of the remaining nitrogen dioxide. The system has to be calibrated under standardized conditions.

Ion chromatography

Under this heading chromatographic techniques are summed up. They consist of a chromatographic column specially prepared for anions. Either a conductivity detector as in the ion- chromatograph or a UV-detector as used in High-Performance- Liquid-Chromatography units. After scavenging HNO_2 in impingers or on filters the accumulated amount is determined. For the ion-chromatograph a lower limit of detection of about 5 ppbv of NO_2^- in solution is cited [2], with an HPLC-UV- detector combination a similar detection can be obtained

[28]. This method and the previous one will be more limited by the stability of the sample then by the sensitivity of the method.

gas-phase

Optical absorption spectroscopy
The absorption bands of nitrous acid at 354.1 nm and 368.1 nm have successfully been used to detect and quantify HNO_2 in the troposphere. This method is described in more detail at the section of NO_2. By using a lightpath of 5 km length Perner and Platt [29] have a lower detection limit of about 25 pptv. Limiting conditions of this technique will be discussed when dealing with the measurement techniques of NO_2.

Fourier-Transform-Infrared Spectroscopy
Infrared absorption bands at 853 cm^{-1} of the nitrous acid molecule can be used to measure the amount of this compound in air. With the FTIR technique Pitts et al.[4] give a lower limit of detection of 5 ppbv when using a lightpath of 1 km length.

2.5. Measurement methods for NO_2

A compilation of used determination methods for nitrogen dioxide is given in table 5. The discussion will first deal with liquid analysis, then with gas-phase techniques.

Table 5

Detection methods for NO_2	detection limit [ppbv]	interference
1.) liquid - phase		
Saltzman, Jacobs-Hochheiser	1	oxidants
2.) gas-phase		
GC	10^4	H_2O
optical absorption spectroscopy (323-347, 354, 357, 365, 435-450nm)	10^{-1}	SO_2, fog
ESR	10^{-1}	(radicals)
photofragmentation (O + NO)	1	lum. reactions
chemiluminescence (NO_2 + O)	1	N_2O
chemiluminescence of NO after conversion and subtraction	10^{-3}	see text

Saltzman and Jacobs-Hochheiser

These are the classical procedures for the determination of NO_2. They make use of the absorption of the gaseous species into the liquid phase, and after a color development, the amount of NO_2 is determined photometrically. The methods differ mainly in the choice of the absorption medium. The Saltzman method uses the color forming reagent directly [30], whereas the Jacobs-Hochheisser takes an alkaline solution [31]. The efficiency of absorption differs in these procedures, and correction factors must be used. Attempts to improve the efficiency of absorption have also been made, sodium arsenite [32], triethanolamine and substituted naphthol [33] and guaiacol [27] were used. The procedures are calibrated by prepared nitrite solutions.

The lower limit of detection is about 1 ppbv depending on the sampling time used. These methods can tend to give higher results due to interferences from other oxidants [34].

Gas-chromatography

The gas-chromatographic method for the determination of NO_2 is described by Lawson and McAdie [26]. The use of this method, which has a lower limit of detection of about 10 ppmv, is limited by the interference of water, which will react on the chromatographic column with the compounds to be determined. Accordingly, this method is more restricted to emission measurements.

Optical absorption spectroscopy

This remote sensing technique makes use of the absorption bands of NO_2 in the UV and visible spectrum. Several bands may be used to derive column densities, and hence mixing ratios from the air volume investigated. Platt et al. used the 323 - 335 nm and the 336 - 347 nm region [35], Platt and Perner the bands at 354, 357, and 365 nm [36], whereas Brewer et al. [37], Kulkarni [38] and Noxon [39] took the region of 435 - 450 nm. While Brewer et al. and Kulkarni modified Dobson spectrometers, which were originally intended for ozone measurements, the others made use of rapid scanning spectrometers. Numerous scans are superimposed to obtain the spectrum of the region employed and the spectra are handled with computers. The mixing ratios are derived from the amplitude of the absorption bands. It is important that in this differential absorption technique the mixing ratios are calculated from ratios of measured light intensities, and provided the mixing ratios of other atmospheric constituents do not change, they do not affect the measurement of NO_2 [40]. Incandescent lamps or sun- or moonlight is used as a light source. With sun- or moonlight different zenith angles have to be used in the measurement in order to differentiate the tropospheric and the stratospheric portions of the determined column densities.

The method is sensitive to fog and to high amounts of SO_2, since the spectrum of SO_2 covers the bands of NO_2. The lower limit of detection depends on the length of the light path and on the visibility. With an incandescent lamp and a light path of about 5 km a limit of 0.1 ppbv can be attained. Thus the method is useful for determining NO_2 in continental background air. For more contaminated air the length of the light path must be reduced correspondingly.

Electron spin resonance

The ESR-technique is a powerful tool for the detection and measurement of atmospheric free radicals. Mihelcic et al. [41] describe the development of a cryogenic sampling device and the meaurement of NO_2. The air is sampled through a critical orifice into a vacuum. The radicals are directed in a molecular beam on a cold finger kept at 77 K. Here they are trapped in a matrix of water or CO_2, which is provided from the atmosphere. The matrix is rigid and the concentration is low enough to prevent recombination. The sticking probabilities of the trace components to be measured in the matrix have to be determined from calibration measurements.

The method has successfully been used in tropospheric air [42]. The detection limit is about 100 pptv for NO_2, making the method suitable for continental background air. Other radicals can interfere, but by an appropriate choice of spectrometer conditions at least one band of NO_2 does not overlap with bands of other molecules.

Photofragmentation

Photolysis of the NO_2 molecule can be achieved by appropriate light sources and leads to the formation of NO and of O-atoms. Both products can be measured. The determination of NO_2 via the NO will be discussed below. Here the evolving O-atoms are used. They are known to react with NO giving rise to a chemiluminescence, which extends from 400 nm into the infrared with a maximum intensity at approximately 650 nm. McClenny and others [43] used a Hg-Xe-lamp to photolyse the NO_2, added pure NO, and monitored the chemiluminescenct emission with wavelength lower than 600 nm by use of a filter. In this way the system discriminated against the luminescence of nitric oxide-ozone. Basically this system can measure ozone also when the photolysis lamp is switched off.

McClenny et al. [43] state to have a linear response up to about 1 ppmv NO_2 and a lower limit of detection of 1 ppbv. Accordingly, this method might be useful for urban air, if there were not the possibility of self-contamination by the NO needed for detection. Interferences might arise from luminescence in the same spectral band from other reactions.

Chemiluminescence of NO_2 + O

Instead of producing oxygen atoms from NO_2 as in the previous method, Black and Sigsby [44] added oxygen atoms, which were produced by thermal decay of ozone, to the sample air. Here the NO_2 of the sample air will be reduced to NO and the excess O- atoms will give rise to the luminescence from the reaction NO + O. Thus the luminescence glow is proportional to the total NO_x. Black and Sigsby [44] give a detection limit of 1 ppbv, the linear response is shown to exist up to about 200 ppmv. An interference may arise from the reaction of the O-atoms with atmospheric N_2O producing NO also. Nevertheless it seems to be difficult to calibrate the system as varying CO and hydrocarbon levels may influence the O-atom mixing ratio and with this the chemiluminescence.

Chemiluminescence of NO after conversion

The high sensitivity of the NO + O_3 luminescence reaction has been used to measure NO_2 in the troposphere. This is done by means of suitable converters, which reduce NO_2 to NO. High temperature conversion, chemical reduction, or photolysis may be used. The NO_2 molecule is known to undergo thermal decay at temperatures above about 450 K [45], but 1200 - 1300 K is required to make the reaction sufficiently fast.

The chemical reduction is effected either by molybdenum metal, where the high affinity of Mo to oxygen is utilized, or by carbon, where CO is formed, both kept at about 600 K. While high conversion rates can be obtained with these converters, their selectivity is questionable. The response of these converters to several organic NO_x compounds and HNO_3 has been reported [46,47].

Another chemical converter is crystalline $FeSO_4 . 7\ H_2O$. The "brown ring compounds" found in the common qualitative analysis for nitrate with $FeSO_4$ are known to all chemists. Once the NO_2 is reduced to NO, the NO is easily driven out of the iron-complex by N_2 [48]. The ferrous sulfate should not be used in dry air, as the $[Fe(H_2O)_6]^{+2}$-complex seems to disintegrate then and the conversion efficiency diminishes. This converter will have positive interference to those compounds, whose thermal decay produces NO_2 during the contact time like PAN or pernitric acid.

Photolytic converters for the reduction of NO_2 to NO by photodissociation at wavelengths shorter than 400 nm have been constructed [49,50]. The selectivity of this type of converter is good, but there exists an interference by traces of water inside the photolysis cell leading to varying conversion rates.

The very low detection limit of the NO + O_3 chemiluminescence monitors when used in conjunction with converters make then measurement systems appropriate for background air.

2.6. Measurement methods for nitric acid

In this section, again, a separation is made between techniques, whose detection is done in the liquid phase, and gas phase measurement methods. A short summary is given in table 6.

liquid phase

While in the coulometric method the gaseous compound is led into the measurement cell, where a wet chemical method is employed, the other methods collect the nitric acid and have a separate detection step.

Table 6

Detection methods for HNO_3	~ detection limit [ppbv]	interference
1.) liquid - phase		
coulometry	5	HCHO
colorimetry after reduction	10^{-1}	
ion - chromatography	5 (in sol.)	
GC with ECD	5	
2.) gas - phase		
FTIR (896 cm^{-1})	4 (1 km)	
chemiluminescence of NO after conversion and subtraction	10^{-2}	see text

Coulometry
Spicer et al. [51,52] report the use of a micro-coulometric method to determine HNO_3 in the troposphere. The procedure is based on a Mast micro-coulomb meter adapted to permit detection of acidic gases [53]. The nitric acid mixing ratio is obtained by the difference of two measurements, one which accepts all acidic gases and another one in which nitric acid is excluded by trapping it on nylon fibers. Ethylene is added to the sampled air to destroy the ozone in order to prevent its interference. Other interferences with typical gases present in air have been tested, but exept formic acid, where a slight response was seen, none have been found. Spicer gives a lower limit of detection of 5 ppbv HNO_3 for this method. This may suffice for urban air, but is too high for rural or background air.

Sampling of nitric acid
Gaseous nitric acid has to be differentiated from the aerosol- borne fraction. This is done by filters, a combination of impactor and a filter, or by use of denuder tubes. The first technique uses filters in series, the first filter deposites the aerosol , the second is usually impregnated in order to scavenge gaseous nitric acid [54,55,56]. In place of the first filter one may use an impactor. In the filter-filter set-up, deposits of aerosol may react with atmospheric constituents leading to the formation of artifact gaseous nitrate, which will be measured on the back-up filter. This interference is overcome in the denuder technique, which was described briefly in the ammonia section. Niessner et al. [57] and Stevens et al. [58] have described this procedure.

Colorimetry
The classical method for the determination is colorimetry. The filters or the denuder coatings are eluted with water and the extracts analyzed. Several different procedures have been used.
Direct determination without reduction is achieved with the brucine-

method [59]. In order to obtain reproducible results, it is absolutely essential to adhere rigorously to the quantitative details chosen, as the method is sensitive to changes in temperature, H_2SO_4 concentration and the time of color development. The extinction is measured at 420 nm.
Reduction of nitrate to nitrite can be done with amalgamated cadmium granulate and when using a "Griess"-type reagent [60], an azo-dye is formed, whose extinction can be measured near 540 nm [61].
Sometimes a complete reduction to ammonia is also applied. The "Berthelot"-reaction then provides an azo-dye, which is photometrically determined at 623 nm [62,63].

Ion-chromatography
Aqueous solutions of nitrate can be analyzed by ion-chromatography. Similar to what has been said in the section on nitrous acid, detection limits of about 5 ppbm NO_3^- in solution can be obtained by the ion-chromatography technique [2] or with an HPLC-UV-detector combination [28].

Gas-Chromatography with Electron-Capture-Detector
Instead of forming an azo-dye, benzene [64] or substituted benzenes [65] as electronegative groups have been coupled to nitrate, so that these nitrocompounds can be detected with an electron-capture-detector. The coupling reaction appears to proceed not efficiently at lower concentrations ($<2\times10^{-5}$ M NO_3^-). Also interferences of halide ions are stated by both groups of compounds cited above.

gas phase

Fourier-Transform Infrared Spectroscopy
The FTIR-method, described in earlier sections, has been used to determine the amount of nitric acid in air. The detection limit of 4 ppbv HNO3 at a light-path length of 1 km (band at 896 cm^{-1}, [4]) points to a use of this system in urban air, i. e. the air surrounding the place of the position of the instrument.

Chemiluminescence after conversion
It was noted by Joseph and Spicer [47], that NO_x-chemiluminescence analyzers using high temperature reduction of NO_2 also respond to HNO_3. A quantitative reduction of HNO_3 to NO was found up to about 1 ppmv with a molybdenum converter kept at 625 - 675 K. The amount of nitric acid is determined by the difference between two signals, one due to the sum of NO_x and HNO_3, the other due to NO_x, when HNO_3 is removed with a nylon filter. Kelly et al. [66] have instead chosen to decompose nitric acid in a quartz tube, which was filled with pyrex beads and kept at 625 K, and then measured the products via a ferrous sulfate converter.
In this experimental set-up the detection limit is given by the NO analyzer and the error due to the subtraction. With a sensitive NO analyzer 10 pptv or better can be achieved. Interferences are

expected by those substances, which will release NO on cracking in the hot converter. PAN, alkylnitrites and -nitrates, and nitroalkanes [46] are expected to behave this way.

2.7 Measurement method for the nitrate-radical

Table 7

Detection method for NO_3	~detection limit [ppbv]
optical absorption spectroscopy (623,662 nm)	5×10^{-3} (1 km)

Optical absorption spectroscopy
This technique has been used for the detection and measurement of the NO_3 radical also. Here the bands in the red region of the spectrum at 623 and 662 nm [67,68] have been utilized. The experimental set-up is described at the NO_2 section. The lower limit of detection was given by Platt [67]as about 5 pptv NO_3 for 1 km pathlength. As stated above, atmospheric perturbations may have influence on this technique and thus lift up the detection limit.

2.8. Measurement methods for PAN

Detection methods for PAN are given in table 8.

Table 8

Detection methods for PAN	~ detection limit [ppbv]
GC with ECD	15×10^{-3}
LAS (1150 cm^{-1})	0.3
FTIR (793, 1162 cm^{-1})	5

Gas-Chromatography with Electron-Capture-Detector
The measurement of peroxyacetylnitrate is commonly done with a gas chromatograph in combination with an electron capture detector [69]. This technique may go down to a lower limit of detection of about 5 pptv in a 5 ml air sample [70]. This highly sensitive method is difficult to calibrate as is stated by Penkett [69].

Laser Absorption Spectroscopy
Reid et al. [71] point out that PAN can be measured with laser absorption spectroscopy with use of the bands at 1150 cm^{-1}. They expect for their tunable diode laser system with a pathlength of 200 m a lower detection limit of about 300 pptv. This will be sufficient for urban air, but rarely suffice for background air.

Fourier-Transform Infrared Spectroscopy
Similar to the laser absorption spectroscopy, the FTIR spectroscopy can be used to measure PAN. Pitts et al. used bands at 793 and 1162 cm^{-1} and obtained a detection limit of about 5 ppbv when using a lightpath of 1 km. Thus this technique may be not sufficient sensitive for the expected outdoor PAN mixing ratios.

3. REFERENCES

1) W. J. Mueller, Diplomarbeit, Frankfurt 1972
2) Handbuch der Ionenchromatographie, Biotronik, Frankfurt
3) M. Ferm, Atmos. Environ., 13,1385 (1979)
4) J. Pitts, Jr., B. J. Finlayson-Pitts, and A. M. Winer Environ. Sci. Tech., 11,568 (1977)
5) J. Reid, B. K. Garside, J. Shewchun, M. El-Sherbiny, and E. A. Balik, Appl. Opt., 17,1806 (1978)
6) D. Perner, private communication
7) J. E. Sigsby, Jr., F. M. Black, T. A. Bellar, and D. L. Klosterman, Environ. Sci. Tech., 7,51 (1973)
8) L. P. Breitenbach and M. Shelef, J. Air Pollut. Control Assoc., 23,128 (1973)
9) V. P. Aneja, E. P. Stahel, H. H. Rogers, A. M. Witherspoon, and W. W. Heck, Anal. Chem., 50,1705 (1978)
10) R. Bock und K. Schuetz, Z. Anal. Chem., 237,321 (1968)
11) R.A. Rasmussen, J. Krasnec, and D. Pierotti, Geophys. Res. Lett., 3,615 (1976)
12) W. Seiler, private communication
13) R. F. Weiss, C. D. Keeling, and H. Craig, J. Geophys. Res., 86,7197 (1981)
14) P. S. Connel, R. A. Perry, and C. J. Howard, Geophys. Res. Lett., 7,1093, (1980)
15) B. A. Ridley, K. H. Welge, H. I. Schiff, L. R. Megill and A. W. Shaw, Second Conference on CIAP, 1972 p.146f
16) B. A. Ridley and L. C. Howlett, Rev. Sci. Instrum., 45,742 (1974)
17) D. M. Steffenson and D. H. Stedman, Anal. Chem., 46,1704 (1974)

18) M. McFarland, D. Kley, J. W. Drummond, A. L. Schmeltekopf, and R. H. Winkler, Geophys. Res. Lett., 6,605 (1979)
19) H. I. Schiff, D. Pepper, and B. A. Ridley, J. Geophys. Res., 84,7895 (1979)
20) I. E. Galbally, Spec. Environ. Report 10, p. 10f, World Meteorol. Organ., Geneva, 1977
21) G. Helas, M. Flanz, and P. Warneck, Intern. J. Environ. Anal. Chem., 10,155 (1981)
22) K. Hoeinghaus, H. W. Biermann, C. Zetsch und F. Stuhl, Z. Naturforsch., 31a,239 (1976)
23) F. P. Schwarz and H. Okabe, Anal. Chem., 47,703 (1975)
24) S. C. Bhatia and J. H. Hall, Jr., J. Phys. Chem., 84,3255 (1980)
25) D. Griffith, private communication
26) A. Lawson and H. G. McAdie, J. Chromatog. Sci., 8,731 (1970)
27) T. Nash, Tellus, 26,175 (1974)
28) U. Leuenberger, R. Gauch, K. Rieder, and E. Baumgartner, J. Chromatogr., 202,461 (1980)
29) D. Perner and U. Platt, Geophys. Res. Lett., 6,917 (1979)
30) B. E. Saltzman, Anal. Chem., 26,1949 (1954)
31) M. B. Jacobs and S. Hochheiser, Anal. Chem., 30,426 (1958)
32) A. A. Christie, R. G. Lidzey, and D. W. F. Radford, Analyst, 95,519 (1970)
33) C. Huygen and P. H. Steerman, Atmos. Environ., 5,887 (1971)
34) E. H. Adema, Anal. Chem., 51,1002 (1979)
35) U. Platt, D. Perner, and H. W. Paetz, J. Geophys. Res., 84,6329 (1979)
36) U. Platt and D. Perner, J. Geophys. Res., 85,7453 (1980)
37) A. W. Brewer, C. T. McElroy, and J. B. Kerr, Nature 246, 129 (1973)
38) R. N. Kulkarni, J. Atmos Sci., 32,1641 (1975)
39) J. F. Noxon, Science, 189,547 (1975)
40) U. Platt, Atmos. Environ., 12,363 (1978)
41) D. Mihelcic, D. H. Ehhalt, G. F. Kulessa, J. Klomfass, M. Trainer, U. Schmidt, and H. Roehrs, Pageoph, 116,530 (1978)
42) D. Mihelcic, P. Muesgen, W. Paetz, M. Helten, M. Trainer, H. Fark, D. Kempa und D. H. Ehhalt, BPT-Bericht 2/81 p. 54f (1981)
43) W. A. McClenny, J. A. Hodgeson, and J. P. Bell, Anal. Chem., 45,1514 (1973)
44) F. M. Black and J. E. Sigsby, Environ. Sci. Tech., 8, 149 (1974)
45) F. A. Cotton und G. Wilkinson, Anorganische Chemie, p. 323 2. Auflage, Verlag Chemie, Weinheim, 1970
46) A. M. Winer, J. W. Peters, J. P. Smith, and J. N. Pitts,Jr, Environ. Sci. Tech., 8,1118 (1974)
47) D. W. Joseph and C. W. Spicer, Anal. Chem., 50,1400 (1978)
48) F. A. Cotton und G. Wilkinson, Anorganische Chemie, p. 700 2. Auflage, Verlag Chemie, Weinheim, 1970
49) D. Kley and M. McFarland, Atmospheric Tech. 12,63 (1980)
50) U. Goede, Diplomarbeit, Bonn, 1981
51) C. W. Spicer, G. F. Ward, and B. W. Gay, Jr., Anal. Lett., A11,85 (1978)

52) D. F. Miller and C. W. Spicer, J. Air. Pollut. Control Assoc., 25,940 (1975)
53) D. F. Miller, Wm. E. Wilson, Jr., and R. G. Kling J. Air. Pollut. Control Assoc., 21,414 (1971)
54) B. J. Hubert A. L. Lazrus, J. Geophys. Res., 85,7322 (1980)
55) B. R. Appel, Y. Tokiwa, and M. Haik, Atmos. Environ., 15,283 (1981)
56) T. Okita, S. Morimoto, and M. Izawa, Atmos. Environ., 10,1085 (1976)
57) R. Niessner and D. Klockow, Intern. J. Environ. Anal. Chem 8,163 (1980)
58) R. K. Stevens, T. G. Dzubay, G. Russwurm, and D. Rickel, Atmos. Environ., 12,55 (1978)
59) H. Fadrus and Z. Maly, Z. Anal. Chem., 202,165 (1964)
60) W. Werner, Z. Anal. Chem., 304,117 (1980)
61) A. W. Morris and J. P. Riley, Anal. Chim. Acta, 29,272 (1963)
62) B. J. Huebert and A. L. Lazrus, J. Geophys. Res., 85,7322 (1980)
63) W. Jaeschke, this issue
64) J. W. Tesch, W. R. Rehg, and R. E. Sievers, J. Chromatogr., 126,743 (1976)
65) R. L. Tanner, R. Fajer, and J. Gaffney, Anal. Chem., 51, 865 (1979)
66) T. J. Kelly, D. H. Stedman, and G. L. Kok, Geophys. Res. Lett., 6,375 (1979)
67) U. Platt and D. Perner, Geophys. Res. Lett., 7,89 (1980)
68) J. F. Noxon, R. B. Norton, and W. R. Henderson, Geophys. Res. Lett., 5,675 (1978)
69) S. A. Penkett, F. J. Sandalls, and B. M. R. Jones, VDI Berichte 270,47 (1977)
70) S. A. Penkett, Toxicol. Environ. Chem., 3,291 (1981)
71) J. Reid, J. Shewchun, B. K. Garside, and E. A. Ballik, Appl. Optics, 17,300 (1978)

ANALYTICAL CHEMISTRY OF THE ATMOSPHERIC AEROSOL

Dieter Klockow

University of Dortmund, Department of Chemistry,
P.O.Box 50 05 00, 4600 Dortmund 50, F.R.G.

In the following contribution it is shown, how analytical chemistry can be used for the characterization of the atmospheric aerosol. After an introductory discussion about nature and origin of aerosol particles the techniques for sampling of airborne particulate matter prior to analysis are treated in detail. Particle collection methods without and with size fractionation as well as sampling artifacts are included. In the following section the analytical methodology applied to the investigation of the collected material is summarized with respect to principal aerosol constituents such as elements of crustal origin, organic carbon, anthropogenic trace metals, and conversion products of reactive trace gases (SO_2; NO_x, NH_3). Finally the technical requirements for surface and depth profile measurements of solid aerosol are briefly discussed.

ORIGIN AND NATURE OF AIRBORNE PARTICULATE MATTER

Regarded only superficially our terrestrial atmosphere seems to be just a mixture of several major and minor gaseous elements and compounds. As a matter of fact, however, it behaves like a colloidal system, a so called a e r o s o l , which generally spoken is a mixture of very small droplets and (or) particles suspended in a gas matrix. In case of the a t m o s p h e r i c a e r o s o l (1) the gas matrix consists of N_2, O_2, noble gases, CO_2, water vapor and a wide variety of other gases present at the trace level, and the suspended material (droplets + solid particles = a i r b o r n e p a r t i c u l a t e m a t t e r) exists in a size range between some 10^{-3} and some $10^2 \mu m$ diameter. It is important to note that the atmospheric aerosol is

H. W. Georgii and W. Jaeschke (eds.), Chemistry of the Unpolluted and Polluted Troposphere, 57–90.

characterized by a substantial interaction between the gas phase and the particle phase, the extent of which is dependent on the physical properties and the chemical reactivity of the respective compounds. The subject of this contribution is the investigation of essentially the particulate fraction of the atmospheric aerosol, but because of the interaction just mentioned the gas phase cannot be completely left out.

Everybody can detect quite easily air pollution by particulate emissions because they scatter and absorb light and therefore affect visibility. This property might have been one of the reasons for King Edward I of England to prohibit in 1273 after Christ the use of coal for house heating in London during sessions of the parliament (2, 2a): "Be it known to all within the sound of my voice, whosoever shall be found guilty of burning coal shall suffer the loss of his head". So the year 1273 seems to be the first significant date in the history of air pollution legislation, and the proclamation of Edward I was directed towards a specific emission source which also today contributes substantially to airborne particle mass. There are existing, of course, many other sources and processes which produce particles, and it is important to know about nature and strengths of these productions in order to have an idea about the chemical composition of the particulate fraction of the atmospheric aerosol (1).

Sea spray, soil dust raise by winds, volcanic eruptions, and cement manufacturing are typical processes with *primary particle production*. In these cases material is emitted directly in the particulate state. During fossil fuel combustions or ferrous and nonferrous metallurgical operations *secondary particles* are formed besides primary ones by condensation from vapor after its escape from a high temperature sources. Homogeneous or heterogeneous chemical transformations of airborne trace gases (*gas-to-particle conversion*) also lead to secondary particulate matter and are of special concern for reactive compounds such as SO_2, NO_2 or NH_3.

In 1971 Robinson and Robbins (3) have published global estimates of source strengths of the different particle production processes mentioned. From their data it can be derived (see Table 1) that the production of atmospheric particulate matter is dominated by *natural sources* (e.g. sea spray, NH_4^+ from NH_3). The strength of *anthropogenic sources* (e.g. coal combustion, sulfate from SO_2) amounts to merely 10 - 15 % of the total particle mass production. It must be mentioned, however, that this result holds true only for global considerations. Under local urban conditions the ambient aerosol may be totally dominated by particulate matter of human origin.

	Strength of particle production $[10^{12}$ g/a]		
Source	Primary	Secondary	Total
Natural	1207	1105	2312
Anthropogenic	92	204	296

Table 1. Global estimate of atmospheric particle mass production (3, 4).

Besides its chemical composition the particulate phase of the atmospheric aerosol is characterized by several physical parameters, the most important one being its size distribution (1). One can derive distributions with respect to number, surface, volume or mass of the particles. (Volume and mass are connected to each other by spec. gravity). From the analytical point of view volume or mass distributions, respectively, are the most appropriate ones, because mass concentrations are the parameters to be determined most frequently.

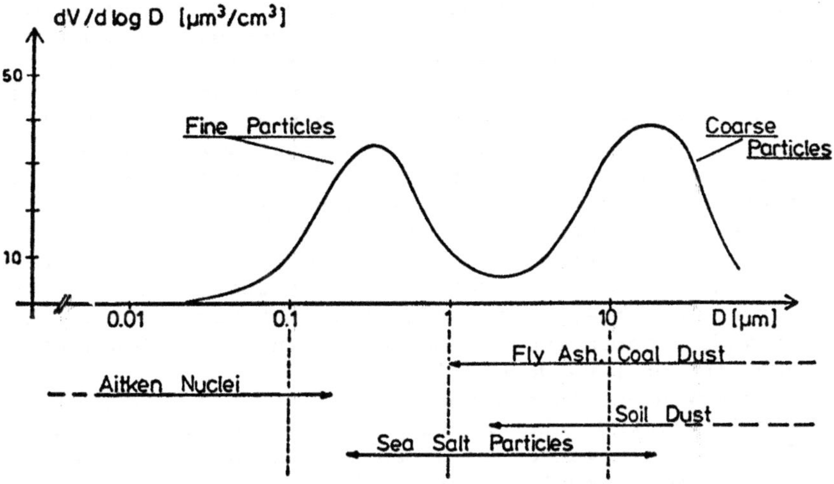

FIGURE 1. Atmospheric particle volume distribution

Results of investigations of urban and non-urban aerosols by Whitby et al. (5, 6) have led to the conclusion that two major modes exist in the atmospheric particle volume (mass) distribution (see Figure 1). The submicron part - the fine particle mode - is largely maintained by anthropogenic sources of small primary and secondary particles and by coagulation of particulate matter below 0.05 µm diameter (transient nuclei) originating from chemical conversions of atmospheric trace gases (gas-to-particle conversion). The supermicron fraction - the c o a r s e p a r t i c l e m o d e - is dominated by natural and anthropogenic sources of coarse primary and secondary particles (sea salt, soil dust, fly ash etc.). The analytical investigation of the submicron mode in ambient air is of special interest, because fine particles,

- exhibit low mobility and gravitational settling and therefore have the highest residence time in ambient air;
- respresent the maximum in particle surface area and therefore are important for heterogeneous conversion processes;
- are responsible (in connection with humidity) for visibility degradation;
- belong to almost 100 % to the thoracic fraction of inspired dust (health effects!).

From what has been discussed until now about sources and size distributions of airborne particles a qualitative conception about their chemical composition can be developed (6, 7, 8):

- In the fine particle mode we should find all the products of gas-to-particle conversion: sulfate and other sulfur compounds, nitrate (which also occurs in the coarse mode), protons of strong and weak acids, ammonium, and organic material (9) from natural and anthropogenic sources. A substantial mass fraction in this mode should consist of liquid water (10) forming submicron droplets with deliquescent compounds such as NH_4HSO_4. Part of the organic matter found as well as elemental carbon has been produced most probably in the form of submicron particles during combustions. The same applies to bromine and lead (auto exhaust), and to some other metals such as cadmium or zinc. These elements and their compounds predominantly occur in the submicron mode because they condense to fine particles from the vapor phase in hot flue gases (11, 12).
- The bulk constituents in the coarse mode, silicon dioxide, silicates, aluminium, and calcium originate from wind blown soil dust and particulate matter of human origin, as for instance emissions of cement and iron works. The same sources are responsible for the occurence of some minor components like iron, titanium, magnesium, sodium, potassium, carbonate, and chloride. The main production of chloride, sodium and magnesium containing particles, however, is by sea spray.

Though all the elements, ions or compounds mentioned can be detected in the particle portion of ambient air, their mass fractions may vary within a wide range depending on the site of collecting the material: Quite different results will be obtained analysing background samples and samples from a polluted area, or samples of marine and continental origin.

A further complication arises from the fact, that single aerosol particles not necessarily consist of well defined chemical compounds. They may have been generated in such a form (e.g. NaCl), but after some time are changed e.g. to a mixed chloride/nitrate salt by conversion and nucleation of NO_2 (13). This transformation of an atmospheric aerosol form the state of a so called *external mixture* to the state of an *internal mixture* (equilibrium mixture) (14, 15) has a high degree of plausibility, but actually little is known about the extent of such processes. What has been shown, however, on particles from combustion sources, is the existence of striking *depth profiles* of certain substances concentrated at the surface (16).

Finally it should be mentioned at this point that besides the "dead" organic and inorganic particulate material we also have *viable particles* present in the atmospheric aerosol such as pollen, spores, bacteria, viruses or small insects (17). These are normally not considered to be of any concern for results obtained in the analysis of airborne dust. But dependent on the season and the sampling site chosen sizable amounts of viable particles may be collected together with *nonviable* ones.

SAMPLING OF PARTICULATE MATTER FOR ANALYSIS

As in the other branches of analytical chemistry the sampling step is the most crucial one also in air analyis. In order to obtain reliable analytical results, we must define a sample of finite dimensions within an extended population, the ambient atmosphere. This sample is to be transferred into a suitable analytical system with as little changes in physical state and chemical composition as possible. Under a few favorable circumstances we may be able to apply an in-situ *remote sensing* technique (18 - 20) to the determination of atmospheric trace compounds without any material manipulation. Here we only have the problem of defining the air volume or the air mass in which the analytical signal is generated. In most cases, however, sampling means extraction and preconcentration of gaseous and (or) particulate trace constituents from ambient air.

The transfer of a representative ambient aerosol sample, containing particles in a wide size range, into an appropriate measurement system gives rise to serious technical problems already at an early stage of the whole procedure. It is not only the sampling inlet which must be designed with care (21), but sometimes also the piping leading to the different particle collectors and measuring devices.

In many monitoring stations the air under study is transported through a central intake duct to which the actual analytical sampling trains are connected by bent pipes. Another possibility is to bend the intake duct and to take out of it a sample stream by a straight pipe (64) (Figure 2). At the sampling points (P in Figure 2) a significant particle fractionation due to different inertia of particles of different size may lead to wrong analytical results. Such an unintended particle segregation is operative as long as the linear velocity of the main stream flow is different from the velocity within the pipe and can be observed with particles above about 3 μm diameter (see Figure 2). Only when working under i s o k i n e t i c conditions, this means using thin walled pipes aligned with the main stream flow and pulling air through them without changing the linear velocity at the sampling point, a particle collection representative of the main stream aerosol can be expected to occur.

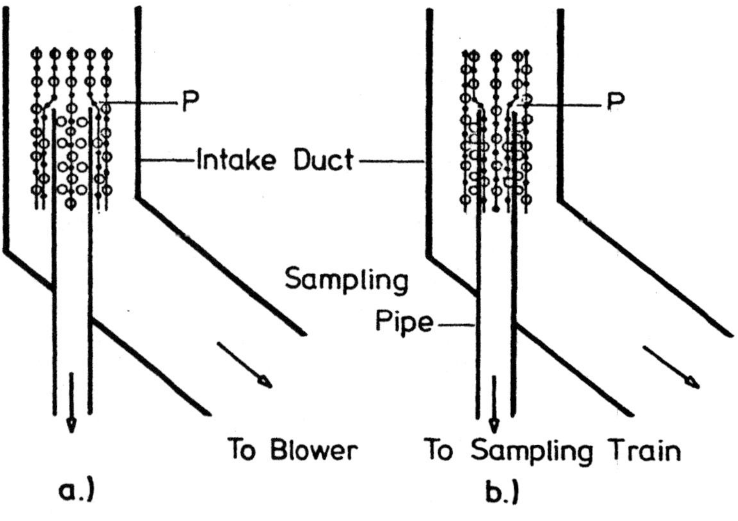

FIGURE 2. Particle fractionation by non-isokinetic sampling, stream lines bent at sampling point P: a.) sample velocity in the pipe too low, too many coarse particles collected; b.) sample velocity in the pipe too high, too few coarse particles collected

The extraction of particulate matter from the suspending gasphase in the sampling train can be accomplished by techniques based on a few physical principles: collision, adhesion, inertia, and diffusion in flow fields under the influence of gravitational and (or) electrical forces. Five of these techniques are treated in the following sections.

Filtration

The most frequently used means to extract and to preconcentrate suspended particles from ambient air is f i l t r a t i o n (22). It can be used as such or in combination with other techniques. Sample volumes usually employed for collection of airborne particulate material by filtration range from a few hundred liters (low volume sampling) up to a few thousand cubic meters (high volume sampling), depending on which information has to be obtained and which analytical methods are available.

For removal of particles from a gas stream by filtration (see Figure 3) collection media are in use, which are made of fibrous fabric (22), porous synthetic polymer membranes, or of membranes which were subjected to charged particle irradiation and chemical etching in order to obtain uniform cylindrical pores (Nuclepore^R filter) (23 - 25). Because different mechanisms such as i n t e r c e p t i o n , i m p a c t i o n , and d i f f u s i o n (to the surface of fibres or walls of pores) are effective in air filtration, particles can be captured o n or i n a filter. The extent to which the processes mentioned contribute to the overall collection efficiency of a particular filtration system is dependent on several experimental parameters: filter type used, size distribution of the particulates to be collected, and linear velocity of the aerosol stream pulled through the filter (face velocity). In addition filter performance is a function of sampling time because particles already collected reduce the effective pore diameter. Therefore the initial collection efficiency of a sampling filter is lower than its efficiency at a later state.

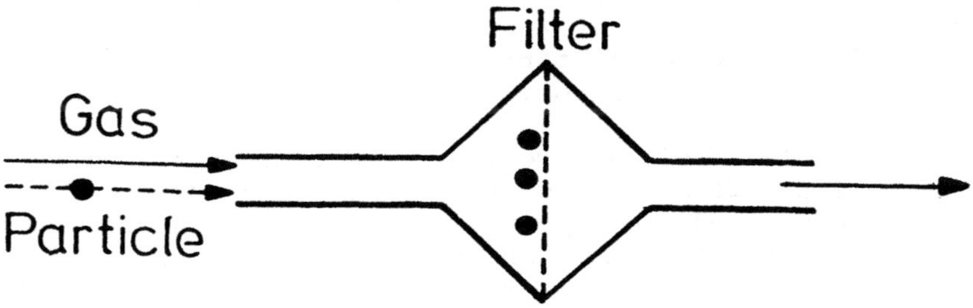

FIGURE 3. Air filtration

In air filtration it is generally not difficult to collect coarse particles (> 1 μm) and very fine particles (< 0.01 μm) with efficiencies near 100 % because in these ranges the mechanisms of impaction and diffusion, respectively, are extremely effective in removing suspended material from a gas stream. In the fine particle mode (0.01 - 1 μm), however, no dominant separation mechanism exists and distinct minima in collection efficiency can be computed and experimentally observed (24 - 26). For efficient extraction of submicron particles from ambient aerosol by filtration it is therefore necessary to select carefully filters with proper collection characteristics (26, 27).

Besides an insufficient sampling efficiency another difficulty may arise during filter collection of aerosol particles, namely a change in composition of the particulate phase on the filter surface. Such a change is of special concern in the case of analysing air polluted by SO_2 and sulfates, or by NO_2 and nitrates.

An increase in sulfate due to a conversion of SO_2 on filter surfaces has been found by several authors (28 - 32). What really happens is not easy to predict, because not only the filter material used but also the type of dust collected seems to be important for a SO_2 conversion (33, 34). In the case of analysing filter samples not for sulfate in general but for sulfate species such as sulfuric acid or acidic ammonium sulfates, the situation is even more complicated. Besides SO_2 conversion, partial neutralization of the acid by NH_3 (35, 36) or irreversible reactions with the filter material used (36 - 39) artifacts due to particle/particle interactions (36, 39 - 41) may have a significant influence on the result. This effect holds for all particle preconcentration procedures and is illustrated by the next two figures.

In the experiment represented by Figure 4 H_2SO_4 droplets and Na_2SO_4 particles were collected on the same polycarbonate filter. Then it was tried to extract the H_2SO_4 with benzaldehyde, because this solvent had been recommended as a selective extractant for H_2SO_4 (39). As one can see, the extraction yield rapidly drops with an increasing molar ratio Na_2SO_4 : H_2SO_4. A similar curve was obtained for the system $(NH_4)_2SO_4/H_2SO_4$ (36), and both findings were explained by the formation of acidic sulfates which are not extracted by benzaldehyde.

What needs to be clarified, however, are the processes leading to such conversions. If one takes into account the surface of the filter used, the mass of the deposited material and the diameter of the collected particles, then one must draw the conclusion from it that there is no chance for the particles to "meet" and to react with each other. If, however, the particles would distribute inhomogeneously on the filter, then such a chance would be much higher.

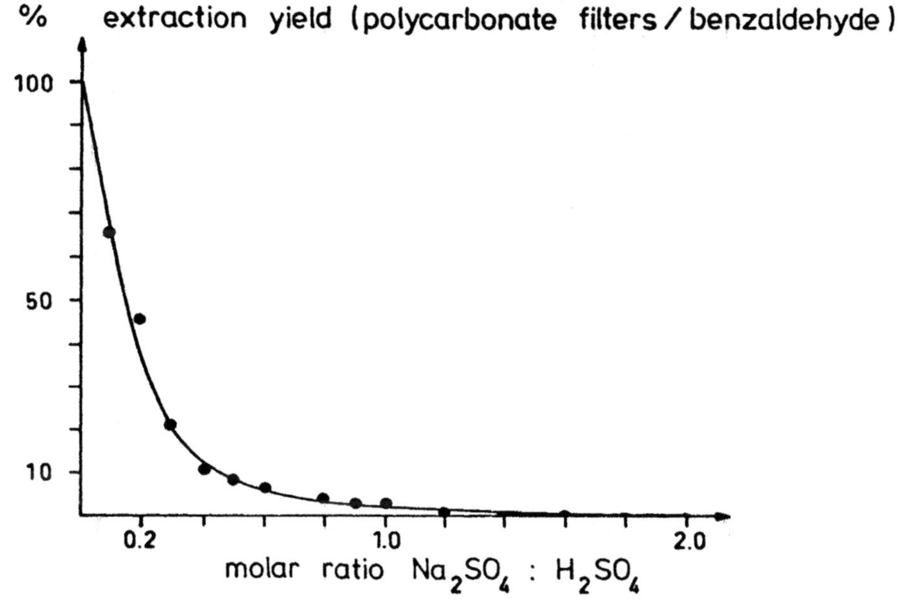

FIGURE 4. Percent H_2SO_4 extractable with benzaldehyde from Nuclepore[R] filter as a function of the molar ratio Na_2SO_4/H_2SO_4. (51×10^{-9} M H_2SO_4 per filter).

An inhomogeneous distribution can result from an improper design of the inlet of the filter holder used and from the separation mechanism of the filter itself. The latter effect can be seen from Figure 5, which shows a scanning electron photomicrograph of the surface of a sponge type PTFE-filter, loaded with $(NH_4)_2SO_4$ particles. These are distributed neither homogeneously nor randomly but accumulate around and in the pores of the filter. This means, there really is a chance that different particles react with each other, and that new compounds can be formed from sulfuric acid and sulfates on a filter.

Such topochemical reactions can be made really evident by collecting H_2SO_4 droplets and halide salt particles on one and the same filter. In this case the reaction products are volatile mineral acids and stable sulfate salts, so that the occurence of a reaction is indicated by the loss of halogen (36, 42). In a similar way an interaction between ammonium salts and alkaline particles may result in a loss of NH_3 (43).

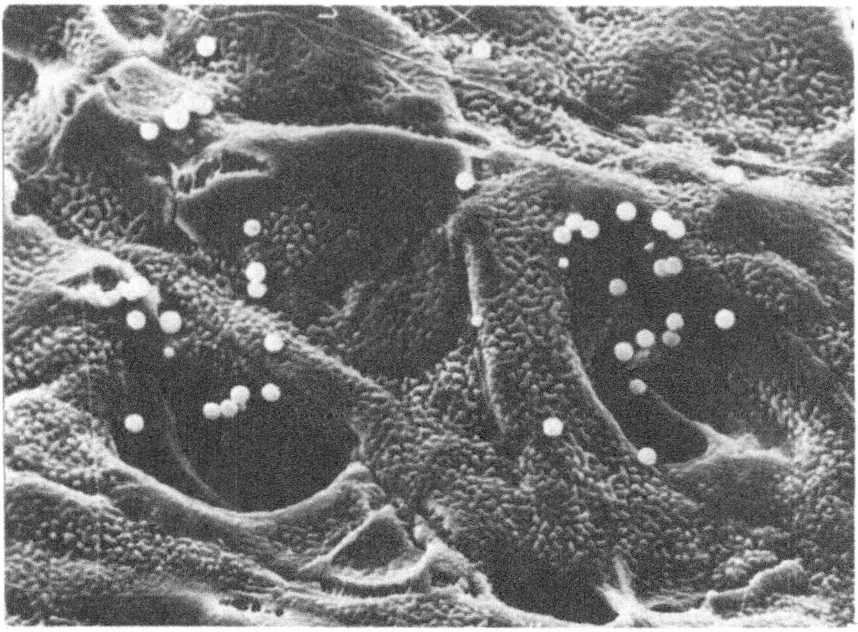

FIGURE 5. Scanning electron photomicrograph showing inhomogeneous distribution of $(NH_4)_2SO_4$ particles (3.4 µm) on Mitex^R filter

Filter sampling of nitrogen containing air pollutants causes problems of similar seriousness. The influence of NH_3 has been mentioned already. Spicer and Schumacher (44) as well as Witz and Mac Phee (31) and Pierson et al. (32) reported about NO_2/filter interactions which lead to the formation of artifact nitrate especially when using glass fibre filters as collection media. On the other hand Harker et al. (45) and Nießner (48) observed a loss of nitrate during collection of H_2SO_4 on filters loaded with particulate matter. In this case partial volatilization of HNO_3 has been most probably the reason for the decrease in nitrate of the collected material. But NH_4NO_3 exhibits also a strong tendency to get volatilized (46 - 48) because of its significant decomposition pressure (49, 50). It can be assumed that airborne gaseous nitric acid strongly interacts with some of the filter types commonly used for gas/particle separation (44, 51). In one case (nylon filter) this interaction is irreversible and quantitative and has therefore been employed for direct and indirect determination of nitric acid in ambient air (47, 52, 53).

Diffusion Separation

A most promising approach to selective collection of reactive trace gases and particles during the sampling step is d i f f u s i o n s e p a r a t i o n . The set up (Figure 6) is very simple but very effective. The air under study is drawn through a tube - called a d e n u d e r - the inner walls of which are coated with a suitable trapping agent for trace gases. At well adjusted conditions for flow rate and tube length (43) the small gas molecules diffuse to the walls and are caught here completely by the absorbent. The less mobile particles, however, are passing through the tube and can be collected on a back-up filter. Examples for the application of a diffusion separator are the stripping of SO_2 (54, 55) or NH_3 (56) from aerosols using denuders coated with PbO_2 (54), $K_2[HgCl_4]$ (55), and H_3PO_3 (56), respectively. Besides an effective gas/particle separation denuders offer the possibility to determine quantitatively the sorbed species. Techniques for the determination of reactive atmospheric trace gases such as NH_3 (43, 60), HCl (55, 57 - 59), and HNO_3 (55, 60) using preconcentration in diffusion tubes have been described in the literature during the past years.

FIGURE 6. Diffusion separation combined with filtration.

It would be really desirable that denuders are introduced to a larger extent into practical air sampling, because by this means the very unpleasant gas/particle and gas/filter interactions can be avoided with small expenditure. As will be shown later in connection with the determination of airborne strong acids and ammonium salts a surprising degree of selectivity can be achieved by using diffusion separators.

Size Fractionation of Particles by Inertial Force

It has been shown already, that different airborne chemical species selectively show up in certain particle size ranges. For studying health effects or atmospheric accumulation processes it is therefore necessary to collect particulate matter in separate size fractions. The most widely used sampling technique for that purpose involves the removal of the particles from an air stream by i m p a c t i o n on a plate (61). The air is directed through a narrow slit or a nozzle onto an opposing surface (see Figure 7) causing the streamlines to be bent abruptly. Particles in the air stream are subjected to the opposing forces of the viscous interaction with the carrier gas and the inertial "compulsion" to move in a straight line. If air velocity and nozzle-plate distance are suitably adjusted, particles above a certain diameter with sufficient inertia overcome the drag force and impact onto the plate while the small particles remain airborne. If a set of such stages with decreasing nozzle diameter and nozzle-plate distance is operated in a series arrangement samples in a sequence of successively smaller particle size ranges, down to submicrometer dimension, are collected.

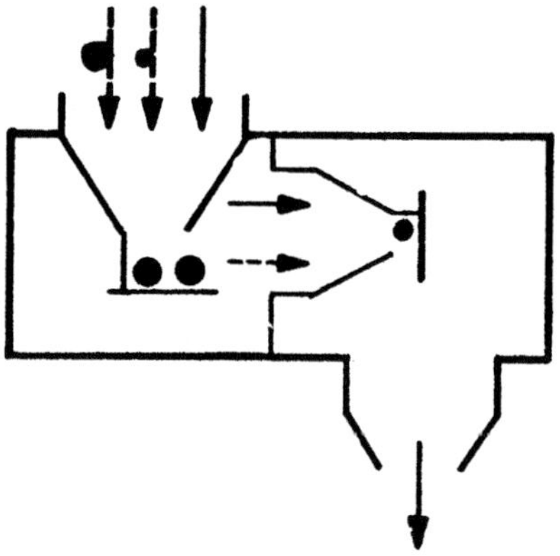

FIGURE 7. Inertial impaction.

In spite of the unquestionable advantages of such a conventional impactor it exhibits some factors of uncertainty:

- Particles may fail to adhere to the surface of the respective impaction plate and become reentrained in the air flow. This "bounce off" phenomenon results in larger particles being collected in a smaller particle fraction.
- The particles are collected on an impactor plate within a small area below the slit or nozzle, resulting in the accumulation of high density deposits. Particles can be "blown off" from the surface of such deposits and be recollected by a later stage.
- Within the deposits the probability for artifact formation by particle/particle interaction is rather high.

- Particle deposition on the walls between the different stages of an impactor can amount to substantial dimensions.
- Sudden pressure drop behind the nozzle in a certain impactor stage may cause losses of volatile or decomposable (NH_4Cl, NH_4NO_3!) substances.

In order to avoid "bounce off" and "blow off" and to circumvent at least partially the other difficulties mentioned, the technique of opposing jets (62) or a so called v i r t u a l i m p a c t o r (55, 63, 64) can be used for size classifying sampling.

In its simplest form (see Figure 8) a virtual impactor is able to divide a polydisperse particle population into two size ranges (d i c h o t o m o u s sampler).

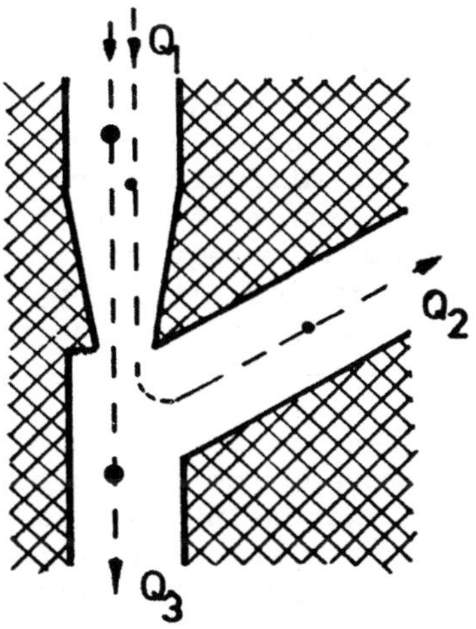

FIGURE 8. Schematic of a virtual impactor (55). Total flow rate [$l \cdot h^{-1}$]: $Q_1 = Q_2 + Q_3$; $Q_2 \gg Q_3$.

An aerosol flow of Q_1 enters the impactor through a conical nozzle. At the end of the nozzle the flow is divided into two portions of Q_2 and Q_3 with $Q_2 \gg Q_3$. Here a virtual impactor plate of stagnant air is built up because the streamlines are drastically turned by an angle of 90° (63, 64) or 120° (55). Only large particles are able to break through this barrier whereas the fine ones follow the direction of the main flow according to the ratio Q_2/Q_1. Both fractions can be collected on backup filters but can alternatively be introduced directly into a suitable analyser system (55). This is a unique advantage of any virtual impactor. If its cut diameter can be set to the minimum of the bimodal mass distribution of the tropospheric aerosol, then the particles of the two modes can be collected as separate samples. For many problems in connection with air pollution studies such a fractionation will be sufficient.

Size Fractionation of Particles by Diffusion Processing

Particles not under the influence of external forces diffuse in a random fashion called Brownian motion. In the size range below about 0.3 μm diameter diffusivity of particles is so high ($>10^{-6}$ $cm^2 \cdot sec^{-1}$) that it can be used for size fractionated sampling.

A sizing device based on diffusion separation is called a diffusion battery (65, 66). It consists of a number of equally spaced narrow plates forming long parallel channels or of a bundle of small-bore parallel tubes of equal diameter and can be regarded as a submicron particle denuder. Alternatively such a battery may contain series of stages made of stainless wire screens (67). The use of screens makes possible a rather compact arrangement (Figure 9) well suited for field measurements. With increasing number of stages the number of screens also increases according to the equation given in Figure 9. Because the diffusional collection mechanism of a set of wire screens with laminar flow is the same as for flow through a round tube, a cumulative equivalent length of such a hypothetical tube can be estimated: It is approximately the product of the actual thickness of a screen, the number of holes in each screen and the number of screens.

Fine particles in such a device diffuse to the surface of the wires of the screen, adhere and are thus removed from the aerosol stream. Because diffusivities decrease with increasing particle size, the extent of penetration of a screen or a set of screens in the battery depends on particle size also. In a staged series, the first screen removes the smallest particles whereas the larger ones are removed in the following stages. Behind each stage the particles still present in the aerosol stream can be detected by a continuous condensation nuclei counter (66) or be collected on suitable filters for chemical analysis (68 - 71). In order to obtain reliable results the battery has to be combined, however, with a preseparator which eliminates particles $\gtrsim 1$ μm diameter. Such coarse particles will otherwise get lost in the screen arrangements by impaction and gravitational settling (70, 71).

The applicability of a diffusion battery as a sizing device is necessarily restricted to the lower submicron particle range. Its size cuts (see Figure 10) are not as sharp as those obtained with an inertial impactor.

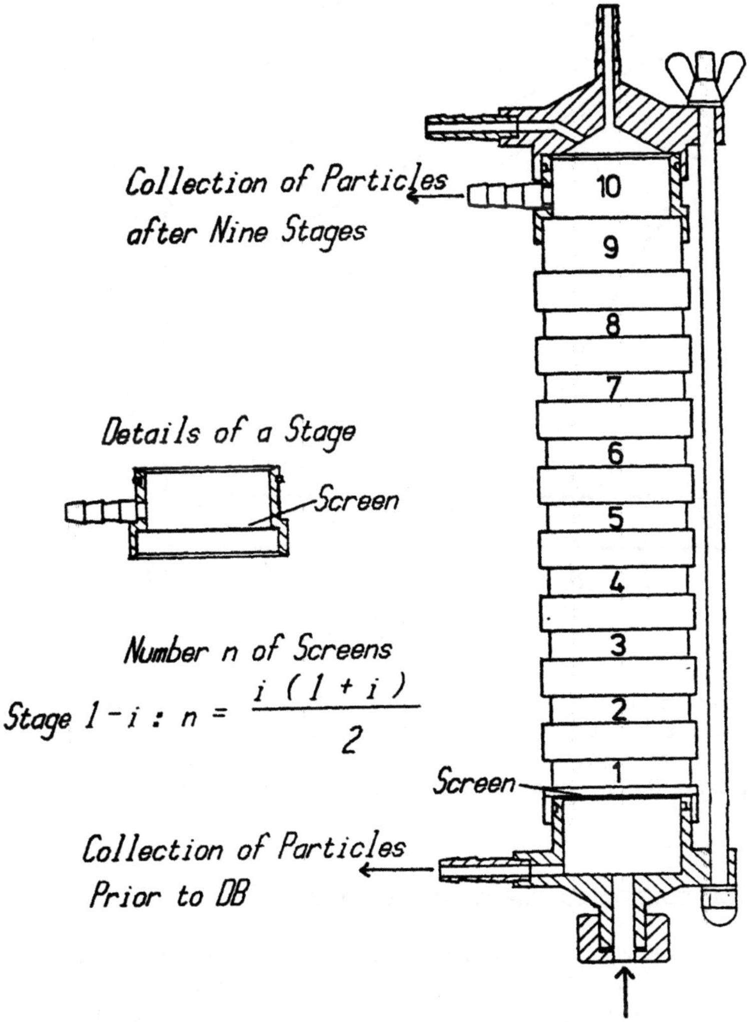

FIGURE 9. Sketch of a screen diffusion battery (DB) according to (67, 48). Total length 32 cm.

On the other hand, however, difficulties inherent in impactor sampling (particle reentrainment, volatilization losses, artifact formation) can be avoided by employing diffusion processing.

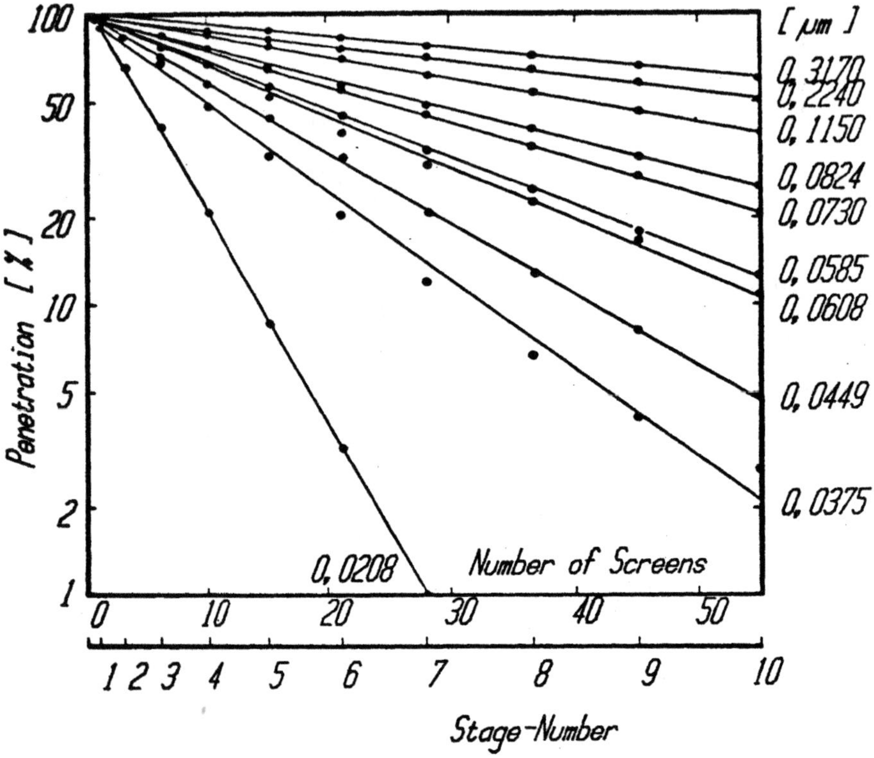

FIGURE 10. Calibration of a screen diffusion battery containing in total 55 stainless wire screens of about 500 mesh corresponding to a cumulative equivalent length of approximately 1100 m (48). Monodisperse NaCl- and DOP-aerosol, respectively; flow rate = 225 $[l \cdot h^{-1}]$.

BULK ANALYSIS OF COLLECTED PARTICULATE MATTER

After having collected airborne particulate matter by one of the methods just described its analytical characterization can be carried out. From the many techniques available for this purpose only those are included in the following condensed survey (Tables 2 - 4) which have been employed already to a larger extent or which may lend themselves to broad-range application in the future, and which are directed towards the elements and compounds identified before as main constituents of atmospheric particles.

The methods listed in Table 2a are just assigned to those elements they can be applied to, without taking into account any

C			Combust./TA/DAA
S	XRF		FPD/TA
Cl, Br	PIXE		
Si			
Al, Fe	NAA	AAS	
Na, K, Mg, Ca	SSMS	AFS	
V, Ni, Pb	ICP		
other trace metals			Voltamm.

Table 2a. Analytical methods for investigation of the elementary composition of air particulates.

change in the operation conditions. Therefore it becomes not so apparent that some of the methods allow multielement survey whereas others are suitable only for few or single element determination. This can be taken better from Table 2b which characterizes the most important spectroscopic approaches to the analysis of airborne particulate matter.

Besides X-ray fluorescence analysis (XRF) atomic absorption spectroscopy (AAS), with flame of flameless, is the most widely used technique for the determination of metallic and a few nonmetallic constituents in collected dust (84 - 86). Though generally time consuming sample preparation steps are necessary, the very low absolute detection limits, medium cost of acquisition, and high reliability of commercially available instruments have made AAS so popular also in the field of aerosol analysis. Atomic fluorescence spectroscopy (AFS) (87) has not yet found that wide application.

The introduction of the inductively coupled plasma discharge (ICP) as a light source (88) has considerably reactivated traditional atomic emission spectroscopy (AES). Because ICP-AES fulfils the requirements for an ideal multielement system to a high degree (89), its use in practical analytical work has drastically expanded during the past years. ICP-AES is essentially a technique for the investigation of solutions. There is no doubt, however, that nevertheless it will find broad application in airborne particulate analysis (82, 83) also. Future research in this context has to concentrate on suitable procedures for sampling, sample preparation, and sample introduction into the plasma.

Of the different ionization processes employed in mass spectrometry (MS) (90) spark ionization is especially well suited for investigation of nonvolatile material.

Method	Consequence of energy transfer to matter	Measurement of	Multielement technique?	Sample preparation?	Detection limit
AAS	outer shell excitation	photon absorption	no	yes	10^{-10} - 10^{-12} g
ICP	outer shell excitation+ ionization	photon emission	yes	yes	10^{-9} - 10^{-11} g/ml
SSMS	outer shell ionization	ion currents (m/e)	yes	yes	10^{-8} - 10^{-7} g/g
XRF PIXE	inner shell ionization	photon emission (X-rays)	yes	no	$\sim 10^{-6}$ g/g
NAA	mass change of atomic nucleus	photon emission (γ-rays)	yes	no	$\sim 10^{-9}$ g/g

Table 2b. Spectroscopic techniques for investigating the elementary composition of air particulates.

Therefore spark source mass spectrometry (SSMS) has been widely used for inorganic trace analysis of solids (91). Its low detection limits (Table 2b) and its applicability to almost every element make SSMS an ideal tool for screening the elementary composition of air particulates (80, 81). On the other hand, however, SSMS data exhibit comparatively poor precision and accuracy though always doubly focussing instruments are used. A way out of this difficulty would be to combine isotope dilution analysis (IDA) with SSMS (91, 92).

Wavelength-dispersive or energy-dispersive XRF is ideally suited for nondestructive multielement analysis of air particulates collected on thin clean substrates like filters (72 - 74). Matrix effects, which are especially harmful in the case of light elements (S, Cl), can be avoided by analysing samples with uniform low weight deposits (<1 mg/cm^2) and employing appropriate calibration and correction procedures. The high standard in reliability and operating comforts exhibited by modern X-ray spectrometers has made XRF a widely used routine technique in aerosol analysis. In particle induced X-ray emission (PIXE) charged particles (e.g. 2 - 4 MeV protons) are employed instead of X-rays to generate characteristic X-ray radiation in the samples to be analysed (72, 75, 76). Compared to XRF the sample size can be much smaller in the case of PIXE analysis, a few mm^2 being sufficient. In addition the detection limits for PIXE are generally lower than for XRF because of the more favorable signal-to-noise ratios achieved. The high technical (accelerator!) and instrumental expenditure necessary is opposed, however, to a common utilization of PIXE as a routine technique.

Every element becomes radioactive when bombarded with neutrons, charged particles such as protons, deuterons or α-particles, or with high-energy photons. The isotope specific radiation - mainly γ-radiation - emitted from an activated sample makes possible its analysis with respect to a large variety of constituents. Irradiation with thermal neutrons (0.1 - 0.2 eV), leading predominantly to (n, γ) nuclear reactions, is by far the most frequently used method in activation analysis. This is because most elements (with the exception of some light ones) have isotopes with sufficiently high neutron-capture cross sections for (n, γ) reactions, and because high thermal neutron fluxes (10^{12}-10^{14} $cm^{-2} \cdot s^{-1}$) are easily obtained at many nuclear reactor facilities. Emitted γ-radiation can be detected and classified according to photon energy at very low intensity levels. Therefore *neutron activation analysis* (NAA) has become an attractive multielement trace technique and has found considerable application also in the analysis of air particulates (77 - 79). For elements with high neutron-capture cross sections detection limits are obtained which cannot be achieved by any other analytical method. The application of NAA to environmental trace analysis is limited, however, to special research projects, because a nuclear reactor as an efficient neutron source and highly sophisticated instrumentation for the set up of an appropriate irradiation/counting schedule and for data treatment are required.

Voltammetric techniques (Table 2a) can generally be employed for detection and determination of electroactive species in solution. Especially *anodic stripping voltammetry*, which includes an electrolytic preconcentration step, has been proved very useful for heavy metal trace analysis and even speciation in aqueous environmental samples (93 - 95). Its broad range application to the investigation of airborne particulate matter should also be possible, though considerable personal skill and comparatively tedious sample preparation procedures are required in this case (95a, b). On the other hand, however, voltammetric techniques offer a uniquely economic tool with high sensitivity and selectivity.

Total carbon in air particulates, i.e. carbonate, elemental (graphitic) carbon, and organic carbon is commonly determined by acidification (for carbonate) and *combustion*, respectively, of the sample and measurement of the CO_2 formed (96). T h e r m o a n a l y s i s (TA, Table 2a) using time programmed heating of the sample in an oxygen atmosphere and relative-conductimetric detection of the CO_2 evolved makes possible the separation of fractions of carbonaceous material of different volatility and reactivity with respect to oxygen (97). Sample amounts of about 50 µg in total are sufficient for such an investigation. An alternative approach to the determination of total carbon is d e u t e r o n a c t i v a t i o n a n a l y s i s (DAA), which makes use of the $^{12}C(d,n)$ ^{13}N nuclear reaction and subsequent detection of the ^{13}N positron emitter (98). A combination of multiple solvent extraction with combustion has been employed for fractionated analysis of particulate organic carbon (99). In all cases carbon-free substrates had to be used for particle collection, of course, namely aluminium impactor plates (96, 97), silver filters (98), or glass fiber filters (96, 99). For selective estimation of the graphitic fraction of total particulate carbon an optical absorption technique (wavelength 630 nm) has been proposed (100). In this case airborne particles were collected on membrane filters.

The TA approach just mentioned in connection with the analysis of carbonaceous material has also been used for the characterization of sulfurous constituents of atmospheric particulate matter (97). In this case SO_2 is formed and measured conductimetrically.

The formation of an excited light emitting S_2^* species in hydrogen rich reducing flames is the basis of the f l a m e p h o t o m e t r i c d e t e c t o r (FPD, Table 2a) for sulfur (101). In combination with g a s c h r o m a t o g r a p h y (GC) it has proved especially useful for the detection of volatile sulfur containing atmospheric trace compounds (102, 103). As far as particulate sulfur analysis is concerned Crider pointed out already 1965 (104) that this can be accomplished by directly introducing an aerosol into the burner. Nevertheless several investigators preferred first to collect ambient air particulates on a suitable substrate and then to volatilize and decompose the sulfur containing constituents and to transfer the resulting gaseous products by a carrier gas stream to the FPD (105 - 107). By this means interferences by airborne sulfurous trace gases (mainly SO_2) are avoided and detection limits in the ng range achieved. The application of low decomposition temperatures (up to 500°C) makes possible the selective determination of H_2SO_4 plus ammonium sulfates (105). At higher temperatures (up to 1200°C) total particulate sulfur is detected (106, 107). A quite similar technique which also makes use of fractionated volatilization and flame photometry is m o l e c u l a r e m i s s i o n c a v i t y a n a l y s i s (MECA) (108). Only during the past four to five years attempts towards direct real-time FPD measurements of particle bound sulfur have been made.

Interferences through gaseous sulfur compounds are avoided by either employing diffusion removal of these gases (109 - 112) or by pulsed electrostatic precipitation of the particles to be measured (113). As will be shown later, the combination of thermal decomposition, diffusion separation, and FPD makes possible speciation of aerosol sulfuric acid (109 - 112).

In Table 3 methods suitable to quantify the most important nonmetal ionic species in the water-soluble part (except carbonate) of ambient aerosol are listed. At present the most versatile technique in this context is *ion chromatography* (IC). It is applicable to trace analysis of mixtures of cations (such as ammonium, alkali, and earth alkaline metal ions) and anions in aqueous solutions, and in essential combines ion exchange chromatography with electrolytic conductivity detection. In order to obtain sufficiently low detection limits high signal/background ratios are required. This can easily be achieved by placing a so called suppressor column between the separator column and the detector, which eliminates the background conductivity of the respective eluent employed and at the same time converts the separated cations (anions) to the corresponding hydroxides (acids). Consequently the measured signal is caused mainly by hydroxyl ions (protons), and therefore cations (anions) of too weak bases (acids) (pk > 5) are not well suited for IC analysis. Since its introduction (114) IC has found already wide

H^+		Titr./RTD
NH_4^+		ISE/Phot
SO_4^{2-}		IDA/Phot
Cl^-, Br^-	IC	IDA/Phot/ISE
NO_3^-		GC/Phot/ISE
CO_3^{2-}		TA/CO_2 measurem.

Table 3. Analytical methods for the determination of nonmetal ions in ambient aerosol particulate matter.

application in environmental analytical chemistry (115 - 118), because it is a sequential oligo-ion method, has rather low detection limits (~ ppb with preconcentrator column), and is characterized by low running expenses.

With the exception of IC any of the methods listed in Table 3 is suitable only for single-ion determination. Protons of strong acids (H_2SO_4, NH_4HSO_4) have been measured by alcalimetric *titration* under carefully controlled conditions using Gran's plot for estimation (119, 120) or by a *radio thermo diffusion technique* (RTD) based on the reaction between strong acids and $Na^{36}Cl$ and yielding $H^{36}Cl$ (36).

A wide variety of wet chemical photometric procedures have been employed to the analysis of ionic aerosol constituents. Quite a number of techniques are known for sulfate, most of them being founded on the precipitation of sparingly soluble $BaSO_4$ (35, 121). Nitrate can be determined by nitration of a suitable aromatic compound (122) or after reduction to nitrite and subsequent formation of an azo dye (123). Direct UV photometric procedures without conversion of the nitrate ion have also been described (124, 125). Widespread photometric approaches to the determination of halide and ammonium ions are the mercury(II)-thiocyanate (126) and the indophenol blue method (127, 128), respectively.

Compared to the important role ion selective electrodes (ISE) play in other fields of analytical chemistry (129) their use in aerosol analysis is limited to a few applications, most of them being related to fluoride determinations. Direct potentiometric ISE procedures for NH_4^+ (130) and NO_3^- (131) have been described. ISE determination of Cl^- in soil extracts (132) or a continuous multiple potentiometric system for Cl^-, F^-, NO_3^- and NH_4^+ (133) could easily be adapted to aerosol investigation.

The application of GC to quantification of inorganic ions in airborne particulate matter requires suitable derivatization and is so far restricted to the determination of NO_3^- (134). There exist, however, quite promising procedures also for traces of halides (135, 136), and quite recently it has been shown that GC could serve as an approach to oligo-ion determination just as IC (137).

Radiochemical IDA employing $^{35}SO_4^{2-}$ and $^{36}Cl^-$, respectively, as tracers have been developed for sulfate and halide ion determinations and have proved especially useful and reliable in the analysis of rain water and aerosol filter samples (138, 139).

Table 4 contains some information about speciation of particle bound compounds in the atmospheric aerosol. Inherent analytical methodology is much more developed for the organic fraction of air particulates than for the inorganic one. The development of selective sampling and detection systems for the most important inorganic species, such as sulfuric acid and ammonium salts of strong acids, has started only a few years ago. In this context it should be pointed out, that a gain in specific information must generally be paid for by a loss in precision and accuracy, and consequently a decision whether a result obtained can be regarded as "quantitative" or as "qualitative" only is sometimes difficult to be made. Therefore qualitative phase identification by X-ray diffraction (140, 141) or electron microscopic techniques (142) may also be used for compound selective bulk analysis.

Because of the complexity of the mixture of organic compounds

Organic comp.			GC/MS/TA
H_2O			MW-DEC
H_2SO_4	TD	TA/FPD, Neph.	
Amm.sulfates	TD	TA/FPD, Neph.	
NH_4NO_3, NH_4Cl	TD		

Table 4. Analytical methods for the selective determination of compounds in ambient aerosol particulate matter.

present in the particulate aerosol fraction *high resolution* techniques have to be employed in this case for speciation. Up to now most work has been done with GC employing glass capillary columns and flame ionization or MS detection (143 - 147), or with high resolution MS using a direct insertion probe technique and different ionization methods (146 - 148). This has proved especially useful when combined with time programmed volatilization of classes of compounds present in the collected dust (146).

In collection of airborne particles for organic analysis filters are usually employed. In order to avoid losses of some of the organic compounds due to volatilization (143, 149, 150) or conversion on the filter surface (147) special attention has to be paid to the arrangement of the sampling train. The loaded filters are extracted and the extracts then subjected to a clean-up and preseparation procedure prior to analysis making use of thin layer or liquid chromatographic techniques (144, 147).

The determination of the free liquid water content of airborne particles is not an easy task because it rapidly changes according to relative humidity. An instrument has been developed for that purpose which measures the *dielectric constant* of continuously collected particles in the microwave range (MW-DEC) (10).

Besides organic matter sulfuric acid and ammonium salts of strong acids belong to the predominant constituents of the fine particle mode, at least in the polluted troposphere. Therefore they are of importance with respect to air chemistry and possible health effects. Most of the techniques proposed for the selective determintation of airborne H_2SO_4 use filters (37, 39, 40, 104, 105, 151 - 159) or impactors (105, 160, 161) for extraction of the acid from ambient air and its separation from gaseous sulfur compounds. In the previous section it has already been pointed out, however, that such procedures may lead to artifact formation during sampling of reactive atmospheric trace constituents. Therefore in situ *nephelometry* (162), droplet mobility measurement as a function of humidity (163), and speciation techniques based either on selective chemical masking prior to filtration (164) or on differences in the physical behaviour of the species to be separated (109 - 113) are a better approach to H_2SO_4 determination.

The combination of thermal treatment of the aerosol, diffusion separation and FPD (109 - 112) has found the greatest interest of the methods just mentioned. Its mode of operation is illustrated by Figure 11: The aerosol under study is drawn through a

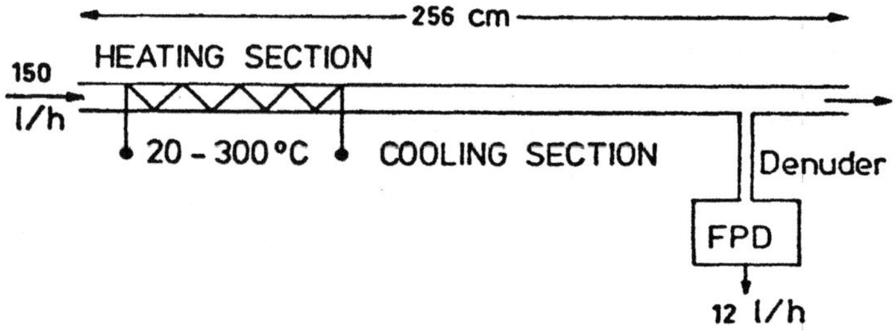

Diagram of the measurement system

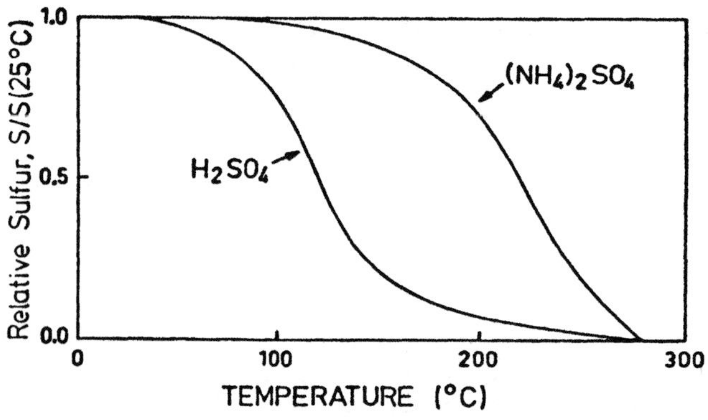

Calibration of the analysis system

FIGURE 11. Thermal analysis/FPD system for speciation of H_2SO_4 in ambient aerosol (110).

stainless steel tube which can be heated periodically from ambient temperature to 300°C. In the heated tube instable components, that is to say H_2SO_4 droplets or ammonium sulfate particles, break down into small fragments. These fragments are trapped - together with concomitant SO_2 - in a coated denuder, so that they cannot reach the sulfur selective FPD. This means, with increasing temperature of the heated tube the signal of the FPD decreases (see Figure 11), and because H_2SO_4 and e.g. $(NH_4)_2SO_4$ exhibit different thermal behavior, the temperature dependence of the corresponding FPD signals is also different.

The instrument just described has the advantage to be a cyclic monitoring system. At the same time, however, it has also two drawbacks: The FPD response depends on a few environmental variables (barometric pressure, humidity) and on the sulfur compound detected (111, 165), and the results derived for H_2SO_4 are founded on the measurement of differences in detector output. For these reasons the basic concept of thermal decomposition and diffusion separation has been modified to work in a discontinuous but direct manner. This alternative thermoanalytical system, called a thermodenuder (TD) (166), is just a heated glass tube the inner wall of which is coated with NaCl or NaF as a trapping agent. If an aerosol is passed through such a tube of about 140°C sulfuric acid droplets are decomposed and the small fragments formed subsequently collected at the wall coating by diffusion controlled deposition. Sulfate and bisulfate salt particles as well as SO_2 remain airborne and can be trapped behind the denuder. Besides H_2SO_4 gaseous strong acids (HNO_3, HCl) and easily decomposable ammonium salts (NH_4NO_3, NH_4Cl) are likewise collected in the heated diffusion separator. If therefore the tube is rinsed with water and the dissolved coating analyzed for SO_4^{2-} and NO_3^- (NaCl-tube) or SO_4^{2-}, NO_3^- and Cl^- (NaF-tube), selective information about the content in H_2SO_4, HNO_3 + NH_4NO_3 and HCl + NH_4Cl of the aerosol under study is obtained. A further

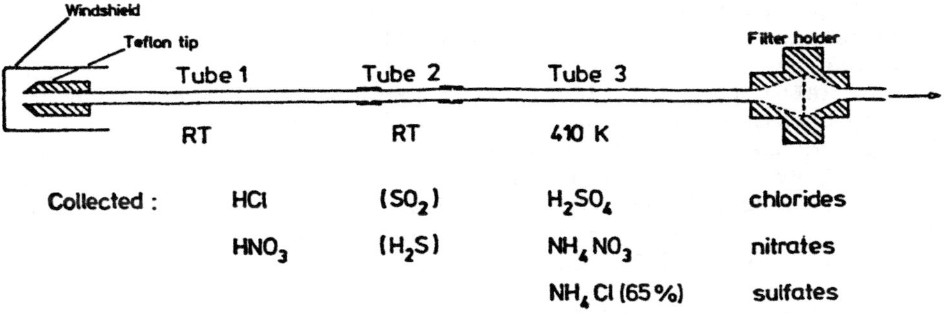

FIGURE 12. Thermodenuder sampling system (55). Inner tube diameter: 6 mm. Tube 1: room (ambient) temperature, NaF-coating, 90 cm length. Tube 2: room (ambient) temperature, tetrachloromercurate coating, 25 cm length. Tube 3: 410 K, NaF-coating, 90 cm length. Flow rate: 40 $l \cdot h^{-1}$

separation of these species is achieved by arranging two NaF-denuders in series, the first one being operated at ambient temperature and the second one at 140°C (see Figure 12). By this means the selective collection and determination of HCl, HNO_3, H_2SO_4, NH_4NO_3 and NH_4Cl is possible (55). The decomposition, however, of NH_4NO_3 even at ambient temperatures (49, 50) still remains a potential source of error. The detection limits so far achieved with the system presented in Figure 12 are 0.2 - 0.3 µg of H_2SO_4, HNO_3, and HCl, respectively. It has been shown (60) that future development of the TD technique might be directed towards a complete detection scheme for ammonia, strong acids and their ammonium salts, including sulfates.

SURFACE AND DEPTH PROFILE MEASUREMENT OF AEROSOL PARTICLES

The characterization of surfaces of airborne particulate matter is important for the following reasons (16):

- The surface of a particle governs its heterogeneous reactivity with respect to gaseous or dissolved species. Therefore information on heterogeneous atmospheric reactions can be obtained by detecting surface bound compounds.
- The surface of a particle is directly accessible to extraction by aqueous solutions in the natural environment. Inhaled particles, for instance, can be attacked by body fluids.
- Potentially toxic trace metals and organic compounds are highly enriched at the surfaces of several types of airborne particles.

Surface characterization of particulate matter generally requires highly sophisticated and expensive instrumentation. *Electron spectroscopy for chemical analysis* (ESCA) has extensively been used for investigation of the chemical states of carbon, sulfur, nitrogen and some trace metals at particle surfaces (167). A complementary technique is *Auger electron spectroscopy* (AES) (16, 168, 169). Depth profile measurements can be made by combining ESCA and AES with Ar^+ *ion etching*. This procedure may lead, however, to artifact formation because of the cleavage of chemical bonds and changes in oxidation state during ion bombardment (48, 170). A direct approach to depth profiling is *secondary ion mass spectrometry* (SIMS) (16, 168, 169).

A quite different method for surface and depth profile investigations, which somehow reflects the fate of aerosol particles in a natural aqueous environment, is *time resolved solvent leaching* (TRSL) (167, 169). Though TRSL has not yet been fully developed it offers an interesting and inexpensive way to elucidate the composition and the possible chemical and physiological behavior of particle surfaces.

CONCLUSIONS

Analytical chemistry as related to air chemistry has to take into account that the bulk material to be investigated is a heterogeneous system. Sampling procedures employed have to be carefully controlled and adjusted in order to avoid severe systematic errors which by far exceed the errors inherent in the determination technique itself. Today so many analytical methods for the investigation of airborne particulate matter are available, that a nearly free choice can be made. For proper chemical characterization of air particulates bulk analysis as well as surface analysis are required.

REFERENCES

(1) Junge, C. E.: 1963, "Air Chemistry and Radioactivity"; Academic Press, New York, London, pp. 111 - 208

(2) Hinch, N.: 1969, J. Chem. Educ. 46, pp. 93 - 95

(2a) Perkins, H. C.: 1974, "Air Pollution"; McGraw Hill, New York, p. 5

(3) Robinson, E., and Robbins, R. C.: 1971, "Emission Concentrations and Fate of Particulate Atmospheric Pollutants"; Amer. Petrol Inst. Publ. No. 4076

(4) Butcher, S. S., and Charlson, R. J.: 1972, "An Introduction to Air Chemistry"; Academic Press, New York, p. 163

(5) Whitby, K. T., Husar, R. B., and Liu, B. Y. H.: 1972, J. Colloid Interface Sci. 39, pp. 177 - 204

(6) Whitby, K. T., and Sverdrup, G. M.: 1980, in "The Character and Origins of Smog Aerosols" (G. M. Hidy, P. K. Mueller, D. Grosjean, B. R. Appel, J. J. Wesolowski, Eds.); John Wiley & Sons, New York, pp. 477 - 517

(7) Hidy, G. M., and Mueller, P. K.: 1980, cf. ref. (6), pp. 17 - 52

(8) Friedlander, S. K.: 1973, Environ. Sci. Technol. 7, pp. 235 - 240

(9) Hahn, J.: 1980, Advances in Environmental Science and Technology 10, pp. 359 - 376

(10) Ho, W., Hidy, G. M., and Govan, R. M.: 1974, J. Appl. Meteorol. 13, pp. 871 - 879

(11) Dannecker, W.: 1982, in "Schwebstoffe und Stäube", VDI-Bericht Nr. 429; VDI-Verlag, Düsseldorf, in press

(12) Müller, J.: 1982, cf. ref. (11)

(13) Martens, C. S., Wesolowski, J. J., Harriss, R. C., and Kaifer, R.: 1973, J. Geophys. Res. 78, pp. 8778 - 8792

(14) Winkler, P.: 1973, J. Aerosol Sci. 4, pp. 373 - 387

(15) Jaenicke, R.: 1978, Atmos. Environ. 12, pp. 161 - 169

(16) Keyser, T. R., Natusch, D. F. S., Evans, C. A. Jr., and Linton, R. W.: 1978, Environ. Sci. Technol. 12, pp. 768 - 773
(17) Jacobsen, A. R., and Morris, S. C.: 1976, in "Air Pollution" (A. C. Stern, Ed.), 3rd Ed., Vol. 1; Academic Press, New York, pp. 169 - 196
(18) Hinkley, E. D. (Ed.): 1976, "Laser Monitoring of the Atmosphere"; Springer-Verlag, Berlin-Heidelberg-New York
(19) Green, A. E. S.: 1979, in "Aerosol Measurement" (D. A. Lundgren, F. S. Harris, Jr., W. H. Marlow, M. Lippmann, W. E. Clark, M. D. Clark, Eds.); University Presses of Florida, Gainesville, pp. 635 - 649
(20) Guillot, P., Bonometti, G., Hasenjaeger, H., van der Meulen, A., Hamilton, P. M., Haulet, R., Laurent, J., Sandroni, S., Cerutti, C., Giovanelli, G., Tirabassi, T., Vittori, O., and Piccinini, P.: 1979, Atmos. Environ. 13, pp. 895 - 917
(21) Liu, B. Y. H., and Pui, D. Y. H.: 1981, Atmos. Environ. 15, pp. 589 - 600
(22) Davies, C. N.: 1973, "Air Filtration"; Academic Press, London, New York
(23) Ref. (22), pp. 111 - 116
(24) Spurný, K. R.: 1972, in "Assessment of Airborne Particles" (T. T. Mercer, P. E. Morrow, W. Stöber, Eds.); Charles C. Thomas, Publisher, Springfield, Illinois, pp. 54 - 80
(25) Spurný, K. R., Lodge, J. P., Jr., Frank, E. R., and Sheesley, D. C.: 1969, Environ. Sci. Technol. 3, pp. 453 - 464 and 464 - 468
(26) Liu, B. Y. H., and Kuhlmey, G. A.: 1977; in ref. (72), pp. 107 - 119
(27) Liu, B. Y. H., and Lee, K. W.: 1976, Environ. Sci. Technol. 10, pp. 345 - 350
(28) Lee, R. E., Jr., and Wagman, J.: 1966, Am. Ind. Hyg. Assoc. J. 27, pp. 266 - 271
(29) Pierson, W. R., Hammerle, R. H., and Brachaczek, W. W.: 1976, Analyt. Chem. 48, pp. 1808 - 1811
(30) Coutant, R. W.: 1977, Environ. Sci. Technol. 11, pp. 873 - 878
(31) Witz, S., McPhee, R. D.: 1977, J. Air Poll. Contr. Assoc. 27, pp. 239 - 241
(32) Pierson, W. R., Brachaczek, W. W., Korniski, T. J., Truex, T. J., and Butler, J. W.: 1980, J. Air Poll. Contr. Assoc. 30, pp. 30 - 34
(33) Liberti, A., Brocco, D., and Possanzini, M.: 1978, Atmos. Environ. 12, pp. 255 - 261
(34) Judeikis, J. S., Stewart, T. B., and Wren, A. G.: 1978, Atmos. Environ. 12, pp. 1633 - 1641
(35) Newman, L.: 1978, Atmos. Environ. 12, pp. 113 - 125
(36) Klockow, D., Jablonski, B., and Nießner, R.: 1979, Atmos. Environ. 13, pp. 1665 - 1676
(37) Dubois, L., Zdrojewski, A., Teichman, T., and Monkman, J. L.:

1972, Int. J. Environ. Anal. Chem. 1, pp. 259 - 266
(38) Maddalone, R. F., Shendrikar, A. D., and West, P. W.: 1974, Mikrochim. Acta, pp. 391 - 402
(39) Leahy, D., Siegel, R., Klotz, P., and Newman, L.: 1975, Atmos. Environ. 9, pp. 219 - 229
(40) Dubois, L., Baker, C. J., Teichman, T., Zdrojeweski, A., and Monkman, J. L.: 1969, Mikrochim. Acta, pp. 269 - 279
(41) Thomas, R. L., Dharmarajan, V., Lundquist, G. L., and West, P. W.: 1976, Analyt. Chem. 48, pp. 639 - 642
(42) Klockow, D., Nießner, R., and Jablonski, B. B.: 1980, Analyt. Letters 13, pp. 1397 - 1408
(43) Ferm, M.: 1979, Atmos. Environ. 13, pp. 1385 - 1393
(44) Spicer, C. W., and Schumacher, P. M.: 1979, Atmos. Environ. 13, pp. 543 - 552
(45) Harker, A. B., Richards, L. W., and Clark, W. E.: 1977, Atmos. Environ. 11, pp. 87 - 91
(46) Forrest, J., Tanner, R. L., Spandau, D., D'Ottavio, T., and Newman, L.: 1980, Atmos. Environ. 14, pp. 137 - 144
(47) Appel, B. R., Wall, S. M., Tokiwa, Y., and Haik, M.: 1980, Atmos. Environ. 14, pp. 549 - 554
(48) Nießner, R.: 1981, Doctoral Dissertation, University of Dortmund, F. R. G.
(49) Stelson, A. W., Friedlander, S. K., and Seinfeld, J. H.: 1979, Atmos. Environ. 13, pp. 369 - 371
(50) Doyle, G. J., Tuazon, E. C., Graham, R. A., Mischke, T. M., Winer, A. M., and Pitts, J. N., Jr.: 1979, Environ. Sci. Technol. 13, pp. 1416 - 1419
(51) Appel, B. R., Wall, S. M., Tokiwa, Y., and Haik, M.: 1979, Atmos. Environ. 13, pp. 319 - 325
(52) Spicer, C. W., Ward, G. F., and Gay, B. W., Jr.: 1978, Analyt. Letters 11, pp. 85 - 95
(53) Joseph, D. W., and Spicer, C. W.: 1978, Analyt. Chem. 50, pp. 1400 - 1403
(54) Durham, J. L., Wilson, W. E., and Baker Bailey, E.: 1978, Atmos. Environ. 12, pp. 883 - 886
(55) Nießner, R., and Klockow, D.: 1981, in "Aerosols in Science, Medicine and Technology - Aerosols in and from Industrial Processes" (W. Stöber, D. Hochrainer, Eds.); Association for Aerosol Research, Schmallenberg, F. R. G., pp. 86 - 91
(56) Stevens, R. K., Dzubay, T. G., Russwurm, G., and Rickel, D.: 1978, Atmos. Environ. 12, pp. 55 - 68
(57) Baily, R. R., Field, P. E., and Wightman, J. P.: 1976, Analyt. Chem. 48, pp. 1818 - 1819
(58) Gregory, G. L., and Moyer, R. H.: 1977, Rev. Sci. Instrum. 48, pp. 1464 - 1468
(59) Bächmann, K., Matusca, P., Vierkorn-Rudolph, B., and Kontos, M.: 1982, Fresenius Z. Analyt. Chem. 310, pp. 89 - 97
(60) Slanina, J., v. Lamoen-Doornenbal, L., Lingerak, W. A., Meilof, W., Klockow, D., and Nießner, R.: 1981, Int. J. Environ. Analyt. Chem. 9, pp. 59 - 70

(61) Marple, V. A., and Willeke, K.: 1976, in "Fine Particles. Aerosol Generation, Measurement, Sampling, and Analysis" (B. Y. H. Liu, Ed.); Academic Press, Yew York, pp. 412 - 446
(62) Pavlik, R. E., and Willeke, K.: 1978, Am. Ind. Hyg. Assoc. J. 39, pp. 952 - 957
(63) Loo, B. W., Jaklevic, J. M., and Goulding, F. S.: 1976, cf. ref. (61), pp. 312 - 350
(64) Dzubay, T. G., Stevens, R. K., and Peterson, C. M.: 1977, in ref. (72), pp. 95 - 105
(65) Thomas, J. W.: 1955, J. Coll. Sci. 10, pp. 246 - 255
(66) Sinclair, D., Countess, R. J., Liu, B. Y. H., and Pui, D. Y. H.: 1979, cf. ref. (19), pp. 544 - 563
(67) Sinclair, D., and Hoopes, G. S.: 1975, Am. Ind. Hyg. Assoc. J. 36, pp. 39 - 42
(68) Marlow, W. H., and Tanner, R. L.: 1976, Analyt. Chem. 48, pp. 1999 - 2001
(69) Tanner, R. L., Marlow, W. H., and Newman, L.: 1979, Environ. Sci. Technol. 13, pp. 75 - 78
(70) Harrison, R. M., and Pio, C. A.: 1981, J. Air Poll. Contr. Assoc. 31, pp. 784 - 787
(71) Nießner, R., Klockow, D., Fißan, H., and Kapitza, F.: 1982, J. Aerosol Sci. 13, in press
(72) Dzubay, T. G. (Ed.): 1977, "X-Ray Fluorescence Analysis of Environmental Samples" - Ann Arbor Science Publ., Ann Arbor, Michigan
(73) Giauque, R. D., Goda, L. Y., and Garrett, R. B.: 1980, cf. ref. (6), pp. 147 - 167
(74) Hammerle, R. H., and Pierson, W. R.: 1980, cf. ref. (6), pp. 255 - 284
(75) Barfoot, K. M., Mitchell, I. V., Eschbach, H. L., Mason, P. I., and Gilboy, W. B.: 1979, J. Radioanal. Chem. 53, pp. 255 - 271
(76) Johansson, S. A. E. (Ed.): 1981, "Particle Induced X-Ray Emission and its Analytical Applications": Proc. 2nd Int. Conf., Lund, Sweden, 1980; Nucl. Instrum. Methods 181, pp. 1 - 546
(77) Dams, R., Robbins, J. A., Rahn, K. A., and Winchester, J. W.: 1970, Analyt. Chem. 42, pp. 861 - 867
(78) Heindryckx, R., and Dams, R.: 1979, Progr. Nucl. Energy 3, pp. 219 - 252
(79) Ragaini, R. C., Ralston, H. R., Garvis, D., and Kaifer, R.: 1980, cf. ref. (6), pp. 169 - 197
(80) Brown, R., and Vossen, P. G. T.: 1970, Analyt. Chem. 42, pp. 1820 - 1822
(81) Briden, F., and Lewis, D.: 1981, Int. J. Environ. Anal. Chem., pp. 249 - 264
(82) Barnes, R. M.: 1978, Toxicol. Environ. Chem. Rev. 2, pp. 187 - 208
(83) Lynch, A. J., McQuaker, N. R., and Brown, D. F.: 1980, J. Air Poll. Contr. Assoc. 30, pp. 257 - 260

(84) Gallay, W., Egan, H., Monkman, J. L., Truhaut, R., West, P. W., and Widmark, G.: 1975, "Environmental Pollutants - Selected Analytical Methods" (SCOPE6), Butterworths, London, pp. 1 - 11
(85) Wegscheider, W., Lorber, K. E., and Müller, K.: 1978, Intern. J. Environ. Anal. Chem. 5, pp. 171 - 187
(86) Geladi, P., and Adams, F.: 1978, Anal. Chim. Acta 96, pp. 229 - 241
(87) Ip, J., Thomassen, Y., Butler, L. R. P., Radziuk, B., and van Loon, J. C.: 1979, Anal. Chim. Acta 110, pp. 1 - 7
(88) Greenfield, S., Jones, I. L., and Berry, C. T.: 1964, Analyst 89, pp. 713 - 720
(89) Fassel, V. A.: 1979, Analyt. Chem. 51, pp. 1290A - 1308A
(90) Smith, S. R.: 1975, Crit. Rev. Analyt. Chem. 5, pp. 243 - 265
(91) Stefani, R.: 1978, Adv. Mass. Spectrom. 7A, pp. 729 - 750
(92) Heumann, K. G.: 1980, Toxicol. Environ. Chem. Rev. 3, pp. 111 - 129
(93) Nürnberg, H. W.: 1978, in "Structure and Dynamics in Chemistry" (P. Ahlberg, L. O. Sundelöf, Eds.); Almquist & Wiksell International, Stockholm, pp. 270 - 307
(94) Nguyen, V. D., Valenta, P., and Nürnberg, H. W.: 1979, Sci. Total Environ. 12, pp. 151 - 167
(95) Wang, J.: 1982, Environ. Sci. Technol. 16, pp. 104A - 109A
(95a) Dennis, B. L., Wilson, G. S., and Moyers, J. L.: 1976, Anal. Chim. Acta 86, pp. 27 - 34
(95b) Ryan, M. D., Siemer, D. D.: 1977, in "Environmental Analysis" (G. W. Ewing, Ed.); Academic Press, New York, pp. 105 - 110
(96) Mueller, P. K., Mosley, R. W., and Pierce, L. B.: 1972, J. Colloid Interface Sci. 39, pp. 235 - 239
(97) Puxbaum, H.: 1979, Fresenius Z. Analyt. Chem. 298, pp. 250 - 259 and 299, pp. 33 - 41
(98) Clemenson, M., Novakov, T., and Markowitz, S. S.: 1980, Analyt. Chem. 52, pp. 1758 - 1761
(99) Appel, B. R., Colodny, P., and Wesolowski, J. J.: 1980, cf. ref. (6), pp. 337 - 352
(100) Rosen, H., Hansen, A. D. A., Dod, R. L., and Novakov, T.: 1980, Science 208, pp. 741 - 744
(101) Dräger, H., and Dräger, B. (Drägerwerk, Lübeck): 1962, German Patent No. 1133918
(102) Stevens, R. K., Mulik, J. D., O'Keeffe, A. E., and Krost, K. J.: 1971, Analyt. Chem. 43, pp. 827 - 831
(103) Bruner, F., Liberti, A., Possanzini, M., and Allegrini, I.: 1972, Analyt. Chem. 44, pp. 2070 - 2074
(104) Crider, W. L.: 1965, Analyt. Chem. 37, pp. 1770 - 1773
(105) Scaringelli, F. P., and Rehme, K. A.: 1969, Analyt. Chem. 41, pp. 707 - 713
(106) Husar, J. D., Husar, R. B., and Stubits, P. K.: 1975, Analyt. Chem. 47, pp. 2062 - 2065

(107) Roberts, P. T., and Friedlander, S. K.: 1976, Atmos. Environ. 10, pp. 403 - 408
(108) Burguera, M., Bogdanski, S. L., and Townshend, A.: 1981, Crit. Rev. Analyt. Chem. 10, pp. 185 - 246
(109) Huntzicker, J. J., Hoffman, R. S., and Ling, C.-S.: 1978, Atmos. Environ. 12, pp. 83 - 88
(110) Cobourn, W. G., Husar, R. B., and Husar, J. D.: 1978, Atmos. Environ. 12, pp. 89 - 98
(111) Tanner, R. L., D'Ottavio, T., Garber, R., and Newman, L.: 1980, Atmos. Environ. 14, pp. 121 - 127
(112) Cobourn, W. G., Djukic-Husar, J., and Husar, R. B.: 1980, J. Geophys. Res. 85, pp. 4487 - 4494
(113) Kittelson, D. B., McKenzie, R., Vermeersch, M., Dorman, F., Pui, D., Linne, M., Liu, B., and Whitby, K.: 1978, Atmos. Environ. 12, pp. 105 - 111
(114) Small, H., Stevens, T. S., and Bauman, W. C.: 1975, Analyt. Chem. 47, pp. 1801 - 1809
(115) Sawicki, E., Mulik, J. D., and Wittgenstein, E. (Eds.): 1978, "Ion Chromatographic Analysis of Environmental Pollutants"; Ann Arbor Science Publ., Ann Arbor, Michigan
(116) Mulik, J. D., and Sawicki, E. (Eds.): 1979, "Ion Chromatographic Analysis of Environmental Pollutants", Vol. 2; Ann Arbor Science Publ., Ann Arbor, Michigan
(117) Smith, F. C., Jr., and Chang, R. C.: 1980, Crit. Rev. Analyt. Chem. 9, pp. 197 - 217
(118) Slanina, J., Bakker, F. P., Jongejan, P. A. C., van Lamoen, L., and Möls, J. J.: 1981, Anal. Chim. Acta 130, pp. 1 - 8
(119) Liberti, A., Possanzini, M., and Vicedomini, M.: 1972, Analyst 97, pp. 352 - 356
(120) Brosset, C., and Ferm, M.: 1978, Atmos. Environ. 12, pp. 909 - 916
(121) Tanner, R. L., and Newman, L.: 1976, J. Air Pollut. Contr. Assoc. 26, pp. 737 - 747
(122) Norwitz, G., and Keliher, P. N.: 1979, Anal. Chim. Acta 109, pp. 373 - 380
(123) Sawicki, C. R., and Scaringelli, F. P.: 1971, Microchem. J. 16, pp. 657 - 672
(124) Slanina, J., Lingerak, W. A., and Bergman, L.: 1976, Fresenius Z. Analyt. Chem. 280, pp. 365 - 368
(125) Rennie, P. J., Summer, A. M., and Basketto, F. B.: 1979, Analyst 104, pp. 837 - 845
(126) Florence, T. M., and Farrar, Y. J.: 1971, Anal. Chim. Acta 54, pp. 373 - 377
(127) Glebko, L. I., Ulkina, Zh. I., and Lognenko, E. M.: 1975 II, Mikrochim. Acta, pp. 641 - 648
(128) Crowther, J., and Evans, J.: 1980, Analyst 105, pp. 841 - 848 and 849 - 854
(129) Koryta, J.: 1979, Anal. Chim. Acta 111, pp. 1 - 56
(130) Eagan, M. L., and Dubois, L.: 1974, Anal. Chim. Acta 70, pp. 157 - 167

(131) Simeonov, V., and Puxbaum, H.: 1977 II, Mikrochim. Acta, pp. 397 - 403
(132) Selmer-Olson, A. R., and Øien, A.:1973, Analyst 98, pp. 412 - 415
(133) Trojanowicz, M., and Lewandowski, R.: 1981, Fresenius Z. Analyt. Chem. 308, pp. 7 - 10
(134) Tesch, J. W., Rehg, W. R., and Sievers, R. E.: 1976, J. Chromatogr. 126, pp. 743 - 755
(135) Belcher, R., Rodriguez-Vazquez, J. A., Stephen, W. J., and Uden, P. C.: 1976, Chromatographia 9, pp. 201 - 204
(136) Vierkorn-Rudolph, B., Savelsberg, M., and Bächmann, K.: 1979, J. Chromatogr. 186, pp. 219 - 226
(137) Faigle, W., and Klockow, D.: 1982, Fresenius Z. Analyt. Chem. 310, pp. 33 - 38
(138) Klockow, D., Denzinger, H., and Rönicke, G.: 1974, Chem. Ing. Techn. 46, p. 831
(139) Klockow, D., and Teckentrup, A.: 1980, Int. J. Environ. Anal. Chem. 8, pp. 137 - 148
(140) Nieman, R. L., and Jenkins, R.: 1977, in ref. (72), pp. 77 - 92
(141) O'Connor, B. H., and Jaklevic, J. M.: 1980, X-Ray Spectrom. 9, pp. 60 - 65
(142) Spurny, K. R., and Stöber, W.: 1981, Int. J. Environ. Anal. Chem. 9, pp. 265 - 281
(143) Grob, K., and Grob, G.: 1971, J. Chromatogr. 62, pp. 1 - 13
(144) Ketseridis, G., and Hahn, J.: 1975, Fresenius Z. Analyt. Chem. 273, pp. 257 - 261
(145) Ketseridis, G., Hahn, J., Jaenicke, R., and Junge, C.: 1976, Atmos. Environ. 10, pp. 603 - 610
(146) Schuetzle, D.: 1980, in "Biochemical Applications of Mass Spectrometry" (G. Waller, O. Dermer, Eds.), Suppl. Vol., Wiley Interscience, New York, pp. 970 - 1001
(147) Schuetzle, D., Lee, F. S.-C., Prater, T. J., and Tejada, S. B.: 1981, Int. J. Environ. Anal. Chem. 9, pp. 93 - 144
(148) Schulten, H.-R., and Schurath, U.: 1975, Atmos. Environ. 9, pp. 1107 - 1112
(149) Beyermann, K., and Eckrich, W.: 1974, Fresenius Z. Analyt. Chem. 269, pp. 279 - 284
(150) König, J., Funcke, W., Balfanz, E., Grosch, B., and Pott, F.: 1980, Atmos. Environ. 14, pp. 609 - 613
(151) Maddalone, R. F., Thomas, R. L., and West, P. W.: 1976, Environ. Sci. Technol. 10, pp. 162 - 168
(152) Mason, D. W., and Miller, H. C.: 1979, cf. ref. (116), pp. 193 - 201
(153) Huygen, C.: 1975, Atmos. Environ. 9, pp. 315 - 319
(154) Dzubay, T. G., Snyder, G. K., Reutter, D. J., and Stevens, R. K.: 1979, Atmos. Environ. 13, pp. 1209 - 1212
(155) Dasgupta, P. K., Lundquist, G. L., and West, P. W.: 1979, Atmos. Environ. 13, pp. 767 - 774
(156) Barton, S. G., and McAdie, W. G.: 1970, Proc. 2nd Intern.

Clean Air Congr., Washington, D. C., pp. 279 - 282
(157) Tanner, R. L., Cederwall, R., Garber, R., Leahy, D., Marlow, W., Meyers, R., Phillips, M., and Newman, L.: 1977, Atmos. Environ. 11, pp. 955 - 966
(158) Appel, B. R., Wall, S. M., Haik, M., Kothny, E. L., and Tokiwa, Y.: 1980, Atmos. Environ. 14, pp. 559 - 563
(159) Penzhorn, R. D., and Filby, W. G.: 1976, Staub-Reinhalt. Luft 36, pp. 205 - 207
(160) Schulten, H. R., and Schurath, U.: 1975, J. Phys. Chem. 79, pp. 51 - 57
(161) Cunningham, P. T., and Johnson, S. A.: 1976, Science 191, pp. 77 - 79
(162) Charlson, R. J., Vanderpol, A. H., Covert, D. S., Waggoner, A. P., and Ahlquist, N. C.: 1974, Atmos. Environ. 8, pp. 1257 - 1267
(163) Liu, B. Y. H., Pui, D. Y. H., Whitby, K. T., Kittelson, D. B., Kousaka, Y., and McKenzie, R. L.: 1978, Atmos. Environ. 12, pp. 99 - 104
(164) Snyder, R. E., Reed, T. J., and McKissick, A. M.: 1977, US Environ. Prot. Agency, EPA-600/2-77-010
(165) D'Ottavio, T., Garber, R., Tanner, R. L., and Newman, L.: 1981, Atmos. Environ. 15, pp. 197 - 203
(166) Nießner, R., and Klockow, D.: 1980, Int. J. Environ. Anal. Chem. 8, pp. 163 - 175
(167) Novakov, T., Chang, S. G., and Dod, R. L.: 1977, in "Contemporary Topics in Analytical and Clinical Chemistry" (D. M. Hercules, G. M. Hieftje, L. R. Snyder, M. A. Evenson, Eds.), Vol. 1; Plenum Press, New York, London, pp. 249 - 286
(168) Natusch, D. F. S., Denoyer, E. R., Keyser, T. R., Kirton, S. E., Taylor, D. R., and Zeller, M. V.: 1980, AIChE Symp. Series 76 (No. 201), pp. 127 - 133
(169) Gardella, J. A., and Hercules, D. M.: 1979, Int. J. Environ. Anal. Chem. 7, pp. 121 - 136
(170) Storp, S., and Holm, R.: 1979, J. Electron Spectrosc. Relat. Phenom. 16, pp. 183 - 193

THE USE OF LONG-PATH FT-IR SPECTROSCOPY IN KINETIC STUDIES OF REACTIVE MOLECULES OF ATMOSPHERIC INTEREST

Jack G. Calvert

Department of Chemistry,
The Ohio State University, Columbus,
Ohio, U.S.A.

ABSTRACT

Long-path, FT-IR spectroscopy has been employed in the kinetic study of chemical changes induced both thermally and photochemically in a variety of simulated atmospheric mixtures. Recent studies of the mechanisms of the formation and decay reactions and the kinetic properties of several interesting reactive species are described to illustrate the great utility of the IR-FTS systems for such studies. Specifically results are presented from our recent studies of peroxynitric acid ($HOONO_2$), hypochlorous acid (HOCl), and dimethylnitrosamine ($(CH_3)_2NNO$). The limitations of the FT-IRS systems as well as their usefulness in these kinetic studies are discussed.

I. INTRODUCTION

The long-path, prism and grating infrared spectrometers have been used successfully for many years to follow the complex chemical changes which occur in simulated atmospheres at low concentrations. For examples, see the work of Calvert [1], Tuesday [2], and Pitts [3] and their coworkers. The unique IR absorption bands characteristic of each molecular species, the non-destructive, in situ analysis of the reactants and products of the reactions, and the relatively high sensitivity of these spectroscopic methods, all add to the attractiveness of the use of long-path IR for atmospheric kinetic studies. However there have been several major problems which have restricted the usefulness of these conventional systems for kinetic studies; chief amoung these are the very slow wavelength scan required to achieve resonable peak resolution with most conventional

H. W. Georgii and W. Jaeschke (eds.), Chemistry of the Unpolluted and Polluted Troposphere, 91–116.

IR equipment, and the real difficulty of making an unambiguous and accurate separation of overlapping absorption bands of different products and reactants. As a result of the first problem it has been necessary to restrict observations in a given experiment to the study of the time dependence of the absorbance of a single species at one wavelength, or to stop the reaction when this is feasible as in the case of photochemically induced changes, and determine the complete absorption spectrum of the system at specific exposure times.

The commercial Fourier transform infrared spectrometers which have become available in recent years provide for kinetic studies all of the advantages of the conventional IR instruments, but more importantly, they overcome many of the dificulties which are inherent in the conventional systems. The most significant of the advantages is the relatively rapid data collection of the FT-IRS systems. Data for the entire absorbance spectrum is collected simultaneously as the mirror of the Michaelson interferometer moves through its cycle. The range of wavelengths sampled is, of course, limited to that for which the IR source is of sufficient energy and to which the detector is sensitive. Typically the entire scan is completed within a few seconds on the commercially available instruments. The interferogram which is collected and processed later by computer, provides the entire spectrum, or any desired portion of it, in any chosen format (transmittance, absorbance, selected wavelength range, degraded resolution, etc.).

An additional feature of the FT-IRS systems which is extremely valuable is the precise instrument control of wavelength and the very reproducible character of the data collection and processing systems. Thus one may ratio the absorbances of one spectrum against those of another or subtract some fraction of one from another to eliminate absorbances due to known compounds and obtain clean spectra of other known or unknown compounds which have overlapping features.

Several research groups have been active in the use of FT-IRS systems in kinetic studies of atmospheric systems. For examples see the work of Hanst [4], Niki [5], Calvert [6], and Pitts [7], and their colleagues. In this work we will illustrate the major advantages and some of the problems associated with the use of the Fourier transform spectroscopy in kinetic studies. Our examples will be drawn from our current research efforts related to three highly reactive species of current interest in atmospheric chemistry, namely, peroxynitric acid ($HOONO_2$), hypochlorous acid (HOCl), and dimethylnitrosamine ($(CH_3)_2NNO$). First we will consider the general methods and problems associated with the collection of kinetic data using the FT-IRS systems.

II. PROBLEMS INHERENT IN THE FT-IRS ESTIMATES OF THE CONCENTRATIONS OF REACTING SPECIES

Signal-to-Noise Ratio and Data Collection Time.

The time required for the collection of one scan with the conventional commercial FT-IRS systems is on the order of a few seconds. However the signal-to-noise ratio which is inherent in such spectra may be enhanced by the averaging of the interferograms collected from multiple scans. If the noise is random and independent of signal level, which is the common case when the detector noise dominates, then the magnitude of the signal-to-noise ratio is expected to be proportional to the square root of the number of scans used in determining the interferogram [8]. This may be seen in Figure 1 where the infrared spectrum of carbon monoxide in the 2050 to 2200 cm^{-1} region is shown, and the interferograms from which the spectra were calculated have been determined using data points from a single scan up to 100 scans. The main features of the CO absorption are evident even in the single scan spectrum, but note that small absorption features which are not observable above the noise in the single scan spectrum become readily visible in spectra obtained at the high number of scans. From the point of view of facilitating product identification and optimizing sensitivity of detection of given compounds, it is apparent that one benefits greatly from an increase in the number of scans used in the calculation of the spectrum.

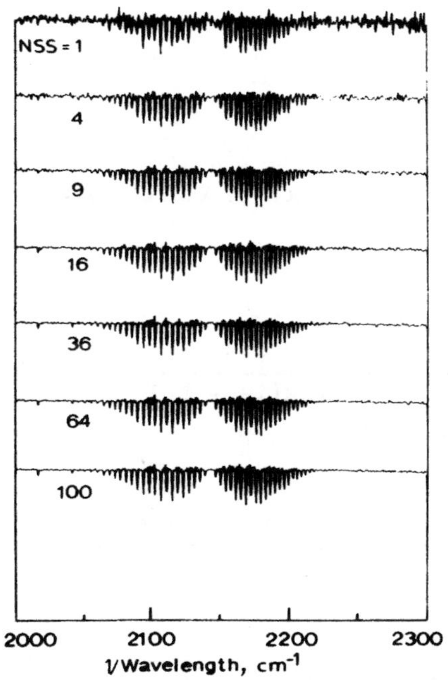

FIGURE 1. Spectrum of carbon monoxide obtained using different numbers of scans; illustration of the effect on signal-to-noise ratio.

However, in kinetic studies one is concerned with the time dependence of the concentration of products and reactants as well as their unambiguous identification. A spectrum which rquires a relatively long collection period, although of low signal-to-noise ratio, has a limited value in establishing the concentration of the reactant at a specific time. In our use of the FT-IRS system in kinetics, it has been convenient, and hence it has become common practice in our group to estimate the average concentration of a given compound over a given scanning time interval and to assign this concentration to the time at the midpoint of the scan period. Obviously if there is a near linear decay or growth of a signal during the scanning period such a procedure is fully justified, and no error arises from the averaging procedure. However if the signal builds or decays by some process which is non-linear in the time, as is often the case, it is important to recognize the magnitude of the error introduced by this adopted concentration estimation procedure. In illustration of this point consider the estimation of the concentrations of a given reactant X which decays by a second order process (1):

$$2X \rightarrow \text{Products} \tag{1}$$

The ratte of change of [X] with time will be given simply by (2):

$$\frac{d[X]}{dt} = [X]^2 k \tag{2}$$

If the concentration of X is $[X]_1$ at the start of our scan period, the concentration of X at any later time t will be given by the familiar integrated second order rat law, equation (3):

$$[X]_t = \left[\frac{1}{[X]_1} + kt \right]^{-1} \tag{3}$$

If we use a number of scans which we average over a collection time extending from t_1 to t_2, the average concentration of X will be given by relation (4):

$$\overline{[X]} = \int_{t_1}^{t_2} [X]_t dt/(t_2 - t_1) = \{\ln \left\{ \frac{\dfrac{1}{[X]_1} + kt_2}{\dfrac{1}{[X]_1} + kt_1} \right\} \} \frac{1}{k(t_2 - t_1)} \tag{4}$$

The ratio of the true value of $(X)_t$, where $t = (t_2 - t_1)/2$, to that estimated from the average concentration during the scan from t_1 to t_2 is given by (5):

$$\frac{[X]_t}{\overline{[X]}} = \frac{[X]_1 k \,\Delta t}{(1 + [X]_1 k \,\Delta t/2)\{\ln(1 + [X]_1 k \,\Delta t)\}} \tag{5}$$

The $[X]_t/\overline{[X]}$ ratio, or the percentage error in estimation of $[X]_t$ from FT-IRS multiple scans, is a function only of the product of $[X]_1$ at the start of the scan, the second order rate constant k, and the time

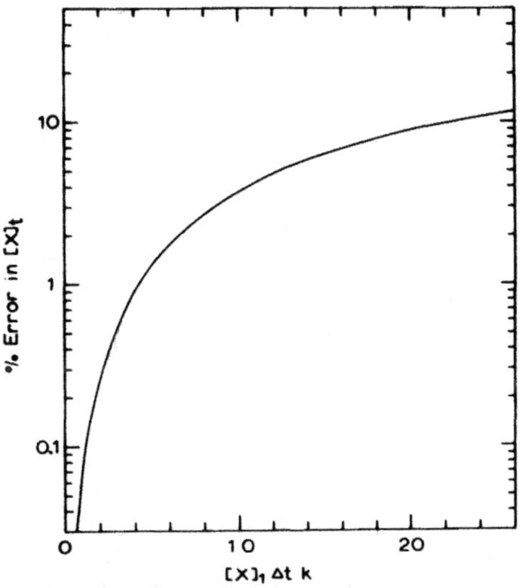

FIGURE 2. Plot of the percentage error introduced in the estimation of the concentration of a species undergoing second order decay on the product of initial concentration of the species, the second order rate constant, and the scan time used to collect the spectral data; see the text.

required for the scan period, $\Delta t = t_2 - t_1$. The magnitude of the error in $[X]_t$ which is introduced by this signal averaging method is shown as a function of the dimensionless parameter, $[X]_1 k\Delta t$ in Figure 2. It can be seen that for a given $[X]_1 k$ product, the error in $[X]_t$ increases rapidly with the data collection time Δt employed. If $[X]_1 k = 0.1\ min^{-1}$ for a given species, then the signal averaging procedure outlined gives $[X]_t$ estimates which are somewhat less than 0.1 % in error when a 1 min averaging time is employed, about 1 % error for 3.5 min, and about 10 % for 10.7 min. Figure 2 can be used as a guide to the choice of scan time for the study of a bimolecular reaction. Thus in our previous study of HONO second order decay, k was estimated to be $1.4 \times 10^{-3}\ ppm^{-1}\ min^{-1}$ [9]. With $[X]_1 = 5$ ppm, we estimate from Figure 2 that for an error less than 0.1 % in $[X]_t$ a data collection time of about 16 min would sufffice. Thus our choice of a 2 min data collection period employed in the previous work with HONO was reasonable, and it did not compromise the accuracy of [HONO] estimates in achieving a reasonably good signal-to-noise ratio. In our later studies of HO_2NO_2 decay in the second order region, an apparent k of the order of $0.15\ ppm^{-1}\ min^{-1}$ was observed with $[HO_2NO_2]$ at about 2 ppm. From the data of Figure 2, we estimate that a one percent error in $[HO_2NO_2]_t$ would arise using .t of about 1 min. Thus our choice of 0.32 min for .t provided a suitable accuracy in (HO_2NO_2) estimates. With the Digilab FT-IRS system employed in our studies, the minimum scan time of 0.08 min restricts the estimate of reactant concentrations at the 1 % accuracy level for a loss process which is second order in reactant, to those systems for which $k < 4\ ppm^{-1}\ min^{-1}$ or $3 \times 10^{-15}\ cc\ molec^{-1}\ sec^{-1}$, when [X] is at the ppm level. Other sources of error in the estimation of $[X]_t$ arise in the measured absorbance of X and extinction coefficient of X. These errors are often greater than 1 %, so that the averaging method employed here for $[X]_t$ measurements is usually sufficiently precise for monitoring those changes in a system which occur on a time scale of a minute or more.

Estimation of Extinction Coefficients of Reactive Molucules.

We have found that the characteristic IR absorbance of most stable molecules follows the Beer-Lambert law well in systems at a fixed total pressure of added N_2 and/or O_2 at 700 Torr. For example, see the data for dimethylnitrosamine given in Figure 3. A linear dependence of the absorbance at the characteristic energies of 1015 and at 1300 cm^{-1} on the concentration of the nitrosamine is seen. In these cases it is a simple procedure to estimate the concentration [X] of the species X from the measured absorbance A_X at some unique band of X for which the extinction coefficient has been determined, and the optical path length l:

$$A_X = \ln (I_0/I) = kl[X] \qquad (6)$$

Of course it is important that the absorbance be determined using the same resolution, apodization, and other adjustable instrumental parameters and the same applied total pressure and temperature as were employed in the determination of the extinction coefficient of X. Certain molecules with very narrow spectral features such as HCl and H_2O show apparent deviations from the simple absorption law when the effective resolution of 1 cm^{-1} or so is employed. In these cases, analysis can still be effected although calibration curves which apply to a given instrument may not apply well to another; thus the calibration for such compounds must be carried out using the same equipment and experimental conditions peculiar to the given study.

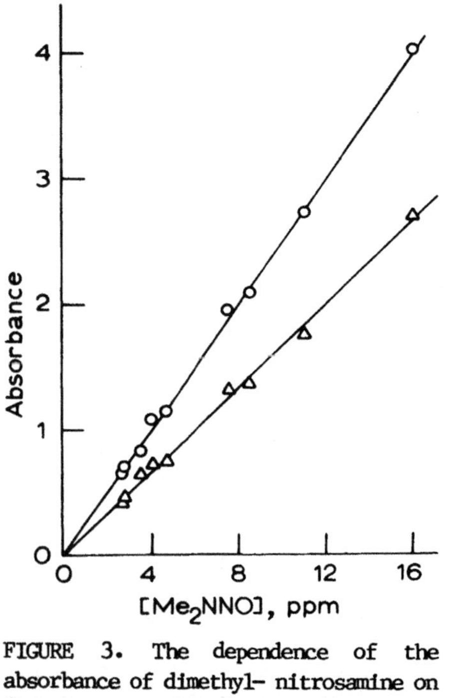

FIGURE 3. The dependence of the absorbance of dimethyl- nitrosamine on its concentration for two characteristic absorption regions.

In the study of highly reactive compounds for which standard pure samples cannot be readily prepared and introduced into the system, less direct means of calibration must be employed. Thus in our studies of the HONO (9), HO_2NO_2 (10), and HOCl (11) systems, we have used various mass balance methods to establish the concentration of a given transient species which should be asociated with an observed absorbance of this species. These methods will be illustrated for certain cases in the discussion of the several systems which follows.

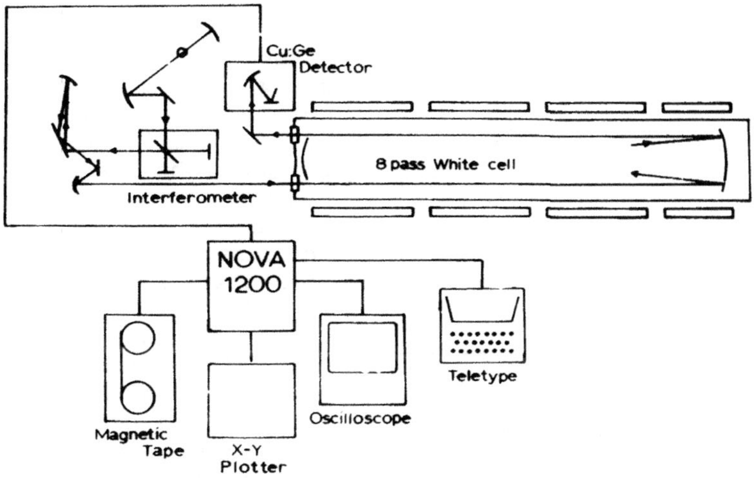

FIGURE 4. Diagram of the FT-IRS system and photochemical reactor used in the studies reported here.

III. THE EXPERIMENTAL FT-IRS AND PHOTOCHEMICAL SYSTEMS

All of the studies described in this work were carried out using a Digilab FTS-20 Fourier transform sprectrometer (model 496 interferometer) coupled to a photolysis cell; see Figure 4. The large, evacuable cell was constructed of 30.5 cm diameter pyrex tubing, 6.3 m in length (445 l volume). It housed a modified white optical system, based upon the design of Hanst [12]; it had a base path of 5.3 m an provided a choice of pathlengths which were multiples of 8-times the base length. All of our experiments reported here were carried out at 170 m (32 traversals) which gave adequate absorption in the ppm range for the compounds studied. The details of the cell construction have been given elsewhere [10] and need not be repeated here. A liquid helium-cooled, Cu : Ge detector (Mullard) was employed for signal detection. Potential problems created by variable background absorption from H_2O and CO_2 in the air path between the transfer optics, Nernst glower IR source, interferometer, cell, and detector, were substantially lowered by enclosing the entire optical system in a plastic housing which was purged continuously with N_2 and contained open dishes of solid NaOH and KOH.

The large photochemical cell was illuminated by black-light fluorescent lamps which surrounded it. An outer reflective shield of aluminum encompassed both the lamps and the cell. Room air could be circulated in the space between the lamps and the cell to help stabilize the cell temperature. Thermocouples and a thermometer placed inside of the cell were used to determine the actual cell temperature at regular intervals during an experiment. Normally the temperature rose about 0.2°C per min or about 1°C during a 5 min irradiation period. The temperature of the cell could be varied within a small range (15 - 27°C) from experiment to experiment by

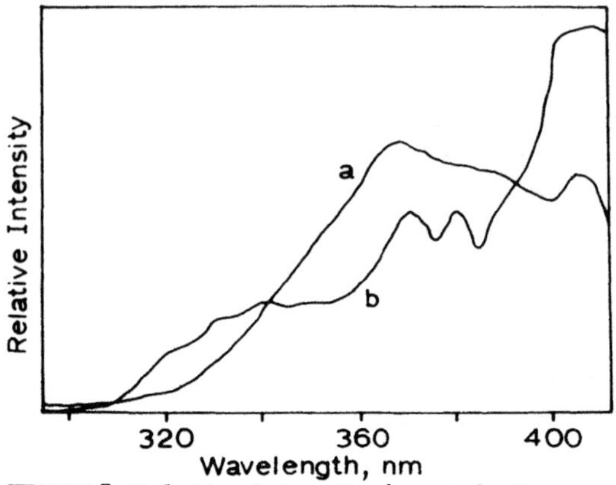

FIGURE 5. Relative intensity (quanta) of the light inside the photochemical reactor versus wavelength (curve a) and a typical distribution of sunlight incident within the atmosphere near ground level (curve b).

regulation of the room temperature. The cell lighting was designed to mimic ground level solar radiation both in distribution and intensity within the photochemically important region, 300 - 450 nm. Direct measurements showed this to be the case. See Figure 5. The NO_2 dissociation rate ($NO_2 + h\nu \rightarrow NO + O$) observed in this system was originally of magnitude, 0.60 min^{-1}, but after three years of use of the system, the value has decreased to 0.35 min^{-1}. These are typical of the values observed in the lower atmosphere for low solar zenith angles. The radiant flux throughout the cell was found to be uniform to within + 10 %.
Reference gases were measured in an all glass, grease- and mercury-free vacuum system equipped with Teflon stopcocks, calibrated volumes, and a quartz spiral gauge employed as a null instrument together with calibrated gauges (Wallace- Tiernan) which were not in direct contact with the gases.
Most of the spectra in kinetic runs were collected using 1 cm^{-1} resolution, with four scans and 8192 data points per interferogram. Extinction coefficients of all reactants and products were determined at several pressures in 700 Torr of synthetic air or N_2, employing 1 cm^{-1} resolution (1 cm optical retardation) and box-car apodization.

IV. THE KINETIC PROPERTIES OF SOME REACTIVE MOLECULES OF ATMOSPHERIC INTEREST AS STUDIED BY FT-IRS METHODS

Peroxynitric Acid Decay.

We have recently studied the rates of formation and decay of HO_2NO_2 in irradiated and dark mixtures of Cl_2, H_2, NO, NO_2, O_2, and N_2 [10]. The system provides new and useful information on this interesting, potential atmospheric contaminant, and some of the results should provide as useful example of the unique information which the FTS-IR systems can provide. A typical spectrum of the products of this system is shown in Figure 6. The characteristic bands of HO_2NO_2 may be separated easily from overlapping spectral features of HONO$_2$ and $ClNO_2$ as seen in the lower spectrum of this figure. Using these bands and those of the other products, the time dependence of the reactants and products from many different mixtures were determined; for

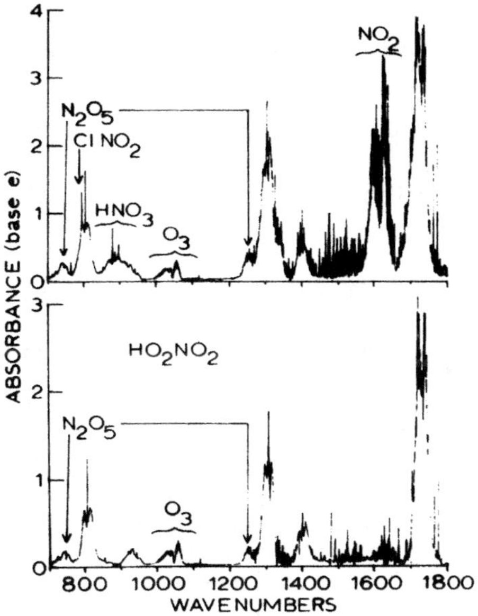

FIGURE 6. Absorbance of products of the photolysis of a dilute mixture of Cl_2, NO_2, H_2 in O_2 and N_2 (total pressure 700 Torr), top spectrum; lower spectrum, remaining absorbance after sutraction of absorbances due to $HONO_2$ and $ClNO_2$ using standard spectra of these compounds; the major unlabeled peaks which are shown result from peroxynitric and acid absorbance.

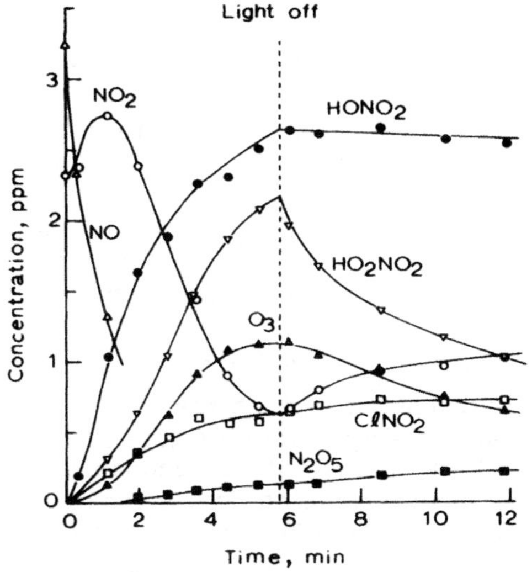

FIGURE 7. The time dependence of the major products of the photolysis of Cl_2 [7.5], H_2[8026], NO_2[2.32], and NO[3.25 ppm] in synthetic air (700 Torr, 23°C); the irradiation was stopped at 5.8 min (vertical dashed line).

example, see Figure 7. We have treated the detailed mechanism and kinetics of the reactions involving HO_2NO_2, $ClNO_2$, $HONO_2$, N_2O_5, O_3, NO, NO_2, and the other minor products elsewhere [10], and these will not be discussed here. However let us consider the kinetic information which one can derive from the HO_2NO_2 dark data. In previous room temperature studies of HO_2NO_2 dark decay by Niki, et al. [5a], Levine, et al. [13], and Hanst and Gay [4b], first order decay kinetics of HO_2NO_2 was observed with $\tau_{1/2} \cong 10$, 8, and 10 - 15 min, respectively. The mechanism of the decay was either unexplained [4b] or attributed to an undefined loss at the wall [5a, 13]. In our recent work [10] we have followed $[HO_2NO_2]$ decay during the dark period following the photolysis of dilute mixtures of Cl_2, H_2, and NO_2 in synthetic air (700 Torr). The $\ln[HO_2NO_2]$ versus time plots of the data from two such experiments carried out near 24°C, are shown in Figure 8. At long times the decay of HO_2NO_2 follows well a first order decay law as was observed and reported previously. However, note that there is a significant curvature to the first order plots at short times for these runs in which $[HO_2NO_2]$ was reatively high (about 5 ppm) and

$[NO_2]$ was relatively low (about 1 ppm). The complete set of data from a given dark decay experiment can be described reasonably well using an empirical rate law of the form (7):

$$-d\,[HO_2NO_2]/dt = [HO_2NO_2]^2\, k_{II} + [HO_2NO_2]\, k_I \qquad (7)$$

The first term in (7) is consistent with the occurrence of the reaction sequence (8), (9), and (10):

$$HO_2NO_2\ (+\ M) \rightarrow HO_2 + NO_2\ (+\ M) \qquad (8)$$

$$HO_2 + NO_2\ (+\ M) \rightarrow HO_2NO_2\ (+\ M) \qquad (9)$$

$$HO_2 + HO_2 \rightarrow H_2O_2 + O_2 \qquad (10)$$

It can be shown that for our conditions the significant removal of HO_2 radicals by reactions with NO ($NO + HO_2 \rightarrow HO + NO_2$), O_3 ($O_3 + HO_2 \rightarrow HO + 2O_2$), and any other reactants than NO_2 and HO_2, can be excluded here. We have concluded from a variety of evidence that the first order term in relation (7) has its origin in a well removal reaction (11):

$$HO_2NO_2\ (+\ \text{Wall}) > \text{Products}\ (+\ \text{Wall}) \qquad (11)$$

The reaction mechanism (8) - (11) can be tested using the kinetic data from HO_2NO_2 decomposition studies given in Figure 8 and other similar data from experiments at several temperatures. If this mechanism is operative and $R_8 = R_9 >> R_{10}$, then relation (12) should apply:

$$\left[\frac{k_8}{k_9}\right]^2 2k_{10} = \left\{\left[-\frac{d\ln[HO_2NO_2]}{dt}\right]_t - k_{11}\right\}\frac{[NO_2]_t^{\,2}}{[HO_2NO_2]_t} \qquad (12)$$

From the $\ln[HO_2NO_2]$ versus time plots (Figure 8 and other similar data) we have measured the slopes at the beginning of the dark period $(d\ln[HO_2NO_2]/dt)_o$, the $[NO_2]_o$ and $[HO_2NO_2]_o$ values for the same time, and the slope of the linear portion of the plot at long times (k_{11}). Coupling these estimates with the literature value for $k_{10} = 5.3 \times 10^3$ $ppm^{-1}\ min^{-1}$ [14,15], values of K_8/k_9 were calculated using relation (12). In Figure 9, our data (circles) are shown in an Arrhenius plot. Within the error of the data, they follow well the dashed line, also shown in Figure 9, which is the locus of pointts derived from the temperature dependent functions for k_8 is the locus of points derived from the temperature dependent functions for k_8 from Graham, et al. [16], k_9 from Howard [17], together with the assumption that $E_9 = -2$ kcal $mole^{-1}$. The two very different kinetic methods lead to results which are in reasonable accord, and some confidence in the methods employed is gained. The least squares fit of the data ponts in Figure 9 gives: $\Delta H_8 \cong 23.8 \pm 2.7\ (2\sigma)$ kcal $mole^{-1}$; $\Delta S_8^o \cong 40.5 + 8.8\ (2\sigma)$ eu (std. state, 1 atm). The FT-IRS system and the data analysis which was

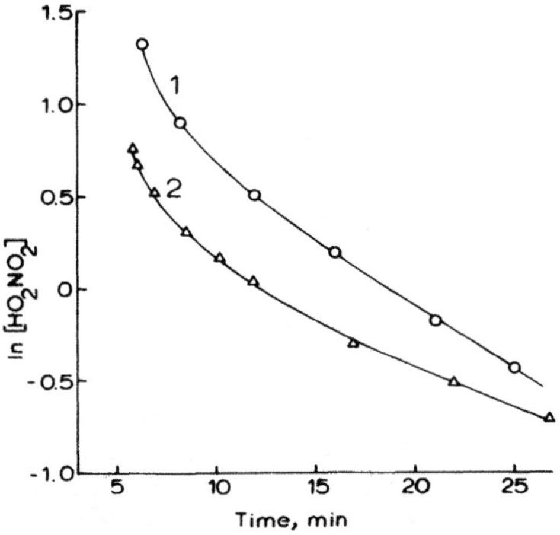

FIGURE 8. Plot of $\ln[HO_2NO_2]$ versus time for two typical HO_2NO_2 experiments during the dark decay period; data from Cl_2, H_2, NO_x, air mixture photolysed; temperatures at the onset of the dark period were near 24°C.

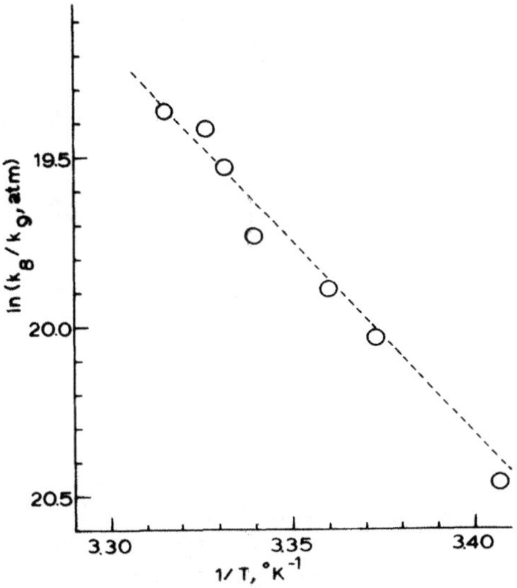

FIGURE 9. Plot of $\ln[k_8/k_9]$ versus 1/T; open circles are from our estimates employing relation [12] and initial rate data for HO_2NO_2 dark decay; the dashed line is the locus of points predicted from the individual rate constant functions for k_8 and k_9 estimated from the results of Graham, et al. [16], and Howard [17].

used in this work appear to offer a convenient, new method of study of the equilibrium constants for the potentially significant, reactive species, peroxynitric acid, and analogous peroxyalkylnitrates (RO_2NO_2).

The Kinetics of HOCl Formation Reaction and HOCl Absolute Extinction Coefficients in thenInfrared Region.

In order to evaluate properly the potential role of HOCl as a temorary sink for Cl atoms in the stratosphere, it is important to establish the absolute IR extinction coefficients of this molecule which will allow direct estimates of its abundance in our atmosphere. We have recently carried out a kinetic study of the HOCl molecule in our system which was designed to establish the needed IR extinction data. It provides an interesting further example of the utility of the FT-IRS system and the complications which are encountered in the kinetic study of reactive species by these methods. HOCl was generated by photolysis of Cl_2 in H_2, O_3, O_2, N_2 mixtures. The spectral changes which result in this irradiated mixture can be seen in Figure 10. The original absorption due to the reactant O_3 can be seen in the spectrum

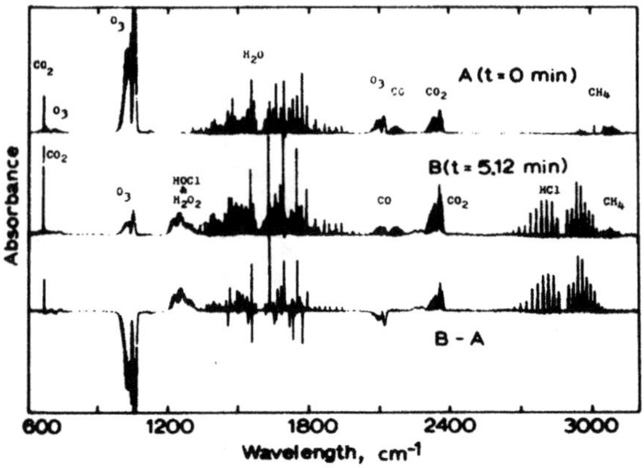

FIGURE 10. Absorbance-wavelength plot for a gaseous mixture of initial composition: Cl_2 (60 ppm), O_3 (23.5 ppm), H_2 (20400 ppm), O_2 (44 Torr), N_2 (641 Torr); the CH_4, CO, CO_2, and H_2O are trace impurities present in the initial reactants; curve A before photolysis; cure B after 5.12 min photolysis; lower curve shows the absorbance difference spectrum (B − A) used to estimate product concentrations.

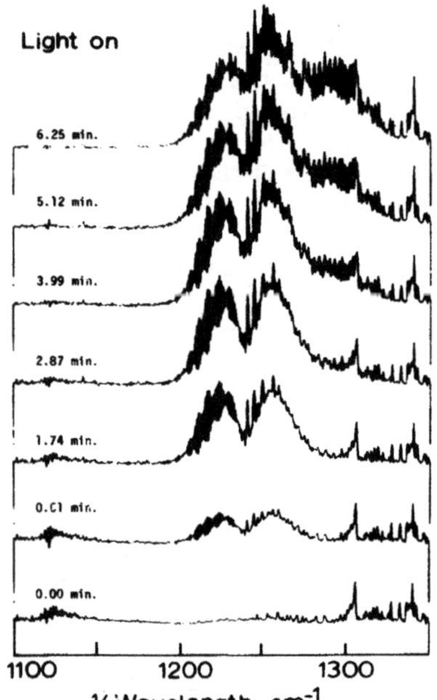

FIGURE 11. Absorbance growth within the HOCl ν_2 band with increasing time of irradiation of a Cl_2, H_2, O_3, O_2, N_2 mixture of initial composition listed in Figure 10.

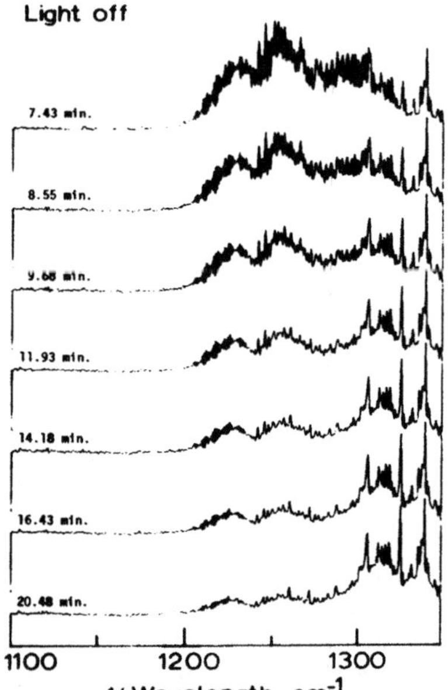

FIGURE 12. Absorbance decay within the HOCl, ν_2 band with increasing length of the dark period; bands beyond 1300 cm^{-1} are from H_2O vapor and $HONO_2$ impurity formed from HCl dark reactions at the cell wall.

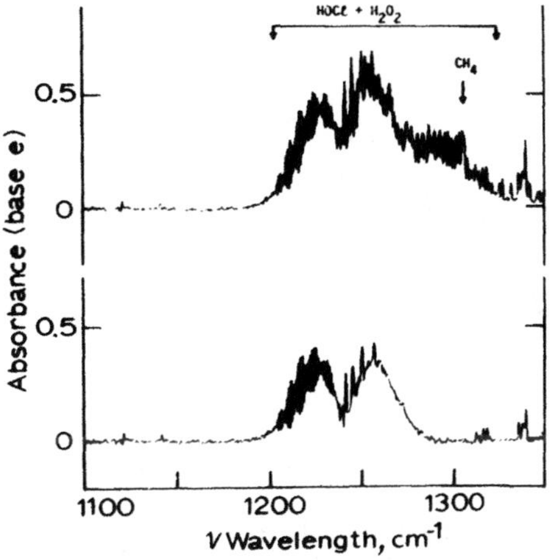

FIGURE 13. Absorbance of HOCl and H_2O_2 products within the 1100 – 1350 cm^{-1} region; upper spectrum is from the products after 5.12 min irradiation of the mixture of composition listed in Figure 10; lower curve, absorbance spectrum of HOCl remaining after absorbance of standard samples of H_2O_2 and CH_4 were used to balance out contributions from these species; remaining absorbance beyond 1300 cm^{-1} result from H_2O.

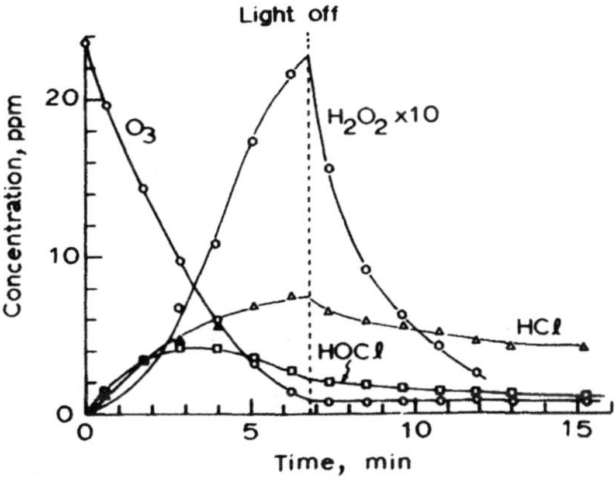

FIGURE 14. The concentrations of products versus time in the photolysis of the Cl_2, H_2, O_3, O_2, N_2 mixture of initial concentrations given in Figure 10; estimated from the absorbance versus time data of Figure 11 and similar data for the other spectral regions.

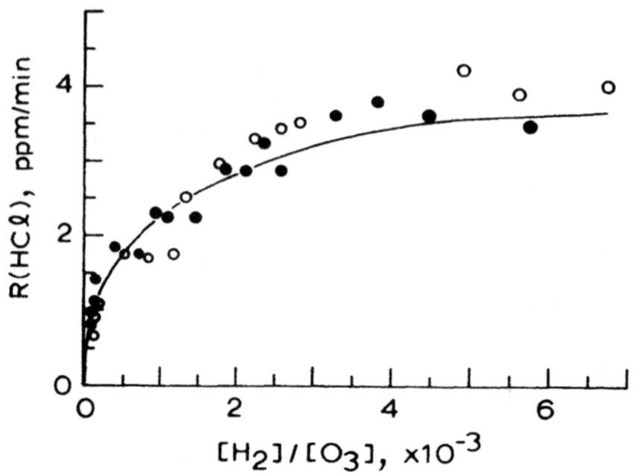

FIGURE 15. Rates of HCl formation in the photolysis of Cl_2, H_2, O_3, O_2, N_2 mixtures at varied $[H_2]/[O_3]$ ratios; data were estimated from the 0.62 (closed circles) and 1.74 min points (open circles); initial concentrations: $[Cl_2]$, 60 ppm; $[O_3]$, 16.1 – 23.5 ppm; $[H_2]$, 570 – 62000 ppm; solid curve represents a computer simulation of rates based upon a complete reaction mechanism including reactions [13] – [19] of the text.

labeled A. Taken after 5.12 min irradiation, spectrum B shows products HCl, HOCl, and H_2O_2, and a decrease in O_3. Spectrum C, the difference spectrum (A-B), shows the product absorption (CH_4 and CO impurities now absent) and negative peaks for the O_3 reactant. Such spectra taken from various times during the photolysis can be used conveniently to follow the products of the reaction. Note in Figure 11 the growth of the HOCl band near 1226 cm^{-1} with increasing time. H_2O_2 absorption centered near 1275 cm^{-1} can be observed to grow in also at the later times shown in Figure 11. As the light is turned off, the characteristic HOCl peak dies as seen in Figure 12. The spectral features of HOCl, H_2O_2 and CH_4 (impurity in O_2 employed) may be resolved by subtraction of some fraction of the standard absorbances for a given quantity of the pure compounds. Thus in Figure 13 absorbances due to CH_4 and H_2O_2 have been removed from the observed spectrum to reveal the absorption in the 1200 – 1275 cm^{-1} band of HOCl alone (ν_2, bend). If the absolute extinction coefficients of the reactants and products are known, then the time

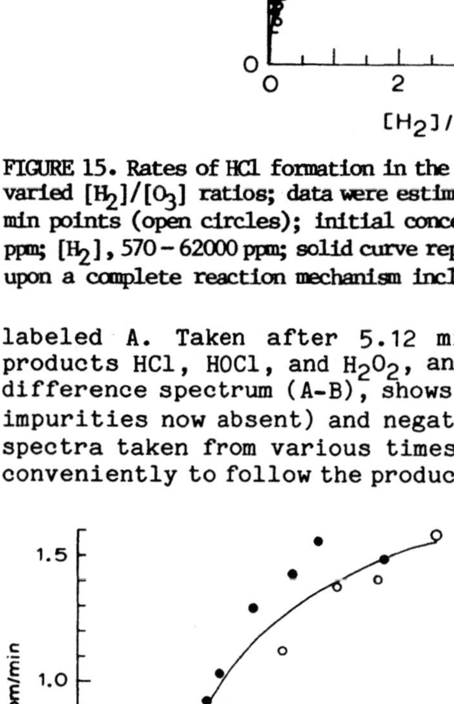

FIGURE 16. Rates of H_2O_2 formation in the photolysis of Cl_2, H_2, O_3, O_2, N_2 mixtures at varied $[H_2]/[O_3]$ ratios; conditions are as outlined in Figure 15.

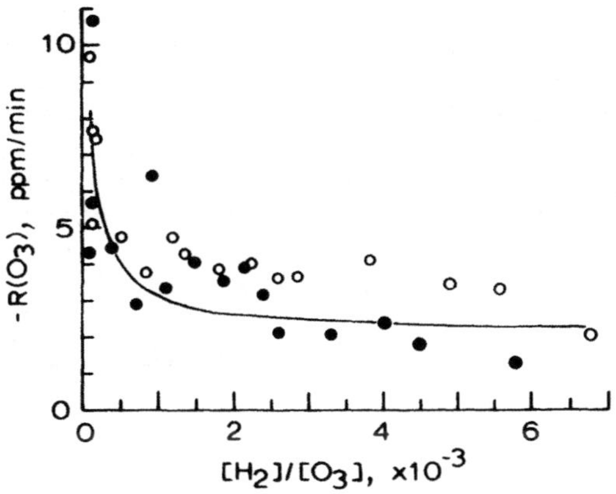

FIGURE 17. Rates of O_3 loss in the photolysis of Cl_2, H_2, O_3, O_2, N_2 mixtures at varied $[H_2]/[O_3]$ ratios; conditions are as outlined in Figure 15.

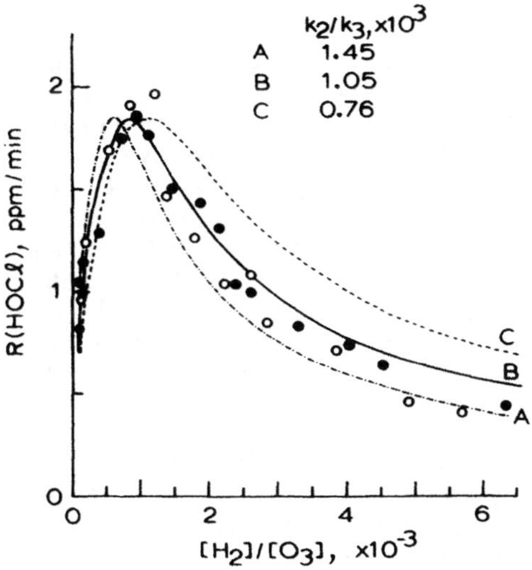

FIGURE 18. Rates of HOCl formation in the photolysis of Cl_2, H_2, O_3, O_2, N_2 mixtures at varied $[H_2]/[O_3]$ ratios; conditions are as outlined in Figure 15, computer simulations are shown for various choices of the k_{14}/k_{15} ratio: 1.45×10^3 (A); 1.05×10^3 (B); 0.76×10^3 (C); the absolute extinction coefficients of HOCl were determined by the matching of the simulation and experiment data points near the maximum on the ordinate scale; note this is insensitive to the k_{14}/k_{15} ratio assumed.

dependence of the concentrations of the individual products and reactants can be estimated as in Figure 14. From such curves we may estimate iitial rates of change of products and reactants for varied reactant concentrations. The initial rate data for the products HCl and H_2O_2 and the reactant O_3 shown in Figures 15, 16, and 17, respectively, were determined in this fashion in experimnts carried out over a wide range of $(H_2)/(O_3)$ reactat ratios.

The reaction mechanism which should describe this system during the early stages of the photolysis is the following:

	k, cc/molec-s, 298°K	Reference	
$Cl_2 + h\nu \rightarrow 2Cl$			(13)
$Cl + O_3 \rightarrow ClO + O_2$	1.16×10^{-11}	18	(14)
$Cl + H_2 \rightarrow HCl + H$	1.68×10^{-14}	18	(15)
$H + O_2\ (+M) \rightarrow HO_2\ (+M)$	1.25×10^{-12} (700 Torr)	9	(16)
$HO_2 + OCl \rightarrow HOCl + O_2$	3.8×10^{-12}	20	(17)
$HO_2 + HO_2 \rightarrow H_2O_2 + O_2$	3.1×10^{-12}	21	(18)
$2OCl \rightarrow Cl + ClOO$			(19a)
$2OCl \rightarrow Cl_2 + O_2$	2.5×10^{-14}	19	(19b)
$2OCl \rightarrow OClO + Cl$			(19c)
$HO_2 + O_3 \rightarrow HO + 2O_2$	1.0×10^{-15}	19	(20)

Most for the elementary rate constants for these reactions which one anticipates to be important in this system have been measured, and they are known with reasonably good accuracy. The rate of reaction (13) determined in a variety of experiments in our photochemical reactor at the time of this study gave $k_{13} = 0.035\ min^{-1}$ [11]. Using this value and the other published rate constant data shown, the computer simulation of the initial rates of HCl, H_2O_2, and O_3 were estimated and are shown as solid lines in Figures 15, 16, and 17, respectively. The predicted magnitudes of the initial rates of HOCl formation for various choices of k_{14}/k_{15} are shown in Figur 18. The kinetic dependence of R_{HOCl} on $[H_2]/[O_3]$ ratio expected theoretically, obviously matches well that

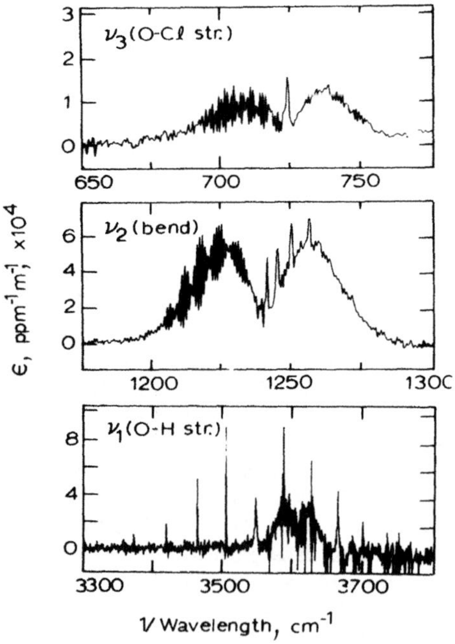

FIGURE 19. The absolute extinction coefficients as a function of wave number (cm^{-1}) for HOCl gas in the infrared region.

observed experimentaly (circles). The choice of the extinction coefficient used in calculating the R_{HOCl} from absorbance- time data for HOCl, was made to procide the best match of these data on the ordinate scale. The choice of k_{14}/k_{15} used affects only the position of the theoretical maximum in the R_{HOCl} versus $[H_2]/[O_3]$ curves and not the magnitude of the maximum. The best fit of our data suggest $k_{14}/k_{15} \cong 1.1 \times 10^3$; this is in resonable accord with the ratio derived from published estimates of the individual rate constants.Thus having established the best kinetic match of the experimental R_{HOCl} versus $[H_2]/[O_3]$ ratios and theoretical predictions based upon the mass balance and reasonable kinetic scheme, our best estimate of the extinction coefficients for HOCl in the three fundamental vibrational bands have been determined; these data are given in Figure 19. They are considered to be reliable to ±20 %.

The Formation Reactions of $(CH_3)_2NNO$ and Relative Rate Constants for the $(CH_3)_2N$ Radical Reactions with O_2, NO, and NO_2

The potential for formation of the nitrosamines, highly carcinogenic compounds, in the polluted atmosphere lies in the rate of generation of the dialkylamino radicals and their subsequent reactions with O_2, NO, and other impurity molecules in the atmosphere [22]. In theory the HO-radical may generate R_2N radicals from secondary amines (R_2NH) by the fast reaction (21), and the nitrosamine (R_2NNO) and nitramine (R_2NNO_2) may be formed by reactions (22) and (23):

$$R_2NH + HO \rightarrow R_2N + H_2O \quad (21)$$
$$R_2N + NO \rightarrow R_2NNO \quad (22)$$
$$R_2N + NO_2 \rightarrow R_2NNO_2 \quad (23)$$

Whether (22) and (23) occur at significant rates in a polluted atmosphere depends upon the relative rates of (22) and (23) compared with the rate of radical loss by undefined reactions with oxygen in the air:

$$R_2N + O_2 \rightarrow PRODUCTS \quad (24)$$

In a recent study we have used the FT-IRS system to establish the needed rate constant data related to the reactions of the dimethylamino radicals [23]. A brief description of some of these results will serve to illustrate the use of the FT-IRS system in relative rate determinations in this interesting system. Our results confirm and extend the qualitative finding reported by Hanst, et al. [24], Glasson [25], and Pitts and coworkers [26]. In one phase of our work, dimethylnitrosamine photolysis was used as the source of the $(CH_3)_2N$ radicals in O_2-rich mixtures (700 Torr) with added small amounts of NO and NO_2. The spectral changes which were observed in irradiation of $(CH_3)_2NNO$ in an experiment with 700 Torr of O_2 are shown in Figure 20. The changes observed can be rationalized quantiatively in terms of absorbances due to $(CH_3)_2NNO$ and its major products: $(CH_3)_2NNO_2$, $CH_3N{=}CH_2$, HONO, NO and NO_2. Spectra for samples of these cmpounds are given in Figure 21. The interesting compound, methylmethyleneamine, $CH_3N{=}CH_2$, was identified as aproduct of the $(CH_3)_2N$ radicals for the first time in this work. The calibration of the extinction coefficients for this compound offered an unusual challenge which is of some interest here.

The methylmethyleneamine trimer was prepared by us, and weighted quantities of it were introduced into our cell. The spectrum of this compound in a mixture with 700 Torr of N_2 was followed as function of time. Note typical spectra for an experiment after 10 min and after 173 min in the dark in Figure 22. There are many changes in absorbance which occur. For example, the characteristic peak at 917 cm^{-1} due to the trimer decreases regularly with time while a new peak at 1025 cm^{-1} increases with time. In terms of these observations and the reported properties of the trimer and its monomer dissociation product (27), it seemed reasonable to attempt a resolution of these spectra in terms of these two components alone. The results of several different experiments are shown in Figure 23 in which weighed samples of the trimer were introduced into our cell, and the absorbances of trimer at 917 cm^{-1} and that attributed to monomer at 1025 cm^{-1} were monitored at selected times. The plot of these absorbances is

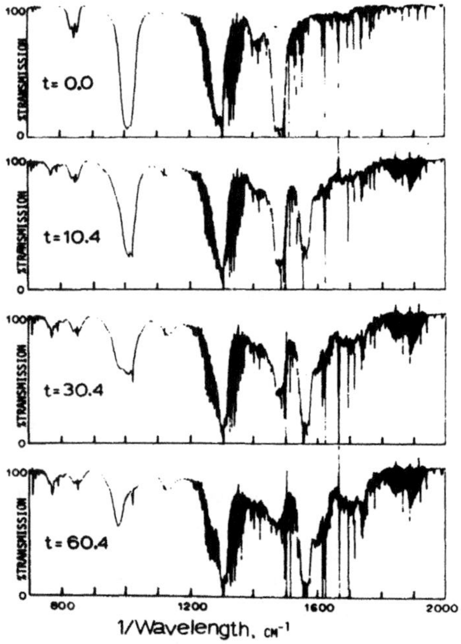

FIGURE 20. Transmission spectra for the photolysis of dimethyl- nitrosamine (5.73 ppm), woth NO (3.93 ppm), NO_2 (1.18 ppm), in 700 Torr of O_2 at various irradiation times shown.

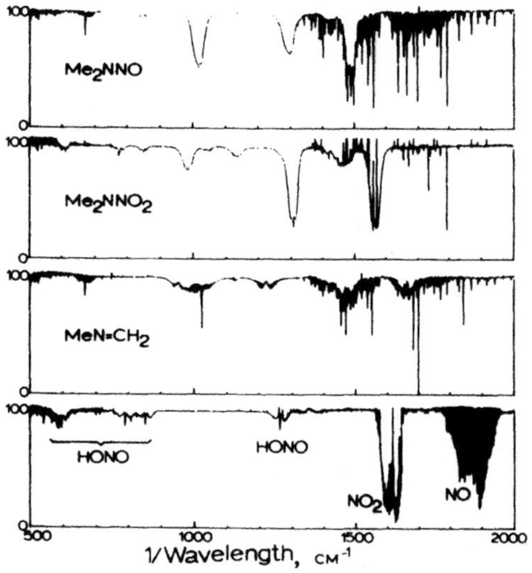

FIGURE 21. Transmission spectra for the major products of dimethyl-nitrosamine photolysis in O_2 containing mixtures.

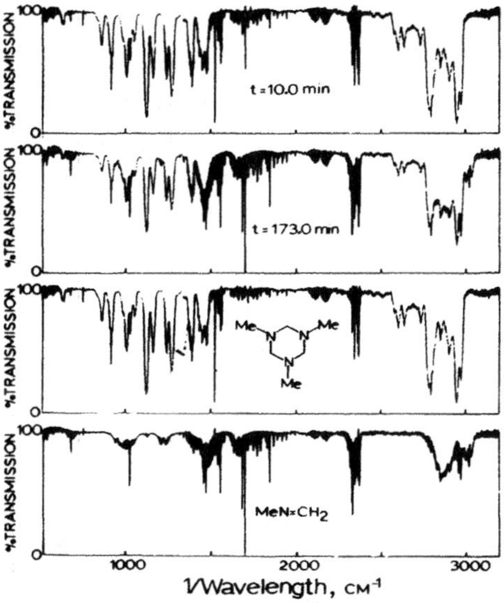

FIGURE 22. Transmission spectra observed after introduction of a sample of methyl-methylene-amine trimer at 10 min and 173 min; all of the features can be rationalized in terms of the spectra of the trimer and the monomer which were derived by suitable subtraction and processing of the mixture spectra at various times.

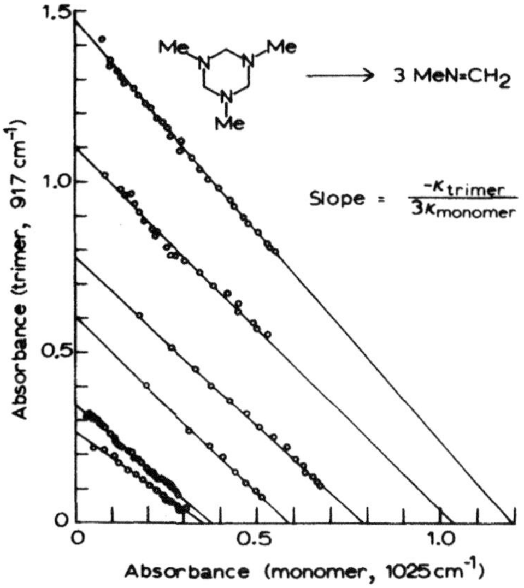

FIGURE 23. Plot of the absorbance of the methyl-methylene-amine trimer at 917 cm^{-1} as a function of monomer absorbance at 1025 cm^{-1}; observations at various times for several different initial concen-trations of trimer; the slope of these lines in theory is equal to the extinction coefficient ratio: $-\kappa_{trimer}$ / $3\kappa_{monomer}$

shown in Figure 23. If the only fate of the trimer is dissociation to monomer then one expects the absorbance plots for the several experiments to consist of parallel lines in whic the ratio of the two intercepts (the slope of the plots) should be related to the ratio of the two extinction coefficients:

$$\text{slope} = -\kappa_{trimer}\ (917\ cm^{-1})/3\kappa_{monomer}\ (1025\ cm^{-1}).$$

This expectation appears to be born out well by the data. The y-axis and the x-axis intercepts of these plots give in theory the absorbance of pure trimer and pure monomer, respectively, at a concentration which can be calculated in each case by the weight of the trimer introduced to the system at the start of the experiment. Beer-Lambert absorbance-concentration plots constructed from these data provided the estimates of the extinction coefficients for the trimer and monomer which were employed in this work; see Figures 24 and 25. A check on the reliability of these estimates can be had from a series of experiments in which trimer and monomer absorbancies were measured at long times at several temperatures. After a given mixture had stood overnight or for longer periods at a near constant temperature, there

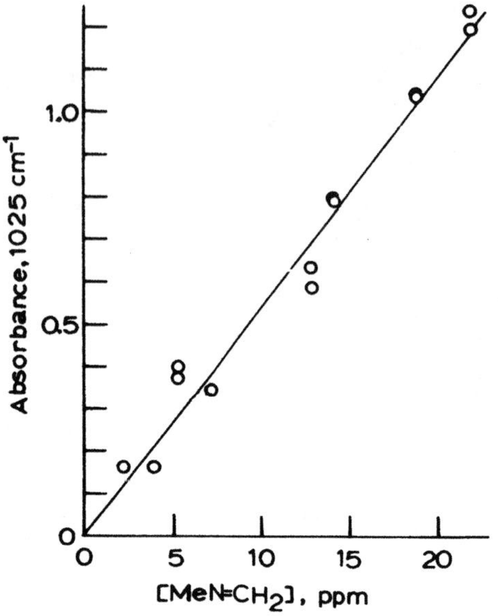

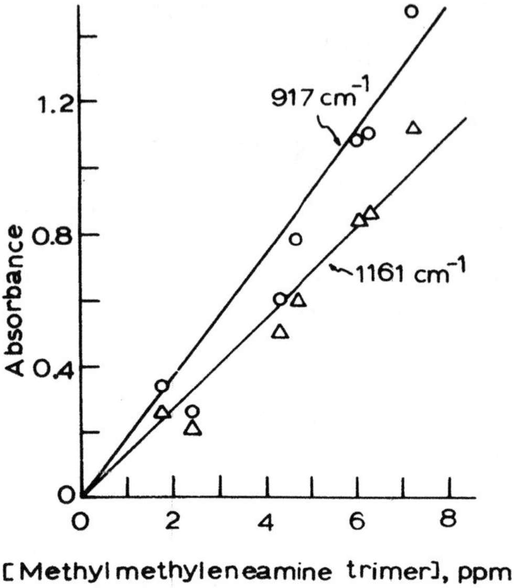

FIGURE 24. Beer-Lambert plot for the absorbance of methyl- methylene-amine monomer at 1025 cm^{-1}.

FIGURE 25. Beer-Lambert plot for the absorbance of the methyl- methylene-amine trimer at 917 (circles) and 1161 cm^{-1} (triangles).

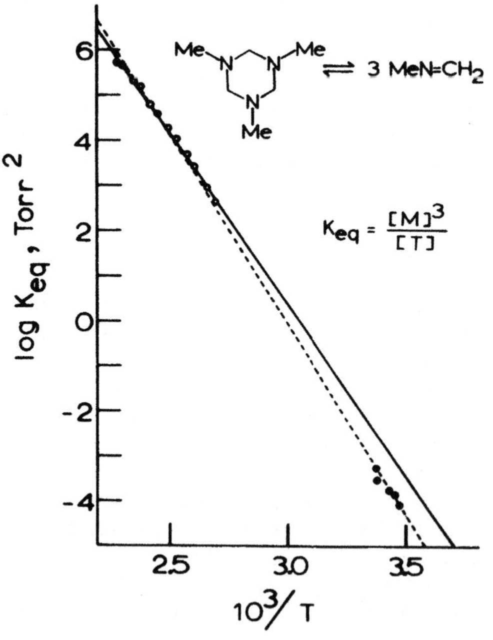

FIGURE 26. Plot of the logarithm of the equilibrium constant for the trimer-monomer equilibrium shown versus the inverse temperature: open circles are from Gowenlock and Thomas [27] who used pressure measurements; closed circles from Lindley et al. [23]; the dotted line is the least squares fit of all the data; the solid line is the fit of the data of Ref. 27 alone.

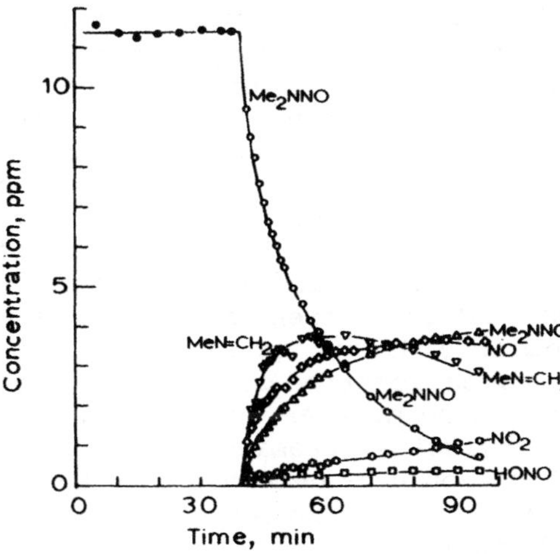

FIGURE 27. Concentration-time profile of the products of dimethyl-nitrosamine photoltyses in 730 Torr of O_2.

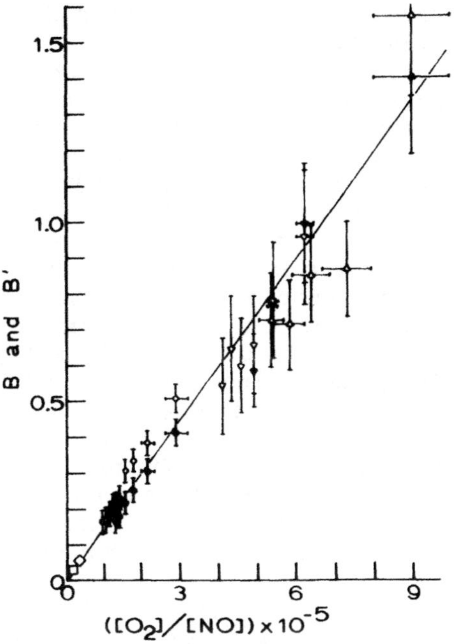

FIGURE 28. Plot of B (open symbols) and B (closed symbols) of the text versus $[O_2]/[NO]$ ratio; data derived from a series of dimethyl-nitrosamine photolyses in various O_2-NO- containing mixtures; in theory the slope of the plot gives k_{25}/k_{26}.

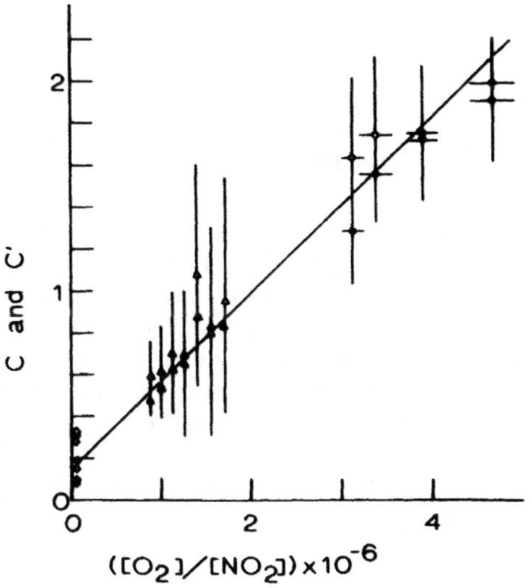

FIGURE 29. Plot of the C and C functions of the text versus the $[O_2]/[NO_2]$ ratio for a series of dimethyl-nitrosamine photolyses in O_2-NO_2-containing mixtures; in theory the slope of the plot gives k_{25}/k_{27a}.

was no significant further change in absorbance of trimer or monomer with time, and it is likely that the trimer-monomer equilibrium was established for these conditions. From these direct spectral observations of trimer and monomer, the equilibrium constant for the dissociation of the trimer to monomer was estimated. These data are shown in Figure 26 together with the equilibrium constants etimated by Gowenlock and Thomas (27) by pressure measurements i experiments at somewhat elevated temperatures. The dotted line is a fit of all of the data, while the solid line is an extrapolation of the Gowenlock and Thomas data alone. It is appaent that the present direct data are in reasonable accord with the older indirect measurements, and additional confidence is gained in our calibration of the extinction coefficients of the methylmethyleneamine monomer and trimer.

Using the extinction data for the methylmethyleneamine and the other products of the $(CH_3)_2N$ radical which could be determined in a more direct fashion, we were able to establish the time dependence of the concentrations of the products of the $(CH_3)_2NNO$ photolyses in NO, NO_2, O_2 mixtures. Those or a typical run at high O_2 concentration are shown in Figure 27. These and our other kinetic data can be rationalized well interms of the following reaction mechanism:

$$[CH_3]_2NNO + h\nu \rightarrow [CH_3]_2N + NO \qquad (25)$$
$$[CH_3]_2N + O_2 \rightarrow CH_2{=}NCH_3 + HO_2 \qquad (26)$$
$$[CH_3]_2N + NO \rightarrow [CH_3]_2NNO \qquad (27)$$
$$[CH_3]_2N + NO_2 \rightarrow [CH_3]_2NNO_2 \qquad (28A)$$
$$\rightarrow HONO + CH_2{=}NCH_3 \qquad (28B)$$

For mixtures at low (NO_2) this reaction mechanism leads to the following rate functions related to the rate constant ratio k_{26}/k_{27}:

$$B = \frac{K_{26}[O_2]}{K_{27}[NO]} = \frac{\bar{R}_{Me_2NNO_2}\ \bar{R}_{Me_2NNO}}{K_{25}[MeNNO] + R_{Me_2NNO}} \qquad (29)$$

$$B' = \frac{K_{26}[O_2]}{K_{27}[NO]} = \frac{R_{MeN=CH_2}}{K_{25}[MeNNO] + R_{Me_2NNO}} \qquad (30)$$

The functions B and B′ have been evaluated from the appropriate rate data derived in experiments at a variety of $[O_2]/[NO]$ ratios. These are plotted versus $[O_2]/[NO]$ in Figure 28; the open and closed symbols represent the B and B′ functions, respectively. The linear relationship between the variables and intercept near zero are consistent with the theoretical expectations based upon the mechanism outlined. The slope of the line gives the estimate: $k_{26}/k_{27} = (1.48 + 0.072) \times 10^{-6}$ (2σ).

In experiments containing significant NO_2, the nitramine, $(CH_3)_2NNO_2$ and HONO are major products. For these conditions we anticipate functions C and C′ to hold:

$$C = \frac{K_{28}}{K_{28}} + \frac{K_{26}[O_2]}{K_{28}\ [NO_2]} = \frac{{}^{-}R_{Me_2NNO}\ {}^{-}R_{Me_2NNO_2}}{R_{Me_2NNO_2}} \qquad (31)$$

$$C' = \frac{K_{28}}{K_{28}} + \frac{K_{26}[O_2]}{K_{28}\ [NO_2]} = \frac{R_{MeN=CH_2}}{R_{Me_2NNO_2}} \qquad (32)$$

The appropriate rate data from a series of experiments were used to evaluate C and C′, and the data are plotted versus the $[O_2]/[NO_2]$ ratio in Figure 29. The slope and intercept, respectively, of this plot provide priliminary estimates of the rate constant ratios: $k_{26}/k_{28a} = (2.90 \pm 0.28) \times 10^{-7}$ and $k_{28b}/k_{28a} = 0.22 \pm 0.06$ (26). The very large rate constants for the $(CH_3)_2N$ reactions with NO and NO_2 compared to that with O_2 suggests that dimethylamino radicals formed in a usual NO-NO_2-polluted atmosphere have a reasonably good chance of forming the nitrosamine and the nitramine even though the NO and NO_2 impurities are at concentrations in the range of a few hundred pphm; concentrations of this magnitude are commonly encountered.

V. SUMMARY

The examples cited in our discussion show that the apllication of long-path Fourier tranform spectroscopy to the study of the kinetics and mechanisms of formation and decay of many reactive molecules may provide useful information concerning atmospheric reactions which is difficult to achieve by other experimental methods. The methods outlined and illustrated here apply to species whose half-lives in the reaction are of the order of several minutes. It seems to be a reasonable expectation that the future refinement of time-resolved FTS systems [28] should ultimately allow direct study the very reactive, highly transient free radical species as well as the reactive molecular species to which our present system are limited. This work was supported by the United States Environmental Protection Agency Grant No. R-806479-03-0. The author is grateful to Dr. R.G. Derwent for his hilp with computer simulations in our early attempts to unscramble the HO_2NO_2 and HOCl reaction mechanisms.

REFERENCES

1. (a) Calvert, J.G., and Hanst, P.L., "The Mechanism of the Photooxidation of Acetaldehyde at Room Temperature", Can. J. Chem., 37, 1671 - 1679 (1959); (b) Subbaratnam, N.R., and Calvert, J.G., "The Mechanism of Methyl Hydroperoxode Formation in the Photooxidation of Azomethane at 25°C", J. Amer. Chem. Soc., 84, 1113 - 1118 (1962); (c) Thomas, S.S., and Calvert, J.G., "The Photooxidation of 2,2′- Azoisobutane at 25°C", J. Amer. Che. Soc., 84, 4207 - 4212 (1962).

2. (a) Tuesday, C.S., "The Atmospheric Photoxidation of trans-Butene-2 and Nitric Oxide", in "Chemical Reactions in the Lower and Upper Atmosphere", Interscience Publishers, N.Y., 15 - 49 (1961); (b) Glasson, W.A., and Tuesday, C.S., "Hydrocarbon Reactivities in the Atmospheric Photooxidation of Nitric Oxide", Environ. Sci. Technol., 4, 916 - 924 (1970); (c) Glasson, W.A., and Tuesday, C.S., "Reactivity Relationships of Hydrocarbon Mixtures in Atmospheric Photooxidation", Environ. Sci. Technol., 5, 151 - 154 (1971).
3. Pate, C.T., Finlayson, B.J., and Pitts, J.N., Jr., "A Long Path Infrared Spectroscopic Study of the Reaction of Methylperoxy Free Radical with Nitric Oxide", J. Amer. Chem. Soc., 96, 6554 - 6558 (1974).
4. (a) Gay, B.W., Jr., Noonan, R.C., Bufaline, J.J., and Hanst, P.L., "Photochemical Syntheses of Peroxyacryl Nitrates in Gas Phase via Chlorine-Aldehyde Reaction", Environ. Sci. Technol., 10, 82 - 85 (1976); (b) Hanst, P.L., and Gay, B.W., Jr., "Photochemical Reactions Among Formaldehyde, Chlorine, and Nitrogen Dioxide in Air", Environ. Sci. Technol., 11, 1105 - 1109 (1977).
5. (a) Niki, H., Maker, P.D., Savage, C.M., and Breitenbach, L.P., "Fourier Transorm IR Spectroscopic Observation of Pernitric Acid Formed via $HO_2 + NO_2 \Rightarrow HO_2NO_2$", Chem. Phys. Letters, 45, 564 - 566 (1977); (b) Niki, H., Maker, P.D., and Savage, C.M., "Relative Rate Constants for the Reaction of Hydroxyl Radical with Aldehydes", J. Phys. Chem., 82, 132 - 134 (1978).
6. (a) Chan, W.H., Nordstrom, R.J., Calvert, J.G., and Shaw, J.H., "A Kinetic Study of HONO Formation and Decay Reactions in Gaseous Mixtures of HONO, NO, NO_2, H_2O, and N_2", Environ. Sci. Technol., 10, 674 - 682 (1976); (b) Chan, W.H., Uselman, W.M., Calvert, J.G., and Shaw, J.H., "The Pressure Dependence of the Rate Constant for the Reaction: $HO + CO \Rightarrow H + CO_2$", Chem. Phys. Letters, 45, 240- 244 (1977).
7. (a) Pitts, J.N., Jr., McAfee, J.M., Long, W.D., and Winer, A.M., "Long-Path Infrared Spectroscopic Investigation at Ambient Concentrations of the 2 % Neutral Buffered Potassium Iodide Method for Determination of Ozone", Environ. Sci. Technol., 10, 787 - 793 (1976); (b) Graham, R.A., Winer, A.M., and Pitts, J.N., Jr., "Temperature Dependence of the Unimolecular Dicomposition of Pernitric Acid and its Atmospheric Implications", Chem. Phys. Letters, 51, 211 - 215 (1977).
8. Bell, R.J., "Introduction to Fourier Transform Spectroscopy", Academic Press (1972).
9. Chan, W.H., Nordstrom, R.J., Calvert, J.G., and Shaw, J.H., "An IRFTS Study of HONO Formation and Decay Reactions in Gaseous Mixtures of HONO, NO, NO_2, H_2O, and N_2", Chem. Phys. Letters, 37, 441 - 446 (1976).

10. Uselman, W.M., Levine, S.Y., Chan, W.H., and Calvert, J.G., "A Kinetic Study of the Mechanism of Peroxynitric Acid Formation in Irradiated Mixtures of Cl_2, H_2, NO_2, and NO in Air", in "Nitrogenous Air Pollutants", Editor D. Grosjean, Ann Arbor Acience Publishers, Inc., 1979, p. 17 - 54.
11. Su, F., Calvert, J.G., Lindley, C.R., Useldman, W.M., and Shaw, J.H., "An FTS-IR Kinetic Study of HOCl and its Absolute Integrated IR Band Intensities", J. Phys. Chem., 83, 912 - 920 (1979).
12. Hanst, P.L., "Spectroscopic Methods for Air Pollution Measurement", Adv. Environ. Sci. Technol., 2, 91 - 213 (1971).
13. Levine, S.Y., Uselman, W.M., Chan, W.H., Calvert, J.G., and Shaw, J.H., "The Kinetics and Mechanism of the HO_2-NO_2 Reactions: The Significance of Peroxynittric Acid Formation in Photochemical Smog", Chem. Phys. Letters, 48, 528 - 535 (1977).
14. Paukert, T.T., and Johnston, H.S., "Spectra and Kinetics of the Hydroperoxyl Free Radical in the Gas Phase", J. Chem. Phys., 56, 2824 - 2838 (1972).
15. Hamilton, E.j., Jr., "Water Vapor Dependence of the Kinetics of the Self-Reaction of HO_2 ind the Gas Phase", J. Chem. Phys., 63, 3682 - 3683 (1975).
16. Graham, R.A., Winer, A.M., and Pitts, J.N., Jr., "Pressure and Temperature Dependence of the unimolecular Decomposition of HO_2NO_2", J. Chem. Phys., in press (1978).
17. Howard, C.J., "Kinetics of the Reaction of HO_2 with NO_2", J. Chem. Phys., in pres (1978).
18. These numbers represent averages of estimates summarized in Watson, R.T., "Rate Constants for Reactions of ClO_x of ClO_x of Atmospheric Interest", J. Phys. Chem. Ref. Data, 6, 871 - 917 (1977).
19. Hampson, R.F., Jr., and Garvin, D., "Reaction Rate and Photochemical Data for Atmospheric Chemistry - 1977", N.B.S. Special Publication 513, May, 1978.
20. This value represents an average of the recent estimates summarized in R.T. Watson, J. Phys. Chem. Ref. Data, 6, 871 (1977).
21. This represents an average of the previous published measurements of Ref. 14, 15, and the following: (a) Hamilton, E.J., Jr., and Lii, R.-R., "The Dependence on H_2O and on NH_3 of the Kinetics of the Self-Reacttion of HO_2 in the Gas-Phase Formation of $HO_2 \cdot H_2O$ and $HO_2 \cdot NH_3$ Complexes", Int. J. Chem. Kinet., 9, 875 - 885 (1977); (b) Foner, S.N., and Hudson, R.L., "Mass Spectrometry of Inorganic Free Radicals", Adv. Chem. Ser., 36, 34 - 49 (1962).
22. Calvert, J.G., "Formation and Persistence of N-Nitroso Compounds in the Environment" in "Assessment of Scientific Information on Nitrosamines", Report of ad hoc Study Group of U.S. Environ. Protection Agency, August, 1976, p. 22 - 25, Appendix D.
23. Lindley, C.R.C, Calvert, J.G., and Shaw, J.H., "Rate Studies of the Reactions of the $(CH_3)_2N$ Radical with O_2, NO, and NO_2", Chem. Phys. Letters, 67, 57 - 65 (1979).

24. Hanst, P.L., Spence, J.W., and Miller, M., "Atmospheric Chemistry of N-Nitroso Simethylamine", Environ. Sci. Technol, 11, 403 - 405 (1977).
25. Glasson, W.A., "An Experimental Evaluation of Atmospheric Nitrosamine Formation", Environ. Sci. Technol., 13, 1145 - 1146 (1979).
26. (a) Pitts, J.N., Jr., Grosjean, D., van Cauwenberghe, K., Schmid, J.P., and Fity, D.R., "Photooxidation of Aliphatic Amines Under Simulated Atmospheric Conditions: Formation of Nitrosamines, Nitramines, Amides, and Photochemical Oxidant", Environ. Sci. Technol., 12, 946 - 953 (1978); (b) Tuazon, E.C., Winer, A.M., Graham, R.A., Schmid, J.P., and Pitts, J.N., Jr., "Fourier Transform Infrared Detection of Nitramines in Irradiated Amine-NO_x Systems", Environ. Sci. Technol., 12, 954 - 958 (1978).
27. Gowenlock, B.G., and Thmas, K.E., "The Gaseous Equilibrium of 1,3,5 - Trimethylhexahydro - 1,3,5 - triazine and N - Methylen - emethylamine", J. Chem. Soc., B 409 -410 (1966).
28. Mantz, A.W., "Infrared Multiplexed Studies of Transient Species", App. Spectros., 30, 459 - 461 (1976).

PART II

ATMOSPHERE, CLIMATE, AND CARBON CYCLE

ATMOSPHERIC CIRCULATION

F. Fiedler

Meteorologisches Institut
Universität Karlsruhe, Kaiserstr. 12
D 7500 Karlsruhe 1, Bundesrepublik Deutschland

ABSTRACT. Atmospheric motion on all scales contribute to the global distribution of chemical compounds. Transport over long horizontal distances is determined by the general atmospheric circulation features. Vertical mixing over the whole troposphere is due to processes on different scales: microturbulence, up - and downdraft in thunderstorms and vertical motion in cyclones and anticyclones.

In this paper a review is given on the most important processes within the atmospheric circulation which have to be known and taken into account when concentrations of chemical substances in the atmosphere are studied.

1. INTRODUCTION

When investigating the mass budget, the spatial and temporal distribution of trace gas constituents in the atmosphere the concept of residence time plays an importand role. This quantity represents an integral over a great variety of processes going on at the same time in the atmosphere and contributing to the observed distribution and the natural cycle of the atmospheric compounds. Beside the chemical reactivity of trace components and the influences of the time and local variations of sources and sinks the flow structure is of fundamental importance to the distribution of the concentrations of trace gases and other solid constituent components. Atmospheric flow conditions must be known if the enrichment and depletion of the atmosphere with pollutants is to be assessed.

H. W. Georgii and W. Jaeschke (eds.), Chemistry of the Unpolluted and Polluted Troposphere, 119–154.

In experimental investigations it is often the case that measurements can be taken only over a short time span and also over a limited area of the atmosphere. Also frequently a large-scale distribution of a particular substance can only be constructed from the results of different investigations which were taken at different times and differences of the atmospheric conditions from day to day or from season to season or even from year to year cannot be taken into account. These great deficiences in distant-transport and global studies limit the value of those investigations even more when the sources and sinks of the components are of great spatial inhomogeneties. Most of the anthropogenic pollutants are released over very limited areas like cities or highly industrialized and densly populated regions. Most of the sources must also be considered as ground sources and the constituents are diluted in the atmospheric boundary layer. Flow structures of all scales will then contribute to the transport and dispersion of the constituents and determine their global distribution. Additionaly, when the sources vary in time the particles and gases will meander over a rather long time according to the actual large scale flow conditions around the globe so that the concentrations at various positions in the atmosphere will undergo a considerable variation. This strong dependence of concentrations upon flow conditions has in turn very often been used to derive flow trajectories of the atmospheric circulations.

Examples are the detection of exchange processes between the stratosphere and the troposphere from monitoring radioactive material in midlatitudes which showed a much faster rate than was known before the nuclear weapon detonations or the inter-hemispheric exchange of air as detected by monitoring the Ozon concentrations in lower heights of the troposphere.

In this paper it is intended to give a review of features of the atmospheric global circulation which must be understood and known for the explanation of the overall distribution of trace constituents. These global flow structures are more important the more the sources of the constituents are found in limited areas like the highly populated and industrialized regions of the continents in the northern hemisphere. The processes of different scales which contribute to the transport and dispersion of the atmospheric compounds are of extreme significance in developing a model for the transport and dispersion of constituents. Following the discussion of the most relevant transport processes and their interaction which is also known as handover of the mass fluxes, some results obtained from numerical experiments with global circulation models will be presented. These model simulations give valuable insight into the mixing and transport conditions in the different parts of the atmosphere which must be taken into account when chemical reactions are to be determined from in-situ measurements or when measurements of several investigations at different locations carried out at large times apart from each other are to be combined to a global life cycle. Atmospheric circulation patterns will be addressed only under the aspects of transport properties. Other important aspects as energetics, momentum balance which are necessary and helpful for the understanding of the general circulation are left out.

2. ATMOSPHERIC TRANSPORT AND CHEMICAL REACTIONS

All atmospheric constituents with partial density ρ_k in the mixture of air must obey a continuity equation of the form

$$\underset{(1)}{\frac{\delta \rho_k}{\delta t}} + \underset{(2)}{\nabla \cdot \rho_k \mathbf{v}} = \underset{(3)}{- \nabla \cdot \mathbb{J}_k} + \underset{(4)}{J_k} \qquad (1)$$

Here $\mathbb{J}_k$ is the diffusion flux $\mathbb{J}_k = (\mathbf{v}_k - \mathbf{v})\,\rho_k$ with $\mathbf{v}_k$ the velocity of the material ρ_k and $\mathbf{v}$ is the baryzentric velocity of the gas mixture. J_k represents sources and sinks by chemical reactions and other processes as phase fluxes.

The local change of partial density (1) is determined by the advection (2) with the velocity $\mathbf{v}$, by the divergence of the diffusion flux $\mathbb{J}_k$ (3) and by the local sources and sinks J_k (4).

Since atmospheric conditions are changing rapidly in time and space it is necessary and convenient to consider mean quantities ρ_k and fluctuations $\rho_k - \overline{\rho_k} = \rho'_k$ of the partial density ρ_k and as well as of velocity, i.e. $\bar{\mathbf{v}}$ and $\mathbf{v}'$. The equation for the averaged density can then be written as

$$\frac{\delta \rho_k}{\delta t} + \nabla \cdot (\rho_k \bar{\mathbf{v}}) + \nabla \cdot (\overline{\rho'_k \mathbf{v}'}) = \overline{J_k} \qquad (2)$$

where molecular diffusion has been neglected in comparison to the diffusion fluxes $\overline{\rho'_k \mathbf{v}'}$ caused by the fluctuating velocity and density field.

For the interpretation of the different fluxes $\overline{\rho_k}\bar{\mathbf{v}}$ and $\overline{\rho'_k \mathbf{v}'}$ which contribute to the local changes in the partial densities it is important to define their meaning. What is phrased mean motion $\bar{\mathbf{v}}$ and mean density $\overline{\rho_k}$ is the outcome of an averaging process. Although the resulting flow pattern may even not exist in this form for a long time and only fluctuate around this state in a rather disordered manner it is still of great help for the study of exchange processes and the interpretation of observations. It is, however, important to keep in mind, that the mean flow circulations and what is considered as the fluctuating part is very strongly dependent on the averaging time or the averaging area. For small scale turbulent flow close to the ground variations of the velocity of the order of 1 hours are usually dissolved as a mean velocity field and therefore the action of the flow associated with synoptic pressure systems is concerned as large scale mean flow. On the other hand when the atmospheric vertical transport of trace constituents is described in a one-dimensional model, in a quasi steady state condition of a season, then the action of synoptic weather systems, i.e. lows and highs are to be interpreted as fluctuations and have to be parametrized in most cases in the same way as small scale turbulence.

From Eq. (2) it is evident, that for different trace constituents the various terms are not all of equal importance for the overall distribution. For passive tracers which are those of no direct feedback to the transporting flow field it depends on the ratio of time scale of the transporting and diffusion mechanism to the time scale of chemical reactions involved. For slow chemical reactions the time variability of the concentration of constituents will beside from the time variability of the source mechanisms only be determined by the dispersion characteristies of the flow field. In those cases it will be extremely difficult to determine chemical reaction rates by in-situ measurements. In the case of fast reacting constituents the flow field of the atmosphere will be of minor relevance for their occurence. In this case the distribution of sources is dominant for the concentration field.

3. SCALES OF ATMOSPHERIC MOTIONS

Meteorological flow always contains an extremely wide range of scales of flow regimes. At the one end molecular processes with scales of millimeters are involved in the gas exchange e.g. at the interface between two media, air and water. At the other end, the largest possible flow structures with sizes of the circumference of the earth determine the redistribution of constituents in a very effective way. Between these flow structures there is always some interaction which limit the possibility of application of a theory for the transport or mixing of constituents in the atmosphere.

In Fig. 1 the typical length scales of atmospheric flow structures are presented together with their typical velocities which are dominant for a special structure and which are representative for the diffusion process. For the determination of ground level concentration close to point sources only small scale flow structures are in volved. For longer distances of the transport larger flow complexes become involved in the diffusion and transport process. Since long distant transport and diffusion occurs in the vertical and horizontal direction through the

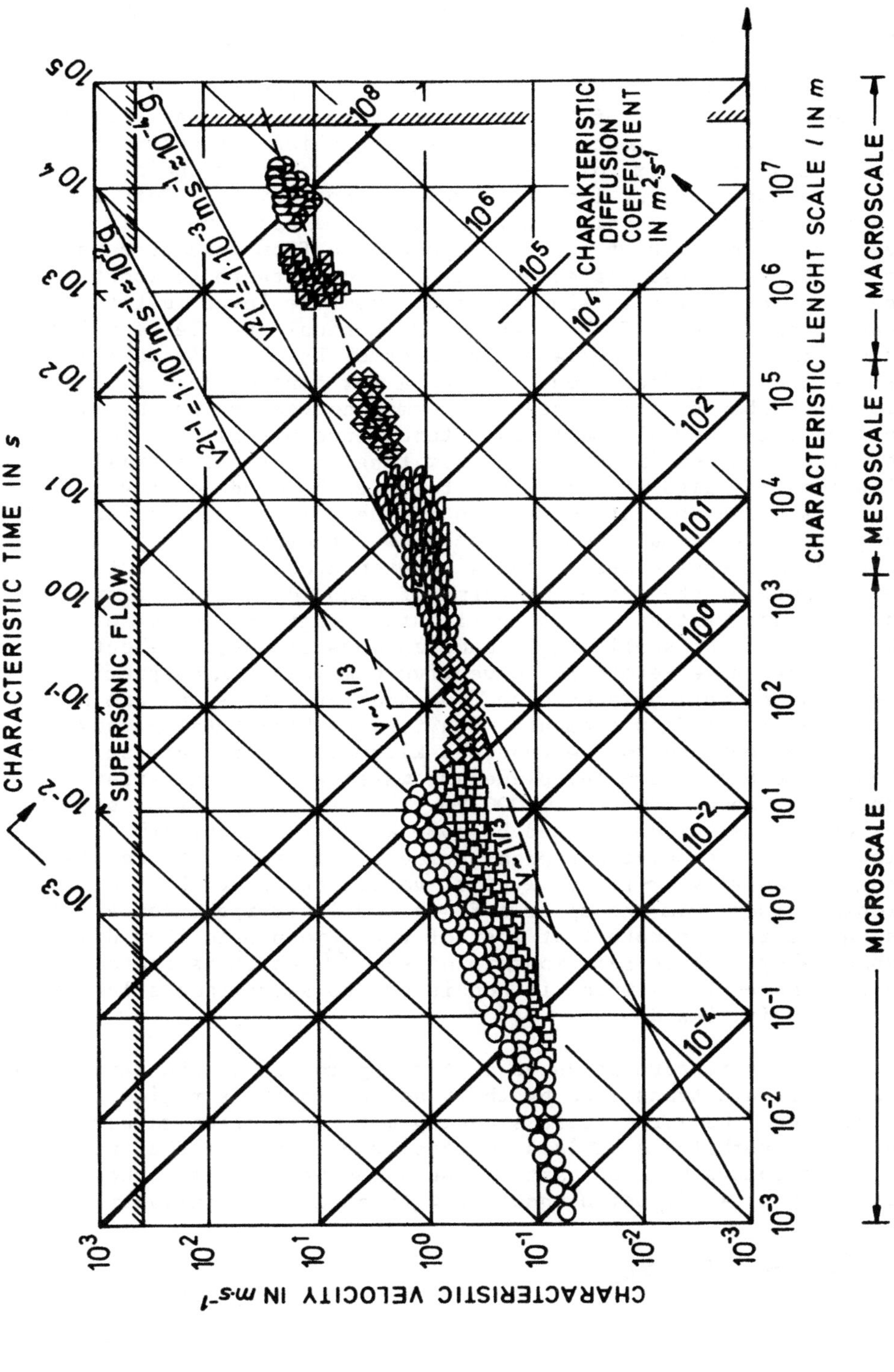

Fig. 1 Characteristic scales of atmospheric flow (adjusted from Wippermann, 1971)

entire troposphere the whole variability of the flow contributes to the ultimate distribution of trace constituents.

In Fig. 1 the length scale of the abscissa spans from millimeter to the circumference of the earth. Microturbulence has scales which go down to almost molecular scales which are determined by the mean free path of the molecules. The corresponding characteristic velocities are given in the ordinate. Both length scale and velocity scale determine the time scale which is a measure of the time of the life cycle of a flow structure which is for the smallest scales on the order of a tenth to about 10 s. The times are given as sloping lines from the lower left to the upper right corner of the diagram. Mechanical turbulence which is determined by the surface roughness or by shear instabilities of the flow span from about 1 mm to a few tens of meters. Larger flow disturbances are caused by thermal convection which may be devided into thermal small scale turbulence with length scales between a few centimeters and a few tens of meters, thermals with length scales of a few tens to a few hundreds of meters which is known as flat convection and deep moist convection. The largest convectional elements are flow structures mainly in the tropical atmosphere associated with pronounced cloud clusters. In the case of the different forms of deep moist convection the ratio of the vertical to the horizontal length scale varies from about 10 : 1 for small single cumulonimbus clouds, to about 1 : 10 for large mesoscale convective complexes. Atmospheric cyclones and anticyclones have even smaller ratios and are almost two-dimensional flow elements with typical horizontal extension from a few hundred to few thousand kilometers. The largest possible flow disturbances are the planetary waves with characteristic lengths of about 3000 to 10 000 km. The size of the flow structures which are all determined by some kind of instability of the flow and the associated typical velocity determine the corresponding diffusivity $K \sim l \cdot v$. Typical values are given by the sloping lines from the upper left to the lower right corner. It is interesting to note that most of the atmospheric flow disturbances obey a close relationship of $v \sim l^{1/3}$ which is indicated by the dashed lines which is consistent to Richardson's law for the diffusivities $K \sim v \cdot l \sim l^{4/3}$.

Every flow structure produces a steering motion in the atmosphere. Pollutants are transfered by their action from one region to the other. Even the large scale steering motion is often rather vagley termed as turbulence. This stems from the fact that the transfer of constituents from on part of the atmosphere to the other is described by a diffusion equation with an eddy diffusivity.

$$\frac{\delta \rho_k}{\delta t} \sim -\nabla \cdot K \nabla \overline{\rho_k}$$

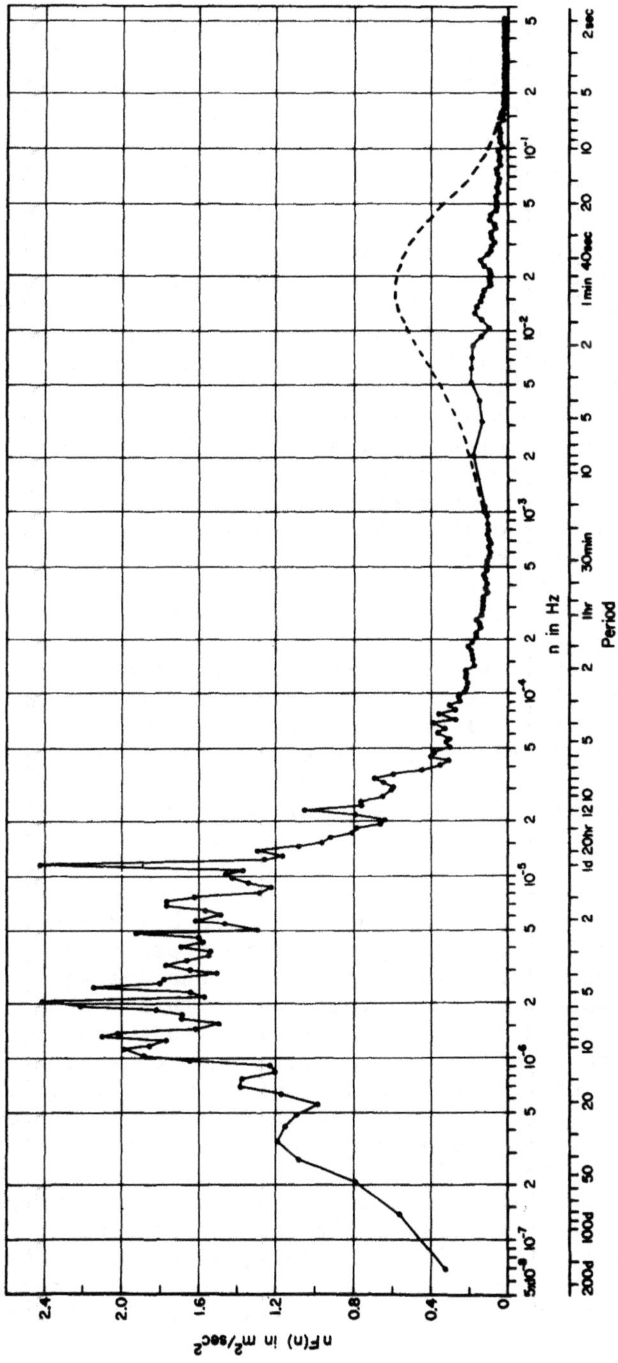

Fig. 2 Spectral distribution of the kinetic energy of flow disturbances in the atmosphere (F. Fiedler, 1971)

in analogy to molecular diffusion. There is however one important difference in this formulation. In molecular diffusion the diffusivities are physical properties of the fluid and only weak functions of temperature and are independent of the structure of the motion field. In application of the K-theory concept to small-scale and large-scale turbulent transport processes the eddy-diffusivities are very crucial functions of the flow structure itself.

Not all structures present in the atmosphere contain the same kinetic energy as can be seen from a variance spectrum of the horizontal wind speed in Fig. 2. Here the power spectrum of the horizontal wind speed measured close to the earth surface at 50 m above ground is shown. It spans a frequency range of seven decades from 0.5 Hz to 5.10^{-8} Hz which is equivalent to time periods from 2 seconds to 200 days. As is seen from this spectrum there are two dominant ranges of higher energy. At the high frequency end we have the small scale turbulence which is almost entirely confined to the atmospheric boundary layer. This small-scale turbulence is three-dimensional, and its maximum energy appears at a time scale of about 1 minute. The prevailing maximum occurs at much smaller frequencies at around 4 to 5 days. This energy of fluctuating velocity is associated with the synoptical eddies, i.e. lows and highs passing the point where the measurements were taken. As is seen from this spectrum meteorological eddy structures are not present with uniform energy content over all time scales but have a rather prefered selection of sizes or time periods. This is completely determined by the inherent instabilities of the flow around the earth. Between the small-scale turbulence and the synoptic variability there is a prefered region with rather low energy known as the mesoscale energy gap.

Because of this broad spectrum one is always forced to introduce some averaging process as has been introduced in Eq. 2 above. It would be desirable to find a clear point where a partitioning between eddying motions of smaller scales than the averaging domain and the mean motion could be introduced. This would be easiest at a point with no energy present, i.e. that now flow structures exist at that range. Unfornately, the atmospheric spectrum of eddies is continuously distributed over all scales. Therefore the part of flow structures which form the mean flow and the part which will be considered as eddying motion is completely dependent on the choise of the averaging domain. This domain is not uniquely definable but is a question of appropriateness or even a question of feasibility of the problem or the degree of complexity of the available models, which determine the scale where the mean and regular motion has be resolved from the smaller scales which is to be treated as statistically irregular and undissolvable motion.

4. IMPORTANCE OF DIFFERENT SCALES IN VARIOUS REGIONS OF THE ATMOSPHERE

Following a gaseous or solid material released at the ground we find from observations that small scale turbulence is the agent close to the ground within the atmospheric boundary for carrying it away from the ground to higher levels. As also can be seen from measurements the turbulence intensity is decreasing with height and dies out at the top of the boundary layer as long as we consider only cloud free air or neglect zones which appear only occasionaly like strong storm cells or frontal zones connected with synoptic lows. Under those conditions it is important to know which larger-scale phenomena take over and act as the next agent to carry away the masses transported upward by the small scale turbulence. Occasionally it is observed that this handover of transport is not very active. Then the lower levels are filled with air pollution for example during smog conditions. It needs then only one of those rather soradic appearing flow structures as a strong front or a storm system which redistributes the masses very efficiently as can often be observed by the clean air in postfrontal zones or after the dissipation of a strong thunderstorm.

In other regions of the atmosphere their is a strong demand to know what types of processes are responsible for the dispersion of tracers. For example in the tropical stratosphere it was believed for a long time that meridional circulations with ascending air close to the equator and downward motion in the high latitude and poleward motion in between are the most important agencies of the transport. Later, it was stressed that eddy-transports are the dominant factors for the poleward transports.

Depending on the location of the sources within the atmospheric regime the dispersion and transport will be caused by flow regimes of remarkable differences. Accordingly the residence time of the same constituents will, independently of chemical reactions, be significantly different.

How the different scales of transport regimes are effective in the various parts of transport cycles can be studied when observing the water vapour transport within the subtropical Headley circulation. Fig. 3 shows schematically the general flow patterns. It is observed that at low levels the air takes up the water vapour evaporating at the sea surface and transported by microscale turbulence up to a level of about 600 m, the cloud base for the trade wind cumuli. With the set up of convection cells the next higher scale comes into play which transports the moisture within its updraughts to the base of the trade wind inversion which is in balance between the large scale subsidence and the mixing of the layer below. Within the low level horizontal

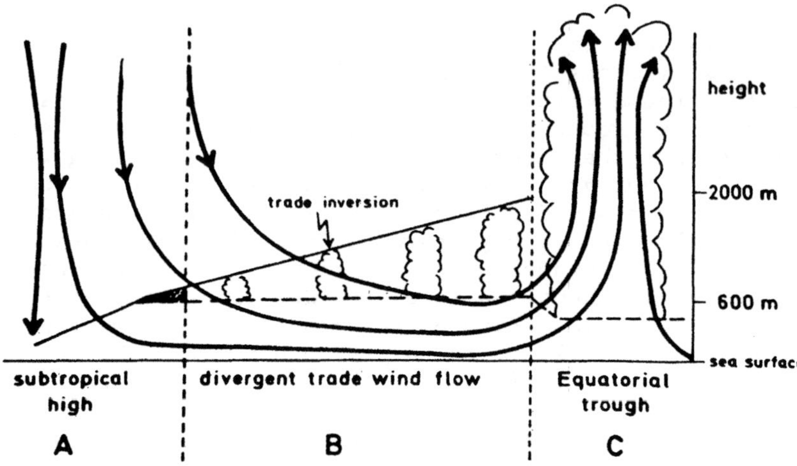

Fig. 3 Schematic flow structure in the over the subtropical ocean as example for handover of moisture transport by different scales (Augstein, 1981)

flow there is an enourmous enrichment with moisture which is condensed in the large mesoscale cumulus towers within the equatorial trough. This energy conversion drives the large scale Headley circulation with polar flow higher up and equatorial flow in the lowest levels. Here turbulent eddy motion is the transport agent in the lowest levels. Partly by the eddying motion of the small cumulus cells which are growing in size with approaching the equator, but mainly by the horizontal mean motion towards the equator the water vapour is taken over.

Similar interrelationships can be observed between small scale eddying motion and the mean large scale flow patterns. In Fig. 4 there is a sketch given of the up-take of contaminants released at industrialized areas or large city complexes. By small-scale turbulent mixing and occasionally by mesoscale structures material is carried upward and by convergence of the material flux in the large horizontal flow the next higher scale transports and diffuses the material over long distances. Thus there is a constant handover of the masses from one scale to the next.

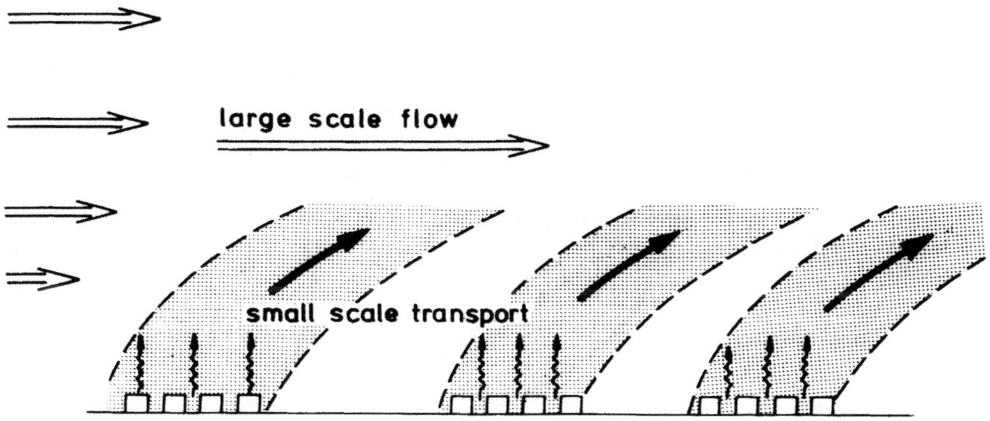

Fig. 4 Sketch of flow and handover of mass transport over urban areas

It is evident, that this handover can only rather poorly be described by the assumption of a K-coefficient.

5. TRANSPORT REGIMES IN THE ATMOSPHERE

We will now proceed and have a closer look at the most effective transport mechanismus which at the same time are mark stones of the general circulations. If we look at the atmospheric motion fields as it is observable from the satellites and compare the information from day to day then we recognize a high degree of disorder and it seems hopeless to find a simple solution for explaining all the observed details. However by averaging the flow field around the globe we can identify with some variability from month to month and even stronger from season to season a more or less pronounced cellular flow pattern for the meridional velocity components (see Fi. 5). Most easily one can detect the direct Headley cell between the equator and

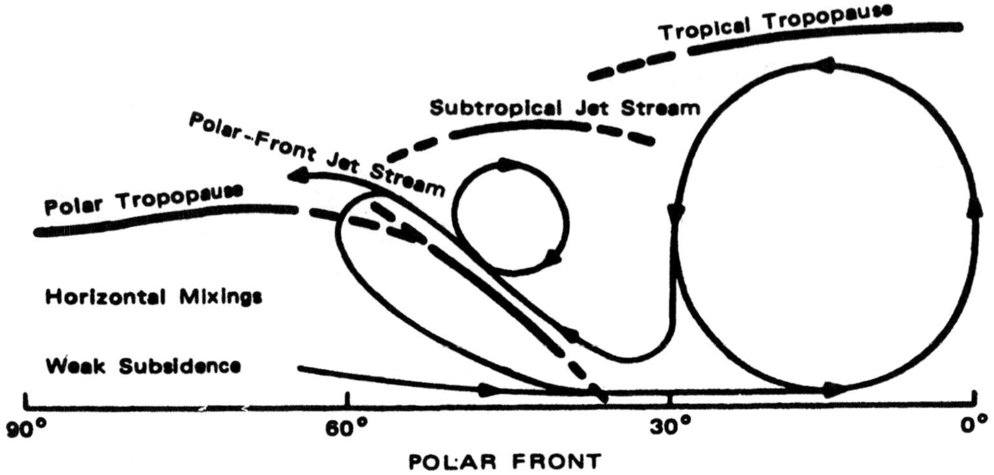

Fig.5 Circulation cells of the meridional flow component

about 30° N or S. This Headley cell can be found only on one side of the equator during the months December to February or June to August and is present on both sides in spring and autumn.

The second indirect circulation cell underneath the subtropical jet is much weaker marked from observations. But it is clear that it steadily looses kinetic energy and generates potential energy. For its survival their must be a continuous supply of energy by eddy transport processes. This structure shows a rather complicated distribution if the meridional energy flow is considered as a function of longitude, especially with respect to zones where the polar jets have their maxima. From an extensive statistical analysis for the northern winter hemisphere the meridional circulations of Fig.6a, b was derived. In this structure it is divided between the upwind and down-

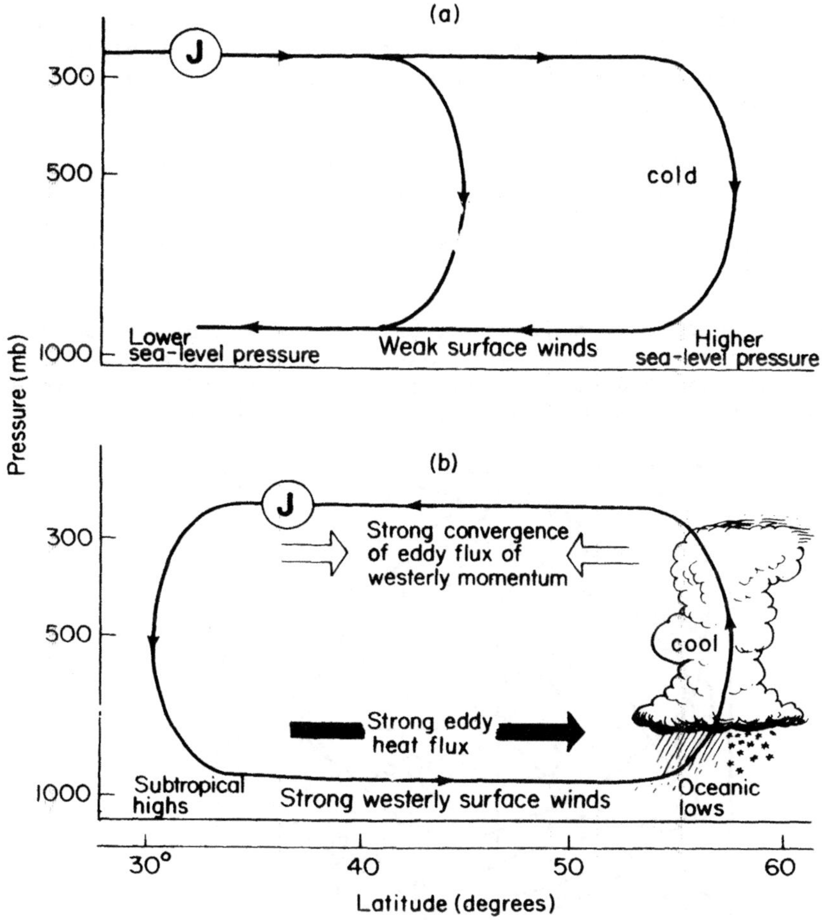

Fig. 6 Secondary meridional flow superimposed on the zonal flow a) at the upstream side b) on the downstream side of a strong jet region (Blackmon et al., 1977)

wind section to the jet stream maximum. Because of longitudional distributions of land and water masses the circumpolar jet does not show everywhere the same intensity. The statistical analysis shows its maxima at the eastern side of the continents. Fig.6a gives the meridional circulation(secondary flow)of the upwind side of the jet maximum and Fig.6b gives the same at the downwind side. The upstream section is governed by a strong thermally direct circulation with ascending air at the warm side and descending motion at the cold side. This direct circulation is required by different reasons (Blackmon et al., 1977).

On the downwind side of the jet stream maximum there is a strong indirect circulation cell. Also this cell is postulated by several reasons, which follow from observations of a balance of momentum and energy in this flow section.

General characteristics of the circulation of the atmosphere are eddies of different sizes which have already been addressed in connection with the scales of motion. We will now study some of them somewhat closer since they are important transport mechanisms for atmospheric tracers.

An outstanding role to maintain a balance of energy and momentum in the atmosphere play the disturbances of the midlatitudes. Since these are strong steering motions they are very effective in distributing heat, moisture and mass constituents. In higher levels

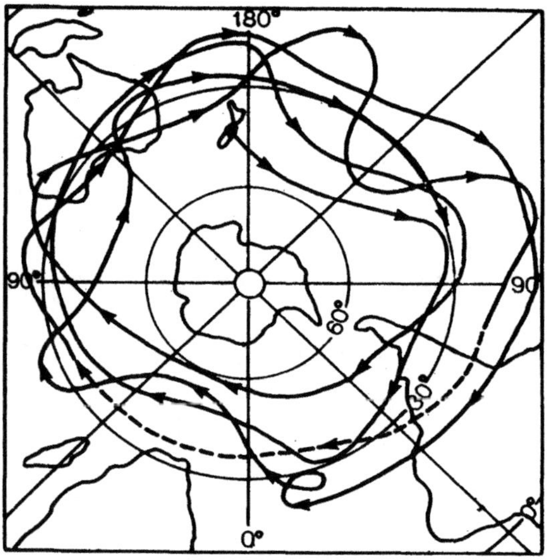

Fig. 7 Trajectories of flow at the 300 mb level in the southern hemisphere (Green, 1979)

they appear in a wave-like pattern as can be seen from a track of a constant level balloon at the 300 mb level (see Fig. 7). If we consider a mass constituent instead of a balloon, then this mass will move several times around the globe until it is distributed over a larger area. If we would take samples at different times and determine average values the concentration would be smeard over a rather broad area because of the changing wave pattern with time although the mass may still be confined to a rather small area within the flow.

Disturbances of the mid-latitudes are baroclinic structures which move generally from west to east. These disturbances receive their energy from potential energy which is stored in inclined temperature surfaces. By lifting warm air in front of the disturbances and subsidence of cold air in the rear potential energy is converted into kinetic energy. This can be understood in a simular way as the processes of small scale thermal convection in the vertical. A small scale thermal will rise vertically upward as long as it is warmer and lighter than the surrounding air. Similarly the synoptic disturbances can be understood as convection cells but with only one important difference. In large scale disturbances motion is quasi two-dimensional and almost horizontal. The production of kinetic energy of a synoptic disturbance is crucially dependent on the slope of potential temperature surface (Θ = const.) given by the angle α (see Fig. 8) in comparison to the slope of the trajectories of air parcels (B to A). In the most common cases the air on the equatorward side is warmer than at the poleward side. That means the isentropes slope towards the south (indicated by angle α). If the trajectory of an air parcel follows the line B to A, then it moves like an air parcel from the pole to the equator in the rear of a disturbance or an air parcel from A to B as in the front of a low. Then potential energy is converted into kinetic energy. When the air parcel follows exactly the isentropic surfaces then no energy exchange is possible. With the exchange of two air parcels across a latitude circle a heat flux along slanting surface is connected. If gradients of traces gases are present there will be also an exchange of mass along slanting surfaces. Cases a) and d) of Fig. 8 are of relevancy for the atmosphere Case a) is the general case observable in the troposphere and Case d) is representative for the motion fields in the stratosphere (see Pedlosky, 1979).

In the troposphere the large scale motions are organized according the Case a) so that potential energy can be converted into kinetic energy and the synoptic disturbances can be supplied with energy to survive against frictional dissipation. Consequently it is important to keep in mind, that large scale disturbances do not transport material directly up-or downward but instead in the vertical direction along slanting surface.

Another special feature of the general circulation are secondary flows which develop along fronts in the vicinity of the tropopause which have recently been studied in detail by Shapiro (1981). In Fig. 9 the dotted lines give the lines of potential temperature which define a strong baroclinic zone in the middle of the Figure 9 a, with the warm side to the left and the cold side to the right. In the range of maximum baroclinicity beneath the corresponding jet core (not shown in the Figure) a direct circulation cell is developing with ascending motion on the warm side and descending motion at the cold side. Fig. 9 b and c

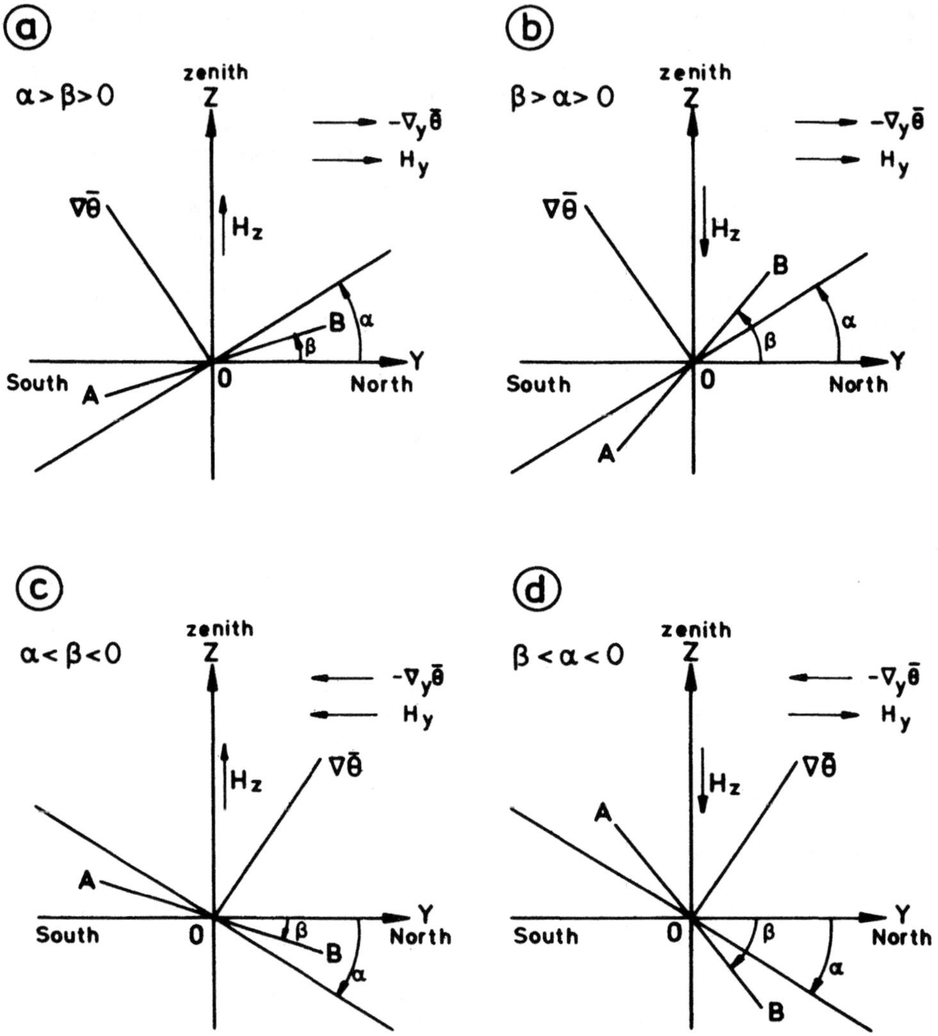

Fig. 8 Schematic conditions for slanting synoptic motions. Air parcels move along the sloping lines AB. Cases a and d are the cases observed in the troposphere and the stratosphere (van Miegham, 1973)

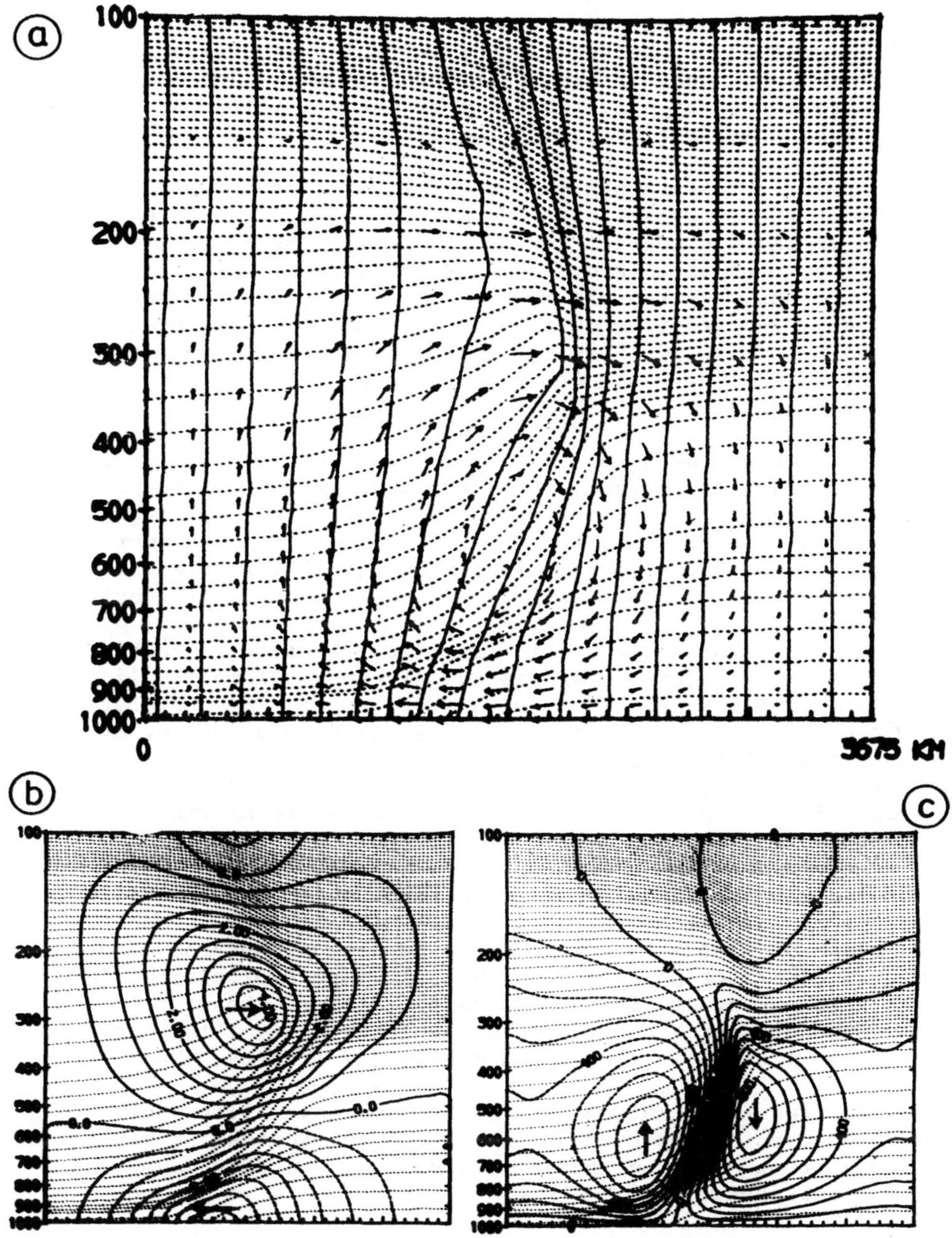

Fig. 9 Secondary flow within a jet stream - frontal zone (a) shows velocity vectors and potential temperature (dottet lines) (b) and (c) horizontal and vertical velocity components respectively (Shapiro, 1981)

show the corresponding horizontal (normal to the front) and vertical components of the secondary flow, which may serve as important mass exchange mechanism across frontal systems. The investigations of these secondary type flows are by far not yet completed and further studies are necessary in this field.

A large step forward has been achieved in recent years in the study of the flow mechanism of large convective cells especially those in the tropics. It is evident that large convective cells are of eminent efficiency to transport and distribute air constituents very rapidly over a large area. A synthesis of the knowledge derived from detailed case studies of Zipser (1977) is shown in Fig. 10. This squall line structure moves from right to left. Relative winds are directed towards the moving cell at all heights.

In the rear of the squall line the flow is directed strongly outward near the surface, it is still outward but with much lower intensity just above the mixed layer and is coming from different directions in the 600 - 900 mb layer. New towers of cumulus clouds develop just above the squall line front. Most of the precipitation falls from the anvil behind the front.

The structure of those squall lines must be three-dimensional in order that up - and downdrafts can alternate along the line. Downdrafts are developed through evaropating cloud droplets within the unsaturated air below the clouds and cooling by that the air so that it is forced to descend. Air from mid-levels is feeding the downdrafts which must be cooled by evaporating drops. This cold air reaches the ground and forms the stable layer behind the front.

For the development of those structures which can have life cycles of several hours the vertical wind profile shows an increase of wind with height. This is often cennected also with wind direction shear, which is an important element for the development of up-and downdrafts. Some of those structures have their upper air outflow to the front side of the convective cell as is shown in the sketch of Fig. 11 (Thorpe, 1981).

The smaller scale motion near the ground have already been addressed as an important link between the sources or sinks at the ground and the concentrations in the middle layers of the troposphere.

The fluxes, caused by small scale turbulence,close to the ground can be described in the same way as the heat flux or moisture flux. If the turbulent diffusion is the dominant process for the distribution of the tracer which has its source at the ground and chemical reactions are slow then it is even possible to

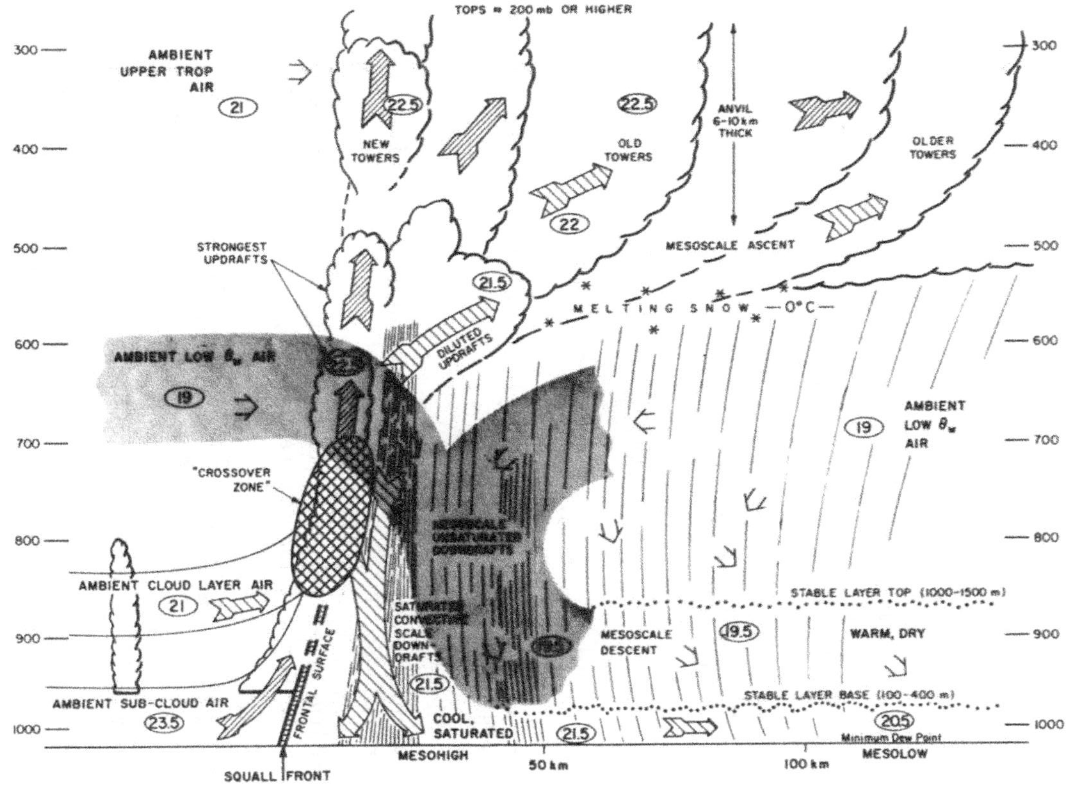

Fig. 10 Schematic flow structure connected with a squall line. The squall line is moving from right to left. Indicated velocities are relative to squall line displacement. (Zipser, 1977)

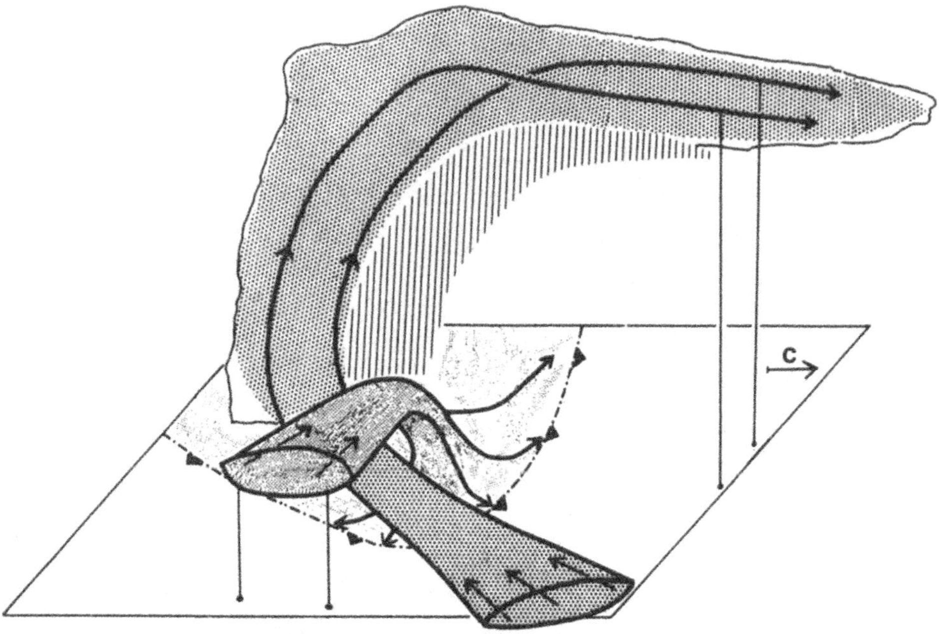

Fig. 11 Flow structure connected with large convective cells (Thorpe, 1981)

determine the source strength at the ground from vertical profiles of the concentration.

Surface boundary layer theory allows the formulation of the vertical profile for the concentrations $c_k = \rho_k / \rho$ in the following way

$$c_k(z_2) - c_k(z_1) = c_*\left[\ln \frac{z_2}{z_1} - \left(\psi_c\left(\frac{z_2}{L_*}\right) - \psi_c\left(\frac{z_1}{L_*}\right)\right)\right]$$

with the characteristic concentration $c_* = \frac{F_{c,o}}{k u_*}$. Here $F_{c,o}$ is the flux density at the ground u_* is the friction velocity of the turbulent boundary layer flow. The friction velocity can be determined e.g. from the vertical wind profile. ψ_c is a known stability function of the stability length L_* (Monin-Obukhov-length).

It is seen that transport mechanisms of all scales are present. These are of different importance depending on the large scale weather conditions and on the layer in the atmosphere where the concentration is released.

6. INTERACTION BETWEEN DIFFERENT RESERVOIRS

In air chemistry the exchange of material between different reservoirs of the atmosphere is of importance. Reservoirs which are most commonly considered are the troposphere and the stratosphere and the air of the northern and southern hemisphere. Some of the mechanisms which are responsible for the mass exchange between the reservoirs will be addressed in the following. It is known for many years that in the vicinity to upper level fronts which is associated with a tropopause break (see Fig. 12) air may penetrate from the stratosphere into the troposphere and vice versa.

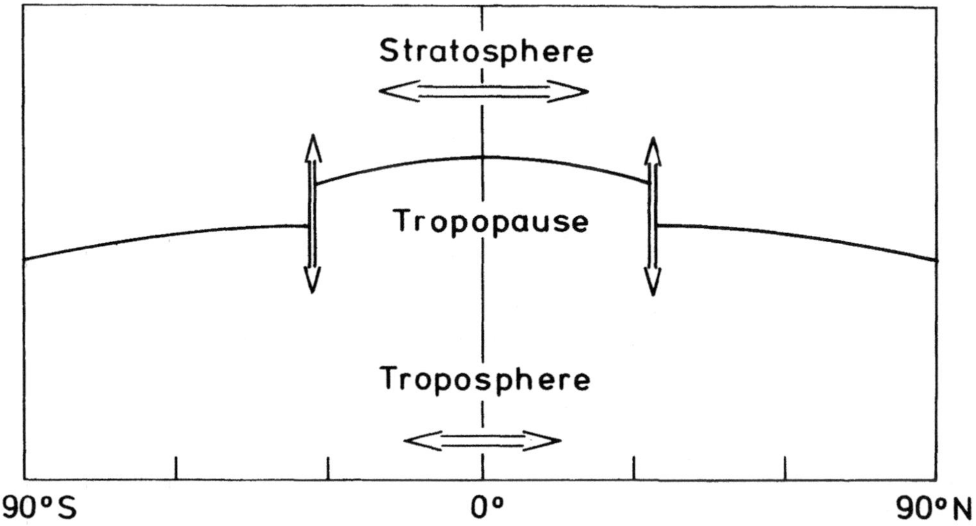

Fig. 12 Sketch of the tropopause break

A recent analysis of air mass exchange processes between the stratosphere and the troposphere has been presented by Shapiro (1980). In this case study detailed information on the mesoscale and synoptic scale flow conditions could be determined from aircraft measurements.

The analysis of the tropopause revealed a remarkable folding associated with a strong jet stream. The cross section analysis of potential temperature and wind speed is reproduced in Fig. 13.

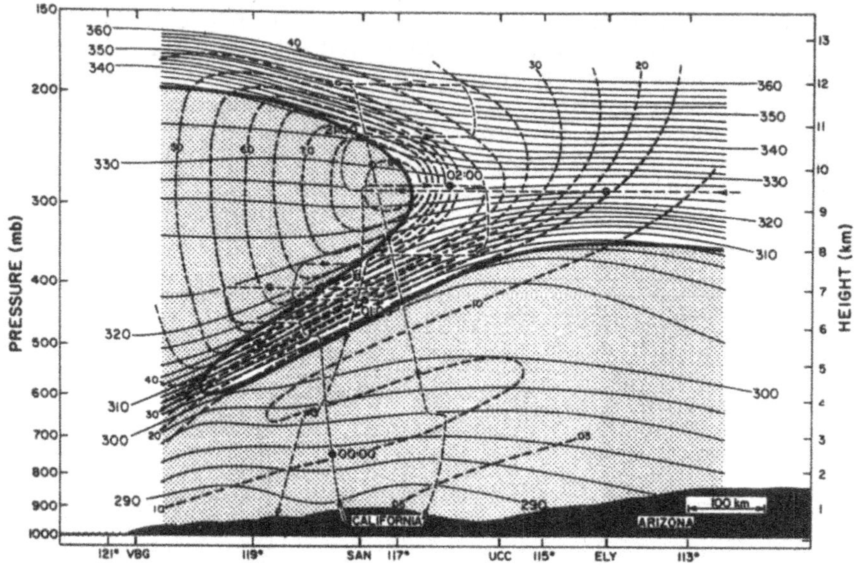

Fig. 13 Vertical cross section through a tropopause fold. Thin lines : potential temperature (K); dashed lines: wind velocity (ms^{-1}). (Shapiro, 1980)

Full lines represent the field of potential temperature as a composite from regular synoptic upper air soundings and the data obtained during the aircraft traverses (indicated by thin dashed lines). The wind field (heavy dashed lines) shows a maximum wind speed in the core of the jet of about 80 ms^{-1}. The position of the tropopause was defined by the field of potential vorticity. The most remarkable result of this analysis is the strong intrusion of stratospheric air deep into the troposphere which shows some resemblance to a waterspout. It can also be seen from this study as was found in previous analyses that a strong wind shear of 70 ms^{-1} /(100 km) to the stratospheric side of the jet core is present.

The important information,that this regions of tropopause folds are very effective mixing zones, could be infered from additional measurements of ozone and condensation nuclei concentrations.

The ozone concentration for that particular case is shown in Fig. 14. It is recognized from these results that within the front region the air from above with high ozone concentrations is penetrating downward into the folded portion of the tropopause.

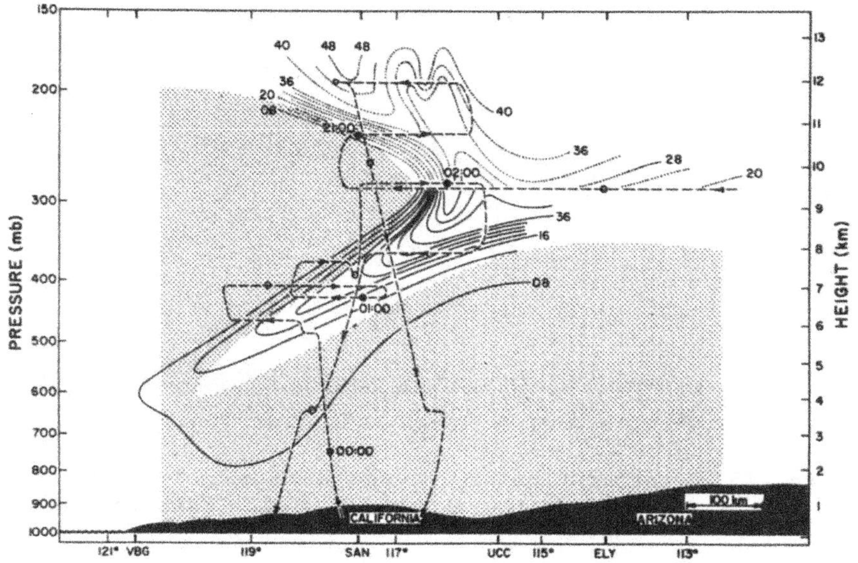

Fig. 14 Concentration of ozone in the vicinity of the tropopause fold (Shapiro, 1980)

The concentration decreases however with increasing penetration of stratospheric air into the fold. Since other important sinks of ozone can be excluded the decrease of ozone concentration must be caused by turbulent mixing in that area.

By simultaneous measurements of condensation nuclei air coming from layers near the earth surface can be traced. It is known that the highest aerosol concentrations are present close to the ground with concentrations of condensation nuclei on the order of 1000 to 10 000 cm^{-3} so air coming from the planetary boundary can very easily distinguished from stratospheric air having concentrations of condensation nuclei on the order 10 to 100 cm^{-3}

Fig. 15 shows the results of the condensation nuclei sampled during the same flights.

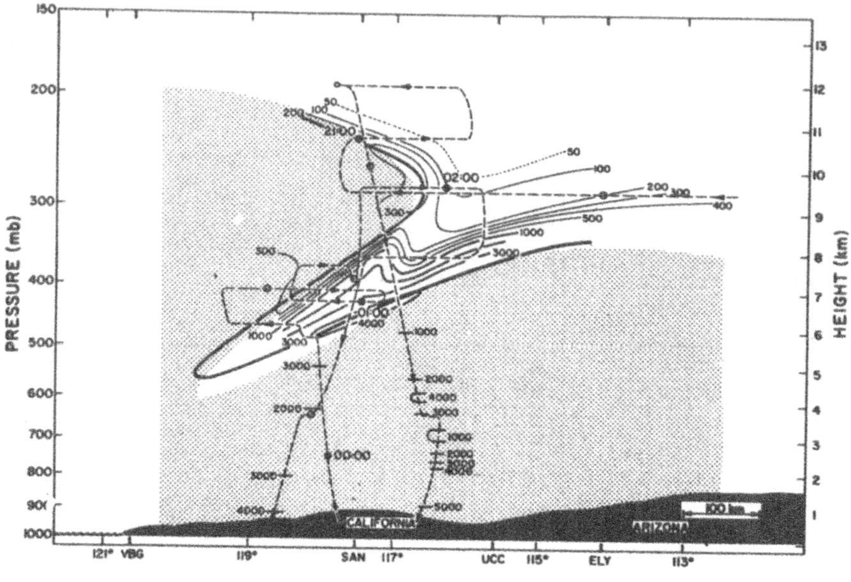

Fig. 15 Concentration of condensation nuclei in the vicinity of the tropopause fold (Shapiro, 1980)

Most surprisingly, the concentration of the condensation nuclei within the fold was found higher than in the troposphere. This could only be explained by the presence of an intermediate transport process which brings the air with high concentrations from the boundary layer close to the tropopause folds and from these by turbulent mixing into the stratospheric air within the folds. It was shown from satellite cloud images that for this case convective cells with their tops reaching the tropopause provided such a transport mechanism which is able to serve as a source for the relative high concentrations within the folds leading to values between those of the mean troposphere and those of stratosphere.

Although other mechanisms are possible for an exchange of trace constituents between the stratosphere and the troposphere the presented case study is an excellent example for the interaction and participation of flow structures of different scales for the ultimate distribution of atmospheric compounds: large scale processes which determine the position of the tropopause and its folds, convective systems which take care for the fast transport of condensation nuclei from the planetary boundary layer to the tropopause and small scale turbulent mixing which is responsible for the mixing and exchange of stratospheric and tropospheric air in the tropopause folds.

With the availability of complex circulation models it became possible to simulate the transport and diffusion of atmospheric tracer constituents. By comparing the final distribution of the tracer with observations one can gain a deeper insight into the transport mechanisms. Only in those complex models the transport processes of different scales can be incorporated at least in a parametrized form.

Also only by this type of investigations the importance of the different transport mechanisms like mean meridional circulation, transport of eddies, transport of mesoscale structures, interaction with boundary layer flow and washout can be compared and assessed for the different atmospheric layers.

The first numerical study which we will consider here is from Hunt and Manabe (1968). Material comparable to radioactive debris were injected into the tropical stratosphere. By following the distribution of the material an assessment was possible what role the different transport processes play for the final distribution.

In Fig. 16 the zonally averaged conditions are shown for the distribution of the material injected at the equator at a height of 20 km after 10, 30, 70, 150 and 180 days after initiation. It is quite apparent from this sequence, that the polward transport occured quasi horizontally with a small downward slope. The lower stratosphere was filled gradually with the highest concentration at a latitude of 30°. Also noticable is the rather fast reduction of concentrations on the equatorial side of the subtropical troposphere which starts after seventy days and leads to a high density of isolines at the tropopause.

From the following Figures (Fig. 17 - 19) the different contributions to the transport mechanisms can be found in form of 10-days averages.

The vertical fluxes are presented in Fig. 17, the action of the large scale eddies are given in the left and that of the meridional circulation is on the right side. It is seen that with continuing model time the range of influence extends to higher latitudes. Shaded areas indicate upward fluxes. Close to the equator the meridional circulation leads to an upward flux and the cellular regime becomes quite clearly active with downward fluxes between about 15 to 40°. Further to the north their is again an upward flux.

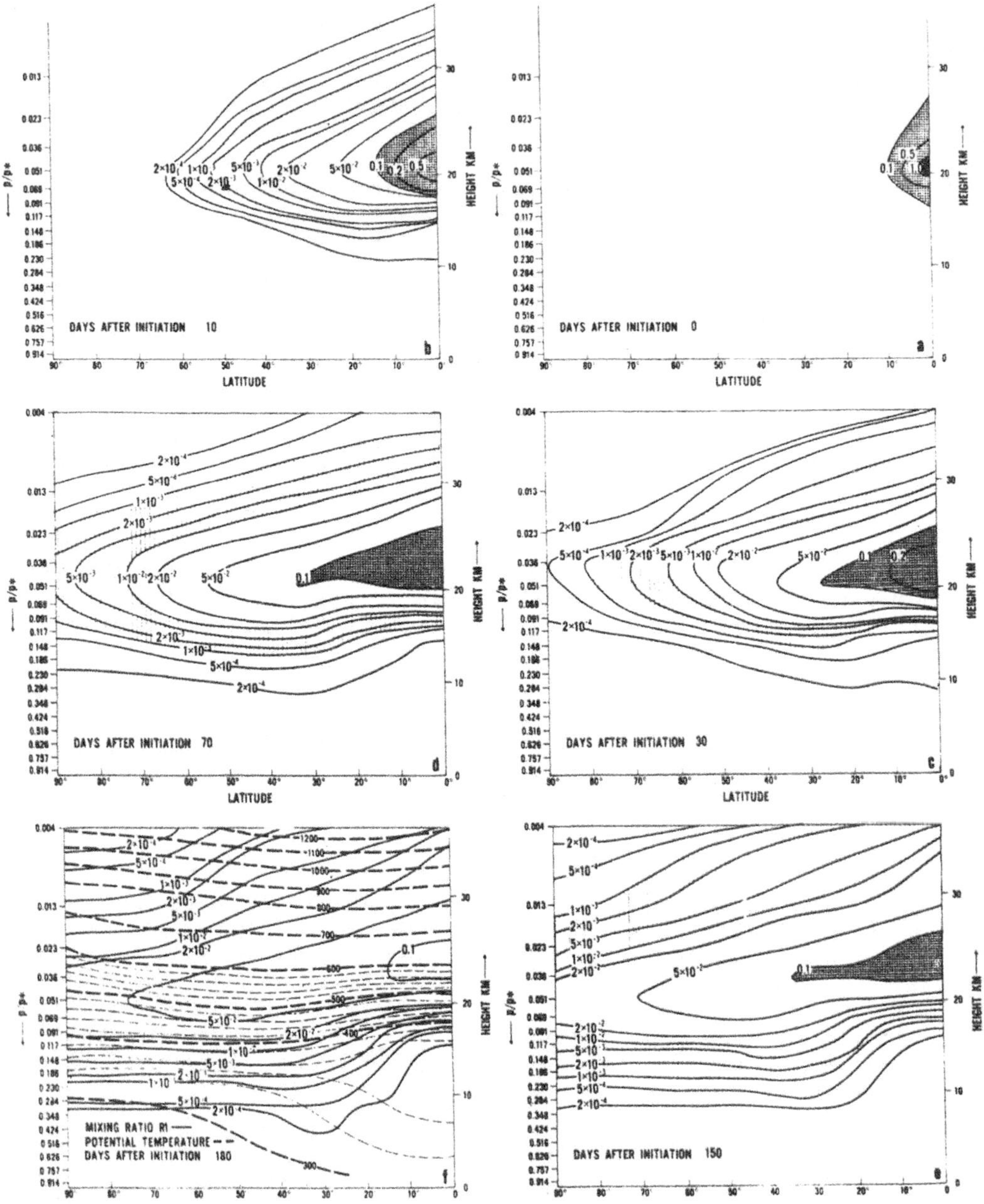

Fig. 16 Zonally averaged concentrations fields of material released at a height of 20 km at the starting time and after 10, 30, 70, 150 and 180 days (Hunt and Manabe, 1968)

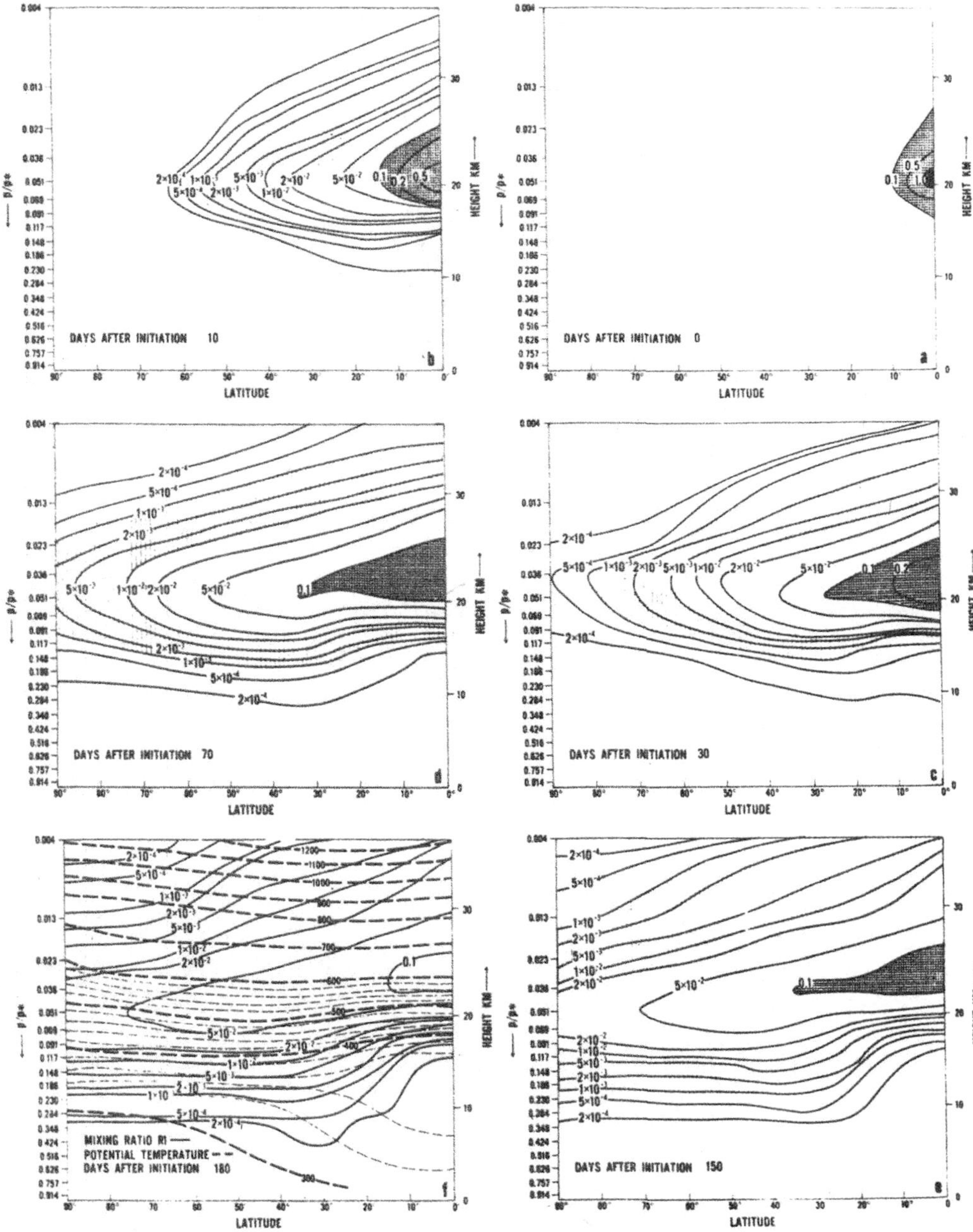

Fig. 17 Vertical fluxes by larger-scale eddies (left) and meridional circulation (right) averaged over different model times. (Hunt and Manabe, 1968)

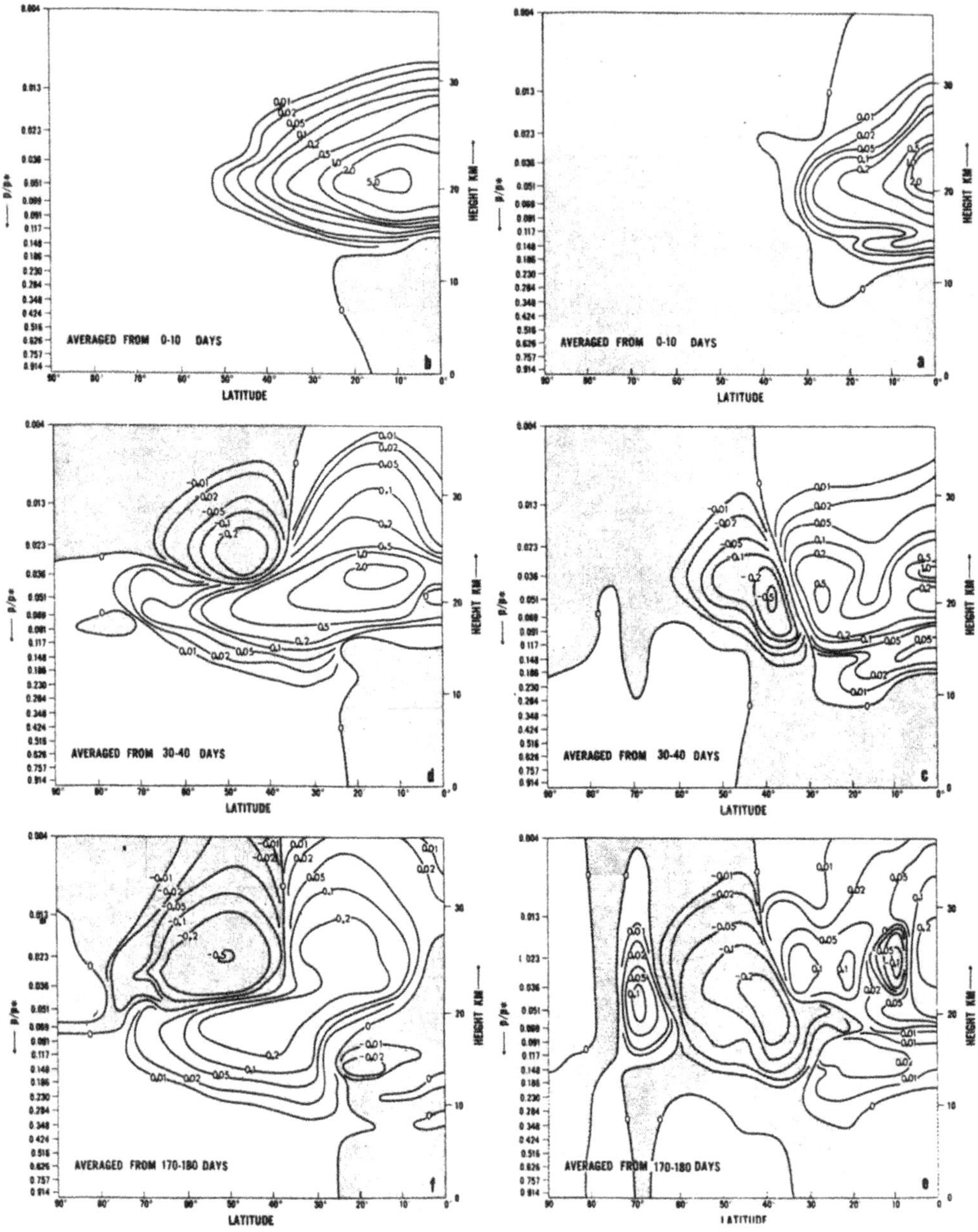

Fig. 18 Horizontal fluxes by large-scale eddies (left) and meridional circulation (right) averaged over different model times. (Hunt and Manabe, 1968)

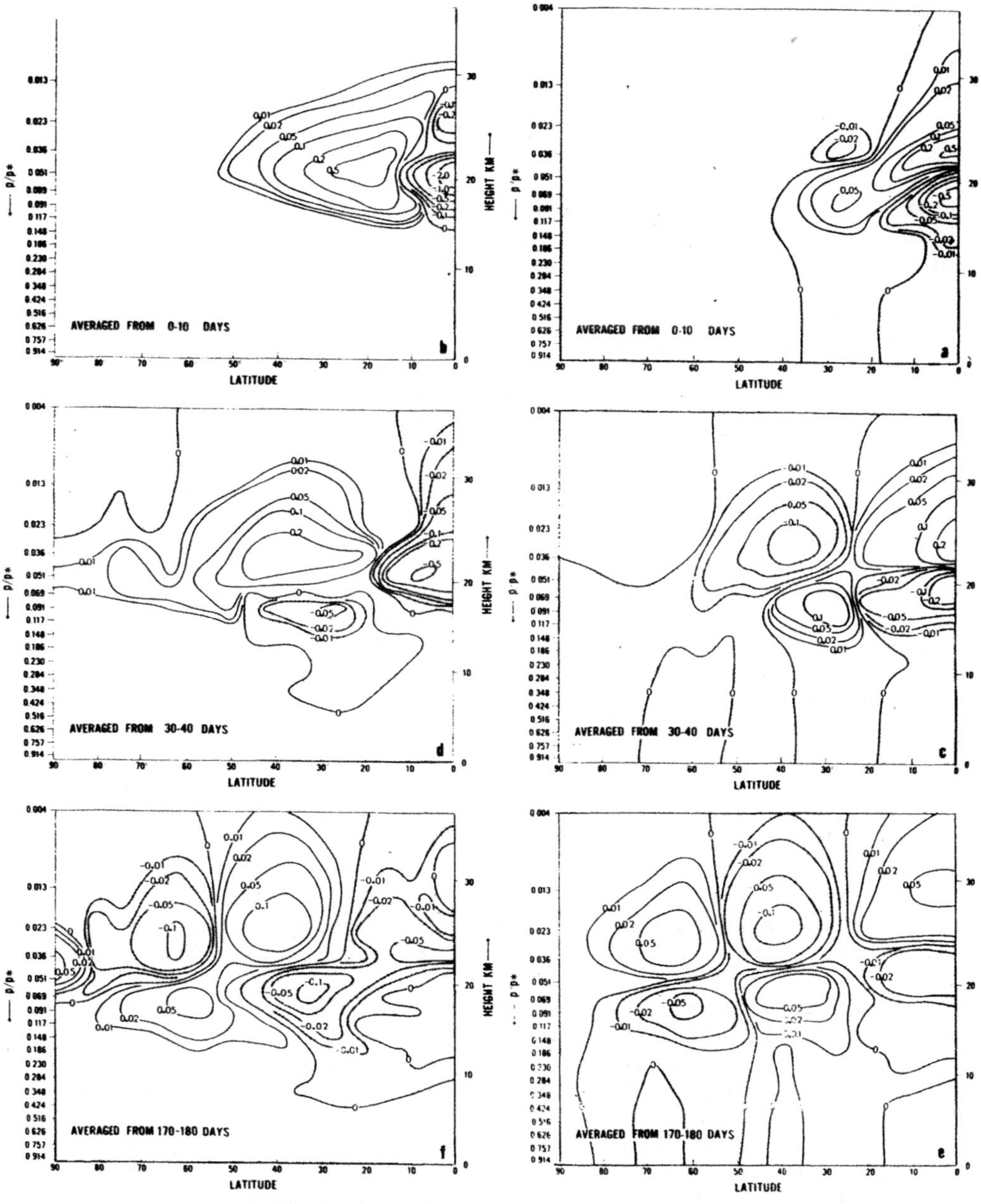

Fig. 19 Combined rate of change of concentration by both vertical and horizontal fluxes separated into the action of large-scale eddies and meridional circulation (Hunt and Manabe, 1968)

In the tropics the eddies transport the tracer on either side of the maximum down the gradient. However in the midlatitudes the resulting eddy flux is countergradient at the lower levels and largely in the opposite direction of the meridional downward flow. With increasing diffusion time the fluxex within the direct cell of the tropics reduce whereas those in the midlatitudes increase.

At the end of the simulation time the eddies produce an upward flux over most of the middle stratosphere. Further down, in the lower stratosphere and the troposphere there was only downward eddy flux with maximum values in the region of the tropopause gap.

In Fig. 18 the corresponding horizontal fluxes are shown. (Shaded areas are equatorward fluxes)
Since horizontal velocities are much larger than the vertical component the corresponding horizontal fluxes have greater values what could already be deduced from the predominant horizontal transport at the beginning of the experiment. The influence of the meridional fluxes as compared to the eddy fluxes is of minor importance. As can be seen from the three averaging periods the horizontal fluxes undergo a considerable change during the simulation. At the end a rather stable countergradient eddy cell, which transports the tracer equatorwards, appear at 50° latitude at about 25 km. A second equatorward eddy cell developed at lower height in the subtropical belt. In the midlatitudes a poleward eddy cell is dominatinating the upper troposphere and the lower stratosphere.

In Fig. 19 the combined rate of changes is presented both for eddies and meridional circulation. As may be expected from the two previous results the horizontal eddy fluxes from the equator to the subtropics dominate over the meridional circulation which only modify the vertical distribution. At the end of the simulation period a cellular structure with symmetry around the height of 20 km appeared both for the eddy fluxes and for the meridional circulation, both fluxes counteracting partly each other.

The combined action of the mean circulation and the eddy flux is sketched in Fig. 20. In this idealized summary the maxima of convergence by the largescale eddies and the meridional circulations are marked by E^{+} and M^{+}. The numbers indicate the order of events as they appear during the simulation. In the first step the meridional cell accumulated the mass in the higher levels of the Tropics. At the same time (step 2) horizontal components of the large scale eddies transfer mass poleward which is then taken over by the downward directed branch (step 3) of the meridional circulation which transported the tracer into the region of the tropopause gap.

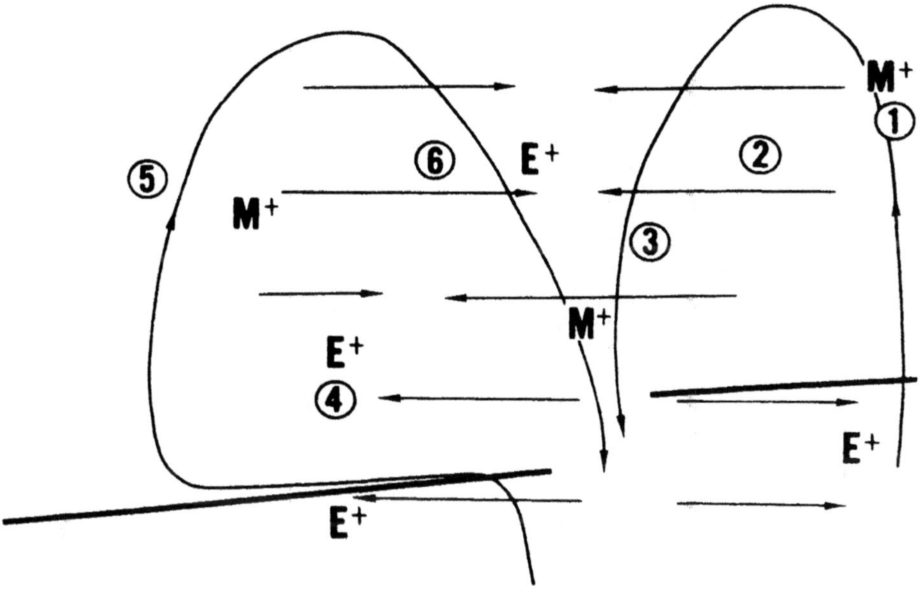

Fig. 20 Sketch of different steps of the mass transport from the stratosphere into the troposphere for mass (Hunt and Manabe, 1968)

From this M^+ region through the further action of the large scale eddies (step 4) the tracer was moved polewards in the lower stratosphere. By the action of an indirect cell (step 5) the material was then shiftet via the upper levels of the stratosphere down to the tropopause gap where it entered the troposphere.

The second simulation demonstrates the transport and diffusion of material injected at a height of 20 km at a latitude of about 40°N (Mahlman and Moxim, 1978). The zonal mean structure for four successive years is shown in Fig. 21 to 24. In the course of the time two important transport directions are evident: (1) the crossequatorial flow in the heights of 15 to 20 km, i.e. in the lower stratosphere and (2) the rather fast intrusion of the material into the troposphere in midlatitudes, notably at around 30°N, in the region of the break of the tropopause. (3)Beginning with April of the first year a zone develops at the equator within the troposphere with low concentrations.

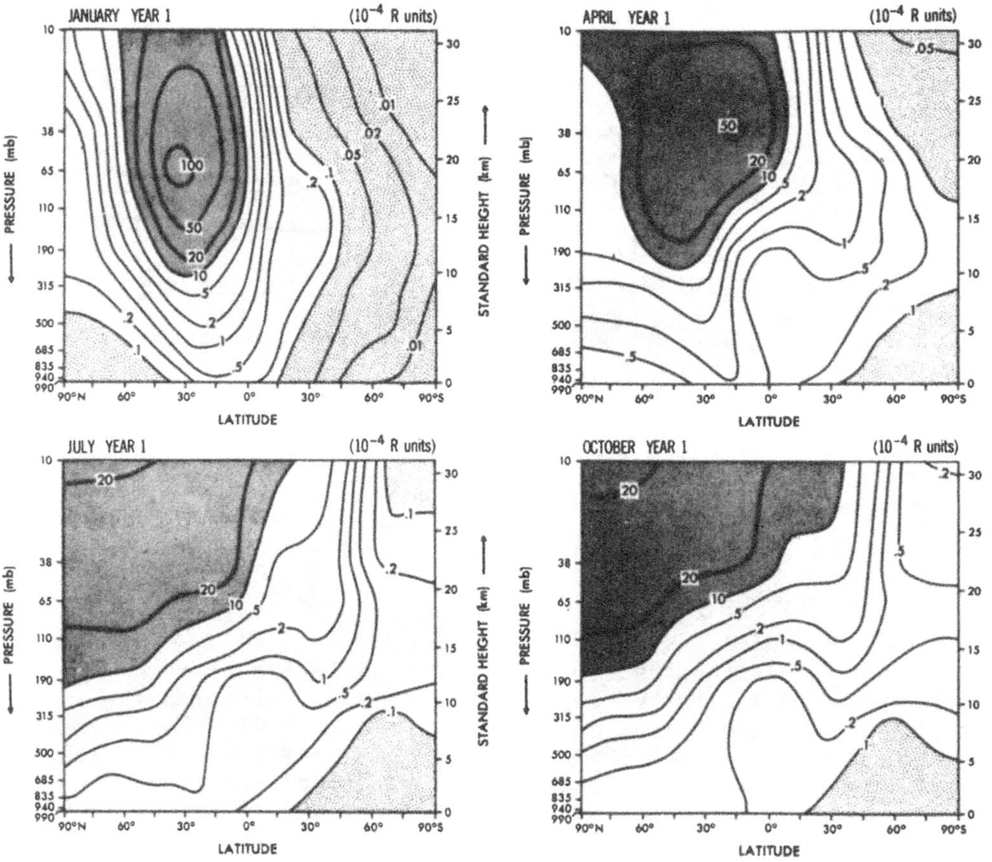

Fig. 21 Zonal mean concentration of mass released at 40°N at a height of 20 km averaged over mouths January, April, July and October of the first year. (Mahlman and Moxim, 1978)

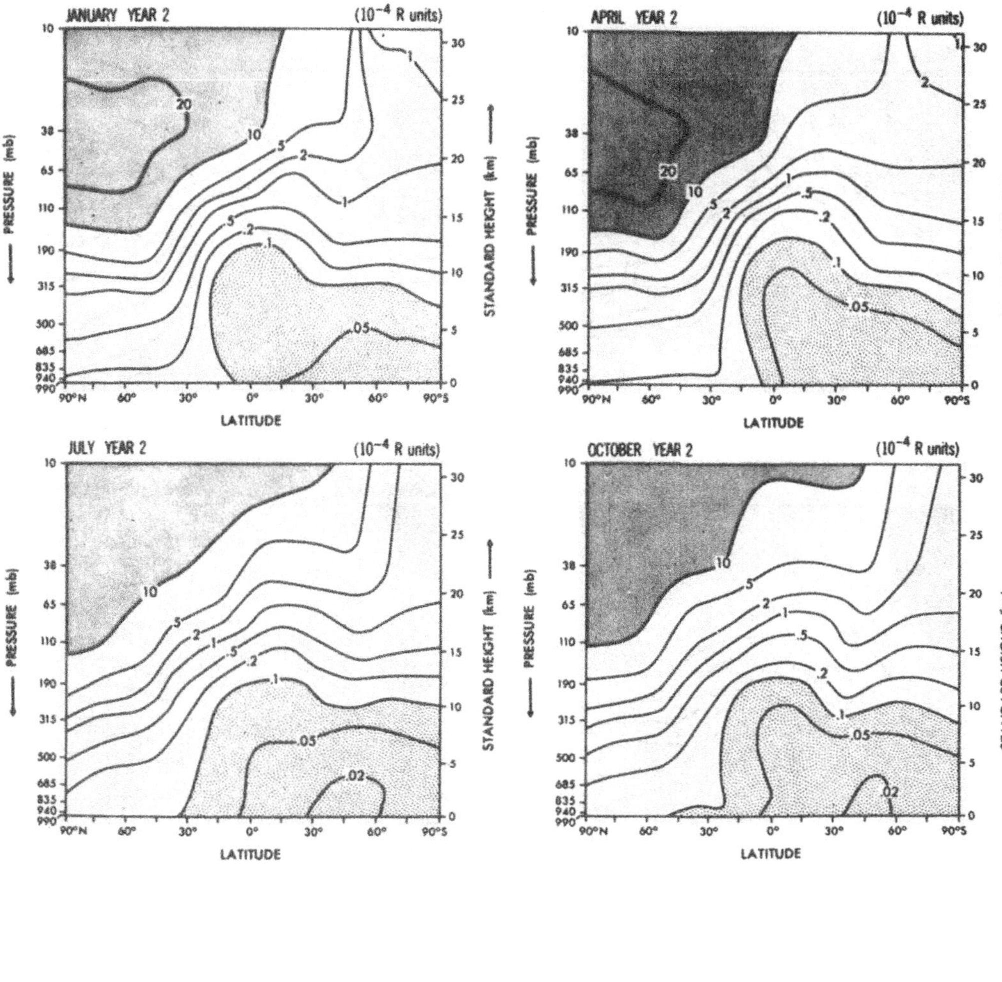

Fig. 22 Zonal mean concentrations as in Fig. 21 but for the second year of model simulation

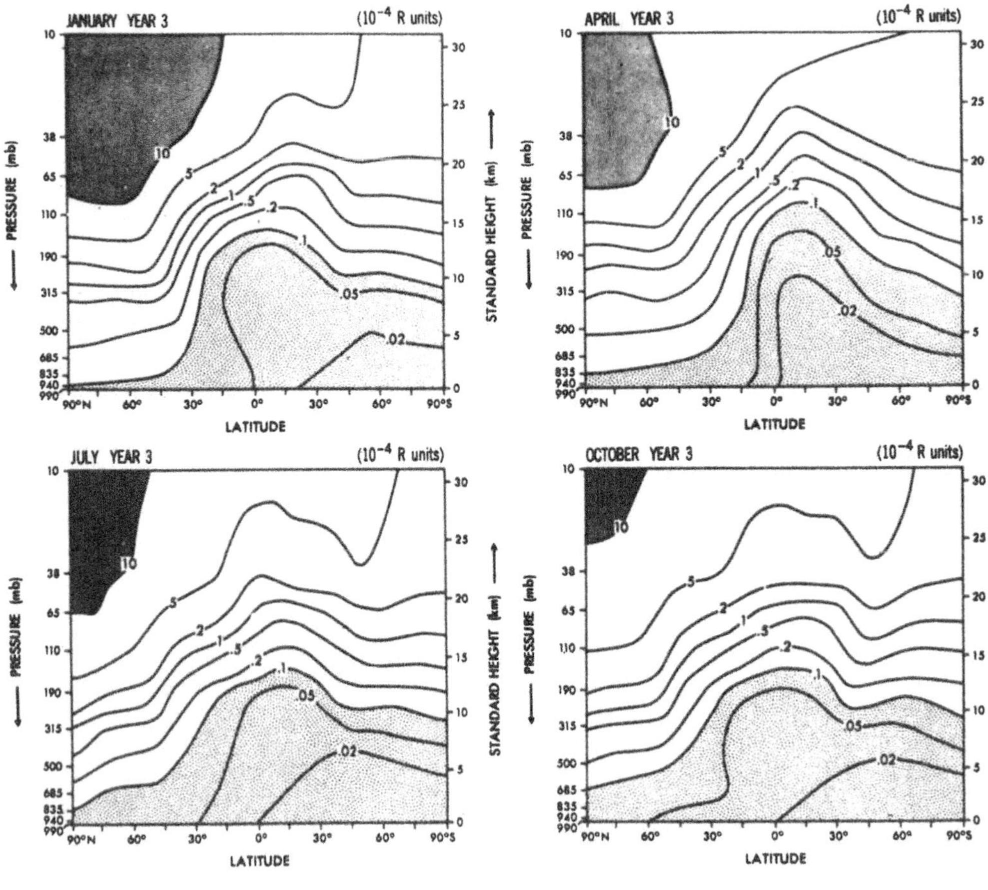

Fig. 23 Zonal mean concentrations as in Fig. 21 but for the third year of model simulation

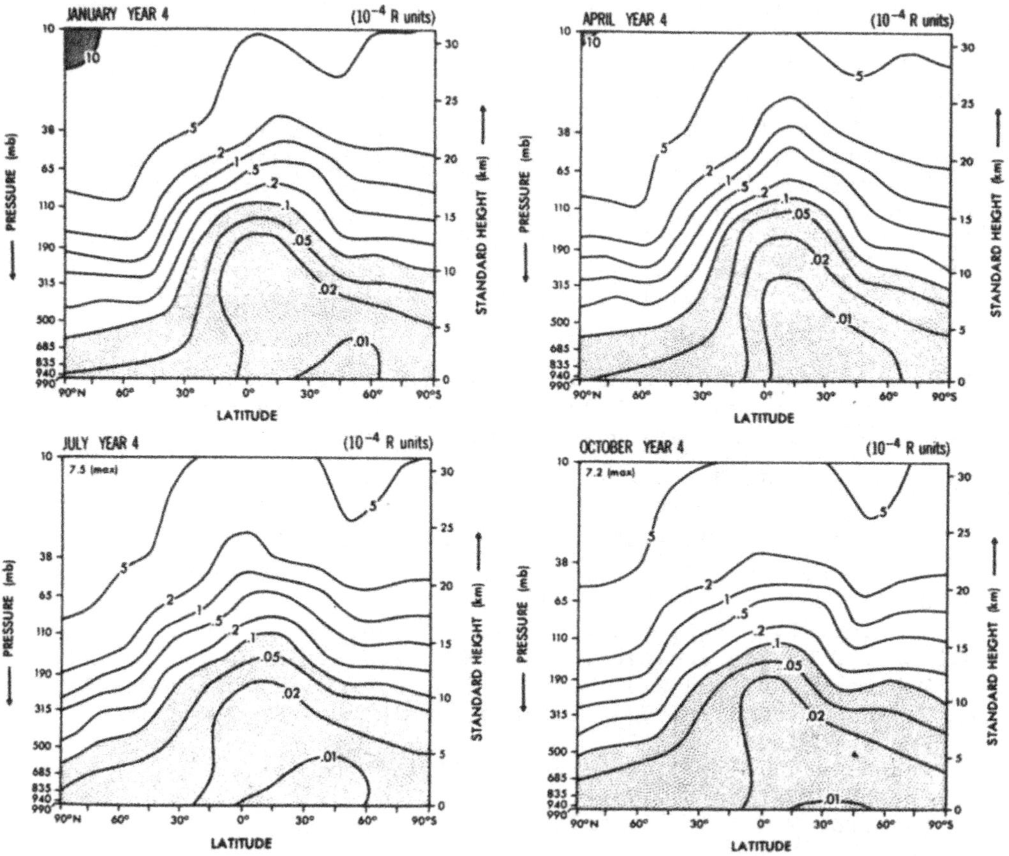

Fig. 24 Zonal mean concentrations as in Fig. 21 but for the fourth year of model simulation

At the end of the simulation, in the fourth year this minimum concentration has reached the top of the model domain.

The main difference in the concentration patterns are dominated by the seasonal difference of the general circulation. The isolines of concentration show a marked slope towards the polar regions, a strong gradient across the tropopause and a inter-hemispheric exchange. This result which is in good agreement with observations reflects the strong influence of the midlatitude cyclones as agents of dispersion.

8. CONCLUSION

It is seen that a broad span of transporting mechanisms are present in the atmosphere which contribute to the overall distribution of tracers. The highly parametrized one-dimensional models will more reflect the skill of making reasonable assumptions on the diffusivities than to give results which are accurate enough in order to assess the action of anthropogenic pollutations. It is believed that only in conjunction with general atmospheric circulation models the mass budjet of chemical reactive constituents can be solved. This will however be a challenge for a longtime afford.

Acknowledgment. The author wishes to thank Mrs. Schnittka, Mrs. Ungeheuer and Mrs. Urbanek for technical and editorial assistence with preparing and typing the manuscript.

REFERENCES

1. Augstein, E.: 1981, Hamburger Geoph. Einzelschriften, 53
2. Blackmon, M.L., J.M.Wallace, N. Lau and S.L.Mullen: 1977, J. Atm.Sci. 34, 1040 - 1053
3. Fiedler, F.: 1971, Beitr. Phys. Atm. 44, 187 - 200
4. Hunt, B.G. and S.Manabe: 1968, Mon. Weather Rev. 96, 503-539
5. Mahlman, J.D. and W.J. Moxim: 1978, J.Atm.Sci.35, 1340-1374
6. Miegham, van J.: 1973, Atmospheric energetics, Clarendon Press, Oxford
7. Pedlosky, J.: 1979, Geophysical Fluid Dynamics, Springer Verlag New York
8. Shapiro, M.A.: 1980, J. Atm.Sci.37, 994-1004
9. Shapiro, M.A.: 1981, J. Atm.Sci.38, 954-973
10. Thorpe, A.J.: 1981, Weather, 36, 108 - 114
11. Wippermann, F.: 1971, promet, 1/2, 16 - 20
12. Zipser, E.J.: 1977, Mon. Weather Rev. 105, 1568 - 1589.

CLIMATIC SYSTEM, CLIMATIC CHANGE AND THE CARBON EFFECT

C.-D. Schönwiese

Institute for Meteorology and Geophysics
University of Frankfurt, Germany (FRG)

ABSTRACT

Climate is produced by the physical and chemical processes acting in the atmosphere-hydrosphere-cryosphere-biosphere-geosphere system some external forcing mechanisms as e.g. solar irradiation included. This climatic system or an actual climatic state can be described by statistics derived from directly measured or proxy data.

Evidently climate has varied in past on all scales of time and will do so in future. An overview of the global climate history is presented and possible causes of climatic change are discussed. The probably most prominent problem of the near future climate (decades and centuries) is the estimation of the effect caused by the anthropogenic rise of the atmospheric CO_2 concentration. The results of the CO_2 measurements and the model calculations of the carbon effect on climate are reported.

INTRODUCTION

In connection with the chemistry of the unpolluted and polluted troposphere it is surely appropriate to discuss the impact of the pollutants on climate. The carbon effect - the expected global rise of temperature near surface due to the anthropogenic rise of the atmospheric carbon dioxide concentration - is a

H. W. Georgii and W. Jaeschke (eds.), Chemistry of the Unpolluted and Polluted Troposphere, 155–184.

very prominent example of such a discussion. Indeed, this CO_2 effect seems to be the most important problem of possible anthropogenic forced climatic change within the time scale of decades and centuries.

It is, however, not possible to separate the problem of anthropogenic forced climatic change from the problem of non-anthropogenic (natural) climatic change which is acting on all scales of time since Earth exists. We may consider one climatic variate, the globally averaged atmospheric temperature near surface for instance. The course of this variate from year to year is forced by a lot of mechanisms, natural and may-be anthropogenic. Apparently, we cannot conclude only from this variate's observations all the variety and complexity of its forcing mechanisms. We need an all-round consideration of climate.

In consequence, the intention of this paper is to discuss the following questions:

- What is climate ?
- What is the climatic system which produces climate ?
- What is going on in the climatic system, as far as we know ? What tracks has climatic change done in history which enable us to get on the track of climatic change ?
- What can we say about future climate and what rôle is carbon dioxide playing in the society of climate forcing factors ?

In an overviewing sense and within limited time it is not possible to answer these questions without selections and simplifications, may-be oversimplifications. The problems can only be touched. For more information the reader is referred to the voluminous literature; for overviews see e.g. (2,3,4,10,14,16, 25,26,33,38,42,44,46,50,53).

CLIMATIC SYSTEM

First, let us have a look on the climatic system which is schematically shown in Figure 1. This system consists of the coupled subsystems atmosphere, hydrosphere (mainly the oceans), cryosphere (ice masses on land and sea ice), geosphere (solid earth; pedosphere and lithosphere emphasized) and biosphere (mainly flora on land and in the oceans). Usually and reasonably those parts of the hydrosphere, which are involved with land (rivers, lakes, ground water) are treated in strong connection with the geosphere.

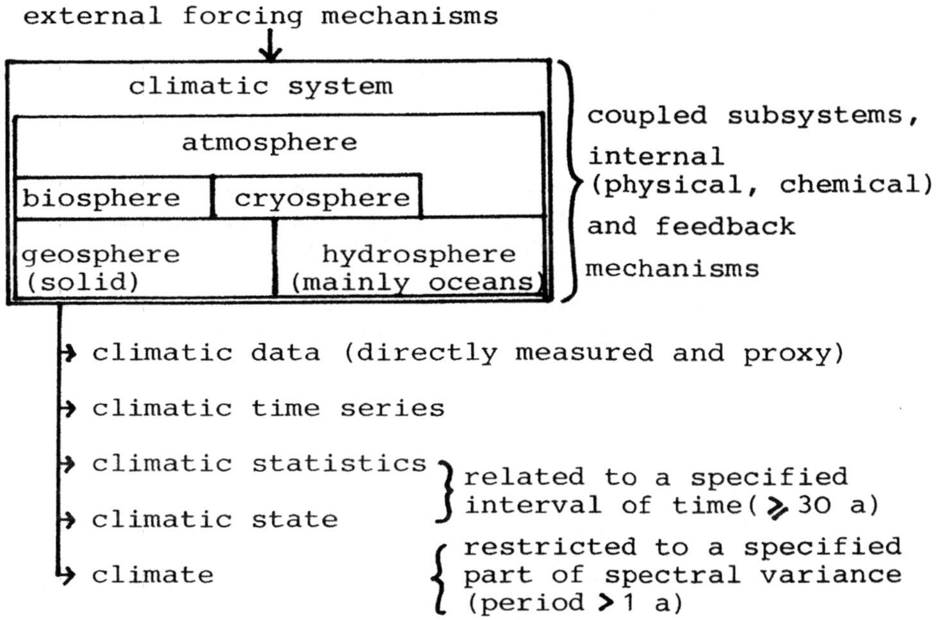

Figure 1 Climatic system and evaluation of climatic properties (a=year).

The challenge, we are far from fulfill, is to understand all physical and chemical processes acting in these subsystems, furthermore to understand the coupling mechanisms from each subsystem to each other, to include the negative and positive feedback mechanisms and finally to take account of the external forcing mechanisms. In contrast to the internal processes of the climatic system the external forcing mechanisms (50) can be treated as isolable factors influencing the climatic system but implying no influence of the climatic system on these factors (at least within the considered characteristical time). Examples of such external forcing mechanisms are solar irradiation (extraterrestrial), volcanism (terrestrial) and, more problematically, human impact on climate.

In order to describe climate we need information of the variates which can be observed in the climatic system. The climatic data obtained by direct measurements and non-direct methods, in this case called proxy data, should cover a time interval as long as possible. By practical and statistical reasons all these data are usually related to time at equal steps of time. It arises a type of information called climatic time series.

The statistical measures of the climatic time series covering a reasonably specified interval of time, according to international statements (19,31) at least 30 years, define a climatic state. Stationary oscillations, most prominent the daily and annual cycle, are included when considering a climatic state.

Adequate 0-dimensional statistical measures of a climatic state are the moments, in detail the mean, variance, skewness, curtosis and other parameters of the frequency distribution. The most important 1-dimensional measures are the frequency distribution, the corresponding empirical or theoretical probability density function and the spectral variance estimate (power spectrum). Measures of correlation and coherence (spectral disintegrated correlation) are assessments of the sharpness of connection of different types of variates or of variates taken from different space coordinates. The all-together set of adequate statistical measures describing the climatic time series leads to the statistics of a climatic state (17,31,34,47) which describe and define it.

When the external forcing factors of the climatic system change, the climatic state will also change; but furthermore it cannot be strictly excluded that constant external forcing factors enable several climatic states. In this sense the climatic state observed today is one realization of climate. And there have been other realizations in past and will be more in future. In consequence, climate includes the real and potential climatic states in history and future produced by the internal and external mechanisms of the climatic system, where the considered characteristical time is restricted to comparably low frequencies. This restriction may be performed in the sense that all phenomena averaged from the individual phenomena of weather are called "climatic" (11) or that the variance spectrum is divided in a turbulent

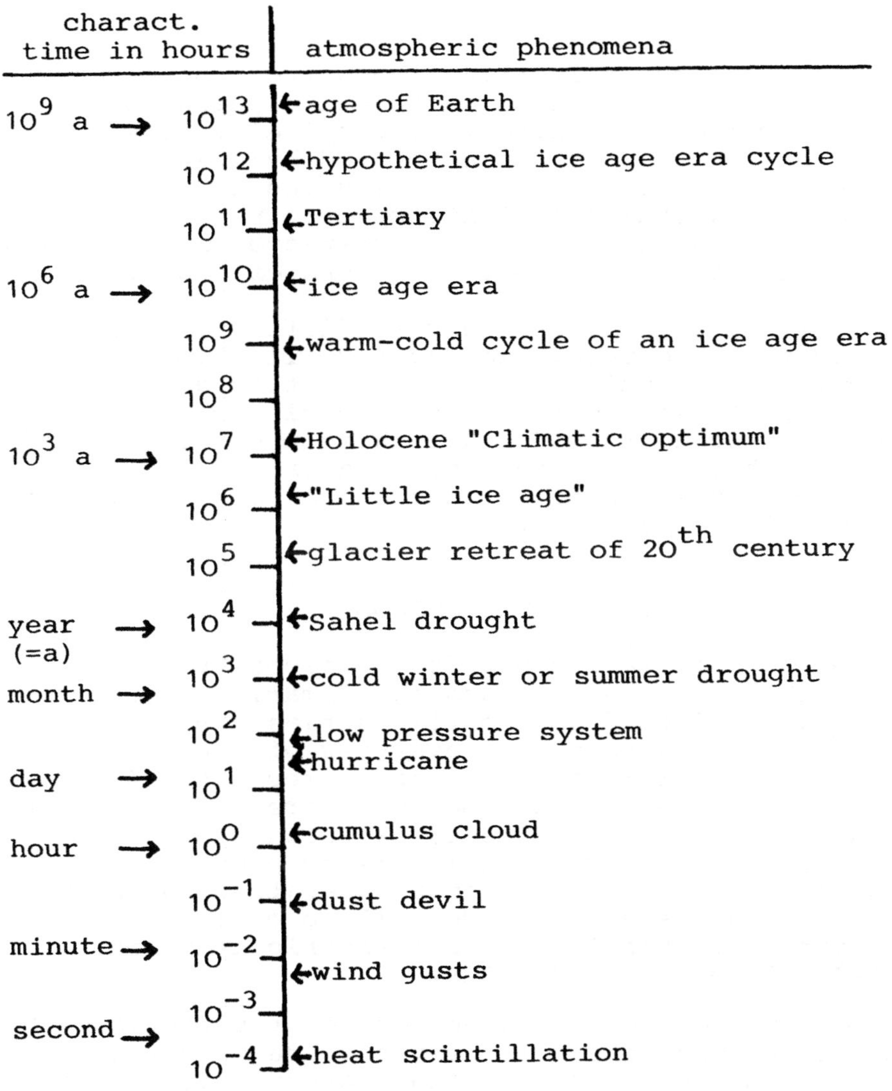

Figure 2 Examples of atmospheric phenomena covering the whole spectrum of characteristical times (periods).

(period $T < 1$ hour), synoptic (1 hour $\leq T \leq 7$ days), seasonal (7 days $< T \leq 1$ year) and climatic ($T > 1$ year) scale (39).

There are not only transitions from one climatic state to another. All climatic state phenomena or observed climatic variates, respectively, vary on all scales of time. This can be proved by spectral variance analysis (23,31). Figure 2 gives in a schematical way a few examples of atmospheric phenomena covering the whole spectrum of characteristical times from parts of seconds up to billions of years.

It is surely one of the most difficult problems of science first to recognize exactly all climatic system phenomena continuously in respect to all characteristical times and in respect to all space coordinates and second to model all these phenomena in order to get a physical and chemical understanding. The second object seems to be of a still higher degree of impossibility than the first.

HISTORY OF GLOBAL CLIMATIC CHANGE

Let us familiarize now from a global point of view with the rough outlines of climatic state variations as derived from the time series available. Table 1 allows a condensed overview of the types of the available climatological data; see also (4,25,42, 50,53). Despite of the immense progress achieved in evaluating proxy (non-direct, paleoclimatological) methods, the data availability is so poor that the simplifications mentioned above are stressed in an extremely intensive way. We will go on in this simplification by considering atmospheric temperature variations near surface only.

Keeping in mind all above quoted restrictions and simplifications, the following picture of the global climate history covering a time interval as long as possible appears, see Figure 3: Approximately 90% of this history there has been a warm climate which enabled no ice to be existent on the Earth's surface. This means a climatic system without cryosphere. The midlatitude continental annual mean temperature as derived for the Mid-Tertiary c. 40 millions of years ago (44) has been 5-10°C higher than today (yearly precipitation rate in Central Europe c. 40-50% higher than today).

Table 1 Overview of climatic data availability (as far as continously recorded). Directly measured data are atmospheric near surface. Time intervals of proxy data are maximal and approximate (a=year).

directly measured	since	proxy	interval
temperature	1659[1)]	lake sediments, varves	$>10^3$ a
precipitation	1697[2)]	tree rings	$\sim10^4$ a
pressure	1740[3)]	polar ice cores	$>10^5$ a
wind	1781[4)]	pollen diagrams, coast lines of oceans	$>10^5$ a
sunshine	1880[5)]	soil sediments	$\sim10^6$ a
snow cover	1881[6)]	marine sediment cores	$\sim10^7$ a
		mineralogical, petrograph. and geomorpholog. signs	$>10^9$ a

today: c. 9000 surface observation stations (weather)
c. 850 radiosonde stations (upper air data)
special networks (e.g. radiation, ozone)
remote sensing techniques (e.g. satellites)

CO_2 data: directly measured since 1958 Mauna Loa (Hawaii); some sporadic early measurements between 1860 and 1940
proxy information from polar ice cores, tree rings, marine and lake sediments

1) Central England 2) Kew near London 3) De Bilt
4) Hohenpeissenberg, FRG 5) Kew 6) Vienna

This warm ice-free climate was several times interrupted by cold intervals, the ice age eras, some of these global and some hemispheric (10,14,25,44,48,50), see Figure 3. The main feature of an ice age era is a climatic state of so low temperature that ice can form on the Earth's surface. The transition from a warm ice-free climatic state to an ice age era climatic state, however, is a very slow progressing process. As soon as c. 40 millions of years ago, that means in the Mid-Tertiary, the last ice formation process started in Antarctica. Approximately 10 millions of years ago the ice shield of West Antarctica was built. Afterwards the southern hemisphere sea ice building and again later arctic icing was initiated. The last few millions of years an ice age era is existent with ice masses in both hemispheres. We still live in this era, which is called Quaternary

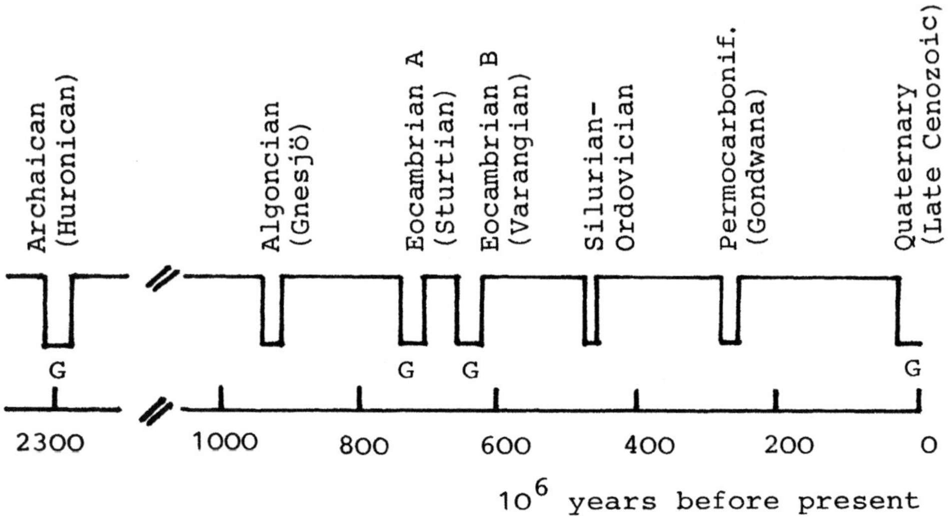

Figure 3 Time related scheme of ice age eras, lower steps, and prevailing ice-free warm climate, upper line; after Tarling (48) and others, simplified. G means global (both-hemispheric) icing.

or late Cenozoic ice age era. It followed the extremely asymmetric climate of the late Tertiary with ice-covered Antarctica and ice-free northern hemisphere (12,13).

The conditions and mechanisms supposed to be responsible for the climatic state of an ice age era cannot be discussed here in detail; see e.g. (10,25, 42,44). They should be briefly mentioned, however, and this can be done as follows:

- Continental drift, where polar situations of landmasses (today Antarctica) or their arrangement nearby the pole (today the big continents Asia and Northern America together with the isle of Greenland) favour snow and ice to be maintained on the Earth's surface.
- Orogenesis creating high altitudes combined with increased volcanic activity whose stratospheric dust layers diminish the effective solar irradiation.
- Ice-albedo positive feedback. (Increased ice cover extent causes temperature decrease because of decreased radiation balance near surface; this causes increased ice cover extent and so on.)

- Astronomical and solar theories, especially variations of the interstellar material affecting the solar system during its movement around the centre of the galaxy (period c. 250 millions of years).
- Oceanic and atmospheric circulation patterns which are, however, probably more responsible for the variability on shorter time scales.

It is very hard to detect signs of climatic change in the time scale of millions and billions of years. These signs are collected in a criminalistic way looking for traces of ice movement on surface (geomorphological phenomena), permafrost soil and other indications pointing to a relatively cold or warm climate (mineralogical, petrographical, geological and biological phenomena). In contrast to that the temperature regime of approximately the last two millions of years is relatively well-known by means of the modern technique of marine sediment core drilling (3,9,33).

The basis of this technique is the fact that the ratio of the oxygen isotopes ^{18}O and ^{16}O is dependent on temperature. Sediment samples of different depth and therefore of different age (additional age assessments by means of determining the radioactive material) which contain limestone components of fauna can be used to apply this technique. Thereby, the oxygen content of the limestone is analysed in respect to the isotope ratio by mass spectrometer (9).

In Figure 4 four results of the sediment drilling projects are brought together. Evidently the considered (Quaternary) ice age era is characterized by an alternative sequence of relatively warm and cold ages (interglacials and ice ages or glacials). This sequence took place rather synchronous all over the world as can be proven by a lot of sediment cores (9,16,25,44,53).

The existence of the last four cold ages (ice ages), labelled here C_1 to C_4, is known since the beginning of this century,where mainly geomorphological, geological and biological (e.g. pollen diagrams of different age) methods have been used. The C_i names are different from region to region; for example C_1 is called "Würm" in the region north of the Alps, "Weichsel" in Northern Germany and the Netherlands, "Devensian" in England, "Valdai" in Russia, "Wisconsin" in the United States and so on. At least

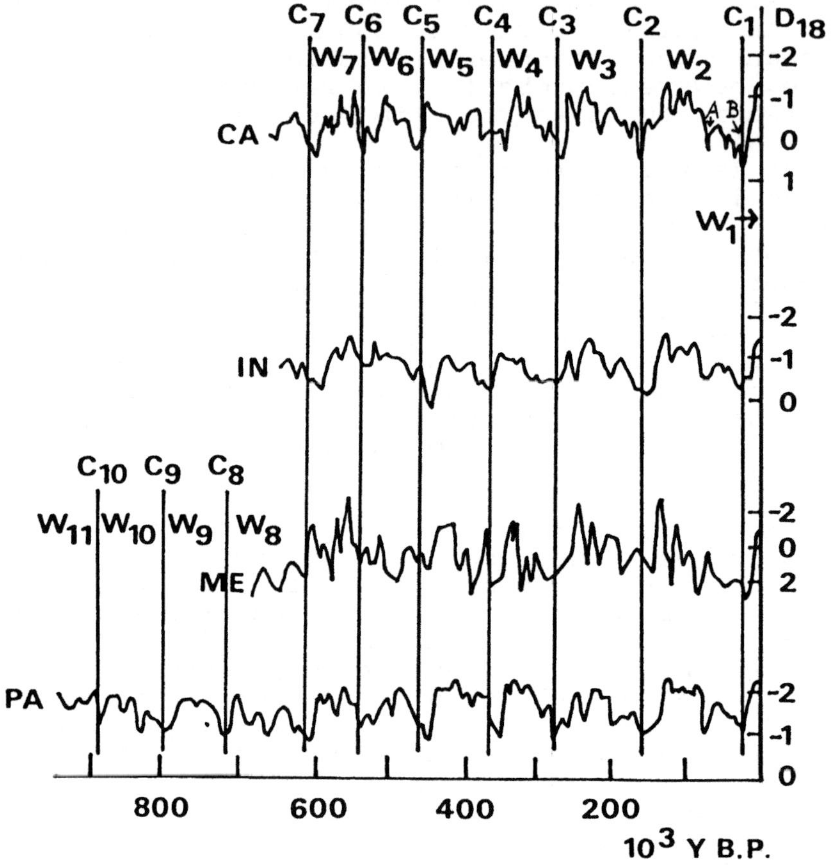

Figure 4 Oxygen isotope ratio (D_{18}) analysis of a Carribean (CA), Indian (IN), Mediterranean (ME) and Pacific (PA) deap sea sediment core, after Duplessy (6). The added C and W symbols indicate the cold and warm ages of the Quaternary ice age era (C_1=Würm or Weichsel, C_2=Riss or Saale; W_1= Neo, W_2=Eem; etc.). Corresponding maximal temperature amplitude in continental mid-latitudes c. 10 °C. (Y=year.)

twenty of these cold ages appeared in both hemispheres. The main epochs of these cold ages were characterized in mid-latitudes by yearly average temperatures c. 10°C lower than today (14,44), whereas in equatorial regions this temperature decrease was not more than 1-2°C. The striking feature were the anormously extended ice masses which covered North America southwards to c. 45°N, total Scandinavia, the British Isles, Iceland and the most parts of the extratropical mountains (but only small parts of North East Europe and North Asia , probably because of very low precipitation rates). These wide-extended ice masses (entailing the name "ice age") caused a sea level drop of c. 140 meters, so that e.g. the British Channel and the adjacent North Sea were not existent.

As soon as 1920 Milankovitch tried to explain the cold-warm age cycle of the Quaternary ice age era by means of the Earth's planetary relation variations (30):

- Variation of the excentricity of the Earth's path around the sun, mean period 95000 years.
- Variation of the tilt of the Earth's axis in respect to the ecliptic plane; mean period 41000 years.
- Variation of the date of aphel and perihel, mean period 22000 years.

Milankovitch computed the solar irradiation variations in mid-latitudes caused by these planetary variations and tried to fit them with the observed C_1/W_1 - C_4/W_4 cycle.

Tens of years this theory was supported to be very doubtful, because the irradiation effects are clearly opposite in different latitudes and seasons; but nowadays, in a more complicated way, the Milankovitch theory is supposed to be a rather probable and effective external forcing factor of climatic change (3,16,42, 44,53). Other important theories considering this time scale are based on orogenesis and volcanic activity, furthermore on oceanic and atmospheric circulation patterns. The corresponding modern astronomical theory (3) predicts the transition from the present W_1 (Neo, Holocene) warm age to conditions corresponding to the C_1 (Würm) B cold age in 50000-60000 years.

Before this will happen, we should have a look on the present W_1 (Neo) warm age climatic state. It seems to be similar to the W_2 (Eem) warm age which preceded the C_1 (Würm) age. (There are two main pecu-

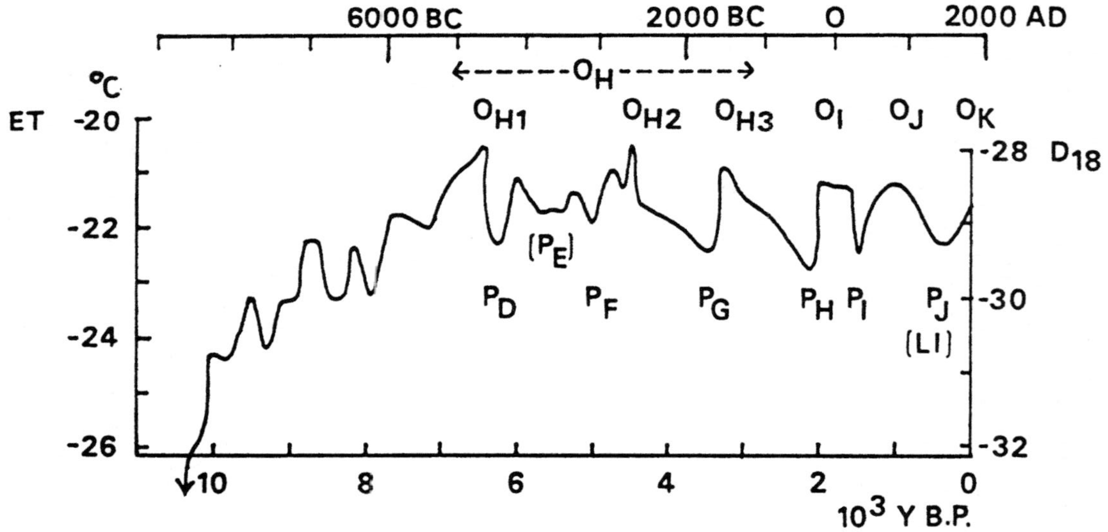

Figure 5 Oxygen isotope ratio (D_{18}) analysis of the Camp Century (Greenland) ice core, after Dansgaard et al. (8). The added symbols mean ET = estimated temperature, O = optimum epochs (relatively warm), P = pessimum epochs (relatively cold), O_H = main optimum or Altithermum, O_J = Medieval warm epoch, LI = Little ice age, O_K = Modern optimum. The whole period covers the Neo warm age (W_1) of the Quaternary ice age era.

liarities of W_2: in mid-latitudes in a yearly average 2-3°C warmer than W_1, but a few rapid sharp coolings on a 100 year scale). W_1 started c. 10000 years ago and reached its "Optimum" (Altithermum) approximately 3000-7000 years ago. This can be read from Figure 5, where oxygen ratio analysis results are presented, taken from the famous Camp Century (Greenland) ice core (8). Drillings in Antarctica (27) show similar results.

According to the established name "Climatic Optimum" a relatively warm epoch of W_1 (Neo) is called here O ("Optimum") and a relatively cold epoch P ("Pessimum"). There are numerous O and P epochs within W_1, as are in comparison numerous relatively (!) warm and cold epochs within C_1 (interstadials and stadials of "Würm"). In Figure 5 the different O and P epochs (of W_1) are indicated by further index characters as used in (42).

The main cold epochs of W_1 - P_G, P_H, P_I and P_J - all occured within the last 3000-4000 years. The "Climate transition" between 1200 and 1400 AD from O_J ("Medieval warm epoch") to P_J ("Little ice age") caused in Europe a dramatical crisis of human economic and social life. This is described in the fundamental work of climatic change written by Lamb (25).

In the 10^3 and 10^2 year time scale the worldwide synchronity of climatic change is anything but definitive. The striking problem is that the ice cores which enable detailed information in this time scale are available only from polar regions and that direct measurements start as late as in 1659, compare Table 1. From 1850 until today there exist c. 100 climatic time series (for the main element temperature near surface). The actual global network of observation stations was established not before the last two decades.

The longest climatological time series, based on direct measurements, the Central England temperature record prepared by Manley (29), starts 1659, see middle part of Figure 8. That means, it covers the last part of P_J ("Little ice age") and the recent warm episode O_K ("Modern optimum"), see Figure 5. And just this series indicates O_K in a very weak form. At many other places O_K has been the most warm epoch since O_J ("Medieval optimum"). The end of P_J is assumed by some climatologists to be around 1700. Most of them, however, prefer the date 1850-1880, because since that time a well-marked retreat of mountain glaciers set in. (The coldest episode of the Central England record was c. 1690-1700; weaker relative minima are found around 1815 and 1880-1890. The warmest episodes were c. 1725-1735 and 1940-1950).

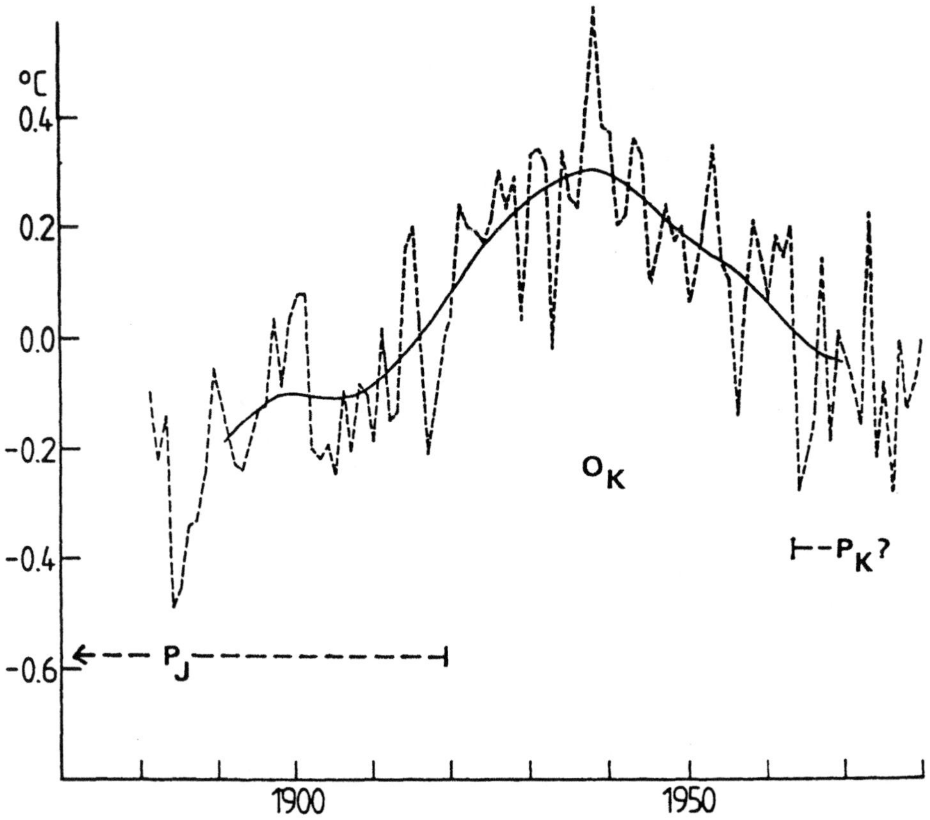

Figure 6 Yearly deviations (anomalies) 1881 - 1980 of the northern hemisphere mean temperature near surface from the 1901 - 1930 average, after Mitchell (55) and others. Data of the years 1976 - 1980 adapted using the Climate Monitor (journal of the Climatic Research Unit, Norwich, U.K.) publications. Solid line Gaussian low-pass filtered data after Schönwiese (42) suppressing the variability of the T < 30 year period range.

It is possible to compute a reliable mean temperature curve for the northern hemisphere since 1881. In Figure 6 the result is shown by means of the yearly deviations (anomalies) from the 1901-1930 average, dashed line, and by means of a low-pass filter, which suppresses variations of periods shorter than c.30 years (42). The culmination of O_K ("Modern optimum") appears 1938, where the greater amplitudes of the polar and subpolar stations cause a weightening effect. It has to be pointed out that there are well-marked regional pecularities so that some regions had a phase shift or were even in anti-phase with the hemispheric trends shown in Figure 6. On the other hand the very most measurements prove this recent climatic change with culmination roughly between 1930 (polar regions) and 1960. The New Zealand record, for example, shows also this type of variation characterized by an uninterrupted temperature increase from c. 1900 until c. 1950 (37).

TOWARDS AN UNDERSTANDING OF CLIMATIC CHANGE

No doubt, the phenomena of climatic change as outlined in the previous chapter (and the very many phenomena not described here!) should be understood as well as possible. This requires
- an exact and exhausting statistical analysis of all data available, both internal and external to the climatic system and
- regional (meso) and global (macro) modelling of the whole climatic system and all mechanisms which cause climatic change in order to realize all physical and chemical properties of the system and to verify the results of statistical analysis.

There are nothing but shortcomings in respect to these requirements. The reasons are too few data, too little man-power in statistical analysis and too drastic theoretical as well as computer-based difficulties in modelling.

The declaration of the World Climate Conference in 1979 (53) found it "urgently necessary for the nations of the world
- to take full advantage of man's present knowledge of climate;
- to take steps to improve significantly that knowledge;
- to foresee and to prevent potential man-made

changes in climate that might be adverse to the well-being of humanity".

Apparently these recommendments can only be fulfilled by the above outlined statistical and modelling requirements.

As ascertained in the first chapter statistical methods are used to define climate and climatic states. Here only some of the most important techniques of climatological time series analysis (1,5,31,34) shall be mentioned:

- spectral estimates (variance spectra) in order to consider the time series variability in relation to frequency (or period, respectively);
- numerical filtering in order to enable a time-related but spectral separated consideration of this variability;
- cross spectrum and coherence analysis in order to consider correlations in a spectral way;
- cross correlations in order to consider correlations phase-related;
- factor and empircal orthogonal function analysis in order to analyse correlations in space and time.

Time series variability can be of the following types:

- Trend, proved by regression and hypothesis decision techniques.
- Periodical, which cause a population spectrum of one (in case of single sinus) or more vertical lines.
- Cyclical, where relatively small variations of period and any variations of amplitude are allowed; this effects a population spectrum of "broadened" relative maxima.
- Random, which causes a population spectrum of a horizontal line.

The finiteness of the available data sets permits only the estimation of sample spectra. Statistical tests have to be used to judge the known sample spectra in respect to their agreement with the theoretical population spectra on a definitive level of significance.

Fig. 7 presents a schematical example of possible applications of statistical spectrum analysis. The upper presentations (RV) combine a stochastic "background" spectrum with signs of cyclic variability found in a lot of temperature variance spectrum estimates. It is dominated by the daily, yearly and c. 100000 year cycle. In addition, however, there are some other signs of cyclical variability established in a varying manner from region to region. The lower presentations (M) indicate spectrum peaks related to potential forcing mechanisms. From this point of view, by comparing the results of variance spectrum analysis, carrying on with comparing filter analysis results and performance of coherence estimation, together with other adequate techniques, some statistical statements of possible climatic mechanisms can arise. There has been done some work in this field of analysis, but by no means in an exhausted way.

Table 2 Possible climatic forcing factors. Abbreviated overview taken from (42).

extraterrestrial	terrestrial
solar radiation	composition of the atmosphere, especially volcanism and anthropogenic emission
solar diameter	atmospheric circulation
other extraterrestr. radiation	cloudiness
meteors	oceanic circulation
interstellar gases and particles	salinity of oceanic water
gravitational effects	vegetation
	ice and snow on surface
	orography and orogenesis
	continental drift
Earth´s planetary relations	
feed back mechanisms	
stochastic transfer of autovariation	

The other approach to the problem comes from the physical and chemical point of view. Table 2 is a list of potential climate forcing factors without spectral consideration. Most of them have already been mentioned; some of them are very questionable, for example the gravitational effects of the big planets. (This may, however, influence solar activity.) Some of them are immanent to the climatic

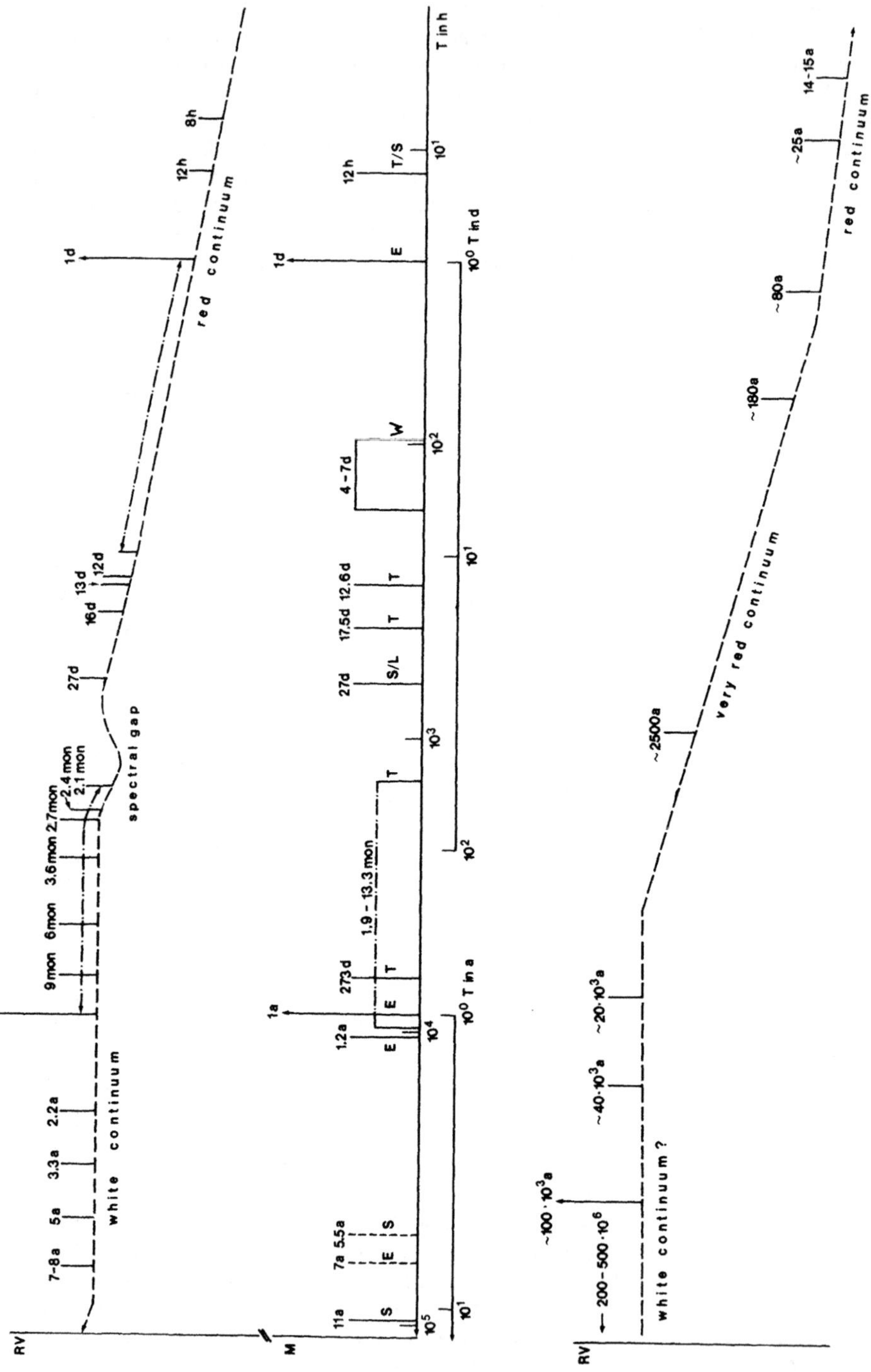
RV
M
RV
white continuum
spectral gap
red continuum
very red continuum
white continuum ?
T in h
10^0 T in d
10^0 T in a
1a
1d
12h
8h
7-8a
5a
3.3a
2.2a
9 mon
6 mon
3.6 mon
2.7 mon
2.4 mon
2.1 mon
27d
16d
13d
12d
11a
7a
5.5a
1.2a
273d
1.9 - 13.3 mon
17.5d
12.6d
4 - 7d
W
T/S
S/L
~2500a
~180a
~80a
~25a
14-15a
~20·10³a
~40·10³a
~100·10³a
200-500·10⁶

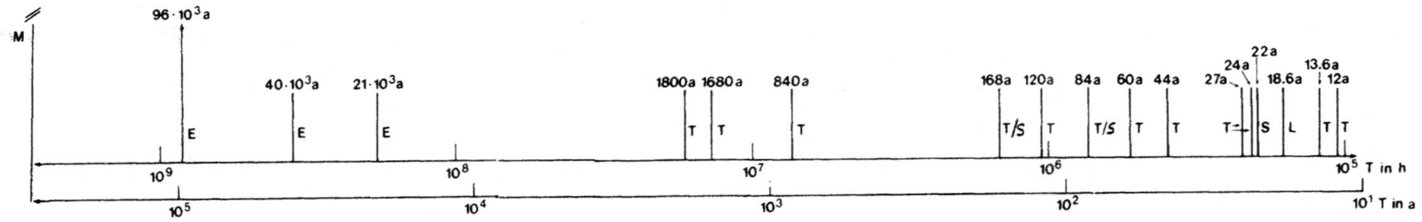

Figure 7 Scheme of measured spectral relative variance (RV), temperature series emphasized, upper graphs, and possible forcing mechanisms (M), lower graphs, after Schönwiese (41). The spectrum peaks are idealized by vertical lines. Except the daily, yearly and warm-cold ice age era cycle ($T \approx 10^5$ years), which clearly dominate, the peaks are not quantitatively shown. Abbreviations: h = hour, d = day, mon = month, a = year; T = period or tidal (gravitational), respectively, S = solar, L = lunar, E = Earth's planetary relations, W = weather systems. The dashed-dotted lines indicate concentrated bands or an extended level of increased spectral variance.

system in a high degree (e.g. cloudiness), so that it is difficult to select cause and effect. The last one is referred to the idea of stochastic variability transfer between the components of the climatic system (20).

The physical-numerical (deterministic) models of the atmospheric circulation, which comprehend in a grid network (characteristical grid-width a few hundred kilometers) the two horizontal coordinates and in vertical direction some isobar planes (characteristical number c. 10) are the most sophisticated simulations of climate. They are called 3D models (D=dimension) or GCMs (general circulation models), where at each grid point the selected set of physical equations has to be solved. They need, however, very much computation time and become not realizable when a too high number of equations is used and especially when the grid width is coniderably diminished. In addition, the time steps used involve numerical instability so that the boundary of progressing time in 3D models amounts to a few weeks, if realistic results shall be expected.

Other difficulties are the lack of corresponding oceanic models (OGCMs), quite apart from including cryosphere and especially biosphere, and the problems of handling negative and positive feedback mechanisms. Nevertheless, 3D models are very adequate and very useful approaches to climatic states when the external conditions (e.g. solar irradiation or composition of the atmosphere) are altered and a steady-state response of the atmospheric data pattern shall be simulated.

In order to mitigate the difficulties, primarily in order to shorten computation time, a step-wise reduction of the model dimensions is necessary: Parameterization of three-dimensional mechanisms by two-dimensional (mostly altitude and geographical latitude related) or one-dimensional (mostly geographical latitude) equations. This leads to the 2D and 1D models (statistical-dynamical models, radiative convective models). The end of this hierachy is a OD model consisting only of one equation which describes the balance of incoming solar radiation and outgoing terrestrial radiation near surface in terms of the effected globally averaged temperature (energy balance models).

In the degree as models are parameterized, in the simplest way by averaging, statistical statements become necessary. These statements obtained by statistical data analysis determine the mechanisms which should be emphasized in the particular considered time scale, time interval (e.g. for W_1 or P_J climatic states), climatic subsystem, geographical region and altitude.

In Fig. 8, medium part, the already mentioned temperature series of Central England is plotted: The dashed line represents annual means and the solid line low-pass filtered data computed as in Figure 6. A simple correlation analysis (43) reveals that this relatively long-term component of the temperature data is correlated with solar activity, parameterized by the observed relative sunspot numbers (+0.56 ± 0.08 confidence region of 95% probability), and volcanic activity, parameterized by Lamb's (24) northern hemisphere stratospheric dust veil index (-0.45 ± 0.09, in each case partial correlation coefficients).

When the sunspot number parameter is exchanged by solar irradiation data derived from this parameter the correlation increases to + 0.64 ± 0.07 (partial). 64% (with 95% confidence an amount between 58% and 71%) of the considered long-term variance (equivalent multiple correlation coefficient r = 0.80 ± 0.04) then can be statistically explained by the two external forcing mechanisms. (86% of the variance within the scale 1 year $< T <$ 30 years is explained by a few components of cyclical variability which have no known deterministic counterparts and may be internal to the climatic system (41)).

A good deal of the residium variance will be due to stochastic transfer of autovariance in the climatic system (20). This explains the increase of variance with period ("red spectrum") as observed in sample spectra.

Now, we may ask whether there are statistical signs or theoretical reasons of a CO_2 signal in the still remaining unexplained variance of the temperature time series. Thereby, of course, the last decades and centuries are emphasized and future projections of the same characteristical time become desirable.

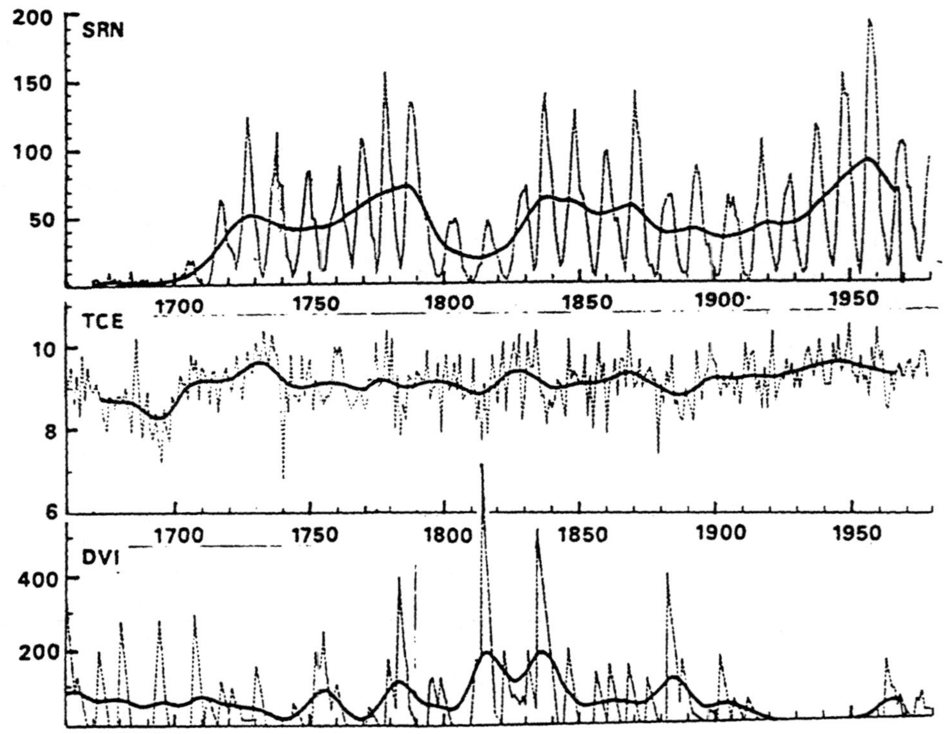

Figure 8 Comparison of SRN = sunspot relative numbers, TCE = temperature (near surface) record of Central England after Manley (29) and DVI = dust veil index after Lamb (24) which parameterizes the northern hemisphere volcanic activity. Dashed lines are yearly data 1660 - 1979, solid lines low-pass filtered data as in Figure 6; after Schönwiese(40,43).

CARBON EFFECT

Carbon dioxide CO_2 (2) is a gas whose atmospheric concentration is influenced by anthropogenic consumption of fossil fuel and cement production. A quantitatively very questionable indirect anthropogen effect is caused by intervening the biosphere; see paper of Stein in this volume. Continuous CO_2 measurements are available since 1958 from Mauna Loa, Hawaii, see Fig. 9. Including some sporadic earlier measurements between 1860 and 1940, the CO_2 concentration level in the atmosphere has increased from c. 290 ppm (= parts per million) in 1860 to 336 ppm in 1980; that means an increase of nearly 16%.

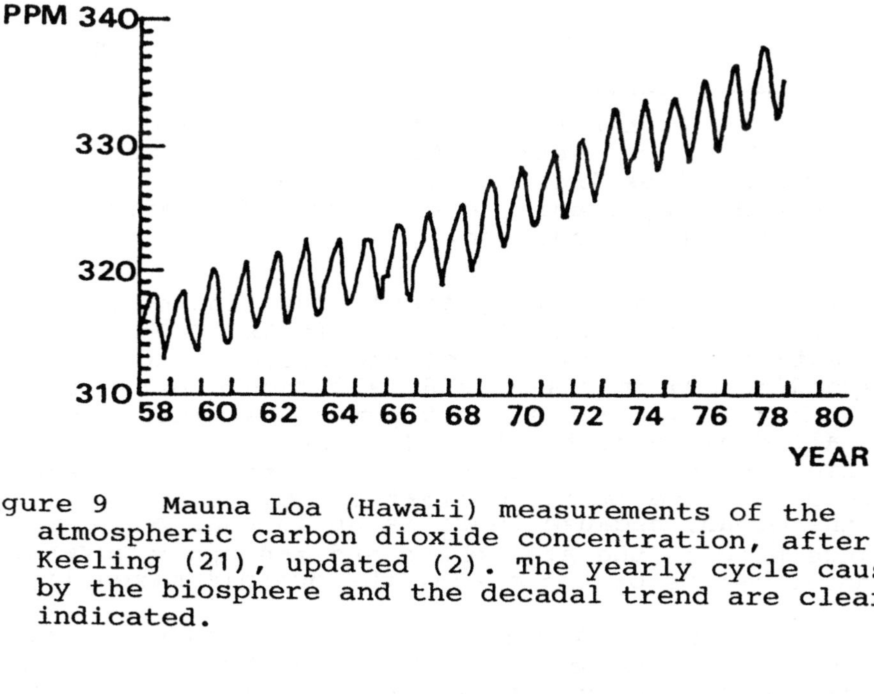

Figure 9 Mauna Loa (Hawaii) measurements of the atmospheric carbon dioxide concentration, after Keeling (21), updated (2). The yearly cycle caused by the biosphere and the decadal trend are clearly indicated.

If this increase is an anthropogenic effect, and this is generally believed, we need first adequate models to explain the CO_2 concentration increase and second models to estimate the response of the climatic system. Fig. 10 shows a "box" scheme, where the estimated carbon reservoirs and fluxes are indicated. Models of this type are used to investigate and assess the atmospheric CO_2 concentration increase.

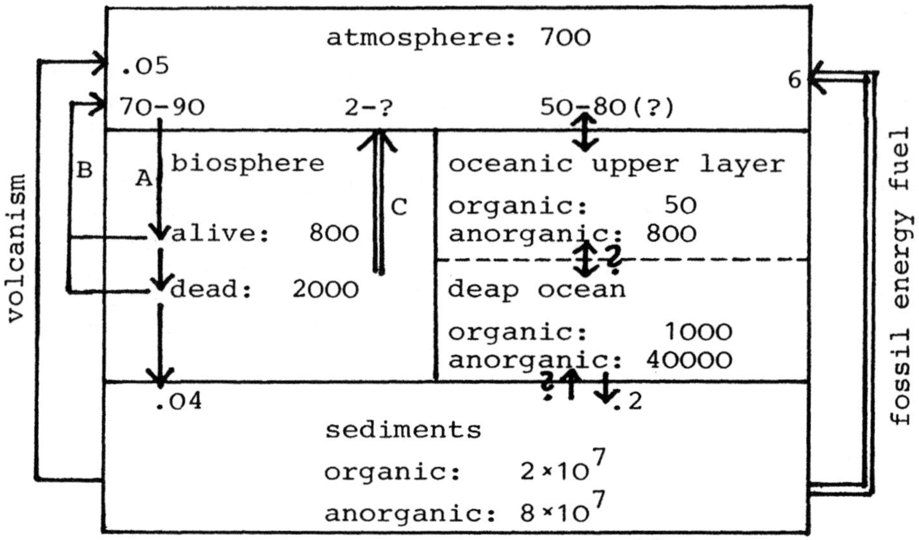

Figure 10 Carbon reservoirs in 10^{15} g and fluxes in 10^{15} g per year, after (2), summarized. The double lines indicate the anthropogenic impact. A = photosynthesis, B = rotting, C = biomass burning and reduction.

The CO_2 production by fossil fuel and cement production in 1980 was c. 6 x 10^9 t (= 6x 10^{15}g) carbon. Neglecting the biosphere for the present, the corresponding CO_2 increase 1958-1978 should be 36 ppm (36), but it was actually 20 ppm. That means, the so-called airborne fraction (atmospheric CO_2 increase divided by CO_2 input to the atmosphere) was only 0.56. The "missing sink" of CO_2 is supposed to be the ocean, because there is an intensive gas exchange between the atmosphere and the upper several hundred meters of the ocean, the oceanic mixed layer.

In another quick acting carbon cycle the biosphere is involved. (Slow acting cycles considering the deep ocean and the sediments are not able to explain effects on a year-to-year scale.) The photosynthesis of the vegetation, concentrated on the northern hemispheric continents, causes a CO_2 decrease during spring and summer, see Fig. 9; during the rottening period in autumn and winter CO_2 is produced which causes an atmospheric CO_2 increase. The anthropogenic deforestation, especially in the tropical regions, the change of lush overgrown to poor monocultur soil and biomass burning causes additional indirect and direct CO_2 input to the atmosphere. The related dramatic assessments of earlier years are now substituted, however, by more realistic comparably small numbers. The realistic modelling of the carbon cycle Kohlmeier is dealing with in this volume is a very hard to solve problem. One of the most recent model results by Viecelli et al. (51) which fits the observations very well is shown in Fig. 11. In contrast to the earlier models this "box model" considers different latitude boxes, this means proceeding from OD to 1D models, and includes upwelling phenomena of deep ocean water. Using one of the energy consumption scenarios ("fossil carbon source"), this model predicts roughly a doubling of the pre-industrial CO_2 level towards the year 2100 and afterwards a slow decrease of this level.

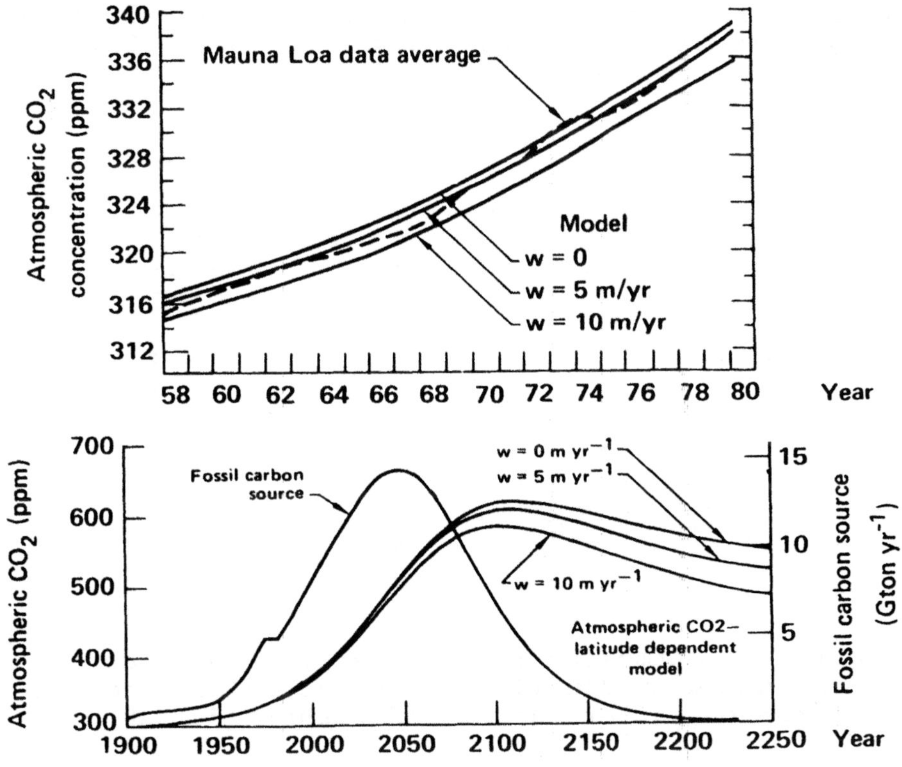

Figure 11 Mauna Loa measurements and model calculations using different oceanic upwelling rates w. Below fossil carbon source based on a consumption scenario and CO_2 predictions; after Viecelli et al. (51), simplified.

The response of the climatic system to an increased atmospheric CO_2 concentration level is a warming of the lower troposphere due to absorption and reflection of IR (infra-red) terrestrial radiation bands, because the short wave solar irradiation is not changed (so-called green-house effect). In compensation, a higher layer of the atmosphere is cooled.

In 1980 Gates has summarized the model calculations which estimate the near surface temperature response of a CO_2 doubling (15). These calculations predict a globally averaged temperature increase between .8 and 4.0°C, where 85% of the models predict a value in the range 1.5-2.6°C. The 3D models, however, tend to higher values, most of them to 2.9-4.0°C. In Fig. 12 the results of such a sophisticated 3D model by Manabe and Wetherald (1980) are presented in respect to latitude and in form of a vertical cross section, where in altitudes higher than c. 15 km a cooling is obtained.

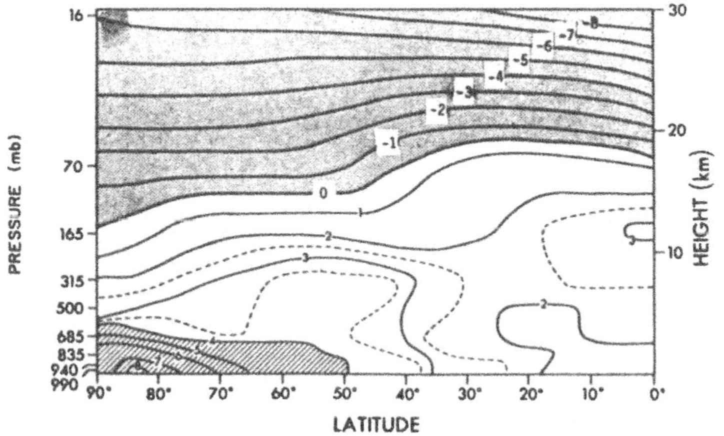

Figure 12 Atmospheric temperature deviation (in °C) effected by doubling the actual CO_2 concentration, after Manabe and Wetherald (28).

It should not be disregarded that some other anthropogenic gases are also absorbers in the IR part of the electromagnetic spectrum.

In Table 3 the most important of these gases are listed up, where NO_2 is produced by fertilizers, the C-F-Cl gases (freones) by the use of spray boxes and the others mainly by industrial activity. The most effective gases in the climatic sense are those which are well mixed-up in the atmosphere and characterized by long residence times in atmosphere. It can be read from Table 3, that these are CO_2, N_2O and the C-F-Cl gases.

Table 3 Anthropogenic trace gases and their estimated global warming effect near surface (= effect) if the actual concentration would be doubled. In abbreviated form taken from (2). Conc. = actual tropospheric concentration, IR abs. = centres of the infra-red absorption bands, res. t. = tropospheric residence time.

gas	conc.	IR abs.	res. t.	effect
CO_2	335 ppm	4.5/14.7 μm	2-10 a	2-4 °C
NO_2	.3 ppm	5.4 μm	20-50 a	.4-.7 °C
NH_3	6 ppb	10.5 μm	10 d	.1 °C
HNO_3	1-10 ppb	7.5/11.3 μm	10 d	.07 °C
SO_2	0-20 ppb	7.3 μm	10 d	.03 °C
$CFCl_3$		9.2/11.8 μm		
CF_2Cl_2	.2-.3 ppb	8.7/9.1/10.9μm	50-100 a	.05 °C

a = year, d = day, ppm = parts per million, ppb = parts per billion (volume).

The temperature effect of the anthropogen absorbers is clearly smaller than that of CO_2, but an combined (non-linear!) effect up to c.50% of the CO_2 effect has been supposed (12,52). This concerns especially NO_2, because the utilization of fertilizers in agriculture has rapidly grown in the last decades (USA from c. 2 x 10^6t in 1920 up to c. 20 x 10^6t in 1980 (38)). Furthermore, the increasing tropospheric ozone may warm the lower troposphere by an additional amount of .5 to 1 °C (54).

To conclude, the following can be said: The time scale which is of the greatest interest in the economic and social life of mankind, is in addition to that of a few days (weather prediction) that of seasons, years and decades, perhaps up to the 100 year scale. Model calculations considering this time scale estimate the natural globally averaged temperature variability of the solar activity signal of c. .2°C (35) and of the volcanic activity signal of .24°C (49). If this is compared with the Central England temperature correlation analysis as described above and if the increasing CO_2 data are involved in this analysis it is possible to explain 84% of the long-term ($T > 30$ years) variability and the CO_2 signal is statistically assessed to be in the order of .1°C.

This is small when compared with the observed northern hemisphere temperature trends, see Figure 6, where an increase of c. .5°C from 1881 to 1940 and a joined decrease of c. .3°C is observed. Just in this 100 year time scale, however, it is not impossible that the anthropogenic signal gets by and by an order of magnitude similar to the natural climatic forcing factors.

The most striking facts of climatological science, however, are the great uncertainties in data recording, statistical analysis of these data and deterministic modelling to explain all the data variability. Therefore, an intensified investigation is needed to proceed significantly in extending the knowledge of climatic change and in diminishing the uncertainties.

REFERENCES

1. Båth, M.: 1974, Spectral Analysis in Geophysics, Elsevier, Amsterdam Oxford New York.
2. Battelle Institute: 1981, Die Auswirkungen von CO_2-Emissionen auf das Klima, Vol. 1-3, BF-R-63.818-1,2,3, Frankfurt.
3. Berger, A.L.: 1981, in 4, pp. 501-525.
4. Berger, A.L. (ed.): 1981, Climatic Variations and Variability: Facts and Theories, Reidel, Dordrecht Boston London.
5. Box, G.E.P., and Jenkins, G.M.: 1970, Time Series Analysis, Holden-Day, San Francisco etc.
6. Bray, J.R.: 1974, Nature 252, pp. 679-680.
7. Budyko, M.I.: 1977, Tellus 29, pp. 193-204.
8. Dansgaard, W.: 1969, Science 166, pp. 377-381.

9. Duplessy, J.-C.: 1978, in 16, pp. 46-67.
10. Flint, R.F.: 1971, Glacial Geology and Quaternary Geology, Wiley, New York.
11. Flohn, H.: 1959, Arch. Met. Geoph. Biokl. Ser.B, pp. 1-13.
12. Flohn, H.: 1980, Possible Climatic Consequences of a Man-made Global Warning, IIASA RR-80-30, Laxenburg (Austria).
13. Flohn, H.: 1981, Geolog. Rdsch. 70, pp. 725-736.
14. Frenzel, B.: 1967, Die Klimaschwankungen des Eiszeitalters, Vieweg, Braunschweig (FRG).
15. Gates, W.L.: 1980, in Bach, W., et al. (eds.): Interactions of Energy and Climate, Reidel, Dordrecht Boston London, pp. 169-190.
16. Gribbin, J. (ed.): 1978, Climatic Change, Univ. Press, Cambridge etc.
17. Ikeda, S., et al. (eds.): 1980, Statistical Climatology, Elsevier, Amsterdam Oxford New York.
18. Johnsen, S.J., Dansgaard, W., et al.: 1972, Nature 235, pp. 429-434.
19. Internat. Meteorol. Org.: 1936, Sekr., Publ. No. 21, Utrecht (Netherlands).
20. Hasselmann, K., 1981, in 4, pp. 481-497.
21. Keeling, C.D., et al.: 1976, Tellus 28, pp. 538-551; updated after 2.
22. Kennet, J. P., and Thunell, R.C.: 1977, Science 196, pp. 1231-1234.
23. Kutzbach, J. E., and Bryson, R.A.: 1974, J. Atm. Sci. 31, pp. 1958-1963.
24. Lamb, H.H.: 1970, Philos.Trans. Roy, Soc. A 266, pp. 425-553.
25. Lamb, H. H.: 1972 and 1977, Climate: Present, Past and Future, Vol. 1 and 2, Methuen, London.
26. Landsberg, H.E., and Flohn, H. (eds.): World Survey of Climatology, Vol. 2, Elsevier, Amsterdam London New York.
27. Lorius, C., et al.: 1979, Nature 280, pp. 644-648.
28. Manabe, S., and Wetherald, R.T.: 1980, J. Atm. Sci. 37, pp. 99-118.
29. Manley, G.: 1974, Quat. J. Roy. Meteorol. Soc. 100, pp. 389-405.
30. Milankovitch, M.: 1920, Théorie mathématique des phénomènes termiques produit par la radiation solaire, Gauthier-Villars, Paris.
31. Mitchell, J. M., et al.: Climatic Change. WMO Tech. Note No. 79, Geneva.
32. Mitchell, J.M.: 1976, Quatern. Res. 6, pp. 481-493.

33. Oeschger, H. et al. (eds.): 1980, Das Klima, Springer, Berlin Heidelberg New York.
34. Panofsky, H.A., and Brier, G.W.: 1958, Some Applications of Statistics to Meteorology, Pennsylv. State Univ., University Park.
35. Pollack, J. B., et al.: 1979, Nature 282, pp. 600-603.
36. Rotty, R.B.: 1980, Experientia 36, pp. 781-783
37. Salinger, M. J.: 1979, Climatic Change 2, pp. 109-126.
38. Schneider, S. H.: 1976, The Genesis Strategy, Plenum Press, New York.
39. Schönwiese,C.D.: 1978, Beitr. Phys. Atm. 51, pp. 139-152.
40. Schönwiese, C.D.: 1978,Arch. Met. Geoph. Biokl. Ser.B. 26, pp. 1-16.
41. Schönwiese, C.D.: 1978, Arch. Met. Geoph. Biokl. Ser. B 26, pp. 105-120.
42. Schönwiese, C.D.: 1979, Klimaschwankungen, Springer, Berlin Heidelberg New York.
43. Schönwiese, C.D.: 1981, Arch.Met.Geoph.Biokl.Ser. B. 30, pp. 1-22.
44. Schwarzbach, M.: 1974, Das Klima der Vorzeit, Enke, Stuttgart.
45. Siegenthaler, U. and Oeschger, H.: Science 199, pp. 338-395.
46. Singer, S. F.(ed.): 1973, Global Effects of Environmental Pollution, Reidel, Dordrecht (Netherlands).
47. Spiegel, M.R.: Statistics, McGraw-Hill, New York etc.
48. Tarling, D.H.: 1978, in 16, pp. 3-24.
49. Taylor, B.L., et al.: 1980, Quart. J. Roy. Meteorol. Soc. 106, pp. 175-199.
50. U.S. GARP Comm.: 1975, Understanding Climatic Change, Nat.Ac.Sci., Washington.
51. Viecelli, J.A., et al.: 1981, Climatic Change 3, pp. 281-302.
52. Wang, W.C., et al.: 1976, Science 194, pp. 685-690.
53. World Meteorl. Org.: 1979, Proceedings of the World Climate Conference, WMO Publ. No. 537, Geneva.

54. Crutzen, P.: Comment during the session (1981)and Fishman, J., Crutzen, P., et al.: 1979, Nature 282, pp. 818-820.
55. Mitchell, J.M.: Private communication (1977).

THE ROLE OF THE TERRESTRIAL VEGETATION IN THE GLOBAL CARBON CYCLE

Stein, Norbert

Chair of Biogeography, University of Saarland,
6600 Saarbruecken, Fed. Rep. Germany

ABSTRACT

The land biota have an important role in the global carbon cycle because of their prominent position concerning a large carbon storage pool and simultaneously concerning a fast carbon exchange flux with the atmosphere and aquatic systems. In addition they can cause considerable net transfers between these reservoirs due to man-induced alterations of the vegetation cover. There are some contradictions in our present knowledge about these transfers and about carbon storage in the terrestrial vegetation of the earth.

After careful consideration of all known facts this paper comes to the conclusion that 2×10^{15}g/yr C are released to the atmosphere due to deforestation and burning in the tropical rain forest belt. The same amount is probably being fixed in the temperate forests and the northern ecosystems. A possible steady balance of all terrestrial ecosystems would, therefore, be the result of a particular historical constellation whereby a destruction of vegetation in the tropics would occur simultaneously to an increase of phytomass and also to a humus and peat accumulation in temperate and northern ecosystems.

INTRODUCTION

In the last few years there has been a great number of research activities, conferences and publications on the global carbon issue. A great percentage of these works with corresponding public discussion has to be seen against the background of hypothetic negative climate change due to an anthropogenically

H. W. Georgii and W. Jaeschke (eds.), Chemistry of the Unpolluted and Polluted Troposphere, 185–202.

increased CO_2 - content in the atmosphere. In conjunction with this, scenarios, based on reputed estimations of the future use of fossil fuels and on the calculated temperature increases, have been interpreted as an actual forecast for a climatic catastrophe for the next century.

In the meantime, however, these types of forecast are now being made more cautiously. Climate is a continuous process with numerous reactions in the interacting system atmosphere-hydrosphere (particularly the oceans) - cryosphere-lithosphere-biosphere. A potential anthropogenically caused climatic change can only be assessed if all these compartments will be taken into consideration.

It is important to emphasize that these climate compartments are simultaneously carbon reservoirs and that carbon on earth is present in different forms and in different quantities in these compartments: as CO_2 in the atmosphere, as carbonates in the earth's crust, as bicarbonate (HCO_3^-) and carbonate (CO_3^{2-}) ions in the oceans, and in the many organic compounds of terrestrial ecosystems. There are permanent metabolisms and flows of these carbon bonds between the reservoirs (Fig. 1).

As BOLIN and others (5) state, the biosphere, as it exists today, has evolved in a complex interplay between carbon and many other elements: primarily hydrogen, oxygen, the basic nutrient elements nitrogen, phosphorus, sulphur, and some metals in minor quantities that are fundamental to the development of life. Therefore the pools and fluxes of carbon in terrestrial biota form important links in the total biogeochemical cycle.

The role of terrestrial ecosystems for the global carbon balance becomes particularly important by the following facts:

1) Primary productivity of terrestrial vegetation as a basic process of ecosystem functioning and herewith as one of the fundamental principles for life requires a permanent and efficient uptake of atmospheric CO_2. Respiration, on the other side, signifies a release of CO_2 from the vegetation to the atmosphere again. Burning of vegetation is an other source for atmospheric CO_2. Fluxes of carbon between atmosphere and vegetation are necessarily fast, in contrast to the slow geological flows.

2) Carbon can be stored in the living phytomass for considerable periods of time. The length of time as well as the intensity of fluxes is dependant on the type of vegetation itself. Even larger quantities of carbon may be stored more effectively and for longer periods of time in the soils beneath the living vegetation body itself, depending on the climatic zone

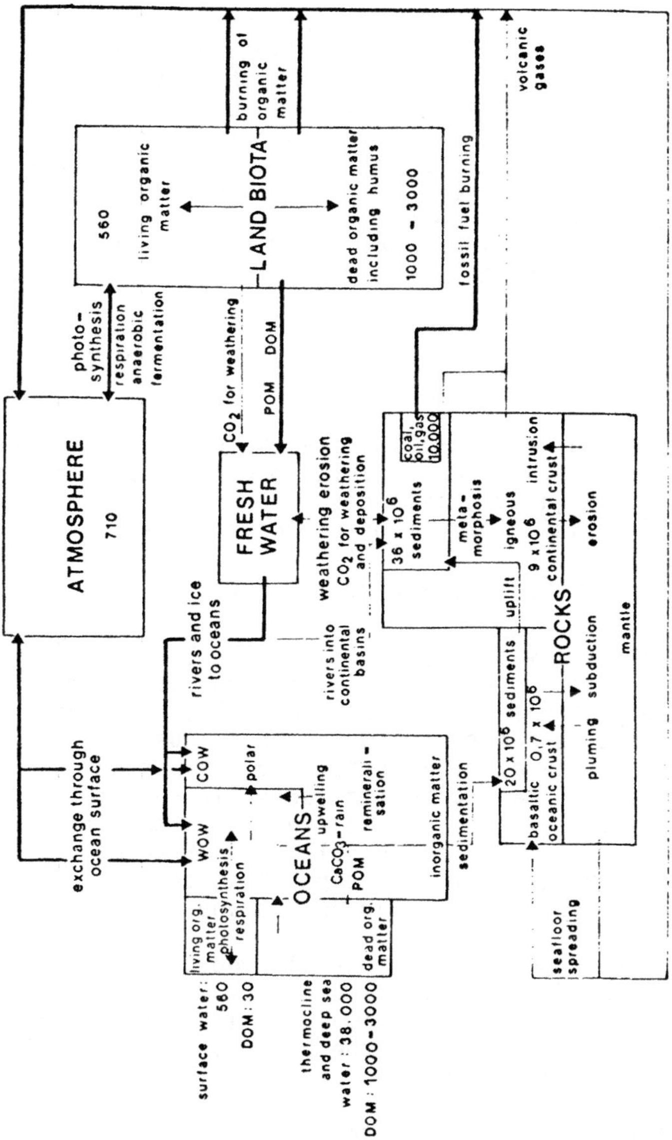

Fig. 1: Principal reservoirs and fluxes in the carbon cycle (WOW = warm ocean water; COW = cold ocean water; POM = particulate organic matter; DOM = dissolved organic matter); after (5); values for the reservoirs from various authors and from own calculations (thick lines signify potential man-induced carbon transfers)

with its respective decomposition characteristics.

3) Man performs important impacts of various intensities on the vegetation of the earth and may thus create significant net transfers between land biota, atmosphere and probably hydrosphere.

There are some contradictions in our present knowledge about the role of the terrestrial vegetation within the global carbon cycle. There is the more traditional assumption that terrestrial ecosystems are accumulating CO_2 from the atmosphere and hence are to be considered as a sink for atmospheric CO_2. More recent analyses suggest that terrestrial ecosystems are an additional source of CO_2. Because of the important role of carbon in terrestrial biota and because of these contradictions mentioned above a re-examination of the vegetation within the global carbon budget is of utmost importance and justifies a careful consideration of all known facts with particular emphasis on some new results.

MAGNITUDE OF PRIMARY PRODUCTION AND CARBON STORAGE IN DIFFERENT TERRESTRIAL ECOSYSTEMS

The vegetation - both terrestrial and aquatic - is the only natural compartment on earth which is able to convert anorganic material (H_2O, CO_2) to organic substances through the process of photosynthesis. The basic formula of photosynthesis is:

$$6\ CO_2 + 12\ H_2O \xrightarrow[2827\ kJ]{PHAR} C_6H_{12}O_6 + 6\ O_2 + 6\ H_2O$$

Whereas the availability of CO_2 for plants is comparatively constant in terms of geographical distribution the other requirements for photosynthesis - H_2O and PHAR (= Photosynthetically Active Radiation, as a distinct wave length spectrum of the solar radiation) - show a characteristic distribution over the earth according to the main climatic zones. Terrestrial primary productivity is therefore very different in the respective climatic zones and is thus responsible for different vegetation types. Each of these vegetation types has its specific phytomass and herewith also its own carbon storage capacity.

The magnitude of the world's terrestrial production and phytomass can be estimated in two different ways, either by classifying the biosphere into ecosystem types and estimating averages and total values for each of these, or by modelling the effects of environmental factors on productivity and phytomass and integrating the results of the model for the earth surfaces (1). Table 1 is derived from the first approach. Details on the second approach can be found in (11, 12).

	Vegetation type	Area 10^{12} m²	Net-Primary Productivity (NPP)	
			g/m².a C (average)	10^{15} g/a C
1	Forests	31,3	700	21,91
2	Tropical rain forests	10,0	1 035	10,35
3	Tropical deciduous forests	4,5	720	3,24
4	Mangroves	0,3	450	0,14
5	Mixed deciduous forests of the temperate zone	3,0	675	2,02
6	Evergreen forests of the temperate zone	3,0	585	1,76
7	Boreal forests (closed)	6,5	338	2,49
8	Boreal forests (open; "forest-tundra")	2,5	293	0,73
9	Afforested areas	1,5	788	1,18
10	Shrubs of the temperate zone	2,0	675	1,35
11	Chapparal, etc.	2,5	360	0,90
12	Savannahs	22,5	787	17,71
13	Shrub savannah	6,0	945	5,67
14	Grasslands	6,0	1 035	6,21
15	Dry forests	3,5	585	2,05
16	Dry thornshrub savannah	7,0	540	3,78
17	Grasslands of the temperate zone	12,5	351	4,39
18	Moist (Meadows, etc.)	5,0	540	2,70
19	Dry ("Steppe")	7,5	225	1,69
20	Tundra and alpine vegetation	9,5	100	0,95
21	Polar desert	1,5	11	0,02
22	High arctic/alpin	3,6	68	0,24
23	Less extreme	4,4	158	0,69
24	Deserts and semi-deserts	21,0	64	1,35
25	with dry vegetation	9,0	90	0,81
26	degraded	12,0	45	0,54
27	Extreme deserts	9,0	7	0,06
28	Hot	8,0	5	0,04
29	Cold	1,0	23	0,02
30	Permanent ice-covered areas	15,5	0	0,00
31	Fresh water systems	2,0	180	0,36
32	Swamps	2,0	1 630	3,26
33	Temperate zone	0,5	1 125	0,56
34	Tropics	1,5	1 800	2,70
35	Bogs	1,5	450	0,68
36	Cultivated lands	16,0	423	6,77
37	Temperate zone, annual plants	6,0	540	3,24
38	Temperate zone, perennual plants	0,5	675	0,34
39	Tropics, annual plants	9,0	315	2,83
40	Tropics, perennual plants	0,5	720	0,36
41	Built-up areas, etc.	2,0	225	0,18
42	Total	149,3	401	59,87 ≅ 60,00

Table 1: Net primary productivity of the more important vegetation types of the earth (after 1).

It must be emphasized that productivity or production rates and carbon storage in the same vegetation are not necessarily correlated (Tables 1 and 2). Net primary production (NPP = gross primary production or apparent photosynthesis minus dark respiration of the autotrophic organisms) is primarily a function of the availability of favourable growth conditions, especially as PHAR, water, and temperature is concerned. Where these favourable conditions are given NPP is high, independant from the vegetation type. This becomes important when successional phases produce a high NPP and simultaneously accumulate carbon for longer periods of time.

As Table 2 shows, the total carbon stored in the various subcomponents of the earth's vegetation as a whole amounts to 2076 x 10^{15}g C. However, this figure may change according to the classification scheme used for subdividing the biosphere in biomes. The phytomass estimates vary between 450 x 10^{15}g and 1080 x 10^{15}g C (1). According to our own estimates based on the studies of deforestation of tropical rain forests the figure of 560 to 590 x 10^{15}g C for the earth's vegetation must be considered as realistic. A further uncertainty consists in the determination of the amount of humus. Estimates vary between 1000 x 10^{15}g C and 3000 x 10^{15}g C. It is therefore possible that the above figure of 2076 x 10^{15}g C for carbon in the whole vegetation system may be even higher.

As can be seen in Table 2, forests, among the standing phytomass, contribute the highest share of the carbon stored (= 450,8 x 10^{15}g).

As stated above, for an efficient photosynthesis plants require a certain level of irradiance, sufficient water supply and favourable temperatures. Consequently the day-length and the duration of the productivity period in the year determine to a high extent the amount of matter accumulation. Trees of the tropical rain forest are photosynthetically active throughout the year. Therefore, they reach higher yields. The productivity period of deciduous trees in the temperate zones is limited by vegetation periods or phenologically determined seasons of the year. Evergreen conifers in the boreal zone achieve an efficient photosynthetic activity only during a short summer period because in these latitudes PHAR and particularly temperature are the very limiting factors for a considerable time of the year.

Therefore, in conjunction with the assessment of the world's vegetation for the global carbon issue it is of the utmost importance to judge the forests of the world according to their photosynthetically determined productivity and according to the carbon storage capacity of the vegetation body itself

Vegetation formation	Area $10^6 km^2$	Living and dead phytomass $10^3 g/m^2$ C	$10^{15} g$ C	Litter $g/m^2 C$	$10^{15} gC$	Humus $10^3 g/m^2 C$	$10^{15} gC$	Total $10^3 g/m^2 C$	$10^{15} gC$
1 Tropical lowland rainforest	8,3	19,85	164,7	325	2,7	9,8	81,3	29,96	248,7
2 Tropical mountain forests	2,5	18,24	45,6	1500	3,8	28,7	71,8	48,48	121,2
3 Tropical deciduous forests	4,0	11,81	47,3	425	1,7	9,8	39,2	22,05	88,2
4 Forests of the temperate zone	7,5	12,85	96,4	1500	11,3	13,4	100,5	27,76	208,2
5 Forests of the boreal zone	9,0	10,76	96,8	1750	15,8	20,6	185,4	33,11	298,0
6 Shrub lands	8,0	6,26	50,1	467	3,7	6,9	55,2	13,63	109,0
7 Tropical Savannahs	19,0	2,32	44,1	175	3,3	4,2	79,8	6,69	127,2
8 Grass lands of the temperate zone	12,5	0,77	9,6	198	2,5	18,9	236,3	19,87	248,4
9 Tundra and alpine areas	8,0	0,76	9,1	1253	10,0	20,4	163,2	22,41	179,3
10 Semideserts	21,0	0,37	7,7	50	1,1	5,8	121,8	6,22	130,6
11 Deserts	10,5	0,04	0,5	8	0,1	0,17	1,8	0,23	2,4
12 Rocks and ice covers	15,5	0,00	0,0	0	0,0	0,0	0,0	0,00	0,0
13 Agricultural lands	18,0	0,38	6,9	39	0,7	6,8	122,0	7,22	129,6
14 Swamps and bogs	3,5	4,56	15,9	1250	4,4	47,1	165,0	52,94	185,3
15 Land masses of the earth	147,3	4,02	591,7	413	60,9	9,66	1423,2	14,09	2076,0
16 Forests	31,3	14,40	450,8	1128	35,3	15,28	478,2	30,81	964,5
17 Others	116,0	1,21	140,9	221	25,6	8,15	945,0	9,58	1111,5
18 Low latitudes, forests	14,8	17,41	257,6	554	8,2	13,00	192,3	30,95	458,1
19 Low latitudes, others	34,5	2,41	83,0	179	6,2	6,11	210,9	8,70	300,1
20 Low latitudes, total	49,3	6,91	340,6	292	14,4	8,18	403,2	15,38	758,2
21 Temperate zone, forests	7,5	12,85	96,4	1500	11,3	13,40	100,5	27,76	208,2
22 Temperate zone, others	25,0	1,60	40,0	256	6,4	13,69	342,3	15,55	388,7
23 Temperate zone, total	32,5	4,20	136,4	548	17,7	13,62	442,8	18,37	596,9
24 Boreal zone, forests	9,0	10,76	96,8	1750	15,8	20,60	185,4	33,11	298,0
25 Boreal zone, others	9,5	1,02	9,7	1242	11,8	28,23	268,2	30,49	289,7
26 Boreal zone, total	18,5	5,76	106,5	1492	27,6	24,52	453,6	31,77	587,7
27 Areas with scarce or missing vegetation	47,0	0,17	8,2	26	1,2	2,63	123,6	2,83	133,0

Tab. 2: The content of organic carbon in different vegetation formations of the earth (after 7)

and the soil under the vegetation.

Summarising our considerations so far the role of the vegetation in the global carbon cycle consists of the following characteristics: (1) Matter production as a result of photosynthesis, that is - as least as matter production is increasing - a net uptake of CO_2 from the atmosphere; (2) storage of carbon in phytomass and soil; (3) permanent transfers between vegetation and atmosphere which can be considered as relatively fast (= fast biological transfers in contrast to the slow geological transfers).

The magnitude and intensity of these processes are dependant on the climatic zones of the earth and, therefore, can be assessed most efficiently by considering the more important belts of forests as shown in Table 2: the tropical rain forest of the low latitudes, the deciduous forests of the temperate zone, the - mostly coniferous - forests of the boreal zone (including the tundra zone on the poleward side).

The Tropical Rain Forest

The role of tropical forest ecosystems in relation to the global carbon issue is associated with the following aspects: (1) their area; (2) their net primary productivity; (3) the amount of organic matter stored in them; (4) the rate of decomposition; (5) the rate of transport out of the system through rivers. All these aspects become very important for the assessment of potential net transfers between atmosphere and biosphere, especially in the case of the unfavourable boundary condition of a large-scale destruction of tropical rain forests.

The first aspects can be assessed in Tables 1 and 2. It can be concluded that tropical rain forests in their natural state have a high net primary productivity and store a great amount of carbon in large areas. The following facts can be derived in accordance to all data available (see a summary in 7):

1. Organic matter accumulates quickly and in some successional stands it takes about 20 years for total phytomass to be as high or almost as high as that of nearly mature forest stands.

2. Litter accumulates quickly in young stands and remains fairly constant as the stands mature.

3. Soil organic matter production is low and not likely to accumulate with time.

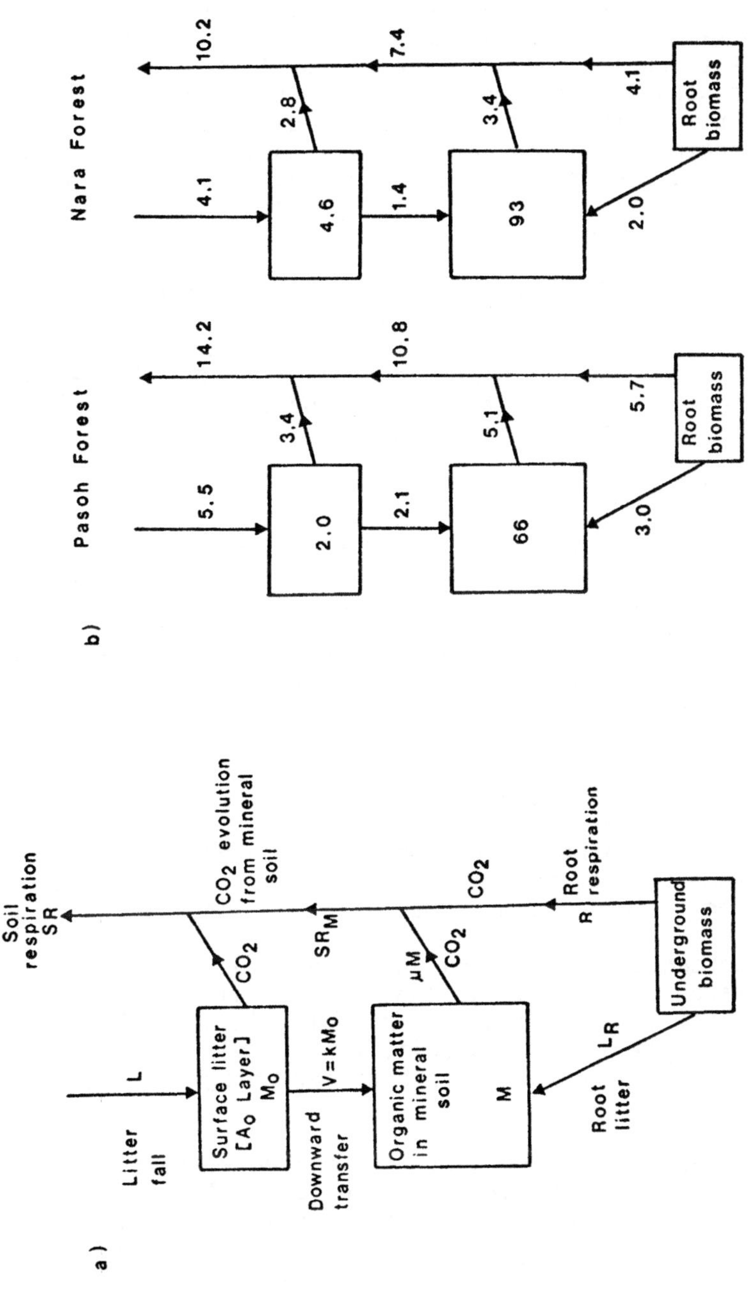

Fig. 2: Carbon cycle in forest soils; a) Model of compartments and fluxes; b) Cycling in soils of Pasoh Forest (W-Malaysia) and of Nara Forest (Central Japan); after (9)

The rate of litterfall is considerably less than in temperate or even boreal ecosystems. In natural ecosystems the input of litter to the forest floor is balanced by losses, either through biotic processes such as microbial decomposition or by physical processes such as leaching. The type of decomposition conditions exerts a greater effect on carbon fixation in the soil than the magnitude of matter production. Therefore, the humic tropics have despite of high matter production only a low accumulation of organic substances in the soil.

A good example for carbon cycling in soils for tropical and temperate forests has been given by (9, Fig. 2). Both forests receive sufficient rainfall, but the annual mean temperature is higher in the tropics (25°C at Pasoh, W-Malaysia) than at Nara (13,5°C, Central Japan). With the rise of temperature, the flows increase and the pools diminish in size.

The Temperate Forests

The temperate forests as a further biotic carbon pool consist mainly of deciduous, mixed coniferous-deciduous and coniferous forest. Most of this vegetation exists in the northern temperate zone between 30° and 60° N latitude, a band of the earth which includes much cultivated land derived from forest areas (2). It is therefore justified to consider this zone within the chapter on anthropogenically caused net transfers between atmosphere and vegetation.

The Boreal and Tundra Ecosystems

The amount of carbon stored in boreal forests is of the same magnitude as that of temperate forests. Table 2 shows that additional storage compartments become important.

The importance of the boreal and tundra ecosystems together is in the large carbon storage pool of the soils. According to (16) three soil orders dominate: Entisols, Inceptisols, and Histosols. Inceptisols and especially histosols have thick surface organic layers containing a high percentage of organic carbon (up to 40%).

Miller (16) could prove that long carbon residence times are valid for tundra ecosystems. Carbon accumulation estimated by assuming that the carbon standing crop is in a steady state, i.e., all net primary production is decomposed indicate turnover rates of about 0,01 gg^{-1} for tundra and 0,001 $gg^{-1}yr^{-1}$ for bogs, indicating that organic carbon takes about 500 years in tundra and 5000 years in bogs to reach 99% of decay.

Miller (16) summarizes the implications of Table 3: (1)

	Standing crop (kg C/m²)	Loss (g C m^{-2} yr^{-1})	Primary production (g C m^{-2} yr^{-1})	Season length (days)	Carbon loss as % of primary production	Carbon accumulation as % of primary production	k, production/std. crp. (g g^{-1} yr^{-1})	Turnover time (year)	k, loss/std. crp. (g g^{-1} yr^{-1})	Turnover time (year)
United States: Point Barrow										
Wet meadow	32	159	--	--						
		37								
		38								
		210								
		13								
		22								
Mean	32	80	130	85	61	39	0.0041	246	0.0025	400
Canada: Devon Island										
Beach ridge	2	--	12.5	--	--	--	0.0063	160	--	--
Wet meadow	14	--	90	--	--	--	0.0064	156	--	--
England: Moorhouse										
Blanket bog										
Sphagnum pools		61	197	--	31	--	--	--	--	--
Sphagnum lawns		35	153	--	23	--	--	--	--	--
Sphagnum hummocks		51	98	--	52	--	--	--	--	--
Mean	83	49	149	--	35	65	0.0039	259	0.0006	1694
Ireland: Glenamoy										
Bog	200	--	156	--	--	--	0.0008	1282	--	--
Norway: Hardangervidda										
Lichen heath	4.5	94	114	140	82	18	0.025	39	0.021	48
Snow bed	10.1	177	--	--	--	--	--	--	0.018	57
Wet meadow	25.7	253	351	133	72	28	0.014	73	0.010	102
Dry meadow	16.8	457	225	140	203	-103	0.013	75	0.027	37
Sweden: Stordalen										
Mire										
Ombrotrophic mounds	--	50	--	134	--	--	--	--	--	--
Ombrotrophic/minerotrophic	--	14	--	--	--	--	--	--	--	--
Minerotrophic depressions	--	16	--	134	--	--	--	--	--	--
Mean	80	27	72		37	63	0.0009	1111	0.0003	2963
USSR (no data)										
Antarctic: Signy Island										
Peat	23	240					0.010	96		
Means										
Mean for tundra		221			105	-4(28)				
Mean for bogs		143			36	64				
Mean for all sites		160			82	18				

TABLE 3: Summary of carbon standing crop, primary production, gaseous carbon loss rate, season lengths, and derived statistics of net carbon accumulation rates as a percentage of primary production and of turnover rates and residence tissues calculated assuming steady state conditions on primary production/standing crop and decomposition/standing crop (after 3).

large amounts of organic carbon accumulate, even though primary production is low, because of the very slow decay rates; (2) any change in primary production takes hundreds of years to work through the decomposition system; (3) any change in the environmental conditions within the soil acts on the decomposition of a large carbon capital which has accumulated over centuries.

THE ROLE OF RIVERS

The ideal steady state of carbon cycling between terrestrial vegetation and the atmosphere may rarely exist on a regional basis, even in mature and undisturbed ecosystems. All ecosystems are open systems and thus can never be expected to be perfectly balanced. The transport of particulate and dissolved carbon by rivers signifies a transfer from one reservoir (= terrestrial ecosystem) to another (= marine ecosystem) even though a changing percentage of the total organic carbon carried by a river may be stored in river sediments or on flood plains or may be en route oxidized.

Meybeck (15) estimated that the total organic carbon load from rivers to oceans is approximately 400 x 10^{12}g C per year. Table 4 shows the input of total organic carbon from rivers to oceans on a worldwide basis. The tropical zone contributes 63% of the organic load. The current estimates of organic export into the oceans suggest that rivers already in undisturbed terrestrial drainage basins may be significant in the global carbon cycle, particularly with regard to increased atmospheric CO_2.

ANTHROPOGENICALLY INDUCED NET TRANSFERS OF CARBON BETWEEN TERRESTRIAL VEGETATION AND ATMOSPHERE

It is a well documented fact that man is able to cause net transfers of carbon between terrestrial vegetation and atmosphere. But there is a great uncertainty about the magnitude of these transfers and also about the regional differences. Two basic possibilities of man-induced net transfer occur: (1) a transfer from the vegetation to the atmosphere as a result of an irreversible destruction of large vegetation areas or as a result of enhanced oxidation of humus in conjunction with vegetation degradation or land use changes; (2) a transfer from the atmosphere to the vegetation through re-afforestions and enlargement of forest areas.

According to the basic principle that photosynthesis, net primary productivity and carbon storage can only be properly judged by considering the different climatic zones the assessment of net transfers has to follow the same line.

A. Studied rivers	Area ($10^6 km^2$)	TOC_{exp} ($t.km^{-2}.year^{-1}$)	Annual Load ($10^{12} gC$)
Siberian rivers	12.6	2.25	28.3
Nelson	1.15	0.62	0.7
MacKenzie	1.8	1.93	3.5
St. Lawrence	1.03	0.81	0.83
Mississippi	3.27	1.23	4.0
Amazon	6.3	8.7	54.8
Murray	1.1	0.28	0.3
Total known sample	27.25	3.4	92.4

B. Extrapolated sample

	Remaining Area ($10^5 km^2$)	Typical TOC_{exp} ($t.km^{-2}.year^{-1}$)	Annual Load ($10^{12} g$ C)
Tundra	4.4	0.6	8.6
Taiga	3.45	2.5	2.6
Temperate	17.7	5	88.5
Tropical	31.0	6	186
Semiarid	14.4	0.3	4.3
Desert	1.7	0.0	0
Total extrapolated sample	72.65	4.0	290

C. Total river discharge

	Total Area ($10^6 km^2$)	Total TOC Load ($10^{12} gC.year^{-1}$)	% TOC Load
Tundra	7.55	4.5	1.2
Taiga	15.85	39.6	10.3
Temperate	22.0	93	24.3
Tropical	37.3	241	62.9
Semiarid	15.5	4.6	1.2
Desert	1.7	0	0
Total	99.9	383	

TABLE 4: Input of Total Organic Carbon from Rivers to Oceans (after 15)

The Tropical Rain Forests

The observed destructions of tropical rain forests in the past and the rapidly expanding conversion of these forests to different land use systems are commonly believed to be a major source for the net transfer of carbon to the atmosphere. The different methods used by various authors to estimate the release of CO_2 from rain forest clearance are frequently not well documented. In order to avoid any misinterpretation it should be stated here that according to the existing observations - that is estimations of the real forest destructions through remote sensing methods and through analysis of reliable historical data - a net transfer of carbon in the humid tropics to the atmosphere appears to be very likely. Some thoughts, however, give reason to believe that this net transfer is not of a magnitude which is assumed by many authors.

There are different kinds of anthropogenic influence on the tropical rain forests, which may signify different degrees of impact on the carbon fluxes.

Shifting cultivation (= swidden agriculture) is a common practice among many indigenous population groups in the humid tropics. The various forms of this land-use system are well known and have often been described. In the course of this land-use system great quantities of dead wood remain on the ground. The decomposition rate of the dead tree trunks in the humid tropical climate is being estimated to be about 0,15-0,44g/m^2/day (18). Probably 40% of the original phytomass would remain as organically fixed carbon in the form of wood and organic soil components (14), provided that successional phases of vegetation development occur, as it is, indeed, the rule within the shifting cultivation system. This means that a temporary release of CO_2 into the atmosphere is possible but certainly no irreversible loss.

A characteristic phenomenon of the tropical shifting cultivation system lies in the fact that only certain areas are being used in the original forest. These areas are allowed to re-grow after a few years. The succession taking place after the period of agricultural use is characterized by rapid growth and, therefore, requires a high net photosynthesis, which ultimately means a net C-uptake from the atmosphere. Quantitative studies on this phenomenon are greatly lacking.

The really decisive problem is, of course, the deforestation of large areas. Since the main function of tropical rain forests within the carbon cycle is the long-term storage of carbon in the living phytomass, the potential land use after forest destruction appears to be the crucial factor.

If there is not a total deforestation and if areas are reserved with the possibility of regeneration, the effect of rapidly growing successional stages could again lead to a high CO_2-uptake and, therefore, also cause a comparatively high C-fixation.

If the area which has been deforested is converted to agricultural land use, than the particular type of agriculture is important. Although little is known about turnover-rates of organic matter in managed cultivation systems, in the case of tree cultivation such as rubber or oil palm plantations, carbon fixation will naturally be again the result of photosynthesis.

In the case of deforestations the rivers might even have a key position in the CO_2-issue. It can be expected that the total organic carbon will considerably increase, especially the particulate organic carbon (POC), including leaf litter, woody debris and soil organics. Since we know that this carbon will reach the oceans to a large extent we must expect a comparatively high net transfer of carbon between forest ecosystems and the oceans, especially for the most unfavourable boundary conditions, i.e. a large-scale deforestation.

The problem of the fate of organic carbon deposited by rivers in the coastal area is very complex and signifies a "research frontier" where detailed investigations should begin as soon as possible.

The conclusions gained by (20) fit to this approach. It appears realistic that 2 x 10^{15}g C/yr are released from the terrestrial biomass to the atmosphere, mainly in the tropics and due to burning. This figure is much lower than in many other estimates.

The Temperate Forests and Northern Ecosystems (Boreal Forests and Tundra)

In order to reach a more precise understanding of the role of the tropical rain forests in the global carbon cycle the importance of the other terrestrial ecosystems should also be considered. Loucks (13) stated that biospheric contributions to CO_2 in the atmosphere represent an asymmetry of opposed effects from 1860 or earlier until the present.

"Asymmetry of opposed effects" refers to the fact that CO_2-source elements, such as temperate forest releases, depletion of soil organic matter, wetland releases, and others, increased or peaked prior to World War II and then decreased, whereas other sources (such as releases from tropical forest conversion) began to increase after these other sources had begun to decrease.

A possible steady balance of all terrestrial ecosystems would, therefore, be the result of a particular historical constellation whereby a destruction of vegetation in the tropics would occur simultaneously to an increase of phytomass and also to a humus and peat accumulation in temperate and northern ecosystems.

It is realistic to assume that in the last 50 years carbon has accumulated particularly rapid in the temperate forests due to the following circumstances stated by (2): 1) the expansion of forests through abandonment of agriculture and by afforestation; 2) net regrowth of forest stands following cuttings in the late 19th and early 20th centuries (in North America); 3) ceasing of forest grazing by domestic animals over the last 50 years (in Europe); 4) decline of the use of wood for fuel (in North America and Europe); 5) systematic control of wildfires; 6) forest management leading to higher growth rates in certain areas.

Miller (16 and personal communication) stated that arctic and tundra regions would continue to be carbon sinks even in the case of a further rise of atmospheric CO_2 level and an assumed temperature increase.

Conclusions

By careful consideration of all known facts which have been dealt with in this paper, one may come to the conclusion that the problem of net transfer from the terrestrial vegetation to the atmosphere is not of that magnitude as frequently believed. One cannot exclude the fact that the tropics have a carbon loss of 2×10^{15}g/yr. If this figure proves to be true, and the perspectives of this article tend in this direction, the overall steady balance of the terrestrial ecosystems would be, indeed, possible. The amount of 2×10^{15}g/yr C could be compensated for by other terrestrial sinks. Even in the case that the release in the tropics would be higher than 2×10^{15}g/yr C the rivers, the flood plains and the delta areas could be an other effective carbon sink.

REFERENCES

1. Ajtay, G.L., P. Ketner and P. Duvigneaud (1979), Terrestrial primary production and phytomass. In: Bolin, B., E.T. Degens, S. Kempe and P. Ketner (eds.), *The Global Carbon Cycle*, Chichester/New York/Brisbane/Toronto, 129-181.
2. Armentano, Th.V. and J. Hett (eds.) (1980), *The role of temperate zone forests in the world carbon cycle-problem definition and research needs*. U.S. Department of Energy CONF-7903105, Washington, D.C.

3. Bliss, Webber, van Cleve, Viereck (1981), Current Carbon Accumulation Rates in Northern Ecosystems. In: (16).
4. Bolin, B. (1977), Changes of land biota and their importance for the carbon cycle. Science 196, 613-615.
5. Bolin, B., E.T. Degens, Kempe, S. and Ketner, P. (1979), *The global carbon cycle*, Chichester/New York/Brisbane/ Toronto.
6. Broecker, W.S., T. Takahashi, H.J. Simpson and T.H. Peng (1979), Fate of fossil fuel carbon dioxide and the global carbon budget. Science 206, 409-418.
7. Brown, S. (1980), Rates of organic matter accumulation and litter production in tropical forest ecosystems. In: Brown, S., Lugo, A.E. and Liegel, B. (eds.), *The Role of Tropical Forests on the World Carbon Cycle*. Gainesville, Florida, pp. 118-139.
8. Hampicke, U. (1980), *Die Rolle terrestrischer Ökosysteme im globalen Kohlenstoff-Kreislauf*. Münstersche Geographische Arbeiten 6, 37-104.
9. Kira, T. (1978), Community architecture and organic matter dynamics in tropical lowland rain forests of Southeast Asia with special reference to Pasoh Forest, West Malaysia. In: Tomlinson, P.B. and Zimmermann, M.H., *Tropical trees as living systems*, Cambridge/London/New York/Melbourne, pp. 561-590.
10. Klinge, H., W.A. Rodrigues, E. Brünig and E.J. Fittkau (1975), Biomass and structure in a Central Amazonian rain forest. In: Golley, F.B. and E. Medina (eds.), *Tropical Ecological Systems*, New York/Berlin, 115-122.
11. Lieth, H. (1972), Über die Primärproduktion der Erde. Z.Angew. Bot. 46, pp. 1-37.
12. Lieth, H. (1975), Primary productivity in ecosystems: comparative analysis of global patterns. In: Van Dobben, W.H. and Lowe-McConnell, R.H. (eds.), *Unifying Concepts in Ecology*, pp. 67-98, The Hague/Wageningen.
13. Loucks, O.L. (1980), Recent results from studies of carbon cycling in the biosphere. In: Schmitt, L.E. (ed.), Proceedings of the Carbon Dioxide and Climate Research Program Conference, Washington D.C., April 24-25, U.S. Department of Energy, CONF-8004 110 UC-11, 3-42.
14. Lugo, A.E. (1980), Are tropical forest ecosystems sources or sinks of carbon? In: Brown, S., A.E. Lugo and B. Liegel (eds.), *The role of tropical forests on the world carbon cycle*. Gainesville, Florida, 1-18.
15. Meybeck, M. (1980), River transport of organic carbon to the ocean. In: Likens, G.E. and others (eds.), *Carbon dioxide effects research and assessment program: Flux of organic carbon by rivers to the oceans*. U.S. Dep. of Energy, Washington D.C. 219-269.
16. Miller, Ph.C. (ed.) (1981), *Carbon balance in northern ecosystems and the potential effect of carbon dioxide induced climatic change*, U.S. Dep. of Energy.

17. Mullholland, P.J. (1980), Deposition of riverborne organic carbon in floodplains, wetlands and deltas. In: Likens, G.E. and others (eds.), *Carbon dioxide effects research and assessment program: Flux of organic carbon by rivers to the oceans*. U.S. Dep. of Energy, Washington D.C., 142-172.
18. Odum, H.T. and R.F. Pigeon (eds.) (1970), *A tropical rain forest*. U.S. Atomic Energy Commission. NTIS, Springfield,Va.
19. Richey, J.E., J.T. Brock, R.J. Naiman, R.C. Wissmar and R.F. Stallard (1980), Organic carbon: oxidation and transport in the Amazon River. Science 207, 1348-1351.
20. Seiler, W. and Crutzen, P.J., (1980), Estimates of Gross and Net Fluxes of Carbon between the Biosphere and the Atmosphere from Biomass Burning, Climatic Change 2, pp. 207-247.
21. Siegenthaler, U. and H. Oeschger (1978), Predicting future atmospheric carbon dioxide levels. Science 199, 388-395.
22. Stein, N. (1981), Is the large-scale destruction of tropical rain forests necessarily crucial for the global carbon cycle? In: Fazzolare, R.A. and Smith, C.B. (eds.), *Beyond the Energy Crisis, Opportunity and Challenge*. (Pergamon Press) Oxford and New York, 715-725.
23. Stuiver, M. (1978), Atmospheric Carbon Dioxide and Carbon Reservoir Changes. Science 199, 253-258.
24. Woodwell, G.M. (1978), The carbon dioxide question. Scientific American 238, 34-43.
25. Woodwell, G.M., R.H. Whittaker, W.A. Reiners, G.E. Likens, C.C. Delwiche and D.B. Botkin (1978), The biota and the world carbon budget. Science 199, 141-146.

MODELLING ASPECTS OF THE GLOBAL CARBON CYCLE

G.H. Kohlmaier, H. Bröhl, U. Fischbach, and G. Kratz

Institut für Physikalische und Theoretische Chemie
Johann Wolfgang Goethe-Universität
Robert-Mayer-Str. 11
D-6000 Frankfurt/Main 1

Since the (man-induced) change of the CO_2-content of the atmosphere may have serious consequences for the climatic system, a prediction of future atmospheric carbon dioxide partial pressures by means of a carbon cycle model seems desirable. Four problems important in this context are dealt with in this publication:

1. CO_2 exchange between atmosphere and oceans and carbon transfer within the oceans
2. carbon exchange between atmosphere and terrestrial biota
3. temperature changes in the past 120 years and their possible effects on accumulation and decomposition of biomass, and
4. development of fossil fuel energy consumption

The first three subjects refer to the efforts to construct a reliable carbon cycle model, whereas the fourth in connection with such a model should yield a probable forecast of the time behaviour of the carbon dioxide concentration in the atmosphere.

1. CO_2 EXCHANGE BETWEEN ATMOSPHERE AND OCEANS AND CARBON TRANSFER WITHIN THE OCEANS

1.1 Atmosphere-Ocean Exchange Flux

In order to determine the quantity of carbon dioxide annually exchanged between atmosphere and oceans essentially three procedures have been used: (a) from the pre-bomb steady-state balance of oceanic C-14 a CO_2 exchange time for the flux

H. W. Georgii and W. Jaeschke (eds.), Chemistry of the Unpolluted and Polluted Troposphere, 203–216.

from the atmosphere to the surface layer of the oceans, τ_{as}, of 5 to 15 ± 5 a with the center of estimates being at 7 - 8 a has been evaluated (1, 2). This method is very sensitive to the natural C-14 abundance in surface waters which has been perturbed by fossil fuel CO_2 dilution (Suess-effect) and by atomic bomb tests. (b) The post-bomb C-14 abundance in atmosphere and surface waters has been employed to estimate τ_{as} which was then found to be in the range of 5 - 12 a (3 - 5). (c) The deficiency of Rn-222 at the ocean surface compared to its concentration at the deeper strata yielded a value of about 9 a for τ_{as} (6).

1.2 Carbon Cycling Within the Oceans

Dissolved carbon species (38000 Gt C inorganic: CO_2, HCO_3^- and CO_3^{2-}, as well as about 2000 Gt C in Organic compounds) are transported in the oceans by water mass transfer. Surface water is moved according to the influence of wind stress, earth rotation and density gradients. In areas where water of high density (low temperature and/or high salinity) is formed one expects this water to sink to greater depths. Thus an overall advective circulation can be imagined which consists of water downwelling mainly in polar regions, propagating in the deep sea towards low latitudes where they may upwell (Preferably at the western margins of the continents) and then reaching higher latitudes as part of surface currents. A second process of vertical exchange of water masses may occur between different strata, generally refered to as eddy or turbulent diffusion (see fig. 1).

Another process of carbon cycling in the oceans consists in that inorganic carbon is taken up by organisms in order to be incorporated as organic compounds into their soft tissue or in order to form organismic $CaCO_3$-parts. These biosynthetic products, when sinking down to the deep-sea as dead biogenic matter, are mostly dissolved in the water column while a small fraction may sediment.

1.3 Vertical Distribution of Carbon Tracers in the Oceans

The depth distribution of total dissolved inorganic carbon (ΣC) and the isotopic ratio ΣC-13/ΣC in the deep-sea originate from the above three processes: advection, turbulent diffusion and particulate transport with dissolution of dead organic matter and $CaCO_3$. Hence, ΣC is greater in the deep-sea than in the surface layers (7) whereas the deeper strata are impoverished with respect to C-13 relative to the upper ones (9, 10), because of biogenic discrimination of C-13 against C-12 (isotopic fractionation). Considering the vertical C-14 distribution, in addition to the above three processes one has to take into account radioactive decay (half-life: 5730 a) which leads to smaller isotopic ratios

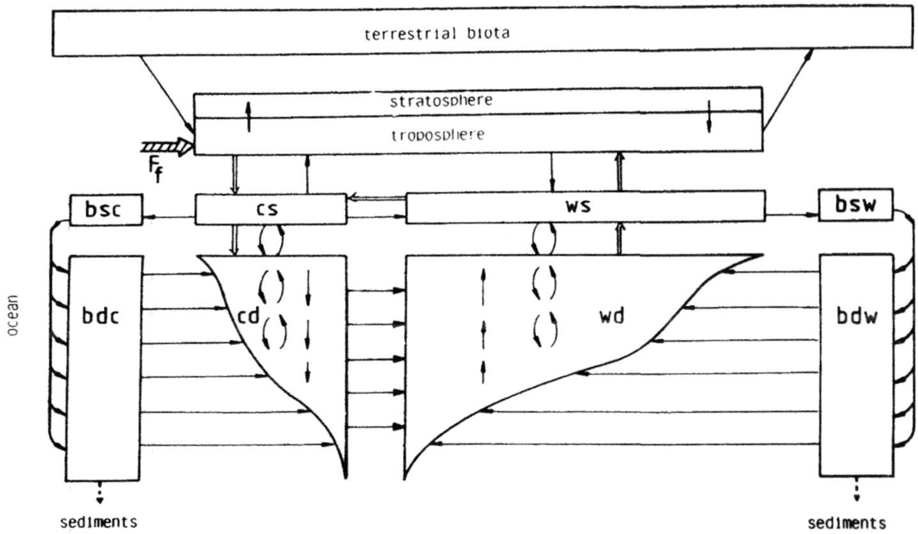

Fig. 1: Illustration of carbon dioxide exchange between atmosphere and oceans as well as carbon transfer within the oceans.
b: marine biomass, c: high latitudes, w: low latitudes,
s: surface layer, d: deep-sea, ⟹ direction of net flux,
↲ ↱ turbulent diffusion, F_f: fossil input

in the deeper layers relative to surface waters (11 - 13, see fig. 2) which are near equilibrium with the atmosphere.

Together with the total atmospheric CO_2 content the ΣC-14/ΣC ratio has been perturbed by the release of fossil fuel CO_2 which contains almost no C-14 (Suess-effect). After this dilution of atmospheric C-14 the atomic bomb tests introduced relatively large quantities of C-14 into the atmosphere by which also an increase of the oceanic ΣC-14/ΣC ratio of surface and intermediate waters was followed (fig. 2.b).

1.4 The Source-Sink Problem of Atmospheric CO_2 and Ocean Modelling

The source-sink problem consists in that model oceans have not been capable to account for fossil fuel plus anthropogenically released biogenic CO_2 (14); if run with these inputs, they produce higher atmospheric CO_2 contents than measured. In the following some ocean models are briefly characterized and the ocean CO_2 uptake capacities they reflect are compared in table 1.

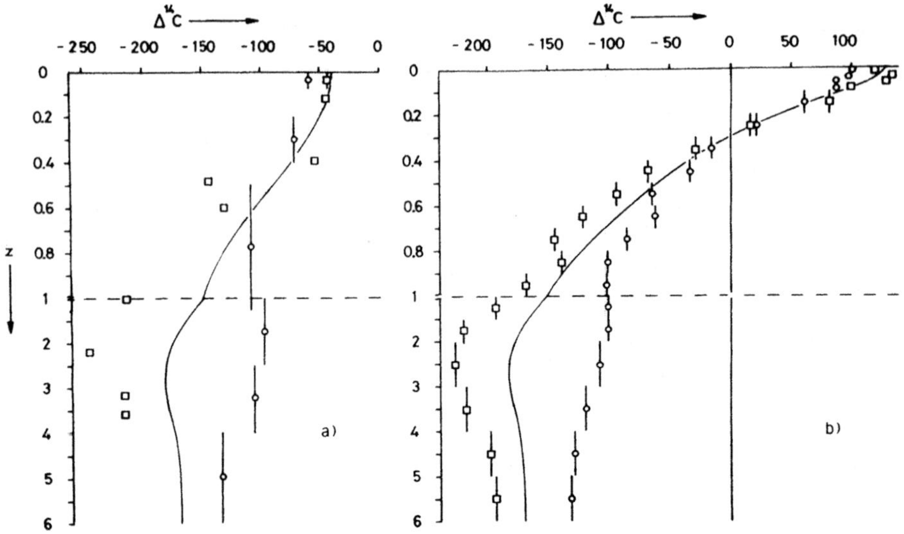

Fig. 2: ΔC-14 distribution (°/₀₀) with depth (km) in the oceans.
a) pre-bomb: o Atlantic (11); □ Pacific (12) (The smooth curve is chosen to illustrate a possible natural distribution.)
b) in 1972 - 1973: o Atlantic (13); □ Pacific (13)

$$\Delta C\text{-}14\ [°/_{oo}] = [(\Sigma C\text{-}14/\Sigma C)_{sample}/(\Sigma C\text{-}14/\Sigma C)_{standard} - 1]\cdot 1000$$

Table 1: Carbon uptake capacities as exhibited by different ocean models

model	part of fossil input CO_2 taken up by respective model ocean
2B	25.0 - 33.3 %
BD	33.3 - 42.2 %
BDU	30 - 35 %
BDK	46.5 %
BDP	25 - 30 %
BDL	49.2 %
3BDU	40 - 50 %

(Only models with an oceanic uptake capacity of more than 44 % (corresponding to an airborne fraction of 56 %) relative to fossil input are able to accomodate a biogenic source.)

The 2-box model (2B) consists of a surface layer and a deep-sea compartment (15, 16). Oceanic response to perturbations of different time constants can be adjusted by variing surface layer depth. The box-diffusion model (BD) simulates turbulent diffusion. A realistic natural C-14 distribution in the oceans is used to determine the atmosphere-ocean exchange time in connection with a global eddy diffusivity (2,17).

Analytically a model with instantaneous transfer of polar waters into the deep-sea and subsequent upwelling with constant velocity (BDU) (18) and one with depth dependent eddy diffusivity (BDK) have been calculated in our group (19, 20). Both models were calibrated with steady-state ΣC-14, the latter one yielding a comparatively high CO_2 uptake (table 1).

A further model incorporates exchange of polar waters with the atmosphere, warm surface, intermediate and deep waters (BDP). Oceanic exchanges simulated are water transport as well as sinking and dissolution of biogenic material. Tracers used for calibration are C-14 and O_2 (21). The steady-state distribution of ΣC and ΣC-14 have been used to determine the exchange coefficients of a model with lateral carbon injection into the bulk of the deep-sea waters (BDL) (22).

A model with three oceanic regions each of which is characterized by a depth independent eddy diffusivity was computed by Viecelli et al. (23). It contains downwelling of polar waters and upwelling with constant velocity and is restricted to the northern hemisphere. H-3 is used for calibration. Initial atmospheric CO_2 partial pressure is 280 ppm instead of 290 ppm as in most of the previous models and for this reason it may accomodate higher CO_2 inputs (3BDU).

As can be seen from table 1, only the more sophisticated models produce a smaller airborne fraction relative to fossil input than is measured and can thus account for a net biospheric source of CO_2.

2. CARBON EXCHANGE BETWEEN ATMOSPHERE AND TERRESTRIAL BIOTA

The terrestrial biota are another great carbon reservoir, containing between 550 and 1000 Gt C in living biomass and between 1000 and 3000 Gt C in dead organic matter. The carbon annually fixed by photosynthesizing plants (net primary production, NPP) is assumed to be between 40 and 80 Gt C/a (for a summary of estimates of biomass and NPP see (24)). Though the biota are capable to store large amounts of carbon, today they may act as a source for atmospheric CO_2 due to human impacts such as burning of wood, forest clearings, humus erosion etc..

Early carbon cycle models usually treated the terrestrial biota in a global way, using one or two compartments for living and dead or long-lived and short-lived biota (8). While all other fluxes within the biota are usually assumed to be linear donor controlled, various non-linear formulations have been proposed for the flux from atmosphere to living biota, i.e. primary production (25). Most of these models assumed the terrestrial biota to be a sink for atmospheric CO_2 in order to account for excess carbon from fossil fuel input not taken up by the respective ocean models.

An important refinement in modelling the terrestrial biota is to distinguish between vegetation units with significantly different dynamic behaviour (biomes) as well as introducing more than two components with different time characteristics. The first approach has been choosen by Moore et al. (26) and by Kohlmaier et al. (27), while the second was employed by Emanuel et al. (28) and Bolin (29), who incorporated a differentiated biota part into a model of the global carbon cycle.

Though the latter two models distinguish between similar components (short-lived biota, further divided into ground vegetation and non-woody parts of trees by Emanuel et al.; long-lived biota or wood; detritus and soil carbon), the structure as well as the dynamic properties of both models are very different, resulting especially in different capabilities to act as a long-term sink for atmospheric CO_2. This is due to differently assumed uptake kinetics and the remarkably different turnover times for some compartments, reflecting the great uncertainties about the behaviour of carbon in different parts of the biota.

Table 2: Comparison of turnover times (in years) for different compartments of the terrestrial biota as given by Bolin (29) and Emanuel et al. (28).

	Bolin	Emanuel et al.	
short-lived biomass	1.2	0.95*	{1.9 ground vegetation {0.5 nonwoody parts of trees
long-lived biomass (wood)	47	15	
detritus	1.1	1.7	
soil carbon	750	100	

* calculated mean value

While no attempt was made by Emanuel et al. and by Bolin to use their models for describing the history of the terrestrial biota, Moore et al. and Kohlmaier et al. presented simulations of the biota from 1860 to 1970 and 2000, resp., using multi-biome models with two-compartment structures for each biome and accounting for different human impacts. Moore et al. distinguish 12 biomes in 10 regions of the world, for each of which linear response curves to perturbations such as forest harvest and land

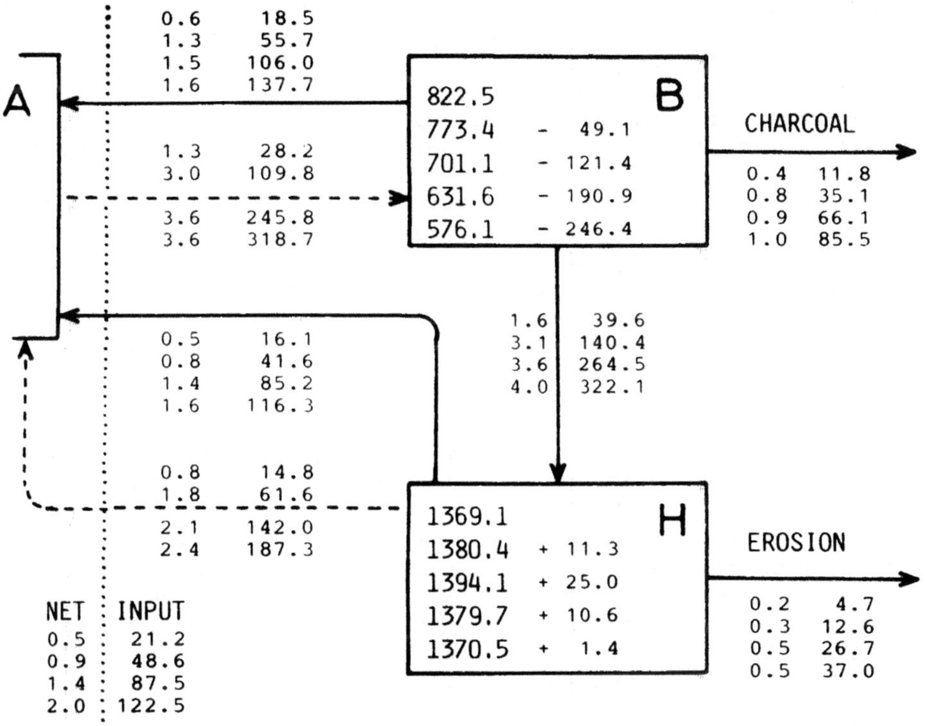

Fig. 3: Global biota carbon pools and fluxes at different times. Pool contents are given for the initial steady state (1860) and for the years 1900, 1940, 1980, and 2000, resp. Small numbers indicate cumulated changes with respect to 1860. Solid arrows indicate fluxes as induced by human impact (per year and cumulated for the resp. years), while the dashed arrows indicate fluxes caused by the response of the system on perturbances. Values at the dotted line give the net input to the atmosphere. (All values in Gt C and Gt C/a, resp.)

transformation for agricultural purposes were established and different usages of wood were incorporated. The amount of human impact in different regions was evaluated from statistical data. The authors found the biota to be a net source for atmospheric CO_2 over the whole time range from 1860 to 1970, the net flux out of the biota being greater than the fossil fuel flux until 1960 and now being between 2.2 and 4.7 Gt C/a.

Kohlmaier et al. used only five biome types, but modelled the response of the biota to perturbations in a non-linear way using power kinetics and accounted for a variety of human impacts such as shifting cultivation, wildfires, forest clearings for wood harvest and agricultural purposes and desertification as well as processes such as charcoal formation and erosion of humus which had not been oxidized. The model was run for total C and C-13 in order to verify the measured isotopic ratios in the atmosphere. The authors also obtained a source function of the biota for the whole time range, the net flux today being 1.4 Gt C/a. Carbon transfers for selected years as obtained by this model may be seen from fig. 3. The increasing amounts of carbon released are due to the assumption of clearing rates in tropical forests increasing in the same way as in the past, while the non-tropical forests in future may act as a sink due to humus accumulation caused by formerly reduced impacts.

3. TEMPERATURE CHANGES IN THE PAST 120 YEARS AND THEIR POSSIBLE EFFECTS ON ACCUMULATION AND DECOMPOSITION OF BIOMASS

The rate constants of fluxes within the biota in general depend on environmental parameters such as temperature, wetness etc. From recent studies by Schleser (30) there is evidence that increasing temperature in the time period from 1860 to 1940 could have remarkably enhanced oxidation of humus and therefore influenced the carbon content of the atmosphere.

According to Schleser the temperature dependence of the flux humus-atmosphere can be described by introducing a Q_{RCR}-value (RCR: rate of carbon release):

$$Q_{RCR} := \frac{k_{T+10}}{k_T}$$

where k denote the rate constants at temperature T and T+10 °C, resp. He found Q_{RCR}-values between 2 and 4, with values up to 6 and greater possible. As shown in a first analytical treatment by Kohlmaier et al. (31), introducing this temperature dependence into common linear donor-controlled compartment models for the humus system will result in a humus-atmosphere flux being proportional to the slope of the temperature function

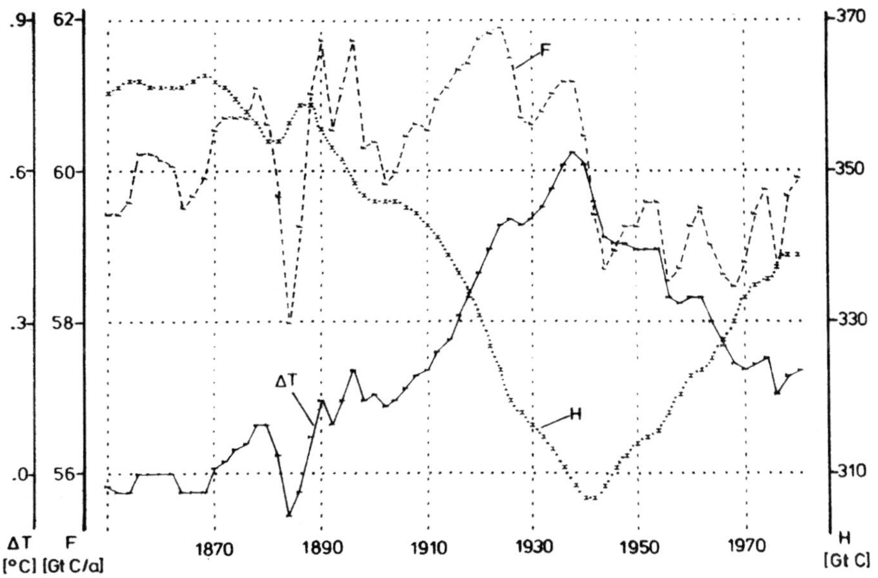

Fig. 4: Response of a 1-box humus model to global mean temperature changes from 1860 to 1980 (32)
H: soil carbon content, H(1860) = H_o = 360 Gt C
F: annual soil carbon release, F_o = 60 Gt C/a
ΔT: temperature variation Q_{RCR} = 4

tional to the slope of the temperature function. Fig. 4 shows the result of a numerical simulation of a 1-box humus model described by

$$\frac{d\,H(t)}{d\,t} = NPP_o - F(t) = NPP_o - k_o \cdot [1 + \frac{Q_{RCR} - 1}{10} \Delta T(t-t_o)] \cdot H(t)$$

where NPP_o is assumed to be constant (60 Gt C/a), $Q_{RCR} = 4$, and $k_o = 1/6\ a^{-1}$ (corresponding to an initial humus reservoir size of 360 Gt C, considering only rapidly decaying humus). As can be seen from this figure, this model yields a global cumulated decrease in humus reservoir size of about 50 Gt C for the time period 1850 to 1940. Though this might be in the right order of magnitude, further studies have shown that total humus decrease could have been between 20 and up to 100 Gt C, the great variation in results being mostly due to our lack of knowledge about the temperature dependence of the formation and decay of different humus substances at different locations.

This enhancement of carbon release out of the biota may have been partly compensated by accumulation of biomass due to temperature increase ture increase. From an empirically derived relation between primary production and temperature given by Lieth (33) Kohlmaier et al. (31) found a Q_{RCU}-value (RCU: rate of carbon uptake) to describe the enhancement of primary production by temperature changes given by

$$Q_{RCU} = 1 + \frac{10\ n}{1 + e^{(nT\ -\ m)}}$$

where n and m are parameters according to the relation of Lieth (n = 0.119/°C, m = 1.315) and T denotes the temperature in °C. For a mean global temperature of 15 °C one finds Q_{RCU} to be 1.5, or a somewhat lower value if one takes an NPP-weighted average for different geographical regions. Introducing this Q_{RCU}-value in a 3-compartment model of the biota-soil system with NPP depending on biomass in a non-linear way as described in the previous section and neglecting anthropogenically induced biomass reduction will yield an accumulation of biomass in the order of 10 to 20 Gt C for the time period 1850 to 1940. Therefore more than one half of the temperature induced carbon output from humus could have been compensated by accumulation of biomass at the same time, the net effect of temperature increase most probably being 10 to 20 Gt C released up to 1940.

4. DEVELOPMENT OF FOSSIL FUEL ENERGY CONSUMPTION

Since the beginning of industrialization the CO_2 content of the atmosphere has been increasing. This fact is mainly caused by the combustion of fossil fuel (coal, crude oil and natural gas). Until now (1980) about 167 Gt C have been transferred to the atmosphere with a fossil input rate of at present 5.4 Gt C/a. The measured increase of atmospheric CO_2 corresponds to a carbon mass of about 93 Gt. This carbon quantity remaining in the atmosphere relative to fossil input is called the airborne fraction α (approximately 0.56).

The future CO_2 input into the atmosphere will certainly be a very complex function of energy supply and demand, of the substitutions of fossil energy by other primary energy sources (e.g. Hydro-, solar, geothermal, wind, biogas and nuclear energy) and of the possibility of energy saving. Table 3 gives a survey of the global primary energy demand and the part of non-fossil energy sources given by Global 2000 (34), IIASA report (35) and Krause et al. (36) for the time from 1975 up to 2030.

IIASA supposes that the non-fossil energy share increases from about 8 % in 1975 to about 33 % of the total energy demand in 2030. The main part of these non-fossil energies is given by

Table 3: Global primary energy by non-fossil sources and total energy demand (in parentheses) in TWa/a

	1975	1985	1990	2000	2030
1	(8.21)	0.44 (12.5)	0.67 (14.9)	3.2 (20.9)	-
2	(8.21)	0.44 (10.5)	0.66 (12.6)	2.6 (17.6)	-
3	0.62 (8.21)	-	-	2.9 (16.8)	10.9 (35.7)
4	0.62 (8.21)	-	-	2.4 (13.6)	7.5 (22.4)
5	(8.21)	-	-	-	8 (15)

conversion coefficients: 1 TWa: $1.078 \cdot 10^9$ t SKE
1 t SKE: 0.7112 t C (coal): 0.5673 t C (oil): 0.4075 t C (gas)
1,2: different strategies as summarised by Global 2000
3,4: IIASA high and low energy scenario
5: low energy path (energy saving) (36)

nuclear energy (fast breeder and light water reactor) followed by hydroelectricity. Krause et al. come to a very low value for the total energy demand (15 TWa/a in 2030) by mainly considering energy saving. 8 TWa/a should be brought up by alternative energy forms. For the year 2000 the various prognoses of Global 2000 and IIASA are in a good agreement. For the far future all scenarios prefer the use of solar energy while fossil fuel and nuclear power should decrease.

To model the fossil fuel input into the atmosphere it is justified to use a logistic function

$$\dot{N}_f = r_f N_f [1 - (N_f/N_{f\infty})]^m$$

because this function describes the decline of consumption when most reserves will have been depleted. Calculations using an adiabatic approximation (19, 20) under the assumption that the growth rate of the fossil input function is slowly variing in time show that in these cases the airborne fraction α is nearly constant within the time range 1980 - 2030 while α decreases for an energy path with constant fossil input from 1980 on, i.e. in this case there remains less CO_2 in the atmosphere. In fig. 5 the annual fossil input rate $\dot{N}_f$ is given as a function of the cumulated input N_f for different energy strategies. Within our present understanding of fossil fuel reserves and technologies curve 1 represents a high estimate of the fossil fuel consumption which lies well above the IIASA high energy path (curve 2 o) and the corresponding IIASA low scenario (curve 3 ◘). Curve 4 shows

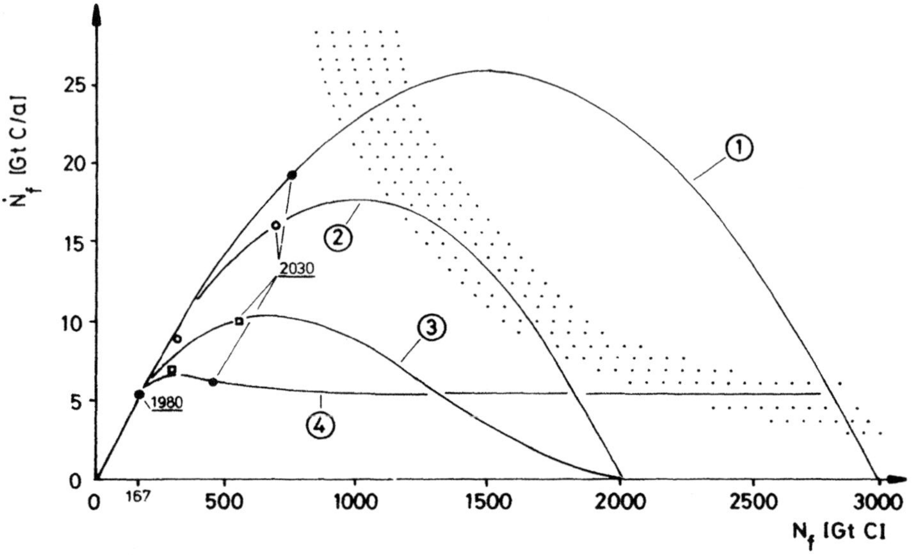

Fig. 5: Annual fossil input rate $\dot{N}_f$ as a function of the cumulated input N_f for different energy strategies

1) logistic function, final cumulated input $N_{f\infty}$ = 3000 Gt C rate constant r_f = 3.4 %/a, m = 1
2) logistic function, $N_{f\infty}$ = 2000 Gt C, f_f = 3.6 %/a, m = 1
3) modified logistic function, same as in 2), m = 2
4) zero-growth path, after a transition time the annual input amounts to 5.4 Gt C/a

a zero-energy-growth scenario as e.g. suggested by Krause et al. Within the validity of the adiabatic approximation a critical area of CO_2 doubling has been estimated (dotted area) beyond which serious climatic consequences may appear.

Taking into account a constant airborne fraction a final cumulated input of at most 1100 Gt C as allowed to avoid CO_2 doubling. On the other hand, following the zero-growth path more than 2000 Gt C can be consumed within entering the critical region. It can be seen that the assumption of a logistic behaviour for future fossil energy paths may lead to a doubling of atmospheric CO_2 (see curve 1 and 2). Only the IIASA low energy scenario gives a chance to prevent the doubling while the zero-growth strategy offers a more riskless future.

REFERENCES

1. Broecker, W.S., and Peng, T.-H.: 1974, Tellus 26, pp. 21-35
2. Oeschger, H., Siegenthaler, U., Schotterer, U., and Gugelmann, U.: 1975, Tellus 27, pp. 168-192
3. Nydal, R.: 1968, J. Geophys. Res. 73, pp. 3617-3635
4. Young, J.A., and Fairhall, A.W.: 1968, J. Geophys. Res. 73, pp. 1185-1200
5. Rafter, T.A., and O'Brien, B.J.: 1970, in: Radiocarbon Variations and Absolute Chronology (I.U. Olsson, ed.) New York, Stockholm, pp. 355-377
6. Peng, T.-H., Broecker, W.S., Mathieu, G.G., and Li, Y.-M.: 1979, J. Geophys. Res. 84, pp. 2471-2486
7. Takahashi, T., Broecker, W.S., and Bainbridge, A.E.: 1981, in (8), pp. 271-286
8. Bolin, B. (Ed.), 1981, Carbon Cycle Modeling. Chichester
9. Kroopnick, P., Weiss, R.F., and Craig, H.: 1972, Earth Planet. Sci. Lett. 16, pp. 103-110
10. Kroopnick, P.M., Margolis, S.V., and Wong, C.S.: 1977, in: The Fate of Fossil Fuel CO_2 in the Oceans (N.R. Anderson, A. Malahoff, eds.) New York, pp. 295-319
11. Broecker, W.S., Gerard, R., Ewing, M., and Heezen, B.C.: 1960, J. Geophys. Res. 65, pp. 2903-2931
12. Bien, G.S., Rakestraw, N.W., and Suess, H.E.: 1960, Tellus 12 pp. 436-443
13. Stuiver, M., Östlund, H.G., and McConnaughey, T.A.: 1981, in (8), pp. 201-226
14. Kohlmaier, G.H., Fischbach, U., Kratz, G., Bröhl, H., and Schunck, W.: 1980, Experientia 36, pp. 769-776
15. Bacastow, R.B., and Keeling, C.D.: 1973, in: Carbon and the Biosphere (G.M. Woodwell and E.V. Pecan, eds.) USAEC Conf. 720510, NTIS, Springfield, Va., pp. 86-134
16. Keeling, C.D., and Bacastow, R.B.: 1977, in: Energy and Climate, pp. 72-95, NAS, Washington D.C.
17. Oeschger, H., Siegenthaler, U., and Heimann, M.: 1980, in: Interactions of Energy and Climate (W. Bach, J. Pankrath, J. Williams, eds.) D. Reidel Publ. Co., Dordrecht, pp. 107-127.
18. Hoffert, M.I., Callegari, A.J., and Hzieh, C.-T.: 1981, in (8), pp. 287-305
19. Kohlmaier, G.H., Siré, E.O., Kratz, G., Fischbach, U., and Bröhl, H.: 1980, Ber. Bunsenges. Phys. Chem. 84, pp. 1076-1081
20. Siré, E.O., Kohlmaier, G.H., Kratz, G., Fischbach, U., and Bröhl, H.: 1981, Z. Naturforsch. 36a, pp. 233-250
21. Björkström, A.: 1979, in: The Global Carbon Cycle (B. Bolin, E.T. Degens, S. Kempe, P. Ketner, eds.) New York, pp. 403-458
22. Killough, G.G., and Emanuel, W.R.: 1981, Tellus 33, pp. 274-290
23. Viecelli, J.A., Ellsaesser, H.W., and Burt, J.E.: 1981, Climatic Change 3, pp. 281-301
24. Hampicke, U., and Bach, W.: 1979, Die Rolle terrestrischer Ökosysteme im globalen Kohlenstoff-Kreislauf. Bericht im Auftrag

des Umweltbundesamtes, FE-Vorhaben Nr. 104 02 513
25. Kohlmaier, G.H.: 1981, ISEM Journal 3, Nos. 1-2, pp. 31-56
26. Moore, B., Boone, R.D., Hobbie, J.E., Houghton, R.A., Melillo, J.M., Peterson, B.J., Shaver, G.R., Vörösmarty, C.J., and Woodwell, G.M.: 1981, in (8), pp. 365-385
27. Kohlmaier, G.H., Kratz, G., Bröhl, H., and Siré, E.O.: 1981. The source-sink function of the terrestrial biota within the global carbon cycle. Paper presented at the international symposium Energy and Ecological Modeling, ISEM, Louisville, Kentucky, USA, April 20-23, 1981 (to appear)
28. Emanuel, W.R., Killough, G.E., and Olson, J.S.: 1981, in (8), pp. 335-353
29. Bolin, B.: 1981, in (8), pp. 315-331
30. Schleser, G.H.: 1982, The response of CO_2 evolution from soils to global temperature changes. Z. Naturforsch. (to appear)
31. Kohlmaier. G.H., Bröhl,H., Siré, E.O., Kratz, G., Fischbach, U., and Jiang Yunsheng: 1982. The response of the biota-soil system to a global temperature change. Report to the CO_2 -climate study group of the Europ. Communities (to appear)
32. Daniel, H.: Man and Climate Variability. WMO, Geneva 1980
33. Lieth, H.: 1975, in: Primary Productivity of the Biosphere (H. Lieth, R.H. Whittaker, eds.) New York, pp. 237-263
34. Council on Environmental Quality and U.S. Foreign Office, 1980: The Global 2000 Report to the President, U.S. Government Printing Office
35. International Institute of Applied Systems Analysis, 1981: Energy in a finite world. Cambridge, Mass.
36. Krause, F., Bossel, H., and Müller-Reißmann, K.-F.: 1980. Energiewende. Frankfurt/Main

PART III

BUDGETS AND CYCLES OF TRACE GASES

THE TROPOSPHERIC CYCLE OF NO_x

D.H. Ehhalt and J.W. Drummond

Institut für Chemie III, Atmosphärische Chemie,
Kernforschungsanlage Jülich, 5170 Jülich, Fed. Rep. of Germany

ABSTRACT

The tropospheric chemistry, budget, and distribution of NO_x are reviewed. For surface air, the chemistry of NO_x can conveniently be separated into daytime and nighttime processes. During the day, NO_x is removed mainly via $NO_2 + OH + M \rightarrow HNO_3 + M$. During the night, some of the NO_2 is converted first to NO_3 then to N_2O_5, which can be removed by moist surfaces thereby forming dissolved NO_3^-. Estimates indicate that this reaction mechanism taking place within nighttime fogs of clouds could remove NO_x as efficiently as the daytime reaction with OH. The final sink for removal of NO_x from the atmosphere is predominately by NO_3^- deposition, mostly by rainout. A smaller fraction appears as dry deposition. The total NO_3^- deposition in 1964 is estimated to have been $(24^{+16}_{-9}) \times 10^{12}$ g N yr^{-1}. The geographical pattern for the measured total deposition is also shown. The sources of NO_x, both natural and man made, totaled $(40 \pm 20) \times 10^{12}$ g N yr^{-1}. Its geographical distribution is also estimated. Existing measurements provide a vertical profile of NO_x for continental midlatitudes; it appears to be dominated by anthropogenic sources.

INTRODUCTION

The nitrogen oxides NO and NO_2 play a key role in the chemistry of the troposphere: Firstly, to a large extent they control the concentration of the hydroxyl radical, OH. Secondly, NO is a necessary participant in the tropospheric formation of ozone, O_3. Both of these functions are effective in polluted as well as in clean atmospheres. Because of the central roles which O_3 and OH assume in tropospheric chemistry, and because of the dependence

H. W. Georgii and W. Jaeschke (eds.), Chemistry of the Unpolluted and Polluted Troposphere, 219–251.

of their concentration on that of the nitrogen oxides, it is important to know the tropospheric fate and distribution of NO_x ($NO_x = NO + NO_2$). Despite this recognition, our knowledge of the tropospheric cycle of NO_x is still somewhat vague. The salient features of what is known about the chemistry, the sinks, the sources, and the distribution of the NO_x are reviewed in that order below.

THE TROPOSPHERIC CHEMISTRY OF NITROGEN OXIDES

Although not all the details of the tropospheric chemistry of nitrogen chemistry are understood, it is possible to identify a number of major reaction pathways which lead to an efficient conversion and removal of nitrogen oxides from the atmosphere. Because of the photochemical nature of many reactions and their dependence on sunlight, it is appropriate to distinguish between daytime and nighttime reaction paths. The daytime reaction paths are summarized in figure 1 and the corresponding reactions are listed in table 1. We first trace out the major reaction paths concerning NO_x itself. As figure 1 indicates, a large fraction of the nitrogen oxides is originally injected into the atmosphere in the form of nitrogen monoxide, NO. NO then reacts rapidly in various ways. In the clean troposphere, reaction [1] with O_3 to form NO_2 (see table 1) is the fastest channel because of the relatively high abundance of O_3. However, reactions with peroxy radicals (reactions [2] and [3]) are also significant. They may even become dominant in polluted environments.

During the daytime, the NO_2 thus formed photolyses rapidly to yield NO and atomic O (reaction [4], table 1). The average lifetime for daytime NO_2 from reaction [4] is about 2.5 minutes. As a consequence of reactions [1] - [4], NO and NO_2 interconvert rapidly, their photochemical stationary state is established in 2-5 minutes, depending on the available sunlight. It is because of this rapid interchange that NO and NO_2 are lumped together as NO_x. Another definition required later is the sum of all odd nitrogen, NO_y ($NO_y = NO + NO_2 + NO_3 + 2 \times N_2O_5 + HONO + HO_2NO_2 + HNO_3 + PAN$).

The next of the major steps in the removal of NO_x is the reaction [5] of NO_2 and OH which produces nitric acid, HNO_3. This step proceeds with a characteristic time on the order of one day near the earth's surface. HNO_3 is fairly longlived with respect to photo-chemistry. It is removed from the atmosphere by heterogeneous reactions, i.e. reactions on the surfaces of liquids and solids. The major removal channels of HNO_3 are rainout and dry deposition at the earth's surface (reaction [20]) with a characteristic time of 2 to 6 days.

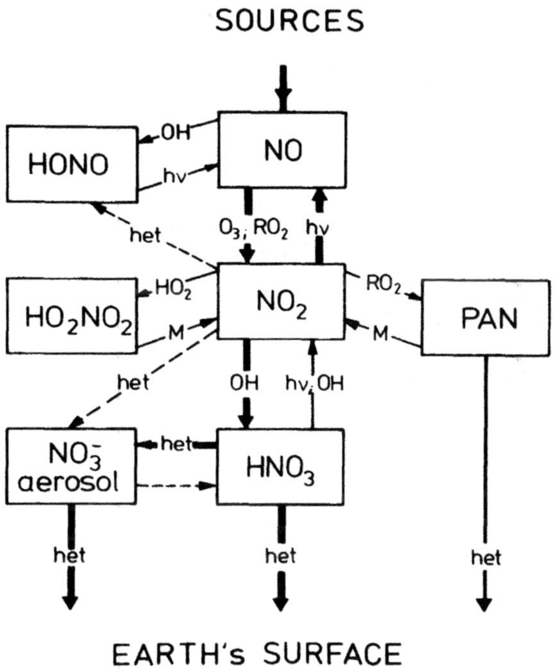

Figure 1. Major daytime reaction paths for nitrogen oxides in the lower troposphere. The symbols in the boxes denote the major oxidized nitrogen compounds present. They are interconverted by reactions, characterized by the symbols within the arrows. The heavy arrows indicate the main paths and the broken arrows show paths that are thought to be likely but are not well established. "M" represents unimolecular decay; "het" refers to heterogeneous processes, for example rainout.

The removal of HNO_3 not only leads to a loss of NO_x but also of OH from the troposphere. In fact, in situations with high NO_x concentration, reaction [5] provides the most important loss mechanism for OH. The fast reactions [5] and [2] exemplify the first important function of NO_x in tropospheric chemistry. It is that of a principal controlling factor for the OH concentration. Reaction [5] exerts control through removal of OH and reaction [2] through the conversion of HO_2 to OH. Thus knowledge of the atmospheric distribution of NO_x is extremely important for the proper modelling of the atmospheric OH-distribution (1, 2).

Table 1. Reactions and rate coefficients for 0 km (T = 285 K and P = 1000 mb) and 6 km (T = 250 K and P = 470 mb). "f" denotes forward, "r" reverse reaction. The 12 h average (no clouds) photolysis rates were calculated using a 1-D model at 45 °N, equinox (Röth, private communication). The units are s^{-1} for unimolecular reactions and photolysis rates, and $cm^3 s^{-1}$ for second order and equivalent second-order three body reactions.

Reactions				Rate coefficients 0 km	6 km	reference
1	$NO + O_3$	$\rightarrow$	$NO_2 + O_2$	1.5×10^{-14}	6.8×10^{-15}	(12)
2	$NO + HO_2$	$\rightarrow$	$NO_2 + OH$	8.6×10^{-12}	9.7×10^{-12}	(12)
3	$NO + CH_3O_2$	$\rightarrow$	$NO_2 + CH_3O$	7.4×10^{-12}	7.4×10^{-12}	(12)
4	$NO_2 + h\nu$	$\rightarrow$	$NO + O$	3.5×10^{-3}	5.5×10^{-3}	model
5	$NO_2 + OH$	$\xrightarrow{M}$	HNO_3	1.3×10^{-11}	1.3×10^{-11}	(12)
6f	$NO_2 + HO_2$	$\underset{\leftarrow}{\xrightarrow{M}}$	HO_2NO_2	2.1×10^{-12}	2.5×10^{-12}	(12)
6r				2.4×10^{-2}	1.5×10^{4}	(13)
7f	$NO_2 + CH_3C(=O)OO$	$\underset{\leftarrow}{\xrightarrow{M}}$	$CH_3C(=O)OONO_2$	1.4×10^{-12}		(14)
7r				6.1×10^{-12}	7×10^{-8}	(15)
8	$NO + OH$	$\xrightarrow{M}$	$HONO$	7.5×10^{-12}	7.2×10^{-12}	(12)
9	$NO_2 + O_3$	$\rightarrow$	$NO_3 + O_2$	2.2×10^{-17}	6.7×10^{-18}	(12)
10a	$NO_3 + h\nu$	$\rightarrow$	$NO_2 + O$	1.5×10^{-1}	1.6×10^{-1}	model
10b	$NO_3 + h\nu$	$\rightarrow$	$NO + O_2$	6.2×10^{-2}	6.7×10^{-1}	model
11	$NO_3 + NO$	$\rightarrow$	$NO_2 + NO_2$	2×10^{-11}	2×10^{-11}	(12)

Table 1. continued

	Reaction			Rate coefficients 0 km	6 km	reference
12f	$NO_3 + NO_2$	$\overset{M}{\rightleftarrows}$	N_2O_5	6.9×10^{-13}	6.8×10^{-13}	(12)
12r				4×10^{-1}	2×10^{-3}	(16)
13	$N_2O_5 + h\nu$	$\rightarrow$	$NO_3 + NO_2$	6.0×10^{-6}	1.2×10^{-5}	model
14	$HONO + h\nu$	$\rightarrow$	$NO + OH$	8.4×10^{-4}	1.4×10^{-3}	model
15	$HONO + OH$	$\rightarrow$	$NO_2 + H_2O$	6.6×10^{-12}		(17)
16	$HNO_3 + h\nu$	$\rightarrow$	$NO_2 + OH$	5.7×10^{-8}	1.5×10^{-7}	model
17	$HNO_3 + OH$	$\rightarrow$	$NO_3 + H_2O$	1.5×10^{-13}	2.0×10^{-13}	(12)
18	$HO_2NO_2 + OH$	$\rightarrow$	$NO_2 + O_2 + H_2O$	4.0×10^{-12}	4.0×10^{-12}	(12)
19	$HO_2NO_2 + h\nu$	$\rightarrow$	$NO_2 + HO_2$	9.3×10^{-7}	2.3×10^{-6}	model
20	heterogeneous reactions, rainout			$2 - 5 \times 10^{-6}$	0	

The other important impact of the NO_x on tropospheric chemistry is given through the reactions [2] - [4]. Every NO molecule which is converted to a NO_2 molecule by reaction with a peroxy radical leads to the production of an oxygen atom by the subsequent photolysis of NO_2, reaction [4]. This O-atom rapidly recombines with molecular oxygen to form ozone:

$$O + O_2 \xrightarrow{M} O_3 \qquad [21]$$

Thus, presence of NO_x leads to formation of O_3. This occurs not only in the polluted atmosphere (3) but also in the remote atmosphere (4, 5), provided the NO_x mixing ratio remains above about 0.01 ppb (5).

There are other reactions besides those in the main path in figure 1 which become more important at lower temperatures. Most of these reactions involve NO_2. NO_2 reacts with HO_2 to form pernitric acid, HO_2NO_2 (reaction [6f]). In the lower troposphere, at temperatures around 285 K, HO_2NO_2 decays within seconds to return the original reactants (reaction [6r]). But at higher altitudes, HO_2NO_2 should become an important reservoir for NO_x, although it has not yet been observed in the atmosphere. NO_2 also reacts with acylperoxy radicals, for example, the peroxyacetyl radical to form peroxy-acetyl-nitrate, PAN (reaction [7f]). Its lifetime against unimolecular decay [7r] is relatively long. PAN thus provides an important reservoir for NO_y even at the surface of the earth. Oxidized nitrogen may be transported in the form of PAN over greater distances than otherwise possible (4). PAN is sufficiently longlived to make its heterogeneous removal possible, although its uptake in liquid water is slow compared to that of HNO_3.

The gas phase reaction of NO with OH (reaction [8]) to form nitrous acid, HONO, is relatively unimportant. However there appear to be unknown heterogeneous processes to form HONO. HONO has been observed during the night, at least in polluted air (6, 7). Whether such a heterogeneous formation process for HONO could also occur during the daytime and in clean environments is not known. If it were to occur, it would be an important process because of the resulting OH production during HONO photolysis (reaction [14]). It would constitute yet another significant contribution by the odd nitrogen chemistry to the OH budget.

It is important to note that some of the gaseous HNO_3 reacts with aerosol to form aerosol nitrate, which in many regions is present in higher concentrations than gaseous HNO_3 (8). There is also the possibility that NO_2 forms aerosol nitrate directly by heterogeneous processes. However, very little solid information on such processes and their rates is presently available. Nitrate aerosol is eventually removed by rainout, washout or dry deposition. A small fraction of the nitrate could possibly evaporate to return HNO_3.

During the night many of the reactions which are most important during daytime virtually stop: there are no photolytic reactions, and furthermore reactions with OH become negligibly slow as a consequence of low OH concentrations. This allows other reactions, unimportant during daytime, to dominate the nighttime NO_x chemistry. They are summarized in figure 2. It should be noted that in the absence of local NO-sources, the nighttime NO concentration also vanishes. The normally more abundant O_3 simply reacts away the NO.

The main feature of nighttime chemistry is the appearance of NO_3. It is formed by the reaction [9] of NO_2 with O_3. Rapid photolysis ([10a] and [10b]) prevents measurable NO_3 concentrations during daytime, but during nighttime significant concentrations are expected and have been observed (9, 10, 11). NO_3 reacts further to form N_2O_5 (reaction [12f]). At 285 K N_2O_5 decays rapidly back to NO_2 and NO_3 (reaction [12r]), and NO_2, NO_3, and N_2O_5 attain equilibrium. Because N_2O_5 is the anhydride of nitric acid it reacts quickly with liquid water (but not water vapor) to form nitric acid. Similarly it should rapidly react upon collision with a wet surface film to form aerosol nitrate.

There is some indirect experimental evidence to support this reaction path. Platt et al. (10), in a number of experiments designed to measure the nighttime variation of NO_3, observed the following behavior in Deuselbach, a rural meteorological monitoring station in West Germany. Whenever relative humidity was below 50 %, there was a steady increase of NO_3 concentration from zero in the evening until its rapid destruction at sunrise. In cases when the humidities were high (above 60 %), no nighttime NO_3 was observed. During a few nights, the relative humidity began low and increased to values above 60 %. Here, NO_3 began to rise as expected. However, when the humidity reached 60 % there was a sudden decrease of NO_3 to below the detection limit. This points to a threshold type of behavior which cannot be explained by homogeneous gas phase chemistry. It leaves a change in aerosol surface characteristic, e.g. the onset of a wet surface film around 60 % humidity as the most likely explanation. Of course, the wetted aerosol surfaces could conceivably react directly with NO_3. However, since N_2O_5 should be more abundant than NO_3 for the observed temperature and NO_2 mixing ratios (5 ppb), and because N_2O_5 is known to dissolve quickly in water, the N_2O_5 loss path is preferred. Due to its equilibrium with NO_3 and NO_2 the loss of N_2O_5 leads to a corresponding loss of NO_3.

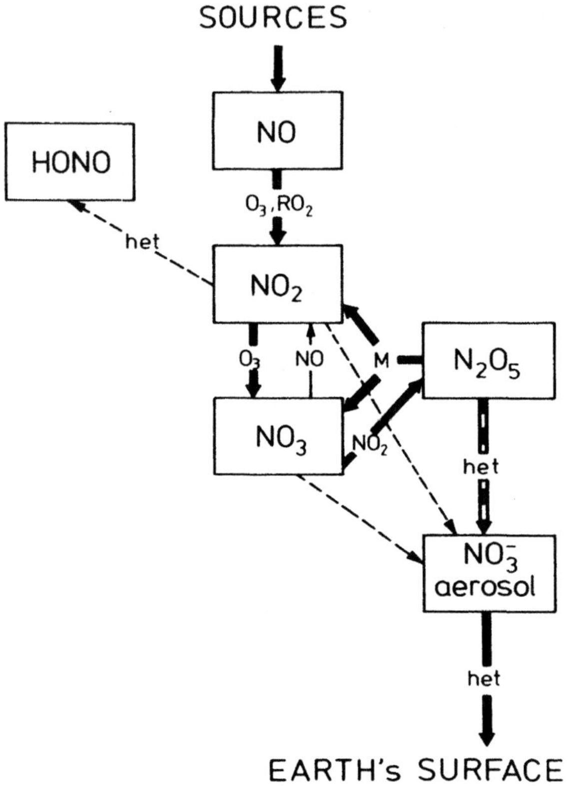

Figure 2. Major nighttime reaction paths for odd nitrogen in the lower troposphere. The symbols in the boxes denote the major oxidized nitrogen compounds present. They are interconverted by reactions, characterized by the symbols within the arrows. The heavy arrows indicate the main paths and the broken arrows show paths that are thought to be likely but are not well established. "M" represents unimolecular decay; "het" refers to heterogeneous processes, for example rainout.

This process of NO_y removal should be particularly efficient in nighttime clouds or fogs, provided there are sufficient concentrations of O_3. A simple estimate assuming 50 % cloud cover, a vertical cloud extent from 0,5 km to 4 km, and standard temperature profile, gives a NO_2-lifetime of 2 days in summer and 2.5 days in winter from N_2O_5 heterogeneous removal alone. This estimate holds for northern temperate latitudes, where the NO_2 concentrations are high (c.f. Figure 10), and reaction [9] is the rate limiting step (see figure 2). On the other hand, the average NO_2 removal through reaction [6] with OH proceeds with a lifetime of about 2 days in summer and 9 days in winter. Thus a large fraction of NO_2 should be removed via the heterogeneous removal of N_2O_5 in clouds at night. It should be noted that this loss process is not considered in present models.

In any case during both day and night most of the injected NO is eventually converted to gaseous HNO_3 or aerosol NO_3^-. Upon removal from the atmosphere, these components are deposited principally as NO_3^- at the earth's surface. This argument is supported by the observation that most of the oxidized nitrogen in precipitation seems to be present as NO_3^- and only a few percent as NO_2^-.

The difference between day and night chemistry of the NO_y causes strong diurnal cycles for most of the odd nitrogen compounds. In figures 3 and 4 model calculated diurnal variations are shown for 0 km and 6 km altitude respectively (1).

Because of its relatively long lifetime, HNO_3 is the most abundant odd nitrogen species at both altitudes. For the same reason and because the model removes HNO_3 by steady average rain, its predicted mixing ratio is also virtually constant with time. Actually, the HNO_3 could vary dramatically as a function of local rain. Near the earth's surface (figure 3), NO_2 is the next most abundant species. It varies diurnally with relatively small amplitudes due to its conversion to NO during the day and to NO_3 during the night. NO and NO_3 are of about equal abundance and their respective diurnal variations are completely out of phase. NO shows a strong diurnal variation with vanishing concentrations during the night. NO_3, and consequently N_2O_5, show a strong variation with vanishing concentrations during the day. Because of the low NO_2 mixing ratios of about 0.03 ppb, and the relatively high temperature, NO_3 achieves higher concentrations than N_2O_5. The other odd nitrogen compounds HO_2NO_2 and HONO, remain quite low in concentration - the latter because the model does not consider heterogeneous production of HONO, the former because of its fast thermal decay.

At 6 km altitude the temperature is lower: 250 K. Since the unimolecular decomposition of N_2O_5, HO_2NO_2, and PAN proceeds with the high activation energies of about 25 kcal, their decay now occurs much more slowly. The longer lifetimes for these compounds permit higher concentrations to be reached, and dampens the amplitudes of their diurnal variations (figure 4). Local equilibrium for these compounds is no longer established because the time for transport is shorter than the time for chemical equilibrium. Due again to the slower decay, N_2O_5 now exists in higher concentrations than NO_3. Because of the colder temperature NO reacts

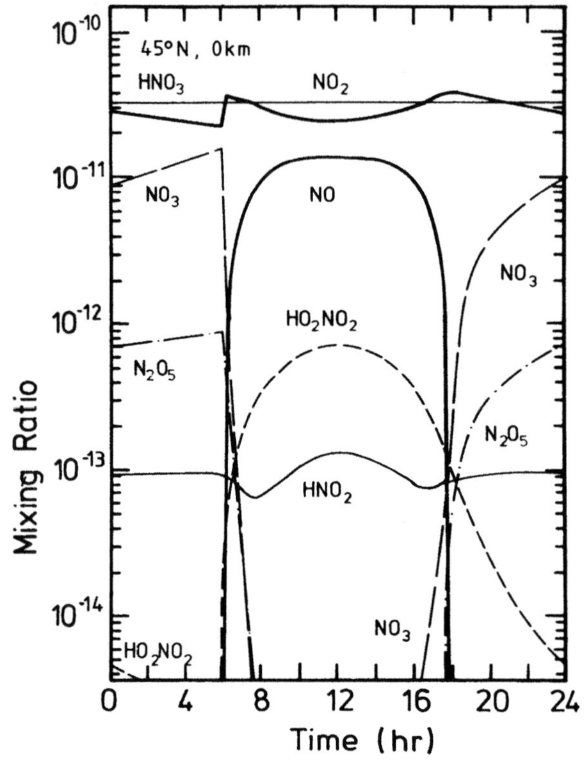

Figure 3. Calculated diurnal variation of odd nitrogen compounds at 45 °N, 6 km altitude, and equinox. The calculations are from the model of Logan et al. (1) and do not include PAN.

less rapidly with O_3 (reaction [1]), giving a daytime mixing ratio for NO that is higher than that of NO_2. By contrast, the daytime NO_2/NO ratio at ground level is 2.5.

Besides diurnal variations, the model predicts also strong vertical gradients in the mixing ratio of the odd nitrogen compounds ((1); cf. figures 3 and 4). The model predictions may change somewhat as our understanding of chemistry improves. Even so, given the predicted strong temporal and spatial variations, simultaneous measurements of the various odd nitrogen compounds should provide a sensitive test for our ability to model the tropospheric nitrogen chemistry. Unfortunately, the techniques with sufficient sensitivity to accomplish this have not been available. As we

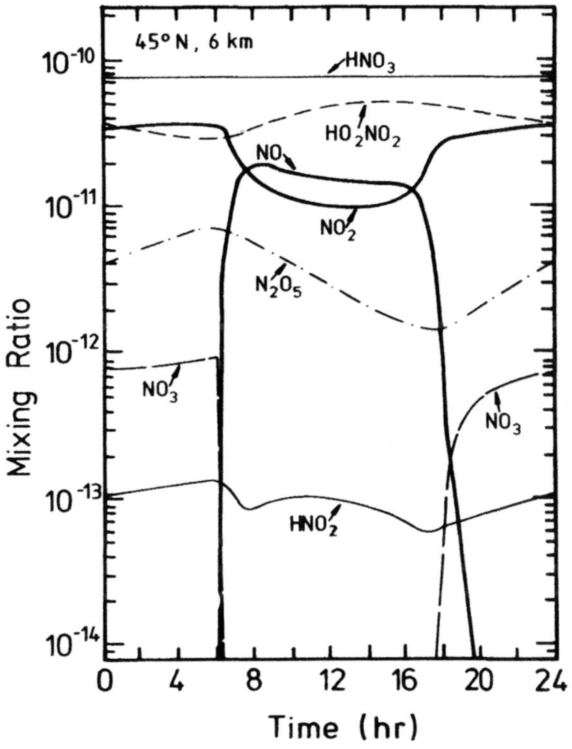

Figure 4. Calculated diurnal variation of odd nitrogen compounds at 45 °N, 6 km altitude, and equinox. The calculations are from the model of Logan et al. (1) and do not include PAN.

shall see in the last chapter, there is even a definite lack of tropospheric data on NO_x itself.

SINKS OF NO_x

The previous discussion of the atmospheric nitrogen chemistry indicated that most of the emitted NO is eventually converted to HNO_3 or nitrate aerosol and removed from the atmosphere by rainout or dry deposition at the ground. This path therefore constitutes the major sink of atmospheric NO_x. The nitrate deposited at the earth's surface can be measured with relative ease, because the NO_3^- concentration in precipitation is greatly enriched over its concentration in air. The measurements are usually made using precipitation samples collected in rain gauges during the course of one to several weeks. Since most such rain collectors are left

Table 2. Published estimates for total NO_3^- deposition in units of 10^{12} grams of elemental nitrogen per year

Author	Total NO_3^- deposition land	ocean	global	basis for estimate, remarks
Eriksson (19)	21	27	48	unreferenced estimate, possibly based on (18)[1]
Robinson and Robbins (22, 23, 24, 66)			75	(18)[1]; assume additional 25 % for dry deposition of NO_3^- [2]
McConnell (25)			20–60	(18), (26), (27), (28); assume 70 additional units for dry deposition of NO_3^- [2]
Warneck (31)	28	10	38	Author gives flux estimate based on (18), (29), (30)
Burns and Hardy (32)	31	18	49	(18), (22), (33), (34), (35), (36), (37); Assume additional 20 % for dry deposition of NO_3^- [2]
Chameides et al. (38)			60–90	(18), (32)

Table 2 continued

Author	Total NO_3^- deposition land	ocean	global	basis for estimate, remarks
Söderlund and Svensson (39)	13-30	5-16	29-46	Land estimate based on (18)[1], (40); Ocean estimate taken from Söderlund (40). Assume additional dry deposition of NO_2 of 19-50 on land and 6-17 on the ocean.
Böttger et al. (20)	17±7	8±6	24±9	new compilation of measurements, centered in time on 1964.
Böttger et al. (20)	"	5	16±6	same, but only for Northern Hemisphere

1) Eriksson's (18) careful analysis of NO_3^- deposition relied on worldwide, mostly continental data centering around 1900. It did not attempt to present a global total.

2) Some authors consider their estimate to represent wet deposition only, and assume an additional dry deposition of NO_3^- between 20 and 30 % of that value. In our opinion most of the data on NO_3^- in precipitation already include the contribution by dry deposition (see text). Thus their value for wet deposition is taken here as total deposition.

open constantly, nitrate deposited on the funnel surface during dry periods is also collected. The deposition thus measured is called total deposition to indicate that it includes both the wet and the dry deposition components - with the tacit assumption that the funnel surfaces collect dry nitrate as efficiently as the earth's surface. There are many measurements of the total nitrate deposition, some from as far back as the middle of the last century (18). One would think that it would be possible to deduce the global nitrate deposition with some certainty. This is unfortunately not the case, as demonstrated by the large variability in the published estimates of the global nitrate deposition summarized in table 2. There is a difference of a factor of three between the lowest and the highest estimates. Moreover, the estimates in table 2 are not always independent. Most are based on Eriksons analysis (18, 19) which summarized measurements between 1864 and 1947 from 90 stations mainly in Europe and North-America. The summary, although carefully done, used a data base that was hardly representative.

To improve this rather unsatisfactory situation Böttger et al. (20) made a careful literature search to generate a larger data base with as much geographical coverage as possible. They found useful data from 500 stations, a factor of five more than before. But even with that large data set, the estimate of a globally averaged nitrate deposition is still problematic. In the following we list the major caveats.

1) To obtain a sufficient number of stations, measurements made in the period between 1950 and 1977 were included during which years the NO_3^- deposition increased.
2) Data records at most stations exist for only a few years.
3) The quality of the measurements is not uniform. For a number of data sets the sampling and measurement techniques are not adequately documented.
4) The geographical coverage still remains extremely non-uniform. This is shown in figure 5. Only Europe with 235 stations, and North-America with 161, have reasonable coverage. Data from the tropics and the oceans are very sparse. The coverage for the Southern Hemisphere is even worse. Moreover, except for a few island stations, the measurements of the oceans were taken for periods considerably shorter than one year.

Despite these shortcomings, one feature clearly emerges when the deposition from all the available stations is mapped: there is a distinct, geographical pattern of total NO_3^- deposition with higher deposition over the continents. The maxima of continental deposition coincide with the heavily industrialized and densely populated areas.

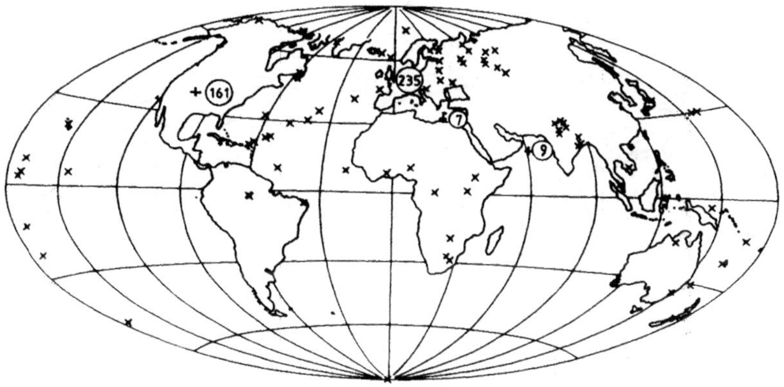

Figure 5. The distribution of stations for measurement of total nitrate deposition (20). The majority of these collection stations are in Europe and the United States as indicated by the encircled numbers.

The estimated isolines of the total NO_3^- deposition are shown in fig. 6. The NO_3 deposition was calculated as weighted mean from the available seasonal records of rainfall and NO_3^- concentration in the precipitation collected at the various stations. The isolines clearly demonstrate the higher deposition over both the industrialized continents and the tropical land masses. There is little deposition over the oceans. A similar pattern was found by H.W. Georgii (private communication) using data from the WMO-network between 1972 and 1979. It is noted that the estimation of these isolines involved large extrapolations, especially over the oceans, the tropics, and Asia, where the network of stations was sparse. There, the extrapolations were guided by two considerations, namely:

1) Wet deposition is generally more efficient than dry deposition; thus the areas with low rainfall were assigned less total deposition.
2) The transport of NO_y and NO_3^- from the source areas is determined by the mean wind field. Thus the contours of the isolines are dilated in the leeward and compressed in the windward directions.

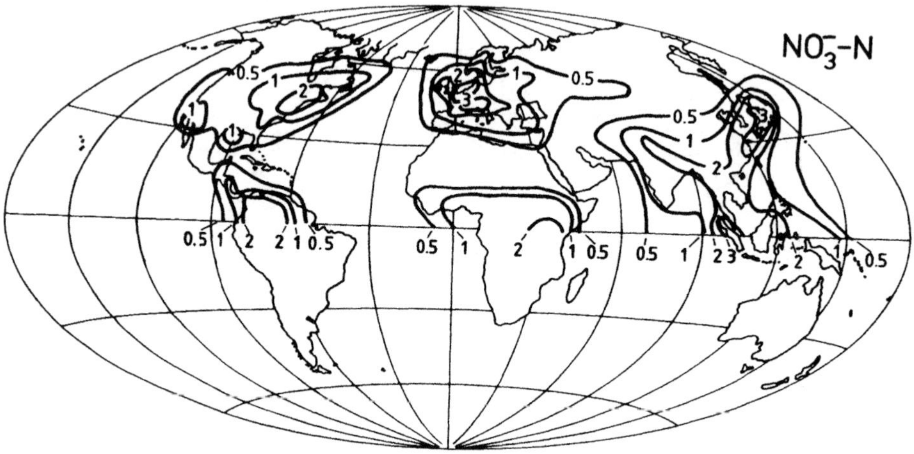

Figure 6. Lines of equal nitrate deposition rates for the Northern Hemisphere (20). The units are in 100 milligrams of elemental nitrogen per square meter per year.

There remains a fair degree of subjectivity in such an analysis, and the isolines, especially in the tropics, should be considered as qualitative. Obviously, considerably more data are needed in the tropics, over the oceans, and from the Asian continent to obtain a reliable estimate of the NO_3^- deposition in these areas. Since the tropical and Asian areas seem to receive a significant part of the global NO_3^- deposition, the uncertainties existing in these areas propagate into the estimate of the global NO_3^- deposition. Nevertheless, figure 6 provided - for the first time - a semi quantitative picture of the total NO_3^- deposition in the Northern Hemisphere.

The hemispheric distribution of the NO_3^- deposition rate in figure 6 was integrated over the longitude to give its variation with latitude (20). The results of this integration are shown in figure 7, for land, ocean and for the area weighted mean.

In addition, Böttger et al. (20) calculated the average total deposition of NO_3^- in the Northern Hemisphere. It is also shown in table 2. Finally including the sparse data from the Southern Hemisphere, they estimated a global deposition of $(24 \pm 9) \times 10^{12}$ g N yr^{-1} of which $(17 \pm 7) \times 10^{12}$ g N yr^{-1} are deposited on land, and $(8 \pm 6) \times 10^{12}$ g N yr^{-1} are deposited on the ocean. These estimates refer to the year 1964, in view of the fact the measure-

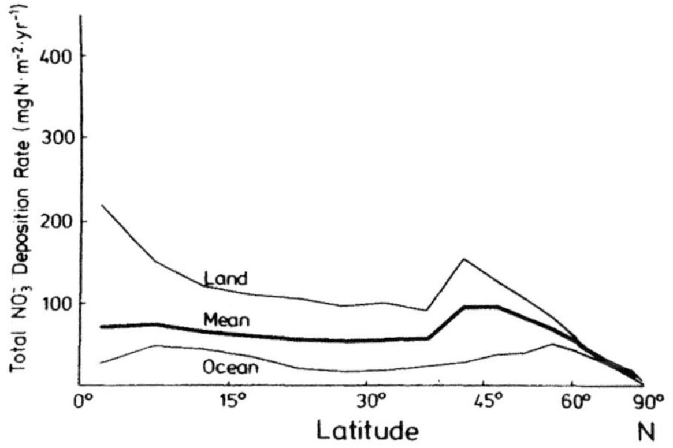

Figure 7. Total nitrate deposition vs. the cosine of latitude for the Northern Hemisphere (20). The results are shown for land, ocean, and the area weighted mean.

ments were made between 1950 and 1977, approximately uniformly with time. They are considerably lower than the previous estimates, owing to the low deposition rates found over the oceans.

We now briefly address the question of whether the estimated total NO_3^- deposition represents the entire sink strength of NO_x removal from the atmosphere. There are two factors which may be not fully included in the total deposition as determined from rain gauges. Both concern the fraction contributed by dry deposition. Firstly, because of the enlarged surface area available for impact, forests or similar vegetation receive a considerably larger contribution from dry deposition than indicated by a rain gauge placed in a nearby open area (21). Thus the deposition in wooded areas may be underestimated. Secondly, odd nitrogen compounds other than HNO_3 or aerosol nitrate also show dry deposition on surfaces, the most important example being NO_2. Dry deposition of NO_2 has not been well studied, but values of the deposition velocity have been published (20). On grass and bare soil the deposition velocity of NO_2 is about 0.3 cm sec^{-1}. Over water it is much slower, about 0.02 cm sec^{-1}.

The dry deposition of NO appears to be negligibly small (20) This allows to roughly estimate the global removal of NO_x by dry deposition of NO_2 alone. Assuming a conservatively high average NO mixing ratio of 1 ppb, and an average deposition velocity of 0.3 cm sec^{-1} over land only, one obtains an upper limit for the global deposition rate of 7×10^{12} g N yr^{-1}. Because of the large uncertainty in this estimate it is included in the error bounds and not in the mean value of the total nitrate deposition. Thus our final estimate for the total removal of odd nitrogen by wet and dry deposition amounts to $(24\ ^{+16}_{-9}) \times 10^{12}$ g N yr^{-1}.

SOURCES OF NO_x

Whereas there is one single major sink for atmospheric NO_x, namely the deposition of NO_3^- at the earth's surface, there are many significant sources. This multitude makes an estimate of the total source strength and of its geographical distribution much more complicated. The known sources are listed in table 3 together with estimates of their respective strengths. Practically all of the estimates are rather uncertain and some represent not more than educated guesses. To take this into account, upper and lower bounds of the source strengths are also given.

We shall briefly discuss the various estimates beginning with the surface sources. Man, owing to the burning of fossil fuels, is responsible for a large share of the total emission. The respective source strengths are estimated as product of the global consumption of a given fuel times an appropriate NO_x emission factor. The latter are usually on the order of 1 g N per kg fuel. Estimates for the emission factors from the various burning of fossil fuel processes have been published (39, 43, 43, 44). For the present estimates we used the means and ranges evaluated by Böttger et al. (20). The fuel consumption was taken from the Statistical Yearbook of the United Nations 1975 (45). The estimates therefore apply to 1975. The various sources from fossil fuel combustion including automobile exhaust add up to a mean NO_x production of 13.5×10^{12} g N yr^{-1}.

Another important surface source is the natural release of NO from soils. Older measurements of exhalation rates have been summarized by Böttger et al. (20). Galbally and Roy (46) measured a NO exhalation rate of (0.1 ± 0.05) g N m^{-2} yr^{-1} from Australian pastures. Similar values were found for forest soils (47). Assuming that similar exhalations occur over all pastures but not any other soils gives the lower bound; assuming that all land surfaces except extreme desert release NO at that rate gives the upper bound. The mean, nothing more than a guess, was taken as 5.5×10^{12} gN yr^{-1}. Clearly NO exhalation from soil could represent a significant source, but at the present time its source strength is extremely uncertain.

A third important NO_x source is the burning of biomass. Here we list only the most important source terms; a more comprehensive list was given by Crutzen et al. (48). Again, the emission factors are rather uncertain. We prefer a mean NO_x production rate of 11×10^{12} g N yr^{-1}. It should be noted that

Table 3. Tropospheric sources of NO_x and their strength.

The units are 1×10^{12} g N yr^{-1}. The estimates refer to the year 1975. See text for references.

	lower bound	mean	upper bound
I. surface sources			
A. fossil fuel burning			
coal	1.9	3.9	5.8
lignite	.8	1.6	2.3
light fuel oil	.5	.7	.9
heavy fuel oil	.7	1.1	1.5
natural gas	.6	1.9	3.1
automobiles	3.7	4.3	4.9
subtotal for A.	8.2	13.5	18.5
B. soil release	1	5.5	10
C. biomass burning			
savanna	1.8	3.1	4.3
deforestation	.8	2.1	3.4
fuel wood	1	2	3
agricultural refuse	2	4	6
subtotal for C.	5.6	11.2	16.4
subtotal for surface sources	14.8	30.2	45.2
II. atmospheric sources			
NH_3 oxidation	1.2	3.1	4.9
lightning	2	5	8
high flying aircraft	.2	.3	.4
NO_y transported from stratosphere	.3	.6	.9
subtotal for atmospheric sources	3.7	9.0	14.2
III. total production	19	39	59

Crutzen et al. (48) admit the possibility of a much larger production rate from biomass burning, 50×10^{12} g N yr^{-1}, which is considerably larger than our upper bound.

Besides the surface sources there are atmospheric sources, i.e. processes which produce NO_x within the atmosphere. They are on the whole weaker than the surface sources. However, most of the NO_x introduced at the surface is relatively quickly converted to NO_3^- within the planetary boundary layer and does not travel very far. Therefore, it is the atmospheric sources, especially those above the planetary boundary layer that determine the NO_x concentrations in the free troposphere and remote areas.

The major free tropospheric source of NO_x is that due to lightning. The estimates for NO production by lightning have decreased considerably over the past years. The most recent estimate (49) gives 1.8×10^{12} g N yr^{-1}, down considerably from the earlier estimates (50, 51) which gave up to 40×10^{12} g N yr^{-1}. To account for the large variability in estimates we choose a range from 2×10^{12} g N yr^{-1} to 8×10^{12} g N yr^{-1} with a mean of 5×10^{12} g N yr^{-1}.

The NO_x produced from NH_3 oxidation is confined mainly to the lower troposphere. NH_3 is destroyed by OH, and both OH and NH_3 have their highest concentration at low altitudes. The global NO_x production from this source averages 3.1×10^{12} g N yr^{-1} (20). There are other organic N-compounds in the atmosphere, e.g. aceto-nitril, CH_3CN, part of which may also be converted to NO_y. Little or nothing is known about their fate or budget.

High flying aircraft are estimated to produce 0.3×10^{12} g N yr^{-1} mostly above 9 km altitude and in the Northern Hemisphere (52).

Finally, the reaction of N_2O with O^1D

$$N_2O + O^1D \rightarrow 2\ NO \qquad [22]$$

produced NO in the stratosphere, of which 0.6×10^{12} g N yr^{-1} are transported to the troposphere largely in the form of HNO_3 (1).

The mean total production of NO_x from all these sources amounts to roughly 40×10^{12} g N yr^{-1}. The lower and upper bounds are 20×10^{12} g N yr^{-1} and 60×10^{12} g N yr^{-1} respectively. There are production estimates in the literature which exceed our upper bound.

In addition to calculating the total source strength it is instructive to consider the distribution of the sources. The strength of many sources varies with altitude, and longitude and

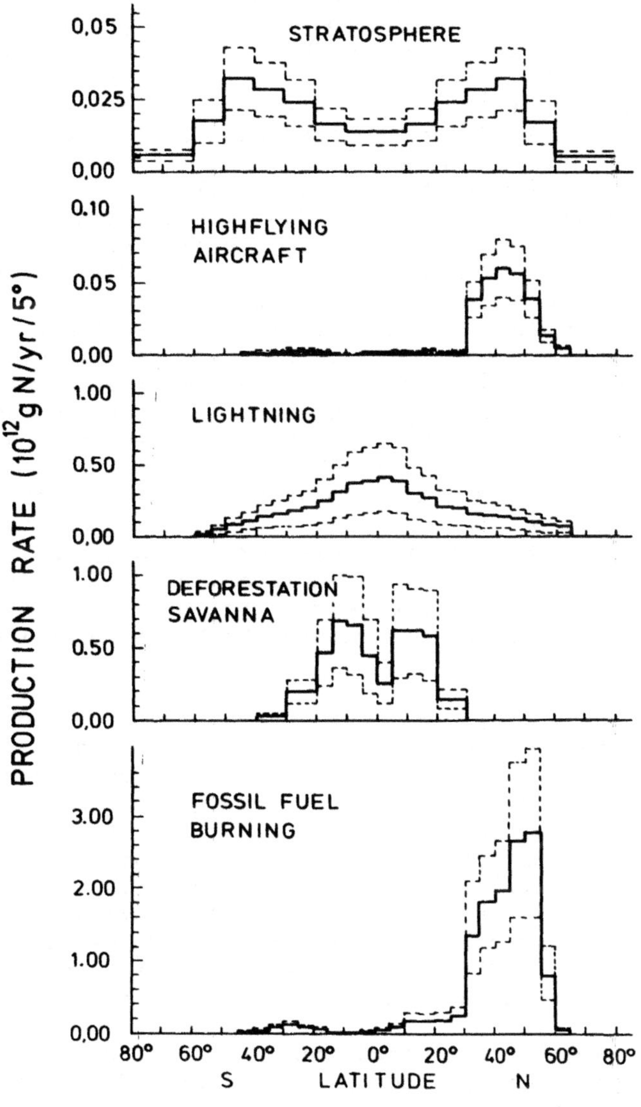

Figure 8. Estimated production rates of odd nitrogen as a function of latitude for five selected sources. The mean, upper and lower bounds were calculated using the detailed latitudinal distributions given by Böttger et al. (20). Note that the scales for the production rates vary up to a factor of 40.

sometimes with season. Since the atmospheric distribution of NO_x concentration depends critically on the source distribution, we provide a few examples for the source distributions. Figure 8 presents the latitudinal distribution of the NO_x production from (a) fossil fuel combustion, (b) from biomass burning in savannas and during tropical deforestation, (c) from lightning, (d) from high flying aircraft, and (e) transport NO_y from the stratosphere. Clearly, (b) is limited to the tropics, and (a) essentially to the industrialized midlatitudes of the Northern Hemisphere. It should also be noted that both are exclusively continental sources. Lightning is somewhat more uniformly distributed with latitude, even though it shows a clear maximum in the tropics with about a factor of three higher production rate than at midlatitudes (20). Although lightning also occurs over the ocean, it is much more common over the continents and is thus mainly a continental source. In addition, it shows a seasonal variation with a maximum during the rainy season in the tropics and during summer at midlatitudes. An undetermined part of the NO_x from lightning is of course produced within the planetary boundary layer. The rest is injected into the lower troposphere normally below about 6 km. For high-flying aircraft we assumed a distribution similar to that of automobile traffic (20). Most of that input is concentrated in the midlatitudes of the Northern Hemisphere. Although most of this input occurs regionally over North America, Europe, the Atlantic Ocean and the Near East, it is expected that at these altitudes, rapid zonal transport should distribute the NO_x more or less uniformally with longitude.

The only truly global NO_y input is the flux from the stratosphere. It takes place at all longitudes but is expected to show a latitudinal variation. The one presented in Figure 8 has been modelled after the measured ^{90}Sr deposition on the ground resulting from stratospheric nuclear bomb debris (53). It should be noted that the transport of stratospheric ^{90}Sr into the troposphere showed also a strong seasonal variation with a maximum downward flux during spring and a minimum flux during late summer and fall (54). The same should be true for the downward flux of stratospheric NO_y. In addition there is the possibility that the ratio of NO_x to NO_y could also change as a function of season.

Fig. 9 contains the estimated total production from all sources as a function of latitude. It is clearly dominated by a maximum at northern midlatitudes caused by the fossil fuel combustion in the industrialized countries. The sources from biomass burning and lighting are not strong enough to generate a second maximum in the tropics. Thus the production decreases evenly between 30 °N and 30 °S. However, it falls sharply south of 30 °S because of the weaker sources in the midlatitudes of the Southern Hemisphere.

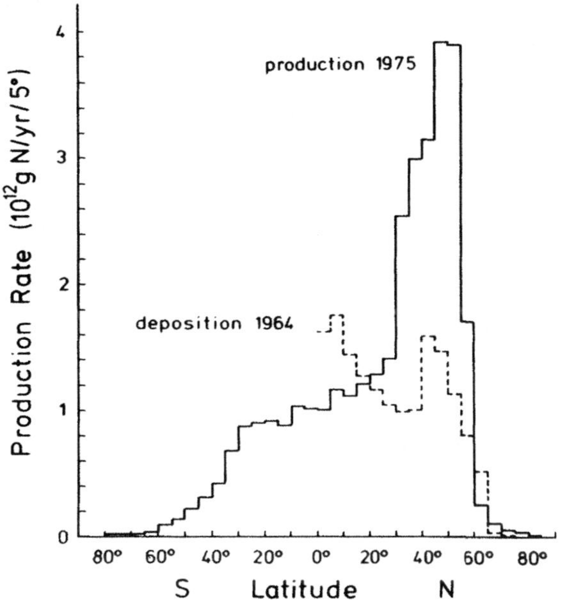

Figure 9. Latitudinal comparison of the estimated total production for 1975 (solid curve) and the estimated deposition in the Northern Hemisphere for the period centered on 1964 (broken curve). The production estimates include both the free tropospheric and ground level sources.

As mentioned above most of the production takes place over the continents. In fact, the only known oceanic surface source releases NO at such a low rate (55) that is was not included in table 3. A few of the atmospheric sources, however, do introduce NO_x over the ocean: lightning, oxidation of NH_3, highflying aircraft, and stratospheric influx inject NO_x in the atmosphere over the oceans at about 5×10^{12} g N yr^{-1} total. This is only about 10 % of the total production. It is more or less evenly distributed with latitude, possibly with a slight maximum in the tropics.

THE ATMOSPHERIC BUDGET OF NO_x, COMPARISON OF PRODUCTION AND DEPOSITION

Total production was estimated to $(40 \pm 20) \times 10^{12}$ g N yr^{-1}, total deposition to $(24 {}^{+16}_{-9}) \times 10^{12}$ g N yr^{-1}. However, production refers to the year 1975, deposition centers around the year 1964, and between those years, the NO_x production must have increased considerably. In fact, fossil fuel consumption increased by about 50 % during that period, and we would expect a corresponding increase in the NO_x production from that source. It is likely that the other anthropogenic emissions such as biomass burning increased as well. If we assume an increase of 50 % for these too, total production in 1964 would have been $(33 \pm 17) \times 10^{12}$ g N yr^{-1}, which would agree with the deposition within the quoted uncertainties.

It is noted that the actual deposition of NO_3^- between 1964 and 1975 in northern midlatitudes may have increased by more than 50 %. The deposition rates over Germany in the winter 1979/1980 (56) exceeded the deposition rates in the period from 1960 to 1970 by about a factor of 2; increases of a factor of 2 were also observed for a number of Swedish and Norwegian stations between the late 50's and early 70's (20). This would point to an increase in deposition at northern midlatitudes of about 60 to 70 % between 1964 and 1975. It thus appears that the global values for NO_x production in 1975 and NO_3^- deposition in 1964 are fairly consistent.

However, a detailed comparison of production and deposition with latitude reveals differences. For that purpose the latitudinal dependence of the deposition in 1964 is also included in Figure 9. It clearly shows a qualitatively different behavior than the production estimated for 1975. The maximum at northern midlatitudes which corresponds to the anthropogenic sources is much weaker and there is a maximum in the tropics where the deposition estimated for 1964 exceeds the production estimated for 1975. Correcting for the secular increase of the anthropogenic NO_x production to give 1964 values would lower the production maximum at northern midlatitudes to about 2.7×10^{12} g N yr^{-1}, and would bring production and deposition at midlatitudes into better agreement. However, a similar correction for an increase of anthropogenic biomass burning with time would clearly worsen the already existing disagreement in the tropics. Here we may have to conclude that either deposition in the tropics has been overestimated or production, probably that by biomass burning, has been underestimated. This is also an example of the importance of having independent estimates of source and sink strengths. The two estimates complement and constrain each other and thereby help to reduce the uncertainty of the budget.

THE TROPOSPHERIC DISTRIBUTION OF NO_x

It is clear that the nonuniform source distribution of the NO_x in conjunction with its short atmospheric lifetime must create large gradients in the NO_x concentration, especially at low altitudes. We expect high concentrations in the atmospheric boundary layer over the industrialized land masses and low concentrations over the remote ocean. This general feature is indeed observed: NO_x mixing ratios above 100 ppb are not rare in densely populated areas. Much lower values on the order of 0.1 ppb are observed in air blowing from the North Atlantic (57, 58). A record minimum of about 0.01 ppb NO_x (59) was measured over the equatorial Pacific (7 °N to 4 °S, 170 °W; July 1978). (In the last case actually 4 ppt of NO were measured from which a total NO_x concentration of 10 ppt were estimated using the measured local O_3 concentration and photolysis rate of NO_2).

The quoted measurements show that the NO_x concentrations at the surface span about 4 orders of magnitude, thus a large number of measurements will be required to establish the horizontal distribution of NO_x. This data base is not yet available, partly due to fact that techniques with sufficient sensitivity to measure the low NO_x concentration in the more remote areas are not widely available.

All we can do at present is to glean a crude outline of the horizontal NO_x distribution from the few existing NO_x measurements and the field of NO_3^- deposition shown in figure 6. The latter also provides an upper bound for the horizontal distances for which NO_x could have been transported from its sources.

The situation is somewhat better when it comes to the vertical profile. There we have enough data, at least at midlatitudes, to construct a full NO_x profile up to the tropopause. The measurements shown in figure 10 were made using two different techniques: Spectrometric, long path absorption around 4500 Å wavelength with the rising or setting sun as light source (60, 61, 62, 63), and chemiluminescence (64). Nearly all measurements were made aboard aircraft. They were collected under different conditions, and thus require some comments. Kley et al. (64) measured NO and NO_2. The authors corrected the actual measurements downward by 0.11 ppb, to account for artifact signals which were uncertain but less than 0.11 ppb. These measurements therefore represent a lower limit for total NO_x.

Brewer et al. (60, 61) and Kerr et al. (62, 63) measured NO_2. However, in the weak light of the rising and setting sun practically all (90 %) of the NO_x is present as NO_2, so that their data are thought to represent total NO_x. For all the measurements the tropopause must have been at around 10 to 11 km altitude. Despite some scatter the data show a clear behavior. There is a strong decrease from about 10 ppb at the earth's surface to about 0.2 ppb at around 6 km altitude, where the mixing ratio reaches a minimum. Above 6 km there is a slight increase to about 0.3 ppb at tropopause height. In the lower stratosphere, between 11 km and 15 km, the fall data (62) indicate a decrease with altitude. The summer data show an increase, but not as large as one would

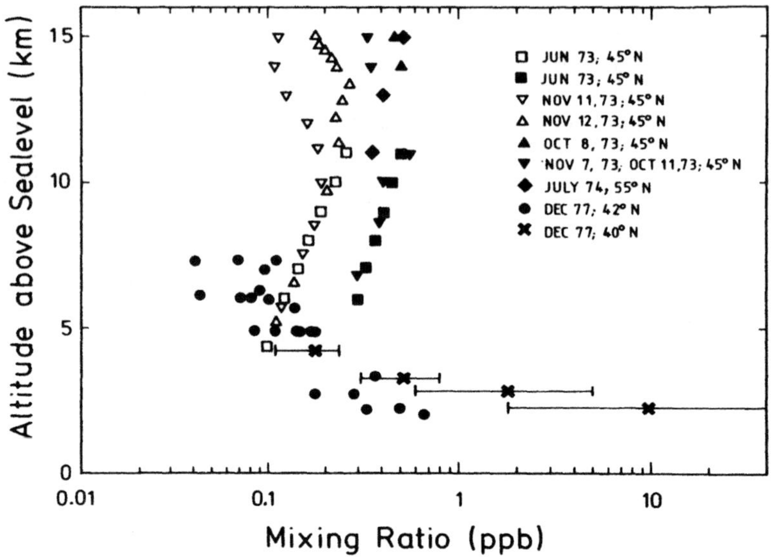

Figure 10. Vertical profiles of total NO_x at northern midlatitudes. The data represented by the full circles were collected over Wheatland, Wyoming, those represented by the crosses over Denver, Colorado, both in the afternoon (Kley et al. (64)). The bars indicate the data range. The other symbols represent data obtained by Brewer et al. (60, 61) and Kerr and McElroy (62, 63) over Ottawa, Ontario (squares), over the Bay of Biscay (triangles) and over Churchill, Manitoba (diamonds). Open symbols denote sunrise, filled symbols sunset measurements.

expect for a stratospheric NO_x source. A significant increase of stratospheric NO_x begins above 15 km (62). It should be borne in mind that most of the stratospheric odd nitrogen is present as HNO_3, and that at the cold temperatures around the tropopause another significant fraction of the NO_y could be present as HO_2NO_2 and N_2O_5. Also, a large part of the annual stratospheric NO_y should be injected in spring at midlatitudes.

The NO profile in Figure 10 should be representative for the continental troposphere at midlatitudes. It relies on the same data base as the profiles given by Kley et al. (64). A close examination of the individual optical measurements (60, 61, 62, 63) gives another interesting detail which should invite further investigation. There appears to be a systematic difference between the morning and evening measurements. The evening values are higher by nearly a factor of three, as was already pointed out by the authors. This is true for all of the profiles: those collected in spring, 1973, over southern Ontario, the profiles in June, 1973, over Ottawa, and in the November flights over the Bay of Biscay. A similar effect was found in the stratosphere. Thus, although there are altogether only 3 profiles at sunrise and 5 profiles at sunset, some incomplete and some with a fair amount of scatter, the difference between the NO_x concentration measured in the morning and that measured in the evening is a consistent feature of all observations. The difference of nearly a factor of three is difficult to explain. N_2O_5 and HO_2NO_2 which act as reservoirs for NO_x are thought to be either too low in concentration or too slow in reaction to cause this large diurnal variation in NO_x. This is clearly seen from the model calculation (1) at 6 km altitude (figure 4), which gives a ratio of 1.3 between sunset and sunrise mixing ratios of NO_x. Only slightly higher ratios are found near the tropopause despite increasing HO_2NO_2 mixing ratios due to lower temperatures. Thus present day chemistry and the model calculations based on it, do not confirm the factor of 3 suggested by the measurements presented in figure 10. If future measurements validate the "measured" factor of three, either some of the rate constants governing the exchange between NO_x and N_2O_5 or HO_2NO_2 are incorrect, or we have to assume a hitherto unknown reaction, possibly with another reservoir species such as organic peroxy nitrate.

It is instructive to compare the NO_x profile measured over the continental midlatitudes (figure 10) with other data on the tropospheric NO_x concentration (figure 11). According to the few available oceanic data (57, 58, 59, 65) the dashed line in Fig. 11 assumed by Kley et al. (64) for a tropospheric NO_x profile without ground sources could represent the NO_x profile over the remote midlatitudinal ocean. It and the continental profile taken from figure 10 could be viewed as the two extremes. They differ by nearly 3 orders of magnitude close to the surface owing to the large differences in the strengths of the NO_x surface sources, but agree well in the upper troposphere. Vertical profiles at intermediate locations in midlatitudes are expected to fall between the two extremes with a transition dependent on the strengths of the regional surface sources of NO_x.

The theoretical NO_x profile shown also in figure 11 is predicted by a 1-D model (1.) This profile describes a zonal average and assumes a downward flux of NO_y from the stratosphere of 1.4×10^8 molec cm^{-2} sec^{-1}, a NO_x column production in the tropopause of 3×10^9 molec cm^{-2} sec^{-1} and a fixed NO_x mixing ratio at the surface of 0.1 ppb (which corresponds to an upward flux of approximately to 8×10^8 molec cm^{-2} sec^{-1}. Most of the column production (e.g. lightning) is concentrated in the lowest 4 km of the troposphere. Although the model calculations were made for 45 ^oN latitude and equinox sunlight conditions and thus matched

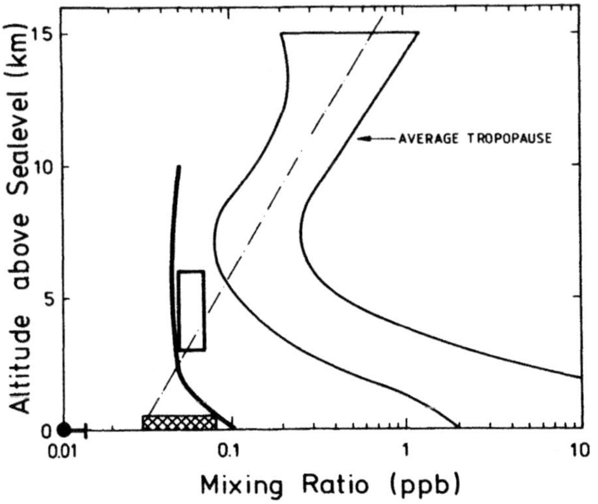

Figure 11. Measured and predicted vertical profiles of NO_x, at midlatitudes. The curved area represents the experimental profile over the industrial continents (Fig. 10). Additionally, oceanic measurements are given for the tropical Pacific (full circle: McFarland et al. (59)), for the Irish West Coast (crosshatched bar: Helas and Warneck, (58)), and for the atmosphere over the Pacific (vertical rectangle). The latter data are derived from NO measurements during GAMETAG, 1978 (Schiff et al. (65)). The dashed line (referenced to about 0.3 ppb at the tropopause, 11 km, and with a scale height of about -4 km) is thought to give the NO_x profile in regions where the stratosphere is the only source of NO_x (Kley et al. (64)). Solid curve: model predicted zonal average at 45 ^{o}N (Logan et al. (1)).

closely season and latitude of the observations in figure 10, the predicted profile does not agree with either the continental profile deduced here or the profile given by Kley et al. (64). Instead, the model predicted profile shows a weak decrease up to 3 km altitude, followed by a more or less constant NO_x mixing ratio of 0.04 ppb at higher altitudes. The discrepancy is probably due to the fact that the actual NO_x source strengths at midlatitudes differ considerably from those assumed in the model. Indeed, we calculate the zonally averaged surface source strength to be

3×10^{10} molec cm^{-2} sec^{-1} between 45° to 50° N (cf. figures 8,9), a factor of 40 higher than implied by the boundary conditions in the model. This high surface source strength certainly explains the stronger gradient and higher concentrations observed in the lower part of the experimental profile over the continents. It also should partly explain the fact that the mixing ratios observed at and above 6 km altitude are higher than those derived by the model. However, the positive gradient observed above 6 km altitude points to additional sources in the upper troposphere. Indeed, taking into account its latitudinal variation, the actual stratospheric influx at midlatitudes, 2.7×10^8 molec cm^{-2} sec^{-1} is about a factor of two higher than the global average flux assumed in the model (cf. Figure 8). However, this is not enough to cause a gradient of the magnitude experimentally observed. In fact, most of it appears to be caused by high flying aircraft which inject about 5×10^8 molec cm^{-2} sec^{-1} above 9 km altitude at 45° to 80° latitude. This source was not included in the model. It appears, therefore, that the difference between the predicted and the measured midlatitude profiles is mainly due to the absence of anthropogenic sources in the model. From this, we conclude that the tropospheric NO_x profile observed at northern midlatitudes is dominated at all altitudes by anthropogenic sources. Measured NO_x profiles at other less polluted latitudes could approach more closely the shape of the model predicted profile in Figure 11.

CONCLUSIONS

It appears that the rough outlines of the atmospheric NO_x cycle are understood. We seem to have a fair grasp of the gas phase chemistry, in the sense that the chemical system in the models includes a sufficient number of fast and important NO_x reactions to model the atmospheric chemistry reasonably well. However, the set of homogeneous gas phase reactions is probably still not complete. Heterogeneous reactions including rainout are poorly understood. The budget for both sources and sinks probably correct to within a factor of 2. The tropospheric distribution of NO_x is poorly known and will require much future research, not only on the distribution of NO_x but also on the gases which interconvert with NO_x, namely PAN, HO_2NO_2, N_2O_5 and HNO_3. Additional knowledge for these gases would also lead to a better understanding of the chemistry and budget of NO_x.

REFERENCES

(1) Logan, J.A., M.J. Prather, S.C. Wofsy, and M.B. McElroy, 1981, Tropospheric chemistry: a global perspective, J. Geophys. Res., 86, pp. 7210-7254
(2) Chameides, W.L. and A. Tan, 1981, The two-dimensional diagnostic model for tropospheric OH: an uncertainty analysis, J. Geophys. Res., pp. 5209-5223
(3) Chameides, W.L., and J.C.G. Walker, 1973, A photochemical theory of tropospheric ozone; J. Geophys. Res. 78, 8571-8760
(4) Crutzen, P.J., 1979, The role of NO and NO_2 in the chemistry of the troposphere and stratosphere, Ann. Rev. Earth Planet. Sci., 7, pp. 43-72.
(5) Fishman, J., S. Solomon, and P.J. Crutzen, 1979, Observational and theoretical evidence in support of a significant in-situ photochemical source of tropospheric ozone, Tellus, 31, pp. 432-446
(6) Perner, D., and U. Platt, 1979, Detection of nitrous acid in the atmosphere by differential optical absorption, Geo. Res. Lett., 6, pp. 917-920
(7) Platt, U., D. Perner, G.W. Harris, A.M. Winer, and J.N. Pitts, 1980, Observations of nitrous acid in an urban atmosphere by differential optical absorption, Nature, 285, pp. 312-314
(8) Spicer, C.W., 1977, The fate of nitrogen oxides in the atmosphere, Pitts, J.N.; Metcalf, R.L.; Lloyd, A.C.; Advances in Environmental Science and Technology, 7, pp. 163-262, John Wiley & Sons
(9) Platt, U., D. Perner, A.M. Winer, G.W. Harris, and J.N. Pitts, 1980, Detection of NO_3 in the polluted troposphere by differential optical absorption, Geophys. Res. Lett., 7, pp 89-92
(10) Platt U., D. Perner, J. Schröder, C. Kessler, and A. Toennissen, 1981, The diurnal variation of NO_3, J. Geophys. Res., submitted 1981
(11) Noxon, J.F., R. Norton, and E. Marovich, 1980, NO_3 in the troposphere, Geophys. Res. Lett., 7, pp. 125-128
(12) World Meterological Organization, NASA, FAA, NOAA, 1981, The stratosphere 1981, theory and measurements, appendix I, December 1981
(13) Graham, R.A., A.M. Winer, J.N. Pitts, 1978, Pressure and temperature dependence of the unimolecular decomposition of HO_2NO_2, J. Chem. Phys., 68, pp. 4505-4510
(14) Cox, R.A., and Roffey, 1977, Environ. Sci. Tech., 11, pp. 900
(15) Wipprecht, V., 1981, Diplomarbeit, Institut für Physikalische Chemie Universität Bonn, June, 1981
(16) Connell, P., and H.S. Johnston, 1979, The thermal dissociation of N_2O_5 in N_2, Geophys. Res. Lett., 6, pp. 553-556
(17) Hampson, R.F., 1980, Chemical kinetic and photochemical data sheets for atmospheric reactions, U.S. Dept. of Trans. Report No. FAA-EE-80-17

(18) Eriksson, E., 1952, Composition of atmospheric precipitation, 1, Nitrogen compounds, Tellus, 4, pp. 215-263
(19) Eriksson, E., 1959, Atmosfärens Kemi, Svensk Kemisk Tidskrift, 71, pp, 15-32
(20) Böttger, A., D.H. Ehhalt, and G. Gravenhorst, 1978, Atmosphärische Kreisläufe von Stickoxiden und Ammonika, Kernforschungsanlage Jülich, Ber. Jül-1558
(21) Höfken, K.D., 1981, Untersuchung über die Deposition atmosphärischer Spurenstoffe an Buchen- und Fichtenwald, Diplmarbeit, Institut für Meteorologie und Geophysik der Johann-Wolfgang-Goethe-Universität, Frankfurt, August
(22) Robinson, E., and R.C. Robbins, 1968, Source, abundances, and the fate of atmospheric pollutants, Final report, Stanford Research Institute, Project PR-6755 for American Petroleum Institute, New York
(23) Robinson, E., and R.C. Robbins, 1970, Gaseous nitrogen compound pollutants from urban and natural sources, J. Air Pollut. Control Ass. 20, pp. 303-306
(24) Robinson, E., and R.C. Robbins, 1971, Sources, abundances, and fate of gaseous atmospheric pollutants supplement, Supplement Report prepared for American Petroleum Institute, Stanford Research Institute, SRI-Project PR-6755, API-Publ. No. 4015
(25) McConnell, J.C., 1973, Atmosphric Ammonia, Journal of Geophysical Research, 78, No. 33, pp. 7812-7821
(26) Angstrom, A., and L. Högberg, 1952, On the content of nitrogen in atmospheric precipitation in Sweden, 2, Tellus, pp. 271-279
(27) Junge, C.E., 1963, Air Chemistry and Radioactivity, Academic Press, New York, London, 382 p.
(28) Jones, M.J., Ammonium and nitrate nitogen in the rainwater at Samaru, Nigeria, Tellus, 23, pp. 459-461, 1971
(29) Junge C.E., 1958, The distribution of ammonia and nitrates in rainwater over the United States, Transactions American Geophysical Union, 39, No. 2, pp. 241-248
(30) Petrenchuk, O.P., and E.S. Seleznewa, 1970, Chemical composition of precipitation in regions of the Soviet Union, J. Geophys. Res., 75, pp. 3629-3634
(31) Warneck, P. 1974, On the role of OH and HO_2 radicals in the troposphere, Tellus 26, pp. 39-46
(32) Burns, R.S., and R.W.F. Hardy, 1975, Nitrogen fixation in bacteria and higher plants, Springer-Verlag, Berlin-Heidelberg-New York, 189 p.
(33) Allison, F.E., 1965, Soil Nitrogen, W.V. Bartholomew and F.E. Clark, eds., Madison, Wisc., Am. Soc. Agron., pp. 573-606
(34) Henzell, E.F., and D.O. Norris, Commonwealth Bur. Pastures Field Crops, Bull. 46, Hurley, Berkshire, U.K. 1962
(35) Köhler, K. H., Biol. Rundschau 6, pp. 167-187, 1968
(36) Martin, D.F., Marine Chemistry, 2, New York: Dekker 1970
(37) Horne, R.A., Marine Chemistry, New York: Wiley 1969

(38) Chameides, W.L., 1978, The photochemical role of tropospheric nitrogen oxides, Geophys. Res. Lett., 5 (1), pp. 17-20
(39) Söderlund, R., and B.H. Svensson, 1976, The global nitrogen cycle, in Svensson, B.H., Söderlund, R. (eds.) N,P,S-Global Cycles, Scope Report 7, Ecol. Bull. (Stockholm) 22: pp- 23-73
(40) Steinhardt, U., 1973, Input of chemical elements from the atmosphere. A tabular review of literature. Göttinger Bodenkundl. Ber. 29: pp. 93-132
(41) Söderlund, R., 1976, Wet deposition of Inorganic Nitrogen Compounds over the Sea, Rep. AC-34, Stockholm, Department of Meteorology, University of Stockholm, International Meteorological Institute in Stockholm, 16 pp.
(42) EPA, 1973, Compilation of air pollutant emission factors (2nd edition), U.S. Environmental Protection Agency, Research Triangle Park, North Carolina, 27711, PB-223 996, Publication AP-42
(43) EPA, 1977, Compilation of air pollutant emission factors (including supplements 1-7), U.S. Environmental Protection Agency, Research Triangle Park, North Carolina 27711, AP-42
(44) NASA, 1981, Chemical kinetic and photochemical data for use in stratospheric modelling, Evaluation number 4: Panel for Data Evaluation, JPL publication 81-3
(45) 1976, Statistical Yearbook of the United Nations 1975
(46) Galbally, J.E., and C.R. Roy, 1978, Loss of fixed nitrogen from soil by nitric oxide exhalation, Nature, 275, pp. 734-735
(47) Marakov, B.N., 1969, Liberation of nitrogen dioxide from soil, Soviet soil science, pp. 20-25, translated from Pochvovedeniye No. 1, pp. 49-53
(48) Crutzen, P.J., L.E. Heidt, J.P. Krasnec, W.H. Pollack, and W. Seiler, 1979, Biomass burning as a source of atmospheric gases CO, H_2, N_2O, NO, CH_3Cl, and COS., Nature, 282, pp. 253-256
(49) Levine, J.S., R.S. Rogowski, G.L. Gregory, W.E. Howell, and J. Fishman, 1981, Simultaneous measurements of NO_x, NO, and O_3 production in a laboratory discharge: atmospheric implications, Geophys. Res. Lett., 8, pp. 357-360
(50) Noxon, J.F., 1976, Atmospheric nitrogen fixation by lightning, Geophys. Res. Lett., 3, pp. 463-466
(51) Chameides, W.L., D.H. Stedman, R.R. Dickerson, D.W. Rusch, and R.J. Cicerone, 1977, NO_x production in lightning Journal of Atmospheric Sciences 34, pp. 143-149
(52) CIAP, 1974, The effects of stratospheric pollution by aircraft, A.J. Grobecker, S.C. Coroniti, and R.H. Cannon, Jr., Dept. of Transportation, DOT-TST-75-50.
(53) Davidson, B., J.P. Friend, and H. Seitz, 1965, Numerical models of diffusion and rainout of stratospheric radioactive materials, Tellus, 18, pp. 301-315

(54) Ehhalt, D.H., and G. Haumacher, 1970, The seasonal variation in the concentration of strontium 90 in rain and its dependence on latitude, J. Geophys. Res., 75, pp. 3027-3031

(55) Zafiriou, O.C., M. McFarland, and R.H. Bromund, 1980, Nitric oxide in sea water, Science, 207, pp. 637-639

(56) Gravenhorst, C., C. Perseke, and E. Rohbock, 1980, Untersuchung über die trockene und feuchte Deposition von Luftverunreinigungen in der Bundesrepublik Deutschland, Institut für Meteorologie und Geophysik der Universität Frankfurt

(57) Platt, U., and D. Perner, 1980, Direct measurements of atmospheric CH_2O, HNO_2, O_3, NO_2, and SO_2 by differential optical absorption in the near UV, J. Geophys. Res., 85, pp. 7453-7458

(58) Helas, G., and P. Warneck, 1981, Background NO_x mixing ratios in air masses over the North Atlantic Ocean, J. Geophys. Res., 86, pp. 7283-7290

(59) McFarland, M., D. Kley, J.W. Drummond, A.L. Schmeltekopf, and R.H. Winkler, 1979, Nitric oxide measurements in the equatorial Pacific region., Geophys. Res. Lett., 6, pp. 605-608

(60) Brewer, A.W., C.T. McElroy, and J.B. Kerr, 1974, Nitrogen dioxide in the atmosphere, Nature, 246, pp. 129-133

(61) Brewer, A.W., C.T. McElroy, and J.B. Kerr, 1974, Spectrophotometric nitrogen dioxide measurements, Third conference on CIAP, U.S. Dept. of Transportation, pp. 257-262

(62) Kerr, J.B., and C.T. McElroy, 1976, Measurements of stratospheric nitrogen dioxide from the AES stratospheric balloon program, Atmosphere, 14, pp. 166-171

(63) Kerr, J.B., W.F.J. Evans, and J.G. McConnell, 1977, The effects of NO_2 changes at twilight on tangent ray NO_2 measurements, Geophys. Res. Lett., 4, pp. 577-579

(64) Kley, D., J.W. Drummond, M. McFarland, and S.C. Liu, 1981, Tropospheric profiles of NO_x. J. Geophys. Res., 6, pp. 605-608

(65) Schiff, H.I., D. Pepper, and B.A. Ridley, 1979, Tropospheric NO measurements up to 7 km, J. Geophys. Res., 84, pp. 7895-7897

(66) Robinson, E., and R. C. Robbins, 1975, Gaseous atmospheric pollutants from urban and natural sources, in: Singer, S.F. (ed.) The Changing Global Environment, D. Reidel Publishing Company, Dordrecht-Holland, pp. 111-123.

IN-SITU PRODUCTION OF TROPOSPHERIC OZONE FROM VARIOUS NO_x SOURCES CALCULATED IN A TWO-DIMENSIONAL ZONALLY AVERAGED PHOTOCHEMICAL MODEL

Louis T. Gidel

University of Miami, Rosenstiel School of Marine and Atmospheric Science, 4600 Rickenbacker Causeway, Miami, Florida 33149

Two dimensional distributions from photochemical model simulations are presented which show the influence of the various NO_x sources from industry, lightning, and the stratosphere on the tropospheric distributions of NO_x, HNO_3, and O_3. It is found that the best agreement with the observation is obtained if the ground level pollution source is included. The influence of the industrial activities in the northern hemisphere may then have led to a doubling of ozone concentrations in the lower troposphere.

1. INTRODUCTION

For over a decade now (Levy, 1971) the photochemical importance of ozone in the atmospheric chemistry has been recognized. The reason for this is that in the presence of ozone and water vapor, the reactions

$$O_3 + h\nu \rightarrow O(^1D) + O_2 \quad \lambda \lesssim 310\text{nm} \qquad \text{(R1)}$$

$$O(^1D) + H_2O \rightarrow 2OH \qquad \text{(R2)}$$

lead to the production of OH, which is by far the most important radical in the troposphere for all pure hydrocarbons, which contain either C-H or carbon double bonds, and many halogen- and sulfur-containing hydrocarbons. Without the presence of ozone in the troposphere, the composition of the Earth's atmosphere would therefore have been totally different from what it is now. If we furthermore realize that tropospheric ozone plays a surprisingly important role in earth's climate (Fishman et al., 1979),

H. W. Georgii and W. Jaeschke (eds.), Chemistry of the Unpolluted and Polluted Troposphere, 253–265.

it becomes clear that knowledge of the processes which determine not only the stratospheric, but also the tropospheric distribution of ozone must be considered as one of the most important goals in the field of atmospheric chemistry and aeronomy. This requires in its turn good knowledge of the chemistry and distribution of NO_x and HNO_3 since NO acts as a catalyst in the production of ozone. Such reactions clearly occur during photochemical smog episodes but they should also take place in the background troposphere during oxidation of CO, CH_4 and other hydrocarbons if only sufficient NO is present (Crutzen, 1973, 1974; Chameides and Walker, 1973; Fishman and Crutzen, 1978).

In the case of carbon monoxide oxidation ozone production occurs via

$$CO + OH \rightarrow H + CO_2 \quad (R3)$$

$$H + O_2 + M \rightarrow HO_2 + M \quad (R4)$$

$$HO_2 + NO \rightarrow OH + NO_2 \quad (R5)$$

$$NO_2 + h\nu \rightarrow NO + O \quad (\lambda \lesssim 400 nm) \quad (R6)$$

$$O + O_2 + M \rightarrow O_3 + M \quad (R7)$$

$$CO + 2O_2 \rightarrow CO_2 + O_3 \quad \text{Net}$$

whenever the ratio of atmospheric concentrations of NO and CO is sufficiently large. Similar relationships apply for the oxidation of methane and non-methane hydrocarbons. As there is a considerable anthropogenic influence on the tropospheric sources of NO_x (e.g., Crutzen, 1979), it is conceivable that tropospheric ozone may have been increasing over the past decades (Fishman and Crutzen, 1978). An increase was indeed reported for some regions of the troposphere by Angell and Korshover (1979) for the period 1969-1976, but is not clear from the analysis of Oltmans (1981). According to the photochemical theory of tropospheric ozone an appreciable portion (maybe most) of the ozone which is present in the troposphere is, therefore, produced in the troposphere in all those air volumes where enough NO is present. This is not always the case, however, as there are probably large areas in the globe with little NO present, e.g. in the Pacific (McFarland, et al., 1979). The processes by which this ozone production takes place are, however, rather slow, so that it is not easy to prove by direct observations that it does indeed occur. As on the other hand, the transport of ozone from the stratosphere to the troposphere is a rather discontinuous process in time and space, it is much more easily detectable than the global in-situ production of ozone. For this reason it was for a long time taken for granted that ozone was mostly produced in the stratosphere from where it is injected into the troposphere where it is destroyed by contact with the earth's surface (e.g. Fabian and Junge, 1970; Danielsen, 1968; Singh et al., 1978; Fabian and Pruchniewicz, 1977). The possibility of an important influence of tropospheric reactions on the tropospheric ozone production is, therefore, still derived from theoretical arguments, although some recent work by Fishman et al. (1980), Seiler and Fishman (1981) of the prevalence of positive correlations between O_3 and CO in the northern hemisphere troposphere may indeed be taken as an indication that tropospheric ozone production occurs in rural and oceanic air.

2. SOURCES OF NO_x CONSIDERED IN THIS STUDY

The following sources of tropospheric NO_x are known to exist:

1) Anthropogenic ground level input: to simulate this we have assumed a source of 20 Tg N/year (20 X 10^{12} g/year), emitted at the different latitudes as a function of industrial activities (Söderlund and Svensson, 1976).

2) Input from lightning: recent studies by Dawson (1980), Hill et al. (1980) and Levine et al. (1981) indicate a possible source strength of about 2-4 Tg N/yr from lightning, about an order of magnitude smaller than earlier estimates of Chameides et al. (1977) and Noxon (1976), but in reasonable agreement with Tuck (1976). Although it is clear from the discussions that the NO_x lightning source is not well established, as a working model, we will adopt a production rate of 3.5 Tg N/yr of odd nitrogen from lightning and we will distribute this globally between the ground and 7 km, adopting the latitudinal and seasonal lightning frequency analysis of Orville and Spencer (1979).

3) Downward flux from the stratosphere: this source of odd nitrogen is always included in the model and is in fact, together with the combustion source, the only reasonably well established source of NO_x. In two recent studies (Levy et al., 1980; Kley et al., 1981) it was hypothesized that this source of NO_x and HNO_3 determines the profile outside polluted areas and Liu et al. (1980) discussed the implications of this hypothesis for the budget of tropospheric ozone. The stratosphere source strength of NO_x which results from the reactions $N_2O + O(^1D) \rightarrow 2NO$ in the current model is estimated to be about 0.3 Tg N/yr (=$8x10^7$ molecules $cm^{-2}s^{-1}$), which is considerably smaller than the ground level combustion source of 20 Tg N/yr. But, as this NO_x is less affected by rainout, the expected residence time of this high altitude source of NO_x is much longer than that of the ground level NO_x.

3. ALTERNATIVE MODELS OF NO_x AND OZONE IN THE TROPOSPHERE

We have performed model calculations in which we separate the effects of the different NO_x sources on the tropospheric distribution of O_3, NO_x and HNO_3. By comparing the calculated and observed distributions of these gases, we hope that we may be able to discuss the relative importance of the different NO_x sources. We have performed computer runs with the following different assumptions:

A) all reactions which affect ozone in the troposphere are neglected, so that only the transport from the stratosphere influences the tropospheric distribution. This computer run demonstrates the "classical" theory of tropospheric origin. Ozone is destroyed by contact with the ground.

B) the stratospheric sources of odd nitrogen are the only ones considered in this run. However, the tropospheric odd nitrogen which is calculated from this source is assumed not to contribute to ozone formation in the troposphere by the "smog" reaction. In the stratosphere the normal reaction scheme is adopted, but in the troposphere only the ozone loss reactions are taken into account. The comparison between computer runs A and B will give a good idea about the importance of the reactions $O_3 + h\nu \rightarrow O(^1D) + O_2$, $O(^1D) + H_2O \rightarrow 2\ OH$, $OH + O_3 \rightarrow HO_2 + O_2$, and $HO_2 + O_3 \rightarrow OH + 2\ O_2$. Rainout of the odd nitrogen species is taken into account in this run.

C) the stratospheric sources of odd nitrogen are again the only ones considered, but the influence of NO_x on tropospheric ozone is now taken into account. The comparison between computer runs B and C is an indication of tropospheric ozone production due to natural odd nitrogen of stratospheric origin. The ozone and odd nitrogen species distribution derived with this model will also reveal how well the present model satisfies the hypothesis of Kley et al. (1980) and Liu et al. (1980) that the "clean" air tropospheric odd nitrogen distribution is dominated by the stratospheric input. Rainout of odd nitrogen is taken into account in this run.

D) the stratospheric source of odd nitrogen and the lightning source are considered. Comparisons between runs C and D will provide information on the possible importance of lightning NO_x in tropospheric photochemistry.

E) in this run we added the ground level industrial source NO_x to all odd nitrogen sources considered so far, in order to study its influence on tropospheric ozone.

4. RESULTS OF MODEL CALCULATIONS OF TROPOSPHERIC NO_x AND O_3

Run A: The only possible source for tropospheric ozone in this run is that supplied by the downward flux of ozone from the stratosphere. In the northern hemisphere the calculated ozone distribution actually agrees reasonably well with the observations. The observed and calculated (Fig. 1) ozone distributions in the troposphere disagree, however, badly especially in the tropics and in the southern hemisphere, where far too much ozone is transported down by the simulated meteorological exchange processes and too little surface ozone destruction takes place because of the low ozone destruction rates at the sea surface.

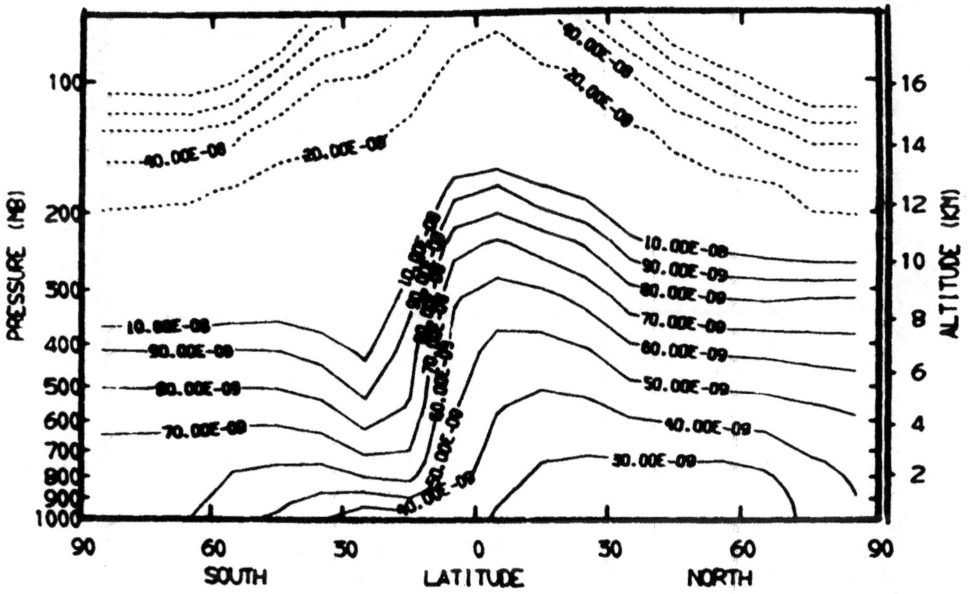

Fig. 1 Calculated meridional distribution of ozone volume mixing ratio for July for run A.

Run B: The neglect of the photochemical ozone production, but inclusion of the destruction terms in the ozone equation in the troposphere leads to a very efficient removal of tropospheric ozone, so that calculated ground level volume mixing ratios are always much smaller than what has been observed (See Fig. 2). Comparisons of the results from runs A and B demonstrate the potential of tropospheric reactions affecting ozone. The results are particularly striking in the southern hemisphere and tropics, where the large surplus of ozone calculated in run A has totally disappeared. It follows clearly that to make up for this ozone loss, production of ozone must take place by "smog-tupe" reactions especially in the northern hemisphere. It seems unlikely that knowledge of the photochemical reactions is so deficient as to invalidate these conclusions.

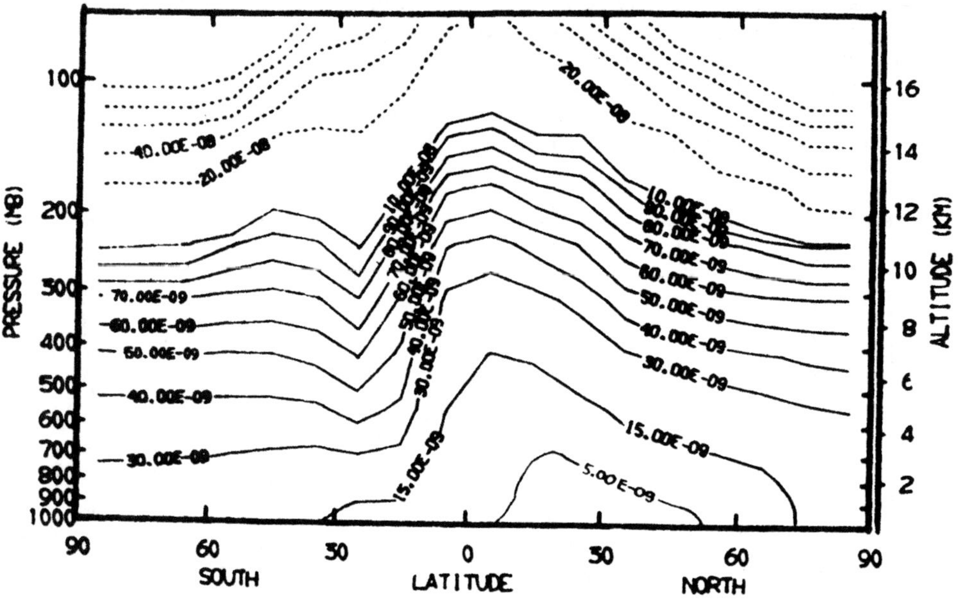

Fig. 2 Calculated meridional distribution of ozone mixing ratio for July for run B.

Run C: This model run does not simulate the observed tropospheric nitric acid distributions as observed during the GAMETAG aircraft field observations of May, 1978. The maximum calculated HNO_3 volume mixing ratios in the mid-troposphere are generally below 10^{-11}, while the measurements predominately give values in the range 5-20 x 10^{-11} northward of about 40°S (Huebert and Lazrus, 1978, 1980). Also the calculated NO_x distributions show less than ten times the concentrations of NO_x which are compiled in the study of Kley et al. (1980) and which includes as the lowest numbers their own set of measurements recorded over Wyoming in December of 1977 covering the volume mixing ratio range 5-10 x 10^{-11}. It is, therefore, clear that this model calculation does not confirm the view of Kley et al. (1980) and Liu et al. (1980) that background concentrations of tropospheric NO_x and HNO_3 are dominated by the downward flux from the stratosphere.

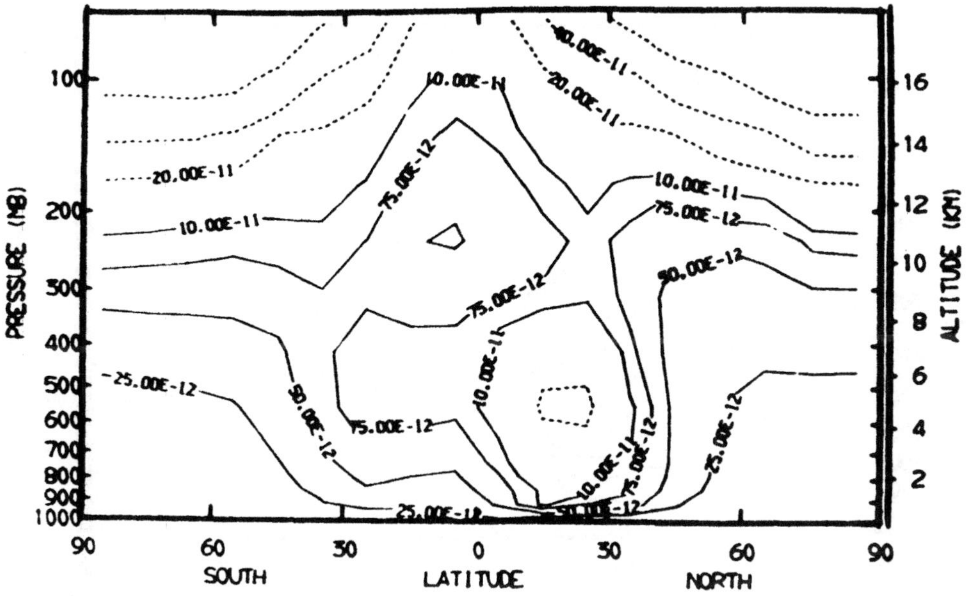

Fig. 3 Calculated meridional distribution of nitric acid mixing ratio for May for run D.

Run D: The inclusion of the odd nitrogen source from lightning as shown in Fig. 3 especially in the tropical troposphere, greatly improves the agreement between the observations of Huebert and Lazrus (1978, 1980) and the theoretical values. However, poleward of 40°N the calculated HNO_3 concentrations remain far below these observed. It is interesting to note from this run how little of the lightning produced NO_x can escape the tropical circulation system to reach latitudes polewards of 30°.

The results from this run could possibly be taken as to represent the composition of the natural troposphere. The calculated ozone distribution for summer is shown in Fig. 4.

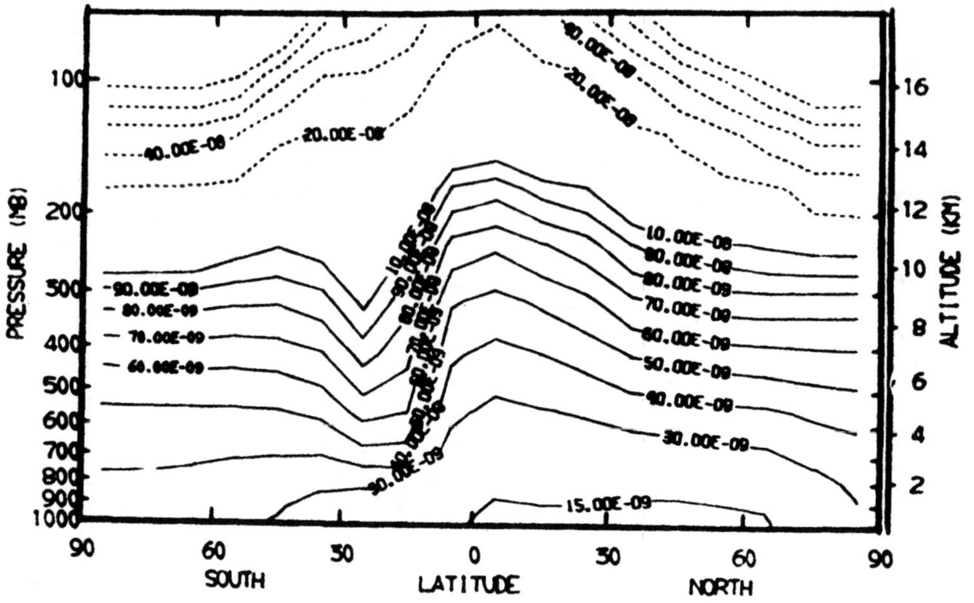

Fig. 4 Calculated meridional distribution of ozone mixing ratio for July for run D.

Run E: With the ground level pollution source of odd nitrogen added to the lightning and stratospheric sources agreement with the observations of the most abundant odd nitrogen species HNO_3 is improved considerably for all cases considered. Most calculated volume mixing ratios of HNO_3 agree roughly within a factor of two with the findings of Huebert and Lazrus (1978, 1980), there is a wall of less than 50 pptv of HNO_3 south of 40°S. Even in this region, however the stratospheric contribution of tropospheric odd nitrogen could be matched by the lightning source. It is interesting to compare the ozone distribution obtained in runs D and E. Run D may be interpreted as a simulation of the natural troposphere without any influence of industrial pollution (compare Fig. 4 with Fig. 5). The inclusion of the ground level NO_x pollution source leads to more than twice as much ozone in the northern hemisphere lower troposphere in summer at mid-latitudes, while there is relatively little change in the southern hemisphere. This computer run tends to confirm, therefore, the hypothesis by Fishman and Crutzen (1978) that tropospheric ozone in the northern hemisphere is substantially influenced by industrial activities. The most important problem with these simulations is that observed and calculated NO_x ($NO + NO_2$) distributions do not agree too well. Compared to the few measurements listed by Kley et al., (1981) we seem to calculate too little NO_x in the summer and too much in the winter (see Fig. 6). Especially during the winter at middle latitudes in the northern hemisphere our calculated total odd nitrogen field is dominated by the industrial inputs up to the tropopause. This must be due to the less efficient removal of odd nitrogen in the model by precipitation processes and can be explained by the inefficient conversion of NO_x to HNO_3 our model which in its turn is due to very low calculated average OH concentrations, where there is an abrupt break from average day-time OH concentrations of about 10^6 cm^{-3} at 35°N to 10^5 cm^{-3} at 45°N. This drastic change in OH concentrations reflects both the much smaller penetration of UV radiation and lower water vapor concentrations. Due to higher OH concentrations in the summer time on the other hand, most odd nitrogen is present as nitric acid which is removed efficiently by precipitation processes. It is interesting to note that in the northern hemisphere summer time, the odd nitrogen field shows everywhere increasing mixing ratios with altitude except below 5 km in regions with strong industrial developments. This means that under these conditions the odd nitrogen distribution in most regions may indeed be determined by

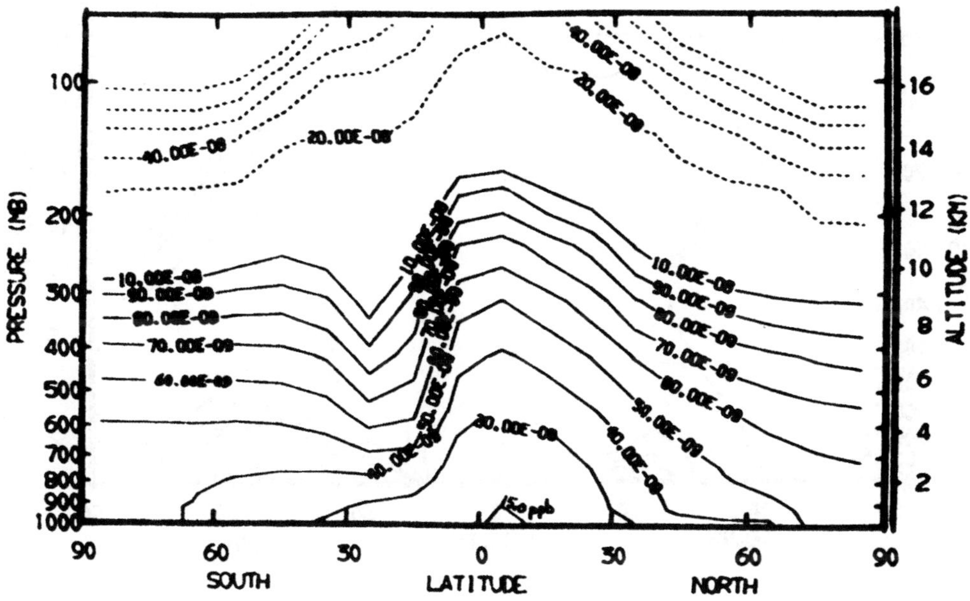

Fig. 5 Calculated meridional distribution of ozone mixing ratio for July for run E.

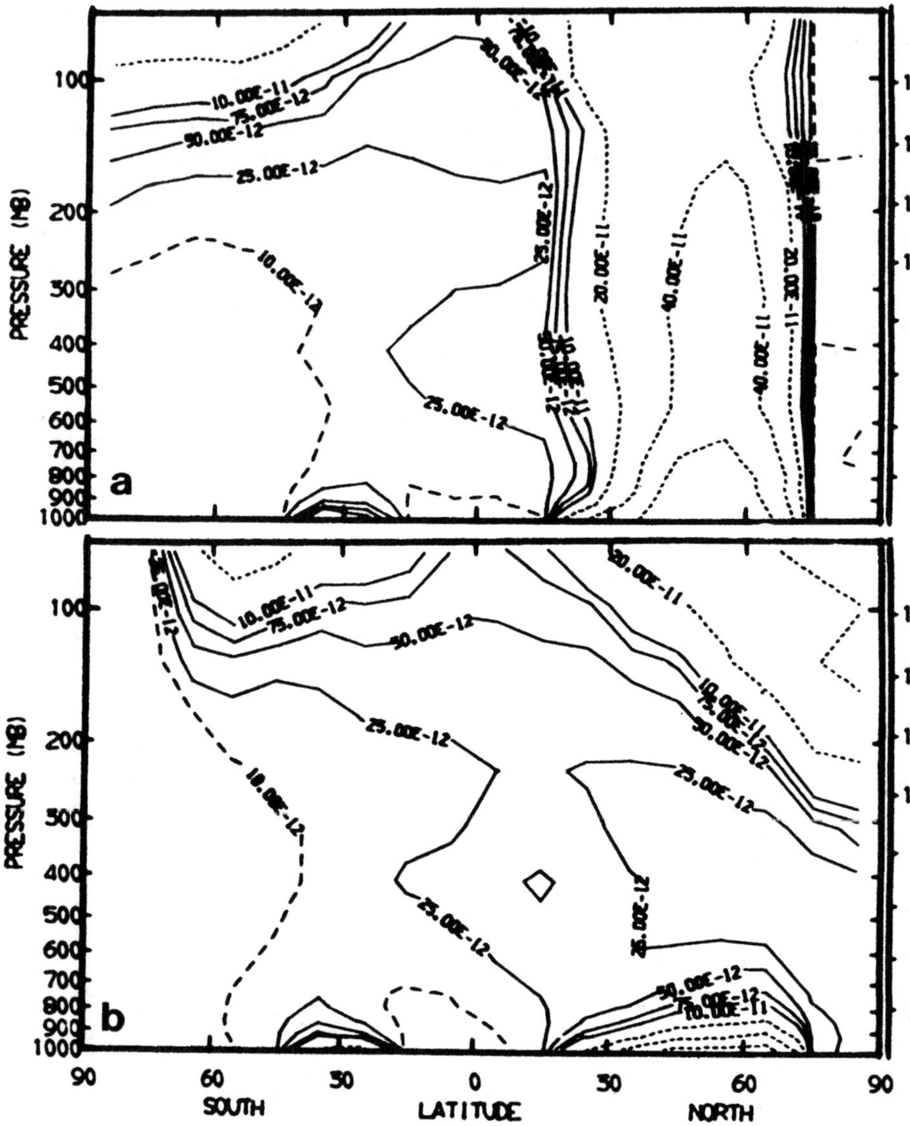

Fig. 6 Calculated meridional distribution of NO_x mixing ratio for (a) January and (b) July for run E.

natural processes, especially lightning, and downward flux from the stratosphere. In the northern hemisphere winter the dominence of the industrial NO_x source is much more pronounced. Interestingly in this model, the largest influence on tropospheric ozone occurs, however, in the summer as a comparison of the results obtained by model runs D and E demonstrate. Clearly, this model run indicates the possibility of a substantial buildup of NO_x at middle and high latitudes in the northern hemisphere in winter.

5. REFERENCES

Angell, J.K. and Korshover, J.: 1979, "Comparison of ozone variations derived from ozonosondes and Umkehr measurements for the period 1969-1976", Mon. Weather Rev. 107, pp. 599-607.

Chameides, W.L., Stedman, D.H., Dickerson, R.R., Rusch, D.W. and Cicerone, R.J.: 1977, "NO_x production in lightning", J. Atmos. Sci 34, pp. 143-149.

Chameides, W.L. and Walker, J.C.G.: 1973, "A photochemical theory of tropospheric ozone", J. Geophys. Res. 78, pp. 8751-8760.

Crutzen, P.J.: 1973, "A discussion of the chemistry of some minor constituents in the stratosphere and troposphere", Pageoph. 106, p. 1385.

Crutzen, P.J.: 1974, "Photochemical reactions initiated by and influencing ozone in unpolluted tropospheric air", Tellus 26, pp. 47-57.

Crutzen, P.J.: 1979, "The role of NO and NO_2 in the chemistry of the troposphere and stratosphere", Ann. Rev. Earth Planet Sci 7, pp. 443-472.

Danielsen, E.F.: 1968, "Stratospheric-tropospheric exchange based on radioactivity, ozone and potential vorticity", J. Atmos. Sci. 205, pp. 502-518.

Dawson, G.A.: 1980, "Nitrogen fixation by lightning", J. Atmos. Sci. 37, pp. 174-178.

Fabian, P. and Junge, C.E.: 1970, "Global rate of ozone destruction at the earth's surface", Arch. Meteorol. Geophys. Biol. climatol., Sec. A, 19, pp. 161-172.

Fabian, P. and Pruchniewicz, P.G.: 1977, "Meridional distribution of ozone in the troposphere and its seasonal variations", J. Geophys. Res. 82, pp. 2063-2073.

Fishman, J., Seiler, W. and Haagenseon, P.: 1980, "Simultaneous presence of O_3 and CO bands in the troposphere", Tellus 32, pp. 456-463.

Fishman, J. and Crutzen, P.J.: 1978, "The origin of ozone in the troposphere", Nature 274, pp. 855-858.

Fishman, J., Ramanathan, V., Crutzen, P.J. and Liu, S.C.: 1979, "Tropospheric ozone and climate", Nature 282, pp. 818-820.

Hill, R.D., Rinker, R.G. and Wilson, H.D.: 1980, "Atmospheric nitrogen fixation by lightning", J. Atmos. Sci. 37, pp. 179-192.

Huebert, B.J. and Lazrus, A.L.: 1978, "Global tropospheric measurements of nitric acid vapor and particulate nitrate", Geophys. Res. Lett. 5, pp. 577-580.

Huebert, B.J. and Lazrus, A.L.: 1980, "Tropospheric gas-phase and particulate nitrate measurements", J. Geophys. Res. 85, pp. 7322-7328.

Kley, D., Drummond, J.W., McFarland, M. and Liu, S.C.: 1981, "Tropospheric profiles of NO_x", J. Geophys. Res. 86, pp. 3153-3161.

Levine, J.S., Rogowski, R.S., Gregory, G.L., Howell, W.E. and Fishman, J.: 1981, "Simultaneous measurements of NO_x, NO and O_3 production in a laboratory discharge: Atmospheric implications", Geophys. Res. Lett. 8, 357-360.

Levy, H. II, Mahlman, J.D. and Moxim, W.J.: 1980, "Stratospheric NO: A major source of reactive nitrogen in the unpolluted troposphere", Geophys. Res. Lett. 7, pp. 441-444.

Levy, H. II: 1971, "Normal atmosphere: Large radical and formaldehyde concentrations predicted", Science 173, pp. 141-143.

Liu, S.C., Kley, D., McFarland, M., Mahlman, J.D. and Levy, H. II: 1980, "On the origin of tropospheric ozone", J. Geophys. Res. 85, pp. 7546-7552.

McFarland, M., Kley, D., Drummond, J.W., Schmeltekopf, A.L. and Winker, R.H.: 1979, "Nitric oxide measurements in the equatorial Pacific region", Geophys. Res. Lett. 6, pp. 605-608.

Noxon, J.F.: 1976, "Atmospheric nitrogen fixation by lightning", Geophys. Res. Lett. 3, pp. 463-465.

Oltmans, S.J.: 1981, "Surface ozone measurements in clean air", J. Geophys. Res. 86, pp. 1174-1180.

Orville, R.E. and Spencer, D.W.: 1979, Global lightning flash frequency", Mon. Weather Rev. 107, pp. 934-943.

Seiler, W. and Fishman, J.: 1981, "The distribution of carbon monoxide and ozone in the free troposphere", J. Geophys. Res. 86, pp. 7255-7265.

Singh, H.B., Ludwig, F.L. and Johnson, W.B.: 1978, "Tropospheric ozone: Concentrations and variabilities in clean remote atmosphere", Atmos. Environ. 12, pp. 2185-2196.

Söderlund, R. and Svensson, B.H.: 1976, "The global nitrogen cycle", Ecol. Bull. (Stockholm) 22, pp. 23-73.

Tuck, A.F.: 1976, "Production of nitrogen oxides by lightning discharges", Q.J.R. Meteorol. Soc. 102, pp. 749-755.

EXPERIMENTAL DETECTION OF OH IN THE TROPOSPHERE

D. Perner and G. Hübler

Kernforschungsanlage Jülich GmbH, Institut für Chemie 3: Atmosphärische Chemie, Postfach 1913, D-5170 Jülich, FRG

ABSTRACT

Recent data on tropospheric OH concentrations are given. Direct measurements by laser induced fluorescence and laser light absorption are presented in more detail but also indirect results from budget consideration are included. The data are compared with model calculations.

1. INTRODUCTION

Bates and Witherspoon /1/ made the first study of the atmospheric photochemistry and cycles of carbon monoxide, CO, and methane, CH_4. They identified reactions involving OH and HO_2 as likely sinks. However, when studying the removal rates they assumed that these radicals were formed by H_2O photolysis and therefore those processes could have been important only in the upper atmosphere.

Based on measurements of the ^{14}CO content of atmospheric CO and on estimates of the ^{14}CO formation by cosmic rays, Weinstock /2/ derived a mean tropospheric residence time of 0.1 year for CO. This value disagreed with earlier estimates of 1-2 years counting mainly anthropogenic sources /3/. Global CO measurements /4, 5/ suggested that significant natural sources for CO must exist as well and the lifetime estimates from those results were equally inconsistent with an exclusive stratospheric sink.

Levy /6/ provided a solution to this problem and extended stratospheric models /7, 8/ into the troposphere and identified the principal source (1), (2) for OH.

H. W. Georgii and W. Jaeschke (eds.), Chemistry of the Unpolluted and Polluted Troposphere, 267–294.

$O_3 + h\nu$	$\rightarrow$	$O(^1D) + O\ (^1\Delta g)$		(1)	$\lambda \leq 310$ nm
$O(^1D) + H_2O$	$\rightarrow$	$2\ OH$	86 %	(2)	
$O\ (^1D) + H_2O$	$\rightarrow$	$O(^3P) + H_2O$	14 %		/9/
$O(^1D) + N_2, O_2$	$\rightarrow$	$O(^3P) + N_2, O_2$		(3)	
$O(^3P) + O_2$	$\overset{M}{\rightarrow}$	O_3		(4)	

He predicted appreciable abundances of OH radicals near the ground and showed that OH could be of importance for atmospheric photochemistry of CH_4 and CO. McConnell et al. /10/ proposed the formation of CO after attack of methane by OH radicals via the intermediate product formaldehyde.

These developments raised much interest in methane and carbon monoxide but also focussed the attention on the OH radical.

2. THE ROLE OF OH

In general the decay rate of a compound y in air is given by the expression (5) which includes photolysis

$$\frac{dy}{dt} = -\ k_y\ [y] \qquad (5)$$

and reactions with reactive species like O_3, O, OH etc. (6)

$$k_y = J_y + k_{O_3}\ [O_3] + k_O\ [O] + k_{OH}\ [OH] + .. \qquad (6)$$

The importance of the OH radical can be checked by comparing the rates for the gasphase homogeneous reaction of various compounds with O_3, O and OH as well as their photolysis rates. Table 1 shows convincingly that OH could be the main initiator of oxidation processes in air. This fact has led to the suggestion that OH rate coefficients may be a good measure for the persistance of chemicals in the atmosphere. If the OH concentration, [OH]*, is known then the atmospheric lifetime τ_y of a substance y can be given with respect to OH. For example a comparison of τ_y with the time for atmospheric transport indicates to what extent a compound is already degraded in the troposphere and how much escapes into the stratosphere.

For such a system however reliable measurements of atmospheric OH under varying atmospheric conditions as well as the determination of rate constants with OH are essential. While the latter data seem to have been well established the OH measurements remain still unsatisfying.

---

* Chemical symbols in square brackets denote concentration in molec cm^{-3}.

Table 1. Loss rate for 1 ppb of C_2H_4, C_3H_8 and CH_2O by different processes

	concentrations in molecules cm^{-3}	loss rate cm^{-3} sec^{-1} C_2H_4	C_3H_8	CH_2O
Unpolluted Air	$[O_3] = 7.5 \times 10^{11}$	3×10^4	-	-
	$[O] = 4 \times 10^3$	62	2.6	16
	$[OH] = 5 \times 10^6$	1×10^6	1.9×10^5	1.8×10^6
	Photolysis	-	-	1×10^6
Polluted Air	$[O_3] = 4 \times 10^{12}$	1.5×10^5	-	-
	$[O] = 5.5 \times 10^5$	8.5×10^3	3.6×10^2	2.2×10^3
	$[OH] = 5 \times 10^5$	1×10^5	1.9×10^4	1.8×10^5
	Photolysis	-	-	1×10^6

Almost all major anthropogenic emissions (e.g. SO_2, NO_2, hydrocarbons) emitted in populated areas react with OH. Provided the [OH] is high enough it could efficiently control the removal of those compounds. While this scheme holds for the remote troposphere, in polluted air ozonolysis (Table 1) and photolysis are sometimes more efficient for selected compounds.

The unpaired electron of the free radical OH is mainly responsible for its reactivity. OH may combine with other radicals (7),

$$OH + NO_2 \xrightarrow{M} HNO_3 \qquad (7)$$

add to unsaturated hydrocarbons (8)

$$CH_2 = CH_2 + OH \xrightarrow{M} CH_2 - CH_2OH \qquad (8)$$

or form water, a preferred reaction of OH, by hydrogen abstraction from alkanes (9).

$$R' - CH_2 - R'' + OH \rightarrow R' - CH - R'' + H_2O \qquad (9)$$

(R', R" denote alkylgroups or H atoms)

However it does not react with the main air constituents N_2, O_2, CO_2, H_2O and noble gases.

To estimate the steady state concentration of radicals like OH, HO_2 etc. is rather difficult in view of the multitude of radical production channels. Compounds like O_3, H_2O_2, HNO_2, CH_2O and other serve as primary radical sources and are discussed in a later section.

Especially the OH production will vary with time and location according to reactions (1) and (2). It depends on the concentration of water which is closely linked to ambient temperature, on the ozone concentration (for a global distribution see /59/) and on the ozone photolysis rate J_1. The latter varies with the stratospheric ozone column density and the solar zenith angle /11, 12/.

In photochemical oxdiation cycles shortlived species are produced which yield radicals as well. This means that the radicals recycle and even may multiply. Only detailed model calculations can assess the combined effects in such a system properly. Unfortunately some of the mechanisms are not precisely enough known. Therefore direct measurements of [OH] are an absolute necessity.

3. MEASUREMENT OF OH

The [OH] may be determined indirectly by its chemical action upon tracers or directly by using physical properties of the OH molecule.

OH concentrations on a longer time scale are obtained from the steady-state relation. If OH removes a tracer by chemical reaction with the rate constant k, the time integrated [OH] is given by (10), which is similar to (5).

$$OH = \frac{\text{source strength (tracer cm}^{-3}\text{sec}^{-1}\text{)}}{k[\text{tracer}]} \qquad (10)$$

The steady-state approach for the measurement of the average tropospheric [OH] was introduced by Weinstock /2, 13, 14/ who used the naturally produced ^{14}CO as a tracer. Volz /15, 16/ following this approach made extensive measurements. He derived a global average (Table 2) and observed as well seasonal and latitudinal variations.

Table 2. OH concentrations - globally and anually averaged as derived from budget considerations

	OH x 10^{-6} cm^{-3}	compound
Warneck, 1975 /17/	0.7	NO_x
Lovelock, 1977 /18/	0.3 - 0.5	CH_3CCl_3
Neely and Plonka, 1978 /19/	0.5 - 1.8	CH_3CCl_3
McConnell and Schiff,1978 /20/	0.5	CH_3CCl_3
Singh et a., 1979 /21/	0.3 - 0.5	CH_3CCl_3
Hanst et al., 1980 /22/	0.4	CO
Volz et al., 1981 /16/	0.45 - 0.95	^{14}CO

Several other authors used methylchloroform as a tracer (Table 2). The original discrepancy between the ^{14}CO and CH_3CCl_3 results for the northern hemisphere is solved when the revised rate constant for the reaction of OH with methylchloroform is used to derive [OH] /22/. Hanst et al. /23/ reconsidered the global CO budget and found a reasonable source to sink balance assuming 4 x 10^5 cm^{-3} OH as a global average value for the OH concentration (Table 2).

As the OH concentrations will vary widely according to air composition and other parameters [OH] more defined with respect to location and time can be obtained only by budget considerations of selected suitable compounds. Calvert /24/ converted the hydrocarbon consumption in Los Angeles plume to an average concentration of (2 $\pm$ 1.8) x 10^6 OH cm^{-3} for the morning hours. The apparent OH concentration depended on the type of compounds. Singh et al. /25/ also stressed this point. Typical results are shown in Figure 1.

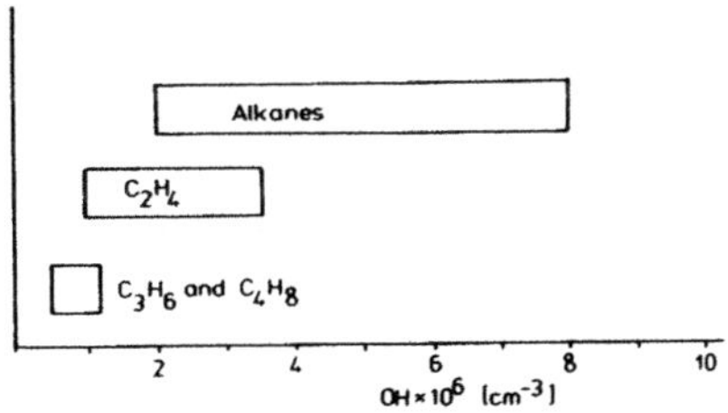

Figure 1. [OH] from hydrocarbon depletion

For example the very reactive C_3H_6 and 1-butene yielded an apparent OH-concentration of only 7 x 10^5 cm^{-3}. This somehow unexpected behaviour may indicate the competition of other processes as well. Neubert et al. /26/ reported a similar observation recently.

Campbell et al. /27/ determined OH-concentrations of 3 x 10^5 to 3.5 x 10^6 cm^{-3} by measuring the local CO oxidation rate using a ^{14}CO tracer method. Interferences may occur with this method through quenching of peroxy radicals by the walls of the teflon bags yielding low [OH]. On the other side light induced conversion of ^{14}CO at the walls could lead to apparently high [OH].

3.1 Direct Measurements

Exact answers about the importance of OH for tropospheric oxidation processes can only be given by physical methods. The results published so far seem to have been hampered by experimental difficulties. Efforts are made to measure OH directly by optical fluorescence and absorption. While the fluorescence method is compact and versatile in application, it seems still to suffer from interferences. These are mainly caused by the high flux density of the laser light. The absorption method uses 3-10 km light paths and offers today the possibility of unambiguous identification and quantitative determination of OH concentrations and does not suffer from interfering OH production by the sounding laser beam.

The structure and spectrum of hydroxyl: The diatomic hydroxyl molecule shows a strong electronic transition between the excited $A^2\Sigma^+$ state and its ground state $X^2\Pi$ near 300 nm. As a diatomic molecule it has a finite moment of inertia about all axes of inertia perpendicular to the join of the nuclei. The angular momentum of the rotating nuclei is Nh and the component of electronic orbital angular momentum about the inter-nuclei axis Λh. N and Λ are integers. If Λ is zero, the molecule is said to be in a Π state, if Λ is 1, the molecule is in a Σ state. The groundstate of the OH radical is a Π state. Its electronic spin angular momentum, Σh, is 1/2 due to the unpaired electron.

The total angular momentum, orbital plus spin, is Jh, where J can take the values $\Lambda + N \pm \Sigma$. If the negative sign is taken, the values of J form the sequence

1/2, 3/2, 5/2.

Those with the positive sign for Σ form the sequence

J = 3/2, 5/2, 7/2

and have the lower energy (Figure 2).

The two sequences of the ground state $^2\Pi$ are indicated by the superscript 2. The sequence is shown by a suffix equal to the last value of J:

$^2\Pi_{3/2}$ or $^2\Pi_{1/2}$

and if need be the actual value of J is added.

$^2\Pi_{3/2}$, J = 5/2 or $^2\Pi_{1/2}$, J = 3/2

The energy of a molecule depends on its electronic state, the vibration of its nuclei and the angular momentum of nuclei and electrons; the Born-Oppenheimer approximation, the basis for the foregoing description, depends on the idea that the velocities of the electrons are very much greater than those of the nuclei. Therefore the latter may be ignored in discussing the former.

However, in a better approximation it is found that all levels of the $^2\Pi$ ground state are splitted into doublets (see Figures 2 and 3). The splitting is known as Lambda doubling and is caused by interactions between the rotation of nuclei and electronic orbital angular momentum. The transition between those two levels has been observed as maser radiation in the galaxy /28/ at 1666 Mhz (~ 0.05 cm^{-1}).

The two levels of a Λ doublet have opposite parity P (Figures 2, 3). In electronic dipole transitions the selection rule for J is: J → J and J $\pm$ 1 → J. It must be strictly satisfied. There is also the selection rule K → K and K $\pm$ 1 → K. Transitions that satisfy both selection rules form strong branches. In addition there is the symmetry selection rule even $\nrightarrow$ odd which also is strict for the free molecule. With these rules 12 possible branches are obtained (Figure 2).

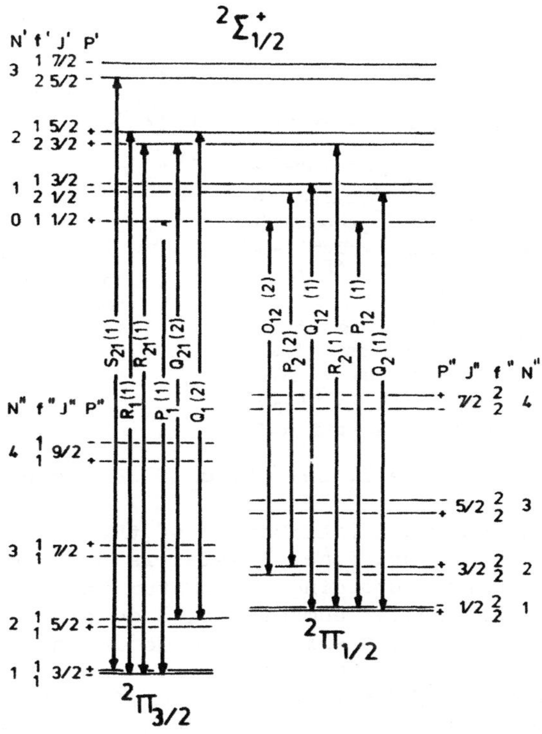

Figure 2. Transitions between OH rotational levels at 308 nm
f"; f': J = K + 1/2
f"; f': J = K - 1/2

There is a further contribution to the energy of the hydroxyl molecule, the hyperfine interaction between the electrons and the proton which depends on the total quantum number F, equal to the vector sum of J and I, where I is the spin quantum number of the proton, namely 1/2. The two possible values of F are thus

$$F_1 = J + 1/2 \text{ and } F_2 = J - 1/2$$

which give rise to two hyperfine levels (Fig. 3).

The spacing of the vibrational energy levels in OH is relatively large so that even the number of molecules in the first vibrational level (v = 1) will be very low at ambient temperature.

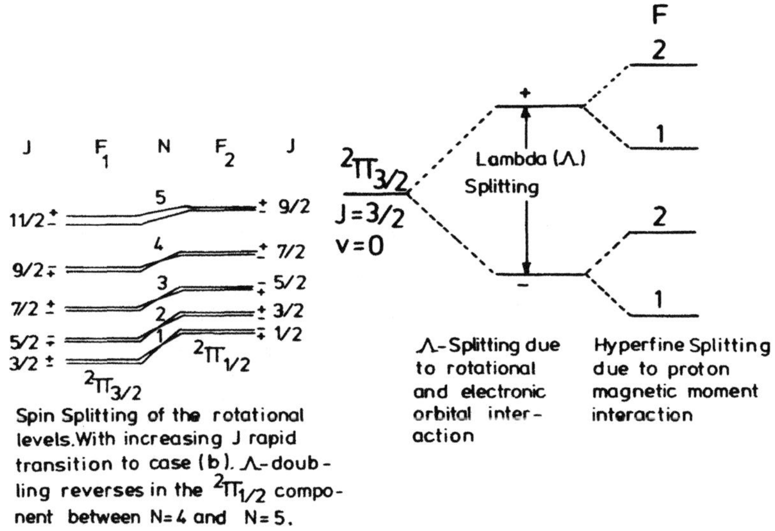

Figure 3. Splitting of OH groundstate rotational level

The thermal distribution of the rotational levels, unlike that of the vibrational levels, is not simply given by the Boltzmann factors. According to quantum theory each state of a system with total quantum number F is degenerated and consists of 2F + 1 magnetic sublevels which coincide in the absence of an external field. The frequency of its occurrence (its statistical weight) is therefore (2F + 1) times that of a state with J = O. The number of molecules N_J in the rotational level J of the lowest vibrational state at the temperature T is proportional to

$$2(F_1 + F_2 + 1)\ e^{-E/kT}$$

The rotational distribution for OH at 293 K is given in Figure 4.

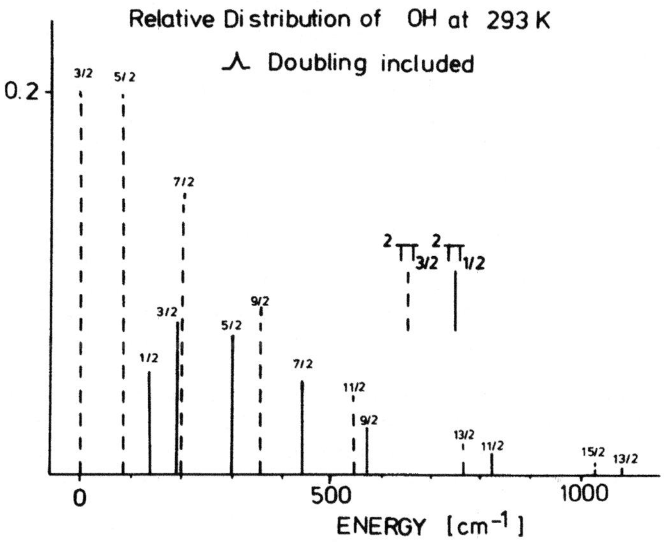

Figure 4. Rotational distribution of OH at 293 K

As each ground state level is Lambda doubled and because the selection rule even $\rightleftarrows$ odd is strict, only one half of the population indicated on the ordinate Figure 4 can be counted for any rotational transition from the X $^2\Pi$ state.

Most important for quantitative spectroscopy is the absolute strength of a line transition which may be conveniently expressed in terms of the oscillator strength f. The absorption cross section at the line center is given by (11).

$$\sigma = a\, r_o\, f \frac{\Delta n}{n} \frac{1}{\nu} \qquad (11)$$

σ = absorption cross section at line center cm^2
a = shape factor Gausz $a = \pi$
Lorentz $a = 2$
f = oscillator strength
r_o = classical electron radius 2.82×10^{-13} cm
$\frac{\Delta n}{n}$ = statistical weight
ν = linewidth at half maximum cm^{-1}

The oscillator strength for OH is obtained from the recently calculated list of Einstein coefficients /28/ which are based on lifetime measurements /30/. For the Q_1 (2) transition from the $^2\Pi_{3/2}$ (v" = 0), J = 5/2 groundstate (Figure 2) the line shows a Voigt profile and the oscillator strength is 7.3×10^{-4}. The Doppler line width at 293 K is 0.1 cm^{-1} and pressure induced broadening in 1 atm air leads to a total line width of about

$\nu = 0.16\ cm^{-1}$ /31, 32/. The absorption cross section at line center is $\sigma = 4 \times 10^{-16}\ cm^2$ taken the statistical weight of 0.1 and a Gaussian shape. It has to be kept in mind that σ varies according to the line width and line shape which are influenced by temperature and pressure. However, a line as it appears in the instrument may be influenced by the spectrograph and the absorption cross section has to be changed according to those effects as well.

Laser induced fluorescence (LIF): Originally proposed by Baardsen and Terhune /33/ the first atmospheric measurements of OH by laser induced fluorescence, LIF, have been reported by Wang and Davis /34/. In these experiments multiphoton water decomposition, one of the many possible interferences, had yielded more than 10^8 OH cm^{-3} /35/. The method is mainly used with the excitation of OH into the first vibrational level of the $A^2\Sigma$ state and the observation of the subsequent OH ($A^2\Sigma^+$, v' = 0 → $X^2\Pi$, v" = 0) emission (Figure 5). This is possible because the vibrational relaxation of OH (12) is very effective and even faster than

$$OH\ (A^2\Sigma^+,\ v' = 1) \xrightarrow{air} OH\ (A^2\Sigma^+,\ v' = 0) \qquad (12)$$

$8.9 \times 10^{-11}\ cm^3\ sec^{-1}$ /36/

physical quenching (13). The latter process causes the

$$OH\ (A^2\Sigma^+,\ v' = 0,1) \xrightarrow{air} OH\ (X^2\Pi,\ v'' = 0,1) \qquad (13)$$

$5.4 \times 10^{-11}\ cm^3\ sec^{-1}$ /37/

percentage of excited particles which emit fluorescent radiation (14) to be rather low in the troposphere.

$$OH\ (A^2\Sigma^+,\ v' = 0) \rightarrow OH\ (X^2\Pi,\ v'' = 0) + h\nu \qquad (14)$$

$1.4 \times 10^9\ sec^{-1}$ /30/

The best sensitivity for the detection of OH fluorescence is obtained by pulsed excitation (frequency doubled dye lasers pumped by Nd-Yag or N_2 lasers). Figure 6 shows the OH fluorescence lifetime compared to the laser pulse duration. Typically light pulses of 1 mJ, a repitition frequency of 30 Hz and beam diameters of 6-8 mm are used. 10^6 OH cm^{-3} yield about one photon per laser pulse. The spectral width of the laser light is smaller than the OH line width and the laser light is tuned on/off the OH wavelength.

A number of interferences and problems occur:
Raman scattering by air
Fluorescence of gases
Fluorescence of aerosols
Stray sun light
OH production by the laser pulse from ozone

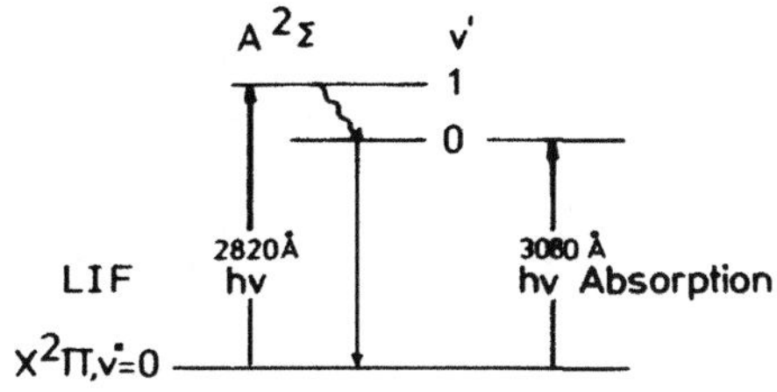

Figure 5. Principal transitions for LIF and absorption measurements of OH concentrations

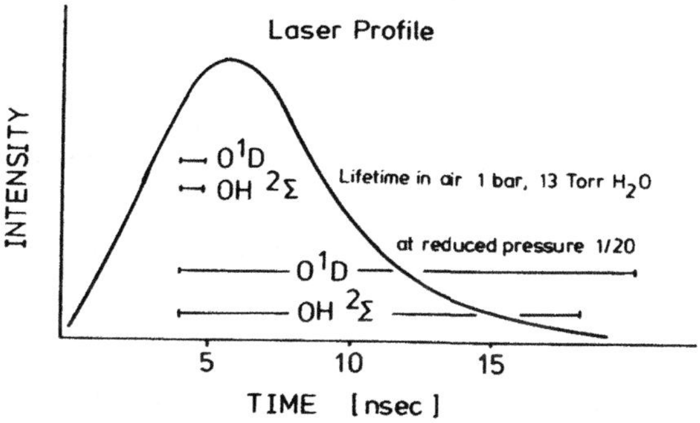

Figure 6. Lifetime of O^1D and OH $^2\Sigma$ for different air pressure compared to laser pulse duration

Accumulation of this selfproduced OH
Saturation of the line transition by the laser light
Calibration of [OH]

The interference due to scattered light of the exciting pulse which is sufficiently separated in wavelength (Figure 7) from the detection wavelength (308 mm) can be eliminated by filtering so that fluorescence detection can begin already during the excitation. Usually a considerable number of pulses has to be used until the desired signal for the OH radical concentration becomes statistically significant. While the problems connected to Raman scattering, fluorescence and stray light can be overcome by a measuring strategy, the interference due to laser general OH is apparently more serious. Ortgies et al. /38/ compared their calculations with previous measurements and showed that the tropospheric OH concentrations to be expected were within the order of magnitude of laser generated OH.

The cause for this interference is seen from Figure 7 where the absorption cross section for O_3 is given. The exciting laser light decomposes ozone according to reaction (1). In the troposhere the lifetime of O^1D is about 1 nsec at ground level (Fig. 6). Since

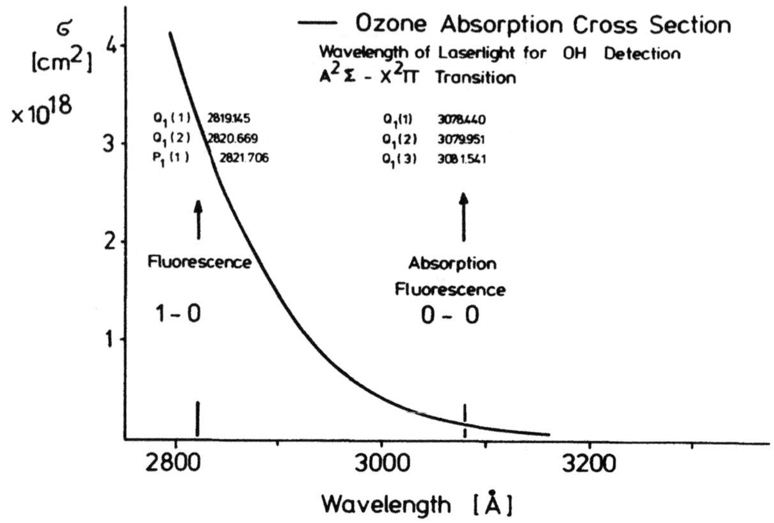

Figure 7.

a certain fraction forms OH within the laser pulse (Fig. 6) the laser produced OH will also be detected by the same laser pulse.

The interference depends crucially on the rotational distribution of the OH initially formed in reaction (2), only thermal OH can be seen by the laser. The argument used before was that this OH is strictly formed in higher rotational levels. However 50 % of the OH initially formed, besides being mostly in the vibrationless state /39/, /40/ exhibits a rotational distribution corresponding to that of about room temperature /40/. After those objections were raised /38/ Davis et al. /41/ carefully reinvestigated their earlier measurements and had to concede that the interference by selfproduced OH was up to 50 %.

A way to circumvent this problem is to reduce the pressure, all interferences related to light intensity like Raman scattering, ozone photolysis and aerosol fluorescence are reduced but not the number of fluorescent photons from OH /42/. Although less OH molecules are excited, the quenching (13) is reduced as well which increases the fluorescence efficiency for the $A^2\Sigma$ state. The interference by laser generated OH through ozone photolysis is less serious as the reaction of O^1D with water becomes slower. In this way the exciting laser pulse passes long before the artificially produced OH has built up (Fig. 6). A quick air exchange avoids accumulation of OH from successive laser light pulses. However the fluorescence has to be observed for a longer time and the possibility of saturation of the OH transition increases at the lower pressure /43/.

The number of OH molecules in the upper state is given by the product of the absorption cross section and the light flux density and by the lifetime. The latter is obviously longer at lower pressure (reaction 13 and Fig. 6). The fraction of molecules in the upper state of the Q_1 (2) transition during the maximum intensity of the laser pulse (7 nsec, 0.1 mJ and 6 mm beam diameter) will be 0.2. This is somewhat high. The problem is eased at high pressure because as only 1/10 of the total OH is found in this rotational level, the depletion can be partly overcome by fast roational relaxation from other levels. At low pressure, however, this saturation problem is intensified by an increase of the cross section at line center and because the linewidth is reduced to the Dopplerwidth and the lifetime of the $A^2\Sigma^+$ state increases also. Consequently the full amount of quanta within a pulse enters the calculation while at the same time rotational relaxation becomes negligible. The laser beam must be widened to 5 cm diameter in order to keep the fraction of excited molecules for pulses of 0.1 mJ below tolerable 0.05 at 293 K.

Absorption: Light absorption allows a more defined direct determination of OH concentrations. Experimentally small differences between almost equally large light levels are determined. Quantitative results are obtained by converting the line absorptions into atmospheric concentrations with the help of the known oscillator strength. Atmospheric OH concentrations were measured by this method in our institute /44, 45, 46/. Determination of atmospheric [OH] by optical absorption was also proposed by other groups /47/. At this time several techniques based on absorption using single or several lines of the OH spectrum are being developed. Light sources used are often lasers but also various others, especially line sources, have been considered.

Basically artificial light is sent through the free atmosphere. The light path has to be sufficiently long to yield detectable absorptions. Lambert-Beer's law (Fig. 8) describes the relation between differential light absorption and concentration.

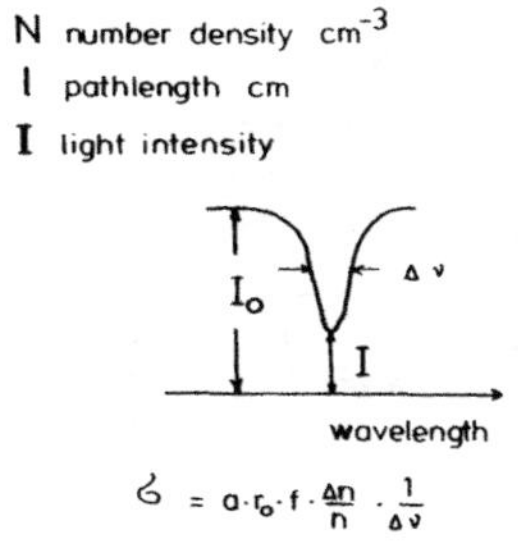

Figure 8. Beer's law $I = I_o \exp - \sigma N l$

σ is obtained from laboratory data as described before (11). Taking $\sigma = 2.7 \times 10^{-16}$ cm^2 for the Q_1 (2) line ($\nu = 0.24$ cm^{-1} as typical in our experiments) and a 10 km light path an optical density ($\log_e I/I_o$) of 10^{-4} corresponds to 3.7×10^5 OH cm^{-3}.

In order to achieve such sensitivity a certain number of photons has to reach the detector. The outgoing laser light-flux ϕ_o is attenuated on its way through air (15)

$$\phi = \phi_o \, e^{-l/b_o} \tag{15}$$

where b_o denotes the distance after which ϕ_o is reduced to 1/e. Under optimal conditions b_o is about 2.7 km. Therefore only 2.5 % of the emitted light will reach the detection system over a path of 10 km and the application of this method depends crucially upon visibility. However during a period of one hour a signal to noise ratio of 10^{-4} may be obtained.

Basically the technique developed before /48/ is adopted and adjusted for the special needs of detecting a weak narrow absorption line. Figure 9 gives the principle of the apparatus. Light at 308 nm is generated by frequency doubling the output of a dye laser which itself is pumped synchronously by an acoustooptically modulated Ar^+ laser. The bandwidth of the laser light obtained with a 3 element birefringent filter is about 0.1 nm. This is broad compared to the useful optical region of 0.045 nm (Fig. 10 and 11). The laser beam is expanded to 25 cm diameter, directed towards a flat mirror in the open field (~ 5 km away) and reflected back into the laboratory. The optical alignment is made be remote control.

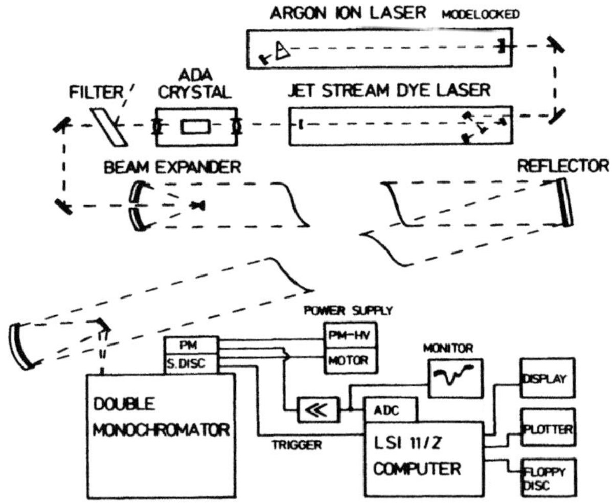

Figure 9. Principle of apparatus for long path optical absorption measurements using laser light

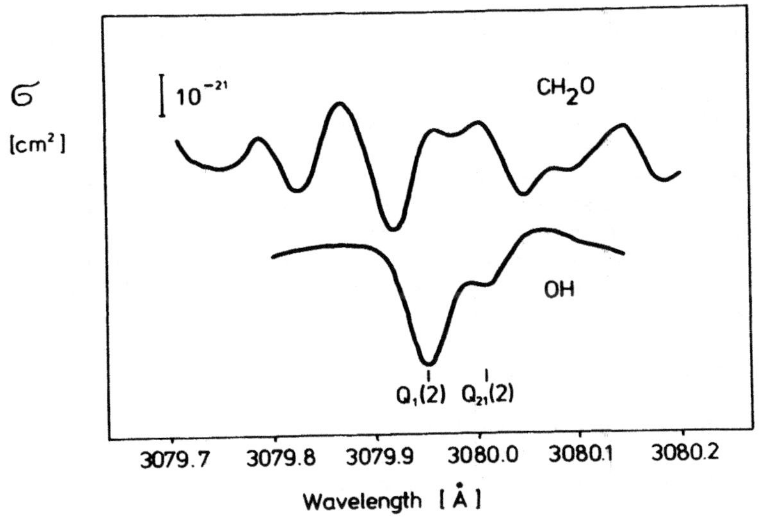

Figure 10. Absorption spectrum of formaldehyde

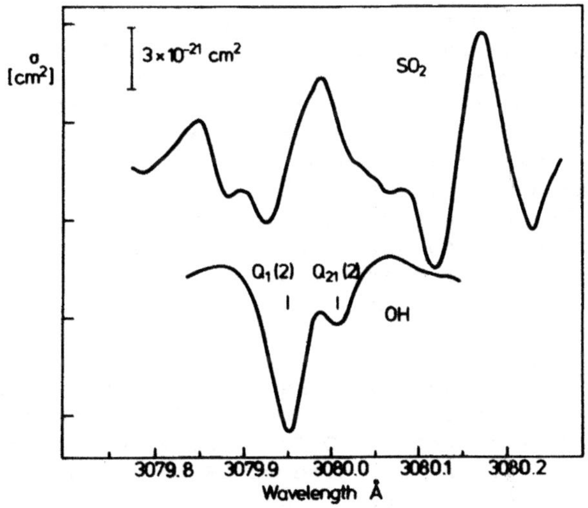

Figure 11. Absorption spectrum of sulfur dioxide

The light after having passed through free air is dispersed by a spectrograph and repetitively scanned (6 khz) by a rotating disc carrying slits of 10 μm width. Thereby the wavelength dependent intensity distribution is converted into a time variable signal which is observed by a single photomultiplier and fed into an analog to digital converter. The data are synchronized and added up for signal averaging by a microcomputer. For the wavelength calibration of the laser and the detection system an acetylene/air flame is placed in the path of the UV light beam. In this flame large [OH] are present. The highest population for ambient OH is found in the lowest three rotational levels (Figure 4). The Q_1 branch (Figure 2) seems to offer the best choice for the OH detection so that in field measurements only the Q_1 (2) line was followed.

Possible sources of error are:
Stray sun light
Saturation of the OH transition
OH production by the laser light
Absorption features of other air constituents.

Stray light from the sun under clear conditions as well as saturation of the OH transition by the laser light is negligible. But the OH production by the laser light and superimposed absorption features of other air constituents have to be discussed. NO_2 and O_3, did not exhibit any structures in this spectral region, however, CS_2, CH_2O and SO_2 did. Figures 10 and 11 show the absorption spectra and the cross for CH_2O and SO_2 respectively. The Q_1 (2) and Q_{21} (2) lines are shown as wavelength reference. Usually in field studies the SO_2 is observed and the data have to be corrected for this by a curve fit. To be able to do so a sufficient number of SO_2 absorption features has to be free from overlap with OH. The same applies for CS_2 and CH_2O but usually the mixing ratio of those two compounds in background air is so low that their absorptions are well below the detection limit. For formaldehyde the spectral overlap (Fig. 10) is such that it can compensate an OH absorption. Therefore formaldehyde has to be considered very carefully in more polluted air. One ppb CH_2O compensates about 1.5 x 10^5 OH cm^{-3} under our conditions.

The OH production by the laser light itself is small. For a 10 mW beam of 25 cm diameter the light flux density is 3 x 10^{13} photons cm^{-2} sec^{-1}. In the presence of 30 ppb ozone 3.4 x 10^6 quanta cm^{-3} sec^{-1} are absorbed to yield O^1D atoms (1). A fraction of 15 % reacts with water (20 Torr) to give a concentration of 1 x 10^6 OH cm^{-3} if the lifetime of OH is one second. However by attenuation of the laser UV beam the average light flux density along the 10 km path is reduced by a factor of 4 so that the interference accounts only for about 2.5 x 10^5 OH cm^{-3}. This is an upper estimate for the worst case e.g. extreme visibility and high water content.

Figure 12 shows a measurement under clear conditions in Deuselbach. In the original air spectrum only SO_2 shows up (upper trace) as compared with the SO_2 reference spectrum. After subtracting the SO_2 spectrum corresponding to an absorption by 5.6 ppb SO_2, the rest spectrum (lowest trace) does show only little OH absorption when compared to the OH reference spectrum. The subtraction of this amount of SO_2 is justified with those SO_2 bands which are free from the OH line. The estimate for the upper limit of [OH] is 5×10^5 cm^{-3}. In Figure 13 an example for high [OH] is shown. Here the absorption is clearly visible and corresponds to 4×10^6 OH cm^{-3}. Table 3 gives the values obtained at Jülich for summer 1979. The average value is $(1.7 \pm 3) \times 10^6$ OH cm^{-3}. In 1980 the apparatus was improved and the measurements obtained had a better time resolution and are given for two measuring sites in Table 4. The values found at Jülich in 1980 are similar to those in 1979 with most values between 1 to 2×10^6 OH cm^{-3}. At Deuselbach, a remote area in the Hunsrück mountains, the highest value of 4×10^6 OH cm^{-3} was observed and the concentrations were found to be highly variable. All those OH values given have not been corrected for formaldehyde absorption, which may raise the numbers by as much as 4×10^5 OH cm^{-3}.

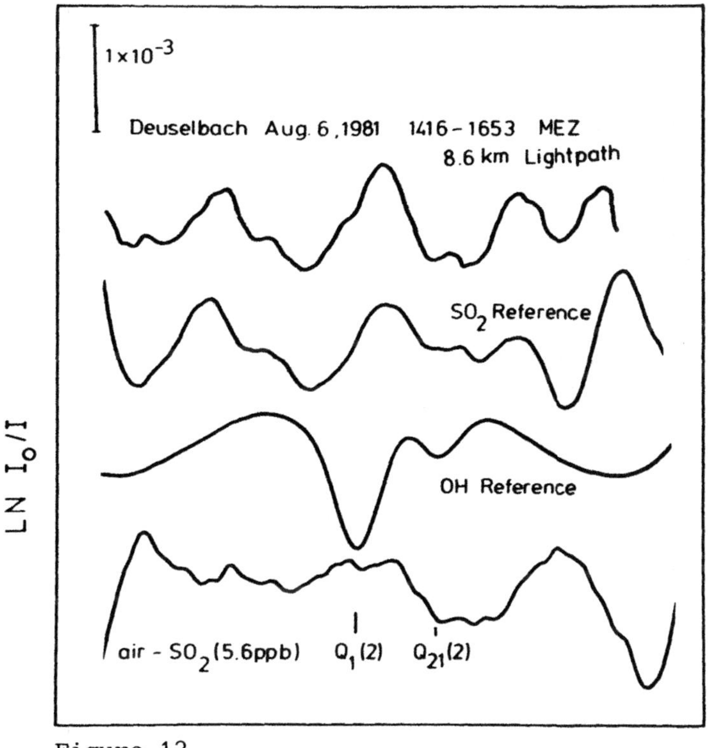

Figure 12.

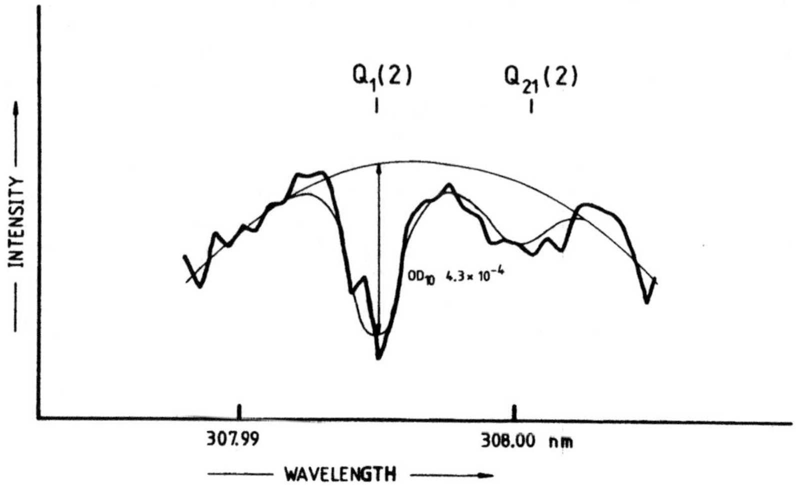

Figure 13. Deuselbach Sept. 24, 1980

Table 3. [OH] measured in 1979 by photon counting. The data are summed up in order to reduce the noise due mainly to low laser light intensity. The results include most and only sunny days.

Date	Time CET	Location	Pathlength km	OH x 10^{-6} cm^{-3}
April	throughout only	Jülich	3	2
May	measurements	"	3	1.8
June	taken between	"	3	1.5
July	1100 and 1600	"	3	1.7
August	are considered	"	3	2
September		"	3	1.7

Table 4. OH concentration measured in 1980 at two locations with an improved machine

Date	Time	Location	Path km	OH x 10^{-6} cm^{-3}
June 6	11.00 to 15.30	Jülich	3	< 3
July 22	11.30 to 15.30	"	3	2.9
July 23	13.00 to 15.30	"	3	< 4
Aug. 7	11.50 to 14.00	"	3	1.8
Aug. 15	11.30 to 15.10	"	3	< 1.2
Sept. 1	15.00 to 16.00	"	8	1.2
Sept. 3	11.00 to 14.30	"	3	0.8
Sept. 8	13.50 to 15.40	"	3	≤ 2
Sept. 23	12.00 to 14.40	Deuselbach	9.7	< 1.8
Sept. 24	11.00 to 12.30	"	9.7	4
Oct. 6	12.45 to 14.20	"	6	2.4
Oct. 15	12.00 to 15.20	"	6	0.8

4. DISCUSSION

Only measurements around noontime will be discussed. Fig. 14 shows all OH measurements reported so far by in situ methods, e.g. laser induced fluorescence, absorption and ^{14}CO. Measurements at higher altitude were exclusively made with LIF techniques and show fair agreement with model calculations /12/. However, those values /49, 50/ need a correction /41/. The same argument probably applies to the measurements /51, 52/ Selzer and Wang /37/ have already conceded that the values /51/ could actually be higher by a factor of 10.

The OH concentrations /44/ have to be raised by a factor of 1.5 due to a change in constants used for evaluation of the absorption coefficient σ /45/. Our recent measurements /46/ show slightly lower values and agree quite well with those in /27/ (see also Tables 3, 4). The measurements made at groundlevel are shown in Figure 15 together with a calculated latitudinal distribution of OH /12/. The variation is large and possible causes for this are aerosol and NO_x chemistry.

In an atmosphere which contains neither NO nor NO_2 (the sum of both is called NO_x) the OH radical reacts mainly with CO and CH_4. The reaction with CO (16) is followed

$$OH + CO \rightarrow CO_2 + H \quad (16)$$

$$H + O_2 \xrightarrow{M} HO_2 \quad (17)$$

$$HO_2 + HO_2 \rightarrow H_2O_2 + O_2 \quad (18)$$

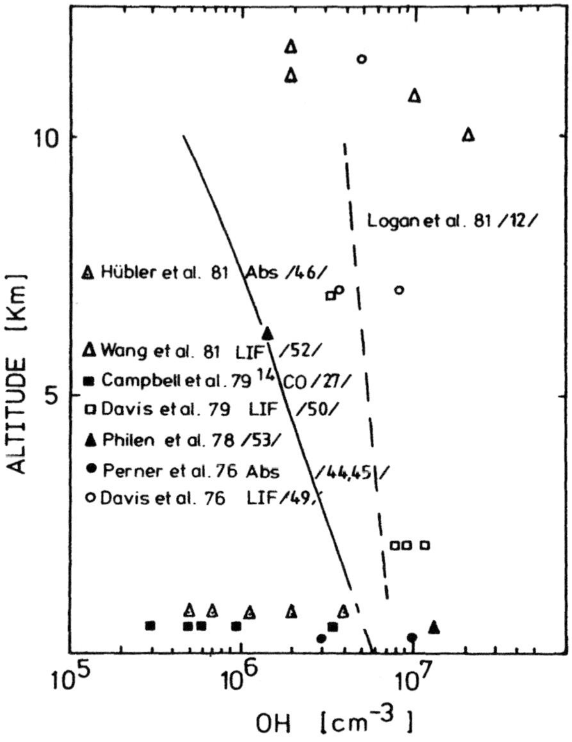

Figure 14.

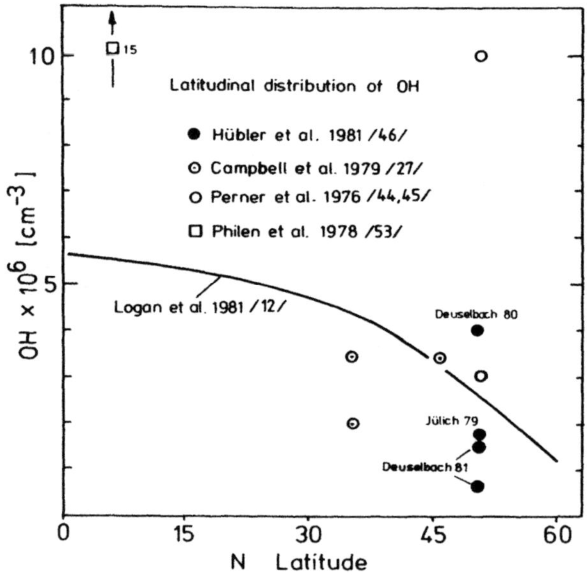

Figure 15.

by reactions (17), (18) and leads to the net reaction

$$2CO + 2\ OH + O_2 \rightarrow 2CO_2 + H_2O_2 \qquad (19)$$

However the HO_2 radical which yields H_2O_2 by disproportionation (18) may react with ozone (20). Therefore the net oxidation path

$$HO_2 + O_3 \rightarrow OH + 2\ O_2 \qquad (20)$$

of carbon monoxide depends upon $[HO_2]$ which is controlled by the primary rate of OH formation. Reaction (19) is found for high OH production rates close to the ground. Reaction (20) prevails in situations of low OH production rates and the net reaction for CO oxidation is then

$$CO + O_3 \rightarrow CO_2 + O_2 \qquad (21)$$

For example at 10 km height the net reaction composed of (19) and (21) is estimated as

$$8\ CO + 2\ OH + 6\ O_3 \rightarrow 8\ CO_2 + H_2O_2 + 5\ O_2 \qquad (22).$$

The reaction of OH with hydrocarbons (9) is an extremely important tropospheric process and is described in detail for methane. The products are methanol, methylhydroperoxide and formaldehyde.

$OH + CH_4$	$\rightarrow$	$H_2O + CH_3$		(23)
$CH_3 + O_2$	$\rightarrow$	CH_3O_2		(24)
$CH_3O_2 + CH_3O_2$	$\rightarrow$	$CH_3OH + CH_2O + O_2$	57 %	(25)
	$\rightarrow$	$2\ CH_3O + O_2$	43 %	(26)
$CH_3O_2 + HO_2$	$\rightarrow$	$CH_3OOH + O_2$		(27)
$CH_3O + O_2$	$\rightarrow$	$CH_2O + HO_2$		(28)

An atmospheric cycle for OH in the boundary layer is outlined in Figure 16 based on the reaction cycles of CO and hydrocarbons. The reaction rate of OH with nonmethane hydrocarbons (NMHC) is assumed to be equal to the reaction rate of OH with methane. The major sinks for radicals are the reactions (18) and (27). Recycling of OH radicals by (20) is inefficient compared to its primary formation rate (2) as expected from the high primary OH production rate and the foregoing discussion.

However other OH production channels have to be considered in detailed models. Depending on the concentrations of the secondary products H_2O_2, CH_3OOH and CH_2O considerable amounts of radicals can be produced by their photolyses (29) - (31).

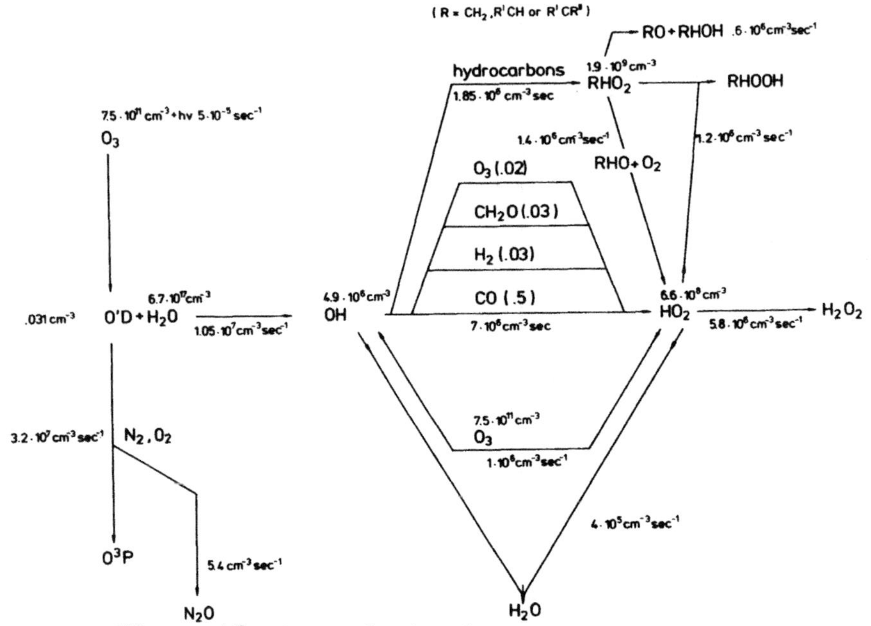

Figure 16. OH cycle in the lower troposphere and in the absence of NO_x

$$H_2O_2 + h\nu \rightarrow 2\,OH \quad (29)$$

$$CH_3OOH + h\nu \rightarrow CH_3O + OH \quad (30)$$

$$CH_2O + h\nu \rightarrow H + HCO \quad (31)$$

$$\rightarrow CO + H_2 \quad (32)$$

$$HCO + O_2 \rightarrow HO_2 + CO \quad (33)$$

However due to their low photolysis rates those species have relatively long lifetimes in the troposphere and they may be lost by deposition or rainout. While OH is directly formed in the photolysis of hydrogenperoxide (29) and methylhydrogenperoxide (30) the photolysis of formaldehyde as representative for carbonyl compounds yields HO_2 (7), (33) first. The HO_2 has to be transformed into OH which could be inefficient. Under the assumed conditions the lifetime of HO_2 is of the order of several seconds so that even interactions with aerosols may become important.

Especially the radical formation by photolysis of carbonyls offers the possibility of increasing the number of radicals in the oxidation cycles of hydrocarbons. This is not possible in the oxidation of CO and H_2 where the radicals are recycled at the most. The uncertainty in the assessment of OH radical concentrations by model calculations is caused in part by the fact that the exact reaction scheme of hydrocarbon oxidation is not known for all hydrocarbons.

The reaction scheme outlined for no NO_x changes dramatically if sufficient NO is present which transfers HO_2 very efficiently into OH (34) and leads to a decreased H_2O_2 and CH_3OOH formation.

$$HO_2 + NO \rightarrow OH + NO_2 \quad (34)$$

Similar reactions occur for alkylperoxides like methylperoxyde (35).

$$CH_3O_2 + NO \rightarrow CH_3O + NO_2 \quad (35)$$

Because of this, the reactions (28) and (33) will lead to increased OH concentrations and to higher oxidation rates. In Figure 15 starting from low [NO] the [OH] increases through the influence of reactions (34) and (35) towards higher [NO]. [OH] goes through a maximum when the rate of the main OH sink reaction (7) comes close to the primary OH production rate through reactions (1) and (2) plus the net OH production from hydrocarbon oxidation. At higher NO_x mixing ratio the [OH] starts to decrease.

The absolute concentration of OH depends crucially upon the hydrocarbon content and upon J_1 as indicated in Figure 17. There the maximum [OH] decreases with latitude because of smaller J_1 values. The highest values for OH calculated are around 1.5×10^7 OH cm^{-3} /12, 54/.

The atmospheric OH cycle including NO_x is given in Figure 18. The rate of OH reaction with hydrocarbons (8), (9) was assumed to be equal to that with CO (16). The steady state concentration for [OH] increases compared to Figure 16 to 1.5×10^7 cm^{-3}. The product pattern is also altered by the NO_x. For example in the case of methane the products from reactions (25) to (27) are reduced in

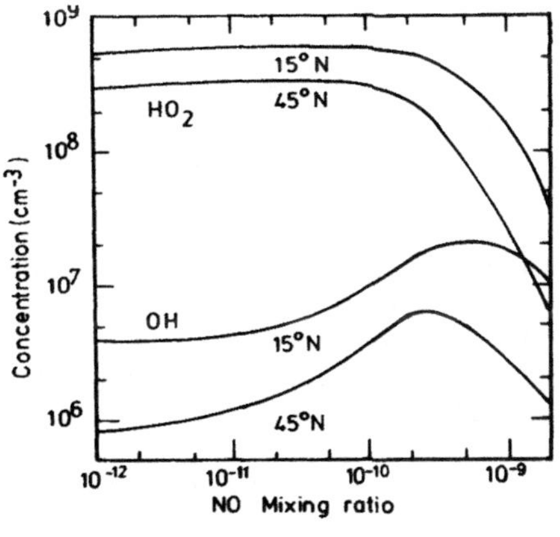

Figure 17. OH concentration as a function of nitric oxide mixing ratio for two latitudes /12/

favour of formaldehyde which becomes the main product (28). The formation of hydrogen peroxide is reduced in favour of nitric acid production (7) which becomes the prevailant radical sink.

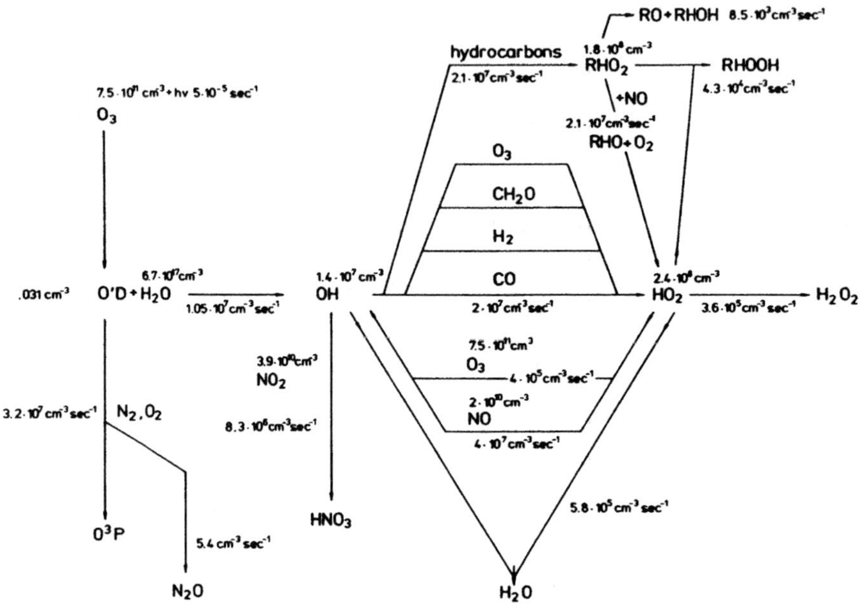

Figure 18. Atmospheric OH cycle in the presence of NO_x

The inclusion of the NO_x chemistry introduces two additional OH sources namely (36) and (37), J_{36} is small so that this

$$HNO_3 + h\nu \rightarrow OH + NO_2 \qquad (36)$$

$$HNO_2 + h\nu \rightarrow OH + NO \qquad (37)$$

OH source is negligible compared to other OH sources even at the highest ambient levels of HNO_3 (30 ppb). HNO_3 mixing ratios are reported in /55/, for example. However several ppb HNO_2 provide an appreciable OH source. This was indeed observed for the morning hours in polluted urban atmospheres /56, 57/. While most primary OH sources depend on previous photochemical formation processes, HNO_2 and probably also HNO_3 are important exceptions. HNO_2 accumulates during the dark to yield appreciable concentrations probably through interactions of NO_x with aerosols /56, 58/. It may be emitted directly also by combustion engines /59/.

Measurements of H_2O_2 are reported by /60, 61/, however these publications are disputed in the moment. Formaldehyde has been measured by many groups /48,62, 63, 64/.

As evident from the measured [OH] given in Tables 3, 4 concentrations of 10^7 OH cm^{-3} have not been observed again recently The average concentration was lower than expected. This could have been caused by a lower J_1 value than used in the calculations and by higher NO_x mixing ratios (Fig. 17). For example the maximum concentration observed so far of 4 x 10^6 OH cm^{-3} under very clear conditions in late September could have well been caused by an optimal NO_x mixing ratio. Surprisingly, during days when strong photochemical action seemed to be anticipated by haze formation, excessively high [OH] were never observed.

In this context the findings /24, 25/ are rather interesting (Fig. 1). Though especially the results for saturated hydrocarbons are subject to considerable error, the OH concentrations are lower in this polluted air than expected for background atmosphere (Figures 15 and 16). The reason could be again a high NO_2 mixing ratio. However Los Angeles smoggy air is always accompanied by high aerosol levels so that also the scavenging of peroxyradicals by aerosol could have lowered the [OH].

$$HO_2 + \text{aerosols} \rightarrow \text{products} \qquad (38)$$

$$RHO_2 + \text{aerosols} \rightarrow \text{products} \qquad (39)$$

A global comparison of directly determined [OH] with that from indirect budget considerations can only be made when the global tropospheric OH distribution is also known which may be obtained from model calculations.

5. CONCLUSION

OH seem to play an important role in the troposphere. Observations indicate that [OH] is lower in polluted air possibly due to the interference of aerosols with peroxyradicals (HO_2) and to an unfavourable NO_x/hydrocarbon ratio. The observation that in heavily polluted air the apparent OH concentrations depended on the type of hydrocarbon may be interpreted to show a depletion of these air constituents not alone involving OH but other processes like gas particle interactions as well. One might speculate on light induced surface reactions as recently observed in smog chamber studies /65/.

REFERENCES

/ 1/ Bates, D.R., and A.E. Witherspoon, Mon. Not. R. Astron. Soc., 112, pp. 101-124, 1952

/ 2/ Weinstock, B., 1969, Science, 166, pp. 224-225

/ 3/ Robinson, E., and R.C. Robbins, 1969, Res. Proj. PR-6755, Stanford Univ., Stanford, Calif.

/ 4/ Seiler, W., and C.E. Junge, 1970, J. Geophys. Res., 75, pp. 2217-2226

/ 5/ Junge, C.E., W. Seiler, and P. Warneck, 1971, J. Geophys. Res., 76, pp. 2866-2879
/ 6/ Levy, H., 1971, Science, 173, pp. 141-143
/ 7/ Hunt, B.G., 1966, J. Geophys. Res., 71, pp. 1385-1390
/ 8/ Nicolet, M., 1970, Aeronimica, A79, pp. 1-100
/ 9/ Amimoto, S.T., A.P. Force, R.G. Gulotty, and J.R. Wiesenfeld, 1979, J. Chem. Phys., 71, pp. 3640-3647
/10/ McConnell, J.C., M.B. McElroy, and S.C. Wofsy, 1971, Nature, 233, pp. 187-188
/11/ Dickerson, R.R., D.H. Stedman, W.L. Chameides, P.J. Crutzen, and J. Fishman, 1979, Geophys. Res. Lett., 6, pp. 833-836
/12/ Logan, J.A., M.J. Prather, S.C. Wofsy, and M.B. McElroy, 1981, J. Geophys. Res., 86, pp. 7210-7254
/13/ Weinstock, B., and H. Niki, 1972, Science, 176, pp. 290-292
/14/ Weinstock, B., and T.Y. Chang, 1974, Tellus, 26, pp. 108-115
/15/ Volz, A., D.H. Ehhalt, R.G. Derwent, and A. Khedim, 1979, Bericht der Kernforschungsanlage Jülich, Jül-1604
/16/ Volz, A., D.H. Ehhalt, and R.G. Derwent, 1981, J. Geophys. Res., 86, pp. 5163-5171
/17/ Warneck, P., 1975, Planet. Space Sci., 23, pp. 1507-1518
/18/ Lovelock, J.E., 1977, Nature, 267, pp. 32
/19/ Neely, W.B., and J.H. Plonka, 1978, Environm., Sci. Techn., 12, pp. 317-321
/20/ McConnell, J.C., and H.I. Schiff, 1978, Science, 199, pp. 174-1
/21/ Singh, H.B., L.J. Salas, H. Shigeishi, and E. Scribner, 1979, Science, 203, pp. 888-903
/22/ Hanst, P.L., J.W. Spence, and E.O. Edney, 1980, Atmosph. Environ., 14, pp. 1077-1088
/23/ Derwent, R.G., and A.E.J. Eggleton, 1981 in press, Quart. J. Roy. Met. Soc., 107
/24/ Calvert, J.G., 1976, Environ. Sci. Tech., 10, pp. 256-262
/25/ Singh, H.B., J.R. Martinez, D.C. Henry, R.J. Jaffe, and W.B. Johnson, 1981, Environ. Sci. Techn., 15, pp. 113-119
/26/ Neubert, E., H.-W. Georgii, and J. Mueller, 1981, Second Europ. Symp. Physico-Chemical Behaviour of Atm. Pollutants, Varese, CEC Cost 61a, 29 Sept. - 1 Oct., 1981
/27/ Campbell, M.J., J.C. Sheppard, and B.F. Au, 1979, Geophys. Res. Lett., 6, pp. 175-178
/28/ Cook, A.H., 1977, Cambridge Univ. Press, Cambridge, London, New York, Melbourne
/29/ Goldman, A., and J.R. Gillis, 1981, J. Quant. Spectrosc. Ratiat. Transf., 25, pp. 111-135
/30/ German, K.R., 1975, J. Chem. Phys., 64, pp. 4065-4068
/31/ Engleman, R., 1969, J. Quant. Spectrosc. Radiat. Transf., 9, pp. 391-400
/32/ Wang, C.C., D.K. Killinger, and C.-M, Huang, 1980, Phys. Rev. A 22, pp. 180-185
/33/ Baardsen, E.L., and R.W. Terhune, 1972, Appl. Phys. Lett., 21, pp. 209-211
/34/ Wang, C.C., and L.I. Davis, 1974, Phys. Rev. Lett., 32, pp. 349-352

/35/ Wang, C.C., and L.I. Davis, 1975, J. Chem, Phys., 62, pp.53-55
/36/ German,K.R., 1976, J. Chem, Phys., 64, pp. 4065-4068
/37/ Selzer, P.M., and C.C. Wang, 1979, J. Chem. Phys., 71, pp. 3786-3791
/38/ Ortgies, G., K.-H. Gericke, and F.J. Comes, 1980, Geophys. Res. Lett., 7, pp. 905-908
/39/ Engleman, R., 1965, J. Am. Chem. Soc., 87, pp. 4193-4194
/40/ Gericke, K.H., G. Ortgies, and F.J. Comes, 1980, Chem. Phys. Lett., 69, pp. 156-160
/41/ Davis, D.D, M.O. Rodgers, S.D. Fischer, and K. Asai, 1981, Geophys. Res. Lett., 8, pp. 69-72; Davis, D.D., M.O. Rodgers, S.D. Fischer, and W.S. Heaps, 1981, Geophys. Res. Lett., 8, pp. 73-76
/42/ Hard, T.M., R.J. O'Brien, and T.B. Cook, 1980, J. Appl. Phys., 51, pp. 3459-3464
/43/ Killinger, D.K., C.C.Wang, and M. Hanabusa, 1976, Phys. Rev., 13A, pp. 2145-2152
/44/ Perner, D., D.H. Ehhalt, H.W. Paetz, U. Platt, E.P. Roeth, and A. Volz, 1976, Geophys. Res. Lett., 3, pp. 466-468
/45/ Perner, D., D.H. Ehhalt, H.W. Paetz, U. Platt, E.P. Roeth, and A. Volz, J. Geophys. Res., submitted
/46/ Huebler, G., D.H. Ehhalt, U. Platt, W. Paetz, A. Toennissen, and D. Perner, 1981, in preparation
/47/ Killinger, D.K., and C.C. Wang, 1977, Chem. Phys. Lett., 52, pp. 374-376
/48/ Platt, U., D. Perner, and H.W. Paetz, 1979, J. Geophys. Res., 84, pp. 6329-6335
/49/ Davis, D.D., W. Heaps, and T. McGee, 1976, Geophys. Res. Lett., 3, pp. 331-333
/50/ Davis, D.D., W. Heaps, D. Philen, and T. McMagee, 1979, Atm. Environ., 13, pp. 1197-1203
/51/ Wang, C.C., L.I. Davis, C.H. Wu, S. Japar, H. Niki, and B. Weinstock, 1975, Science, 189, pp. 797-800
/52/ Wang, C.C., L.I. Davis, P.M. Selzer, and R. Munoz, 1981, J. Geophys. Res., 86, pp. 1181-1186
/53/ Philen, D., W. Heaps, and D.D. Davis, 1978, EOS Trans AGU, 59, pp. 1079
/54/ Weinstock, B., H. Niki, and T.Y. Chang, 1980, Advances Environ. Sci. Tech., 10, pp. 221-258
/55/ Huebert, B.J., and A.L. Lazrus, 1980, J. Geophys. Res., 85, pp. 7322-7328
/56/ Platt, U., D. Perner, G.W. Harris, A.M. Winer, and J.N. Pitts, 1980, Nature, 285, pp. 312-314
/57/ Harris, G.W., W. Carter, A.M. Winer, J.N. Pitts, U. Platt, and D. Perner, 1981, Environm. Sci. Techn., submitted
/58/ Perner, D., 1980, Proc. Int. Workshop Test Methods Assessm. Proced. for Determ. Photochem. Degradation Behaviour of Chem. Subst.; Fraunhofer Ges., Umweltbundesamt (eds.), Berlin, pp. 159-173

/59/ Kessler, C., D. Perner, and U. Platt, manuscript in preparation, 1982
/60/ Kok, G.L., K.R. Darnall, A.M. Winer, J.N. Pitts, and B.W. Gay, 1978, Environm. Sci. Techn., 12, pp. 1077-1080
/61/ Kelly, T.J., D.H. Stedman, and G.L. Kok, 1979, Geophys. Res. Lett., 6, pp. 375-378
/62/ Platt, U., and D. Perner, 1980, J. Geophys. Res., 85, pp. 7453-7458
/63/ Lowe, D.C., U. Schmidt, and D.H. Ehhalt, 1980, Geophys. Res. Lett., 7, pp. 825-828
/64/ Neitzert, V., and W. Seiler, 1981, Geophys. Res. Lett., 8, pp. 79-82
/65/ Kessler, C., U. Platt, D. Perner, P. Bruckmann, and P. Eynck, 1981, Environm. Sci. Techn., submitted

THE ATMOSPHERIC SULFUR-BUDGET

Hans-Walter Georgii

Department of Meteorology and Geophysics
University of Frankfurt/Main, FRG

ABSTRACT

The paper deals with the sources, sinks and reactions of the main sulfur-species in the atmosphere, SO_2 and H_2S. There is also evidence for the presence of organic compounds like $(CH_3)_2S$, CS_2 and COS. SO_2 is converted in the atmosphere to particulate sulfur. This complex process involves gas-phase reactions but also heterogeneous reactions in droplets and on particle-surfaces. In the case of droplet-reactions, catalytic oxidation in the presence of heavy metals is believed to be important in urban areas and stack-plumes. Gas-to-particle conversion leads to the formation of fine sulfate-aerosols. Sulfate aerosols are of regional if not hemispheric importance, they influence the atmospheric radiation budget by absorption and scattering.

INTRODUCTION

Sulfur plays an important role in the tropospheric and stratospheric budget of atmospheric gases. Much attention is being paid to the growing rate of sulfur-components which are emitted into the atmosphere from antropogeneous sources leading to locally and regionally high accumulations of SO_2 and sulfate aerosols. But, on the other hand during recent years, growing attention is being paid to the biogenic sulfur compounds particularly to gaseous organic sulfur-

H. W. Georgii and W. Jaeschke (eds.), Chemistry of the Unpolluted and Polluted Troposphere, 295–324.

compounds like CS_2, COS or DMS which are thought to pay an important role in the atmospheric sulfur budget. Still, the abundances, sources and sinks and the chemistry of atmospheric sulfur compounds are not fully understood although our knowledge has improved during the last years due to intensive research in the field and in the laboratory.

EMISSIONS

It is without question that the global emission of sulfur-compounds mainly in the form of SO_2 has strongly increased during the last 7o years. The anthropogenic input to the sulfur cycle is clearly significant and likely to further increase. Cullis and Hirschler (1979) give the following figures for the global SO_2-production-rate:

19oo:	3o.6 x $1o^6$	to/year
193o:	6o x $1o^6$	to/year
196o:	13o x $1o^6$	to/year
1974:	187 x $1o^6$	to/year = 94 x $1o^6$ to S / year

The huge quantity of SO_2 is mainly due to combustion processes by fossil fuel containing sulfur. However, some industrial activities have also to be taken into consideration. The present estimates of emissions of atmospheric sulfur makes clear that in the northern hemisphere the total man-made sulfur emission exceeds the total natural sulfur emission while in the southern hemisphere the natural emissions exceed the anthropogenic sulfur production by far.

Natural emissions of sulfur take place mainly

- as release of sulfur gases in a chemically reduced form from the decay of biologic material
- as sea salt particles from the sea surface
- as volcanic emissions

The natural sulfur gasemission has until now mostly been calculated as a missing term to close the atmospheric sulfur cycle, the values given by the different authors were highly dependant on the other terms of the sulfur budget and varied very much. The following table shows such estimates:

Table 1

Author		Production rate of volatile reduced sulfur compounds by natural processes in 10^6 to/yr
Eriksson	1960	267
Junge	1963	230
Robinson and Robbins	1970	98
Kellogg et al.	1972	89
Friend	1973	106
Granat	1976	33
Hallberg	1976	37
Cullis and Hirschler	1979	97

While the estimates presented up to 1970 are probably too high the estimates by Granat et al. (1976) may be in the right order of magnitude in spite of the recent detection of the emission of COS, CS_2 and $(CH_3)_2S$. Varhelyi and Gravenhorst (1981) have independantly of the figures given by the other authors tried to estimate the flux of biogenic sulfur compounds into the atmosphere.

They present the following figures:

Table 2

Compounds	Estimated global biogenic sulfur emission 10^6 to S / year
H_2S	0.05 - 1.8
$(CH_3)_2S$	3.0 - 31.6
CS_2	0.17 - 1.8
COS	0.07 - 0.3
Total:	3.3 - 35.5

It is surprising that the upper limit is based on the high production rate of $(CH_3)_2S$ from the oceans and not from H_2S. It must however be considered that the authors extrapolated data gained in restricted areas to global scale. Measurements of H_2S by Jaeschke et al. (198o) in Central Europe showed clearly that the H_2S-emission rate increased with rising temperature. This indicates that tropical regions might be much more interesting source-areas for H_2S. Owing to lack of information on the H_2S-production in the tropics, the global rate may be underestimated. It is necessary to get better estimates of the emissions of volatile reduced sulfur compounds originating from decay of biogenic material in different regions of the world. This is particularly true for tropical areas.

Another uncertainty in the global sulfur budget is the emission of sulfur compounds by volcanoes. The available data for the sulfur content in volcanic exhalations show large variations. Berresheim (198o) could show that an individual volcano like the Etna emits very variable amounts of gaseous exhalations (SO_2, H_2S and sulfates) dependant on the intensity of eruption. Berresheim estimates the total SO_2-production by volcanoes to 7 x $1o^6$ to S/year and the global H_2S-production by volcanoes to 1 x $1o^6$ to S/year which is about 1 to 2% of all natural emissions. The sulfate-production by volcanoes is estimated by Berresheim to 1 to 3 x $1o^6$ to/year. Different authors have estimated the global rate of volcanic emissions to 1 to 4 x $1o^6$ to S/year. The great variation of the estimated amounts of yearly sulfur emission by volcanism shows the difficulty in making these estimations. Varhelyi and Gravenhorst mention that more recently higher emission rates by volcanic activities are proposed, namely in the range of 15 to 3o x $1o^6$ to S/year.

It should be mentioned that 2/3 of the anthropogeneous emissions of sulfur-components take place within about 5% of the earth`s surface, 38% in Europe, on 2% of the global area. The following table 3 shows the present estimates of emissions of sulfur into the atmosphere in Tg S / year according to Cullis and Hirschler.

Table 3 Present Estimates of Emissions of Atmospheric Sulphur ($Tg\ S\ yr^{-1}$)[a]

Sources	1970 N and S Hemisphere	1974 N Hemisphere	1974 S Hemisphere
Natural:			
Volcanoes	5	3 (6o)	2 (4o)
Sea-spray	44	19 (43)	25 (57)
Sub total non-biogenic	49	22 (45)	27 (55)
Biogenic (land)	53	39 (73)	14 (27)
Biogenic (oceans)	44	19 (43)	25 (57)
Sub total biogenic	97	58 (6o)	39 (4o)
Total natural	146	8o (55)	66 (45)
Man made	86	88 (94)	6 (6)
Total	232	168 (70)	72 (3o)
Man-made (% of natural)	59	11o	9
Man-made (% of total)	37	52	8

[a] The figures in brackets represent the percentages in each hemisphere.

TRENDS

SO_2-emissions doubled between 195o and 197o mainly by the increase of use of fossil fuel. Observations of the immission-concentration of SO_2 in industrialized areas show on the other hand a significant decrease of the SO_2-concentration. This decreasing tendency is observed in Frankfurt (Fig.1).

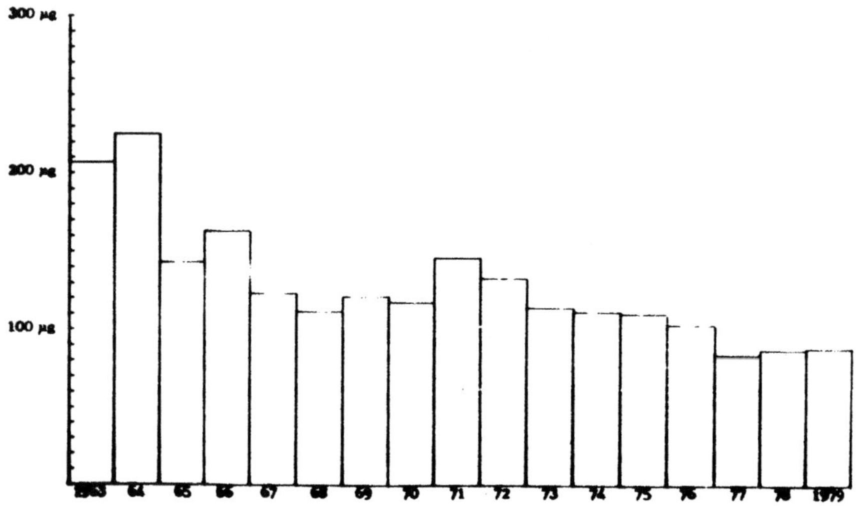

Figure 1 Annual Averages of the SO_2-concentration in Frankfurt (Pilot-Station of the Umweltbundesamt)

Also several measuring-stations in North-Rhine-Westphalia show a tendency of decreasing concentration, mainly a drop of the maximum-immission-concentration between 196o and 197o. Also the weighted mean concentration of the measuring stations in the UK shows a decrease from 1955 to 197o in spite of rising emissions. Salmon, L., Atkins, H.F. et al. (1978) comment the decrease of the SO_2-concentration in England as follows: "A high stack policy in the U.K. for power generation from fossil fuel has resulted in a significant fall in ground level sulfur dioxide concentration though not in its total emission."

In the Federal Republic of Germany there is no unique opinion about the cause for the decrease of the SO_2 -concentration in densely populated areas. Reduction of the emissions, meteorological influences and less numerous inversion situations but also the construction of higher stacks are mentioned. In this connection it is of interest to study the trend of the SO_2-concentration in unpolluted areas in the Federal Republic of Germany on the basis of measurements of the network operated by the Federal Environmental Protection Agency. During the period 1967-1972 the concentration was mainly constant. An evaluation for the period 1973-1978 shows during winter a certain increase of the SO_2-immission concentration. However, the correlation is rather weak. The results of the measuring activity in unpolluted regions does not show a clear trend, but certainly it can be said that we do find a decrease in polluted areas and we do not find a decrease of the SO_2-concentration in unpolluted regions.

One of the most important products of SO_2-oxidation is the sulfate-aerosol. Approximately 2o-5o% of the SO_2 emitted into the atmosphere is transformed into sulfate-particles within hours or days. The particle size distribution of the sulfate-aerosols reveals that a dominant fraction has radii below 1µm, resulting in a size-fraction with a tropospheric residence time of about a week. There exist only few series of measurements of sulfate-aerosols over a longer period of time. An analysis of particulate matter collected during the years 1957-1974 exists for Chilton/Oxfordshire in Great Britain. 27 components of the aerosols were determined. While the concentration of most components remained constant or decreased with time an increase of $SO_4^{=}$, NH_4^{+}, and NO_3^{-} by nearly a factor of 2 was observed. (Fig.2)

The results of the study show that concentrations of $SO_4^{=}$, NH_4^+ and NO_3^- are rising steadily. It is observed that the rise in sulfate-levels is almost entirely confined to summer months with 4,4% /year. The authors explain this phenomenon by the preference of photochemical oxidation during the sunny period of the year. They claim that heterogeneous catalytic oxidation in cloud- and fog droplets is of less importance for the production rate of sulfate-particles. We shall discuss these mechanisms a little later. I want to conclude the results of these measurements in Great Britain with the statement that ammonium-sulfate, ammonium-nitrate and sulfuric acid included 43% of the total mass of the aerosol-samples.

AMBIENT CONCENTRATIONS

SO_2

Man made emissions of gaseous sulfur-components take place mainly in densely populated areas. Since the tropospheric residence time of H_2S and SO_2 is only of the order of days, large variations of the concentration of these components are observed. Before going into details of the SO_2-distribution we want to draw a more schematic picture (Fig. 3). Over the continents the SO_2-concentration is higher than over the oceans resulting in a horizontal gradient of SO_2 from land to sea. Maximum values are found in densely inhabited areas and in industrialized regions which are indicated by a maximum near the ground with 2oo $\mu g/m^3$. However aircraft ascents above industrialized areas show that the dispersion and removal of pollutants is such that the accumulation of SO_2 is largely restricted to the lower troposphere. From a level of 15oo mtrs upwards, the regional concentration-maxima are difficult to distinguish. The production of SO_2 near the ground with sinks and dispersion in the troposphere thus leads to a strong decrease of SO_2 with altitude above the continents. At about 3 - 5 km altitude (changing a little with the season) a background-concentration of approximately o.5 μg SO_2/m^3 is reached (Georgii 1978). On very recent results of the SO_2-concentration at and above the tropopause we shall report later.

Measurements of the vertical distribution of SO_2 have been made at different places by Georgii and Jost (1964), Jost (1974) and in the frame of the OECD-project: "Long range transport of pollutants" (1977).

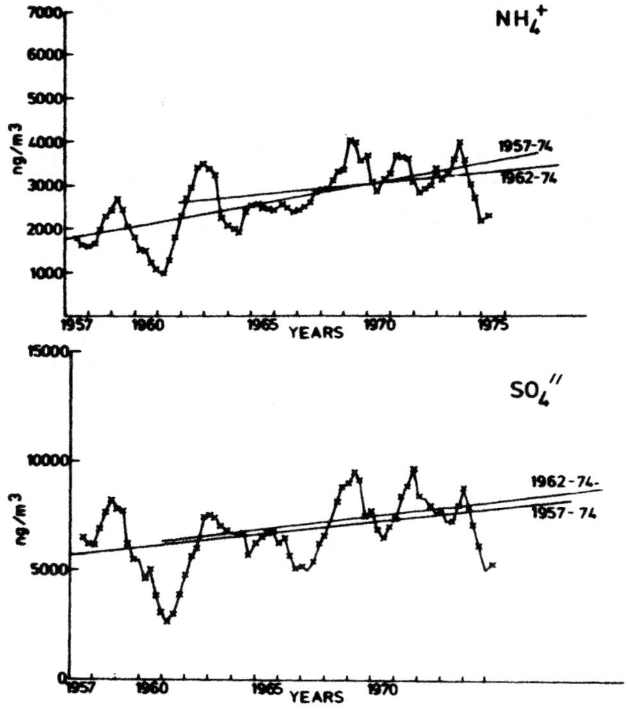

Figure 2 Temporal trend of NH_4^+ and $SO_4^=$ in aerosols at Chilton, Oxon. (By courtery of The Science of the total Environment)

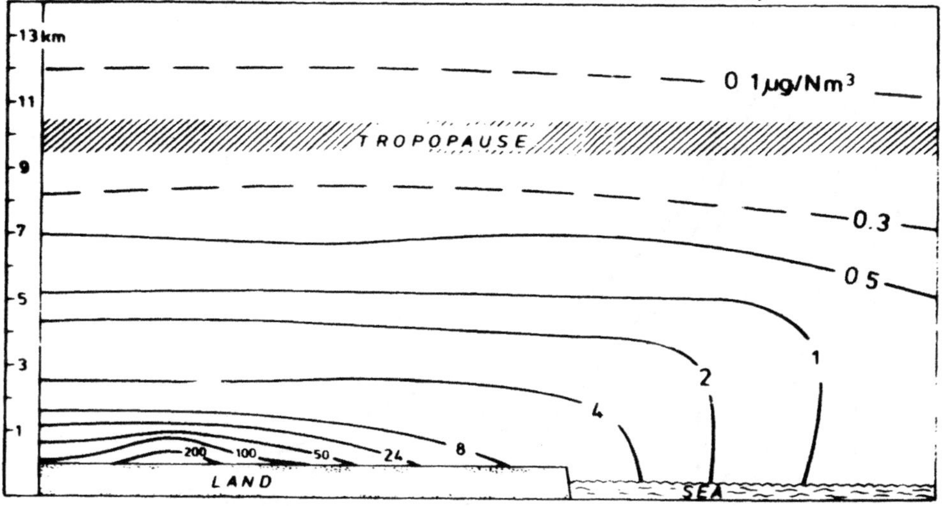

SCHEMATIC DISTRIBUTION OF SO_2 IN TEMPERATE LATITUDES

Figure 3 Schematic distribution of SO_2 in temperate latitudes

As Fig.4 shows over Western Europe the average SO_2-concentrations in unpolluted air decreases with altitude from about 2o $\mu g/m^3$ near the ground during winter and 1o $\mu g/m^3$ during summer reaching 5 $\mu g/m^3$ in 2 - 3 km altitude and o.5 $\mu g/m^3$ in 5 km altitude (Georgii 1978). The half value of the ground concentration is reached at about 1.5 km above ground.

Vertical Distribution of SO_2-S from Aircraft - Ascents

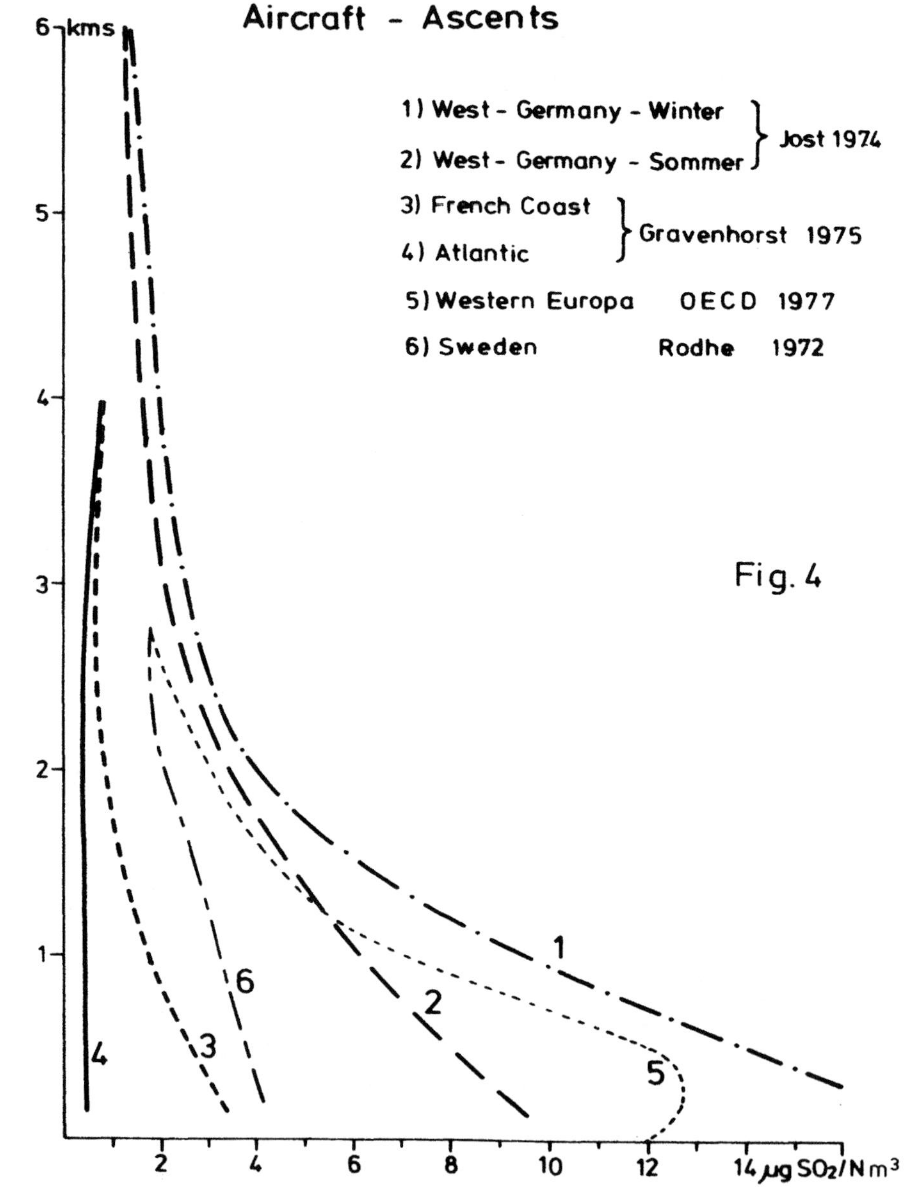

It is of interest to observe that the difference between winter and summer values is only found in the lower troposphere. The higher winter values are probably mainly caused by differences in emission, but differences in the transformation- and removal rate may also play a role. Individual cases of vertical SO_2 distributions are strongly influenced by the vertical thermal stability of the atmosphere, particularly in the case of the formation of inversions. The development of a new chemiluminescence-technique by Stauff and Jaeschke (1975) permitted for the first time the performance of SO_2-measurements in the upper troposphere and lower stratosphere.

Separate sampling and analyzing of SO_2 is performed with a wet chemical filter method. Chemically impregnated filters are used instead of wet scrubbing. The filter is impregnated with o.1 M Tetrachlormercurate. Air samplings are taken by drawing the required amount of air through the filter. Traces of SO_2 are then absorbed in the filter, forming a stable disulfitomercurate complex. Analysis of the complexed SO_2 in the sample is based on the following effect: the complexed SO_2 is oxidized to $SO_4^{=}$ by an acid $KMnO_4$-solution and the oxidizing step is accompanied by chemiluminescence (< 5oo nm). The chemiluminescence effect is proportional to the SO_2 in the sample. This analytical method is 2o-fold more sensitive compared to the usual West-Gaeke method (Georgii and Meixner 198o). Considering all observed and analytical errors a detection limit of o.o1 ± o.oo1 ppbv is obtained. Ascents were performed with chartered jet-planes. Most flights were performed during winter, since the aircraft can reach the lower stratosphere at 5o to 6o°N only during the cold season. The position of the tropopause mainly governed the choice of the flight-track. 12 flights were carried out between 1978 and 198o. Within the troposphere up to altitude of 5 km the SO_2-concentration decreased sharply to values around 1oo ppt. Above 5 km a constant value up to the tropopause is observed. A SO_2-concentration between 2o and 1oo ppt SO_2 seems to be typical for layers above 6 km and can be considered a background value for the higher troposphere. Near the tropopause a very weak vertical gradient is observed. Under the assumption that the SO_2 flux is proportional to the vertical gradient, only a very insignificant SO_2-flux from the troposphere into the stratosphere can be expected.

At the undisturbed tropopause, the SO_2-concentration was observed between 15 and 5o ppt, that means lower than assumed by most authors and only comparable with the minimum assumption by Moortgart and Junge (1974) (Fig.5).

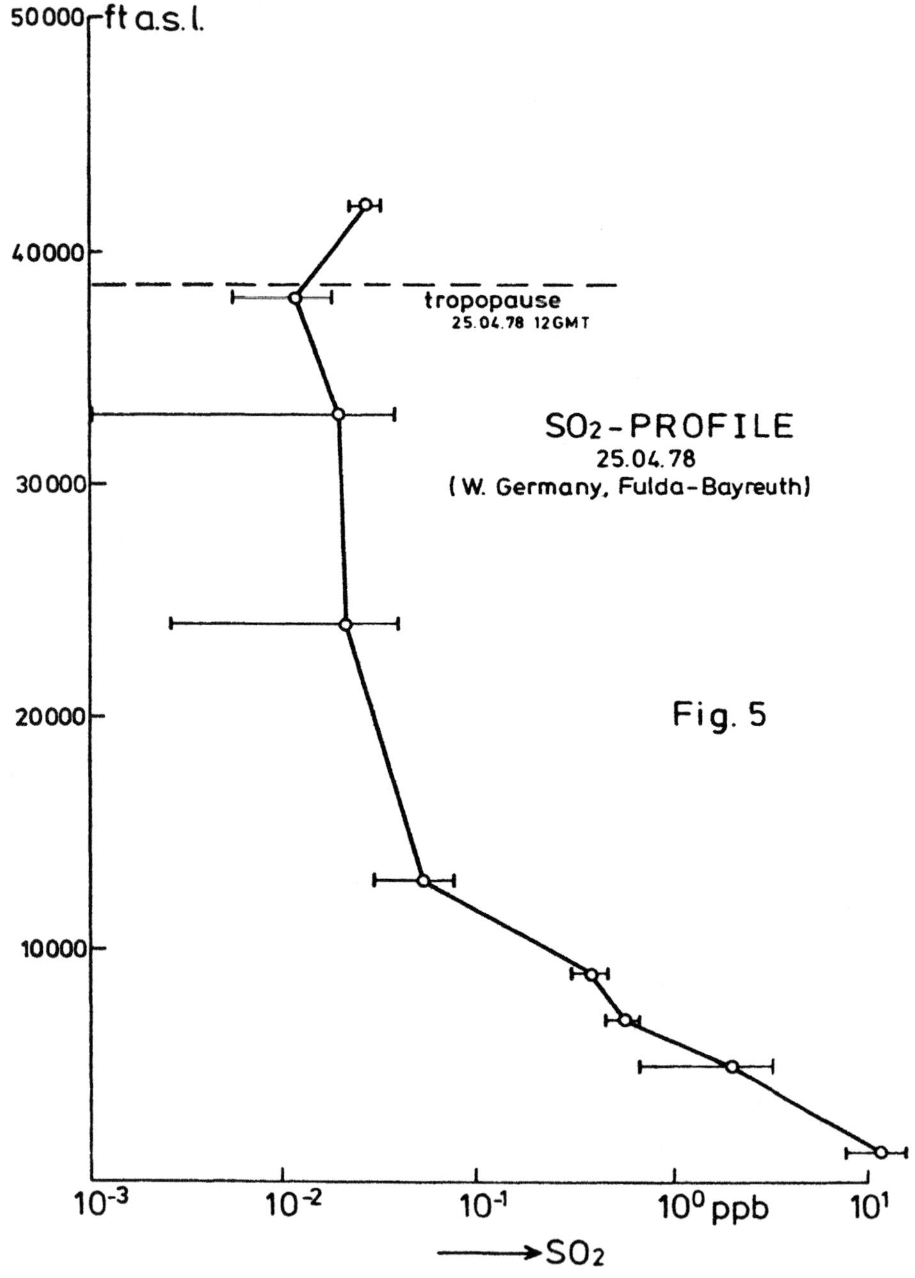

Fig. 5

At a disturbed tropopause near a polar-jet a well structured distribution was found with change of positive and negative gradients, which are caused by stratospheric-tropospheric exchange processes. The results of measurements within the stratosphere show a result which was unexpected and suprising at first sight, namely a slightly <u>increasing</u> SO_2-mixing ratio. This result is in contrast to previous assumptions and considerations on the stratospheric sulfur budget by Harrison and Larson (1974), Junge (1974), Harker (1975), who postulated a strong decrease of the stratospheric SO_2-mixing ratio. Our own measurements, but also recent measurements by Inn demand on the basis of the observed vertical increase of SO_2 in the stratosphere an in-situ source for stratospheric SO_2.

Table 4

Data	vertical gradient of mixing ratio pptv/km	average distance above the tropopause km
25. April 1978	+ 12	o.73
18. Dec. 1978	+ 6	3.59
19. Dec. 1978	+ 5	3.98
2o. Sept. 1979	+ 15	1.83
24. May 198o	+ 24	2.54

Table 4 shows that the vertical gradient of the mixing ratio above the tropopause was positiv, that means increasing SO_2-concentration in the lower stratosphere.

A model of the stratospheric SO_2-distribution by Junge and Moortgart (1977) who reviewed the reaction rates SO_2 + OH describes high tropospheric SO_2-values fairly well but not the stratospheric distributions. New models by Crutzen (1981) and by Turco et al.(1979) include the photodissociation of COS and the oxidation of COS and CS_2 by OH-radicals as important model parameters. These assumptions lead to a rapid decrease of COS and CS_2 in the stratosphere and to an increase of the SO_2-concentration (Fig. 6).

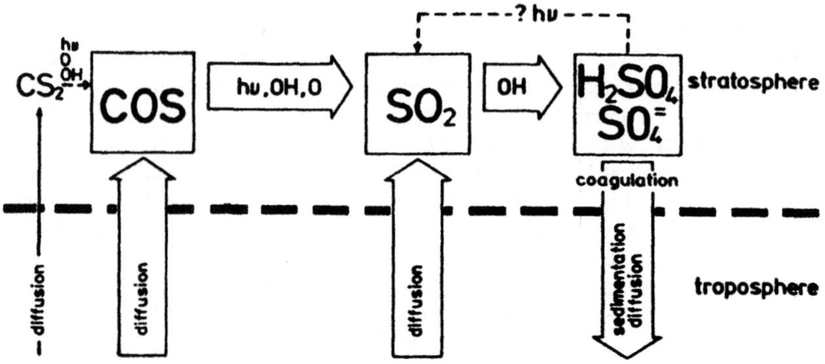

Figure 6 Schematic diagram of the mechanisms of the stratospheric sulfur budget

Without going into details it can be concluded that the flux of tropospheric SO_2 into the stratosphere is insufficient and cannot explain the stratospheric sulfate-aerosol layer. The thesis, postulated by Junge et al. (1961) that the quasistatic diffusion of tropospheric gaseous sulfur components into the stratosphere is the source of stratospheric sulfate-aerosols in times of weak volcanic activity remains valid. However, the fraction of tropospheric SO_2 is only 1o to 45% of the total sulfur flux into the stratosphere, the dominant fraction are organic gaseous sulfur compounds. This is further explained in table 5 giving calculations of the sulfur-flux into the stratosphere.

Table 5

Required sulfur-flux through the tropopause for the conservation of the stratospheric aerosol - layer

Author	Molecules/cm^2sec
Junge (1974)	$5.6 \times 1o^6$
Hofmann et al. (1974)	$2.o \times 1o^7$
Lazrus et al. (1974)	$9.o \times 1o^6$

Calculated sulfur-flux through the tropopause

Component	Molecules/cm^2sec	
	Whitten et al.	Crutzen
CS_2	$1.1 \times 1o^7$	$1.o \times 1o^7$
COS	$1.o \times 1o^7$	$7.o \times 1o^6$
SO_2	$1.2 \times 1o^6$	$4.7 \times 1o^5$
H_2SO_4	$3.2 \times 1o^5$	-

Above the oceans nearly constant SO_2-concentrations in the troposphere were observed during aircraft-flights over the Atlantic. A background concentration of approximately o.2 to 1 $\mu g/m^3$ is suggested and there are indications that in pure maritime air the SO_2-concentration may even increase with altitude. This would not be surprising since ocean water acts as a sink for SO_2 as outlined by Georgii and Gravenhorst (1977). Horizontally, the SO_2-concentration decreases from temperate to tropical latitudes, the background-concentration in the tropical region of the ocean being less than o.3 $\mu g/m^3$ (Büchen and Georgii 1971). G. Ockelmann et al. (1981) measured the SO_2-concentration over the Atlantic during several cruises with the research vessel "Meteor". Mixing ratios range from 3o ppt over the open Arctic ocean to values near the detection limit over the ice-covered polar sea, whereas

3o to 2oo ppt were found over the Atlantic ocean (Fig.7)

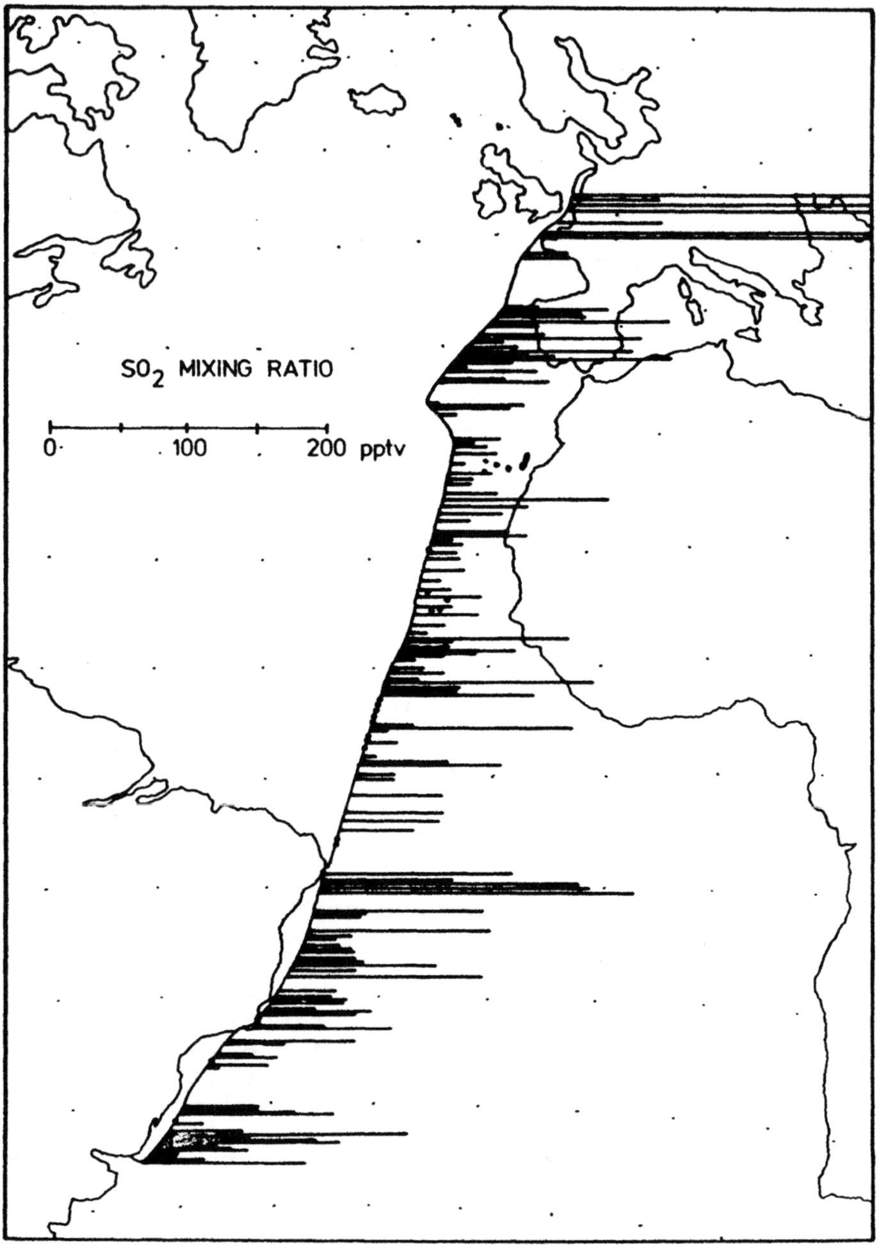

Figure 7 Meridional Distribution of SO_2 over the Atlantic Ocean

This may indicate a meridional dependance of the maritime SO_2-distribution. During all cruises large variations of the mixing ratio could be detected with characteristic periods of the order of days. It can be assumed that enhauced SO_2-concentrations are due to long range transport from the continents.

Earlier measurements of SO_2 over the Atlantic had shown a maximum around 4o°N latitude. Georgii and Vitze(1971) had attributed this maximum to transport from continental sources, mainly from the East coast of North America. However, this conclusion is very tentative. It must also be taken into consideration that dimethylsulfide and carbon disulfide are emitted from the oceans and may be slowly converted to SO_2. Besides long range transport the transformation of ocean produced organic sulfur compounds are responsible for the marine SO_2-distribution.

H_2S

With respect to the large-scale spatial and temporal distribution of H_2S still unsufficient information is available. Although large areas with anoxic soils are assumed to be natural biogenic sources, it is difficult to detect the very low amounts of H_2S released into the atmosphere. An improved analytical method by Grunert , Ballschmitter and Tölg (1968) was adapted for atmospheric measurements by Jaeschke and Haunold (1977) permitting for the first time the detection of H_2S in the concentration range below 1 $\mu g/m^3$. Experiments with different soil samples, typical for Europe showed that some particular soils are sinks for H_2S, only swamps, marshes and tidal flats acted as sources. Aircraft ascents over tidal flats in Northern Germany carried out by Jaeschke, Claude and Hermann (198o) indicated a strong decrease of the H_2S-concentration with altitude. While over tidal flats a concentration of o.1 to o.2 $\mu g/m^3$ was found at ground level the ground concentration in the Rhein-Main area, where H_2S is produced from anthropogeneous sources is much higher, namely around o.8 $\mu g/m^3$. This shows that anthropogeneous sources of H_2S may not be neglected (Fig.8).

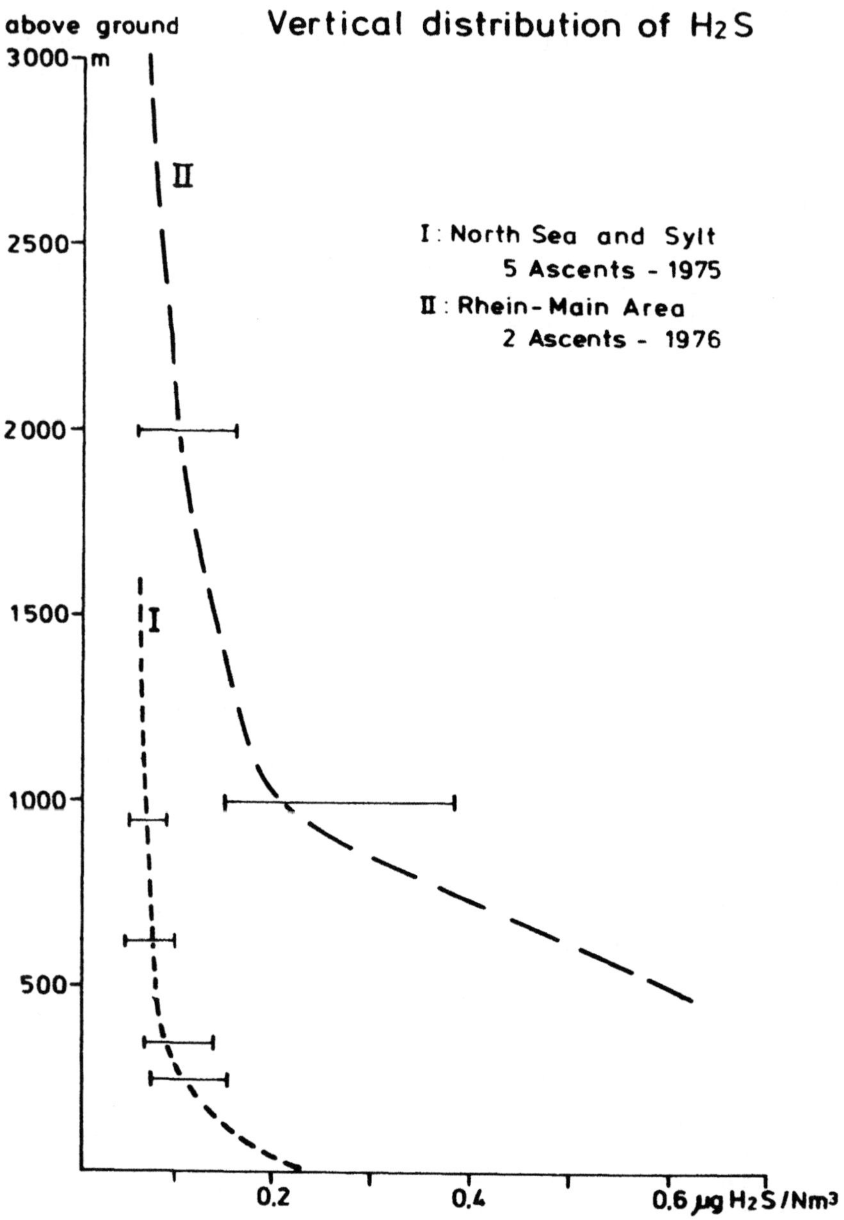

Figure 8 Vertical Distribution of H_2S

Another natural source of local to regional importance is the volcanic activity. Berresheim (198o) measured the source strength of the volcano Etna in Italy and found in September 1978 during a period of very low activity a source strength of 1.3to H_2S/day. In June 1979 a significant increase of the source-strength was registered. At that time the volcano emitted a mean of 1o to H_2S/day.

It is suggested that the oxidation of H_2S with OH-radicals is the main sink of H_2S in the lower troposphere. Jaeschke et al. (198o) have shown the importance of this reaction in an indirect way. They measured the vertical gradient of the H_2S-concentration and calculated the vertical flux by application of a gradient method. On the basis of the decrease of the H_2S-flux with altitude they found a mean removal coefficient of H_2S of $2.3 \times 1o^{-5} s^{-1}$. From the mean removal an average life-time of H_2S of 12.2 hrs was calculated.

COS, CS_2, DMS

Organic gaseous sulfur components like dimethylsulfide and carbon bisulfide are emitted from the oceans into the atmosphere, while COS has swamps and marshes as source areas. However, considerable differences exist with respect to the estimates of the size of flux of reduced biogenic sulfur compounds from the oceans into the atmosphere. The DMS-concentration in surface water seems to be controlled both by the surface temperature and by the biological activity. It can be expected that most oceanic emissions occur in the southern hemisphere, therefore the oceans may be the dominant source of atmospheric sulfur south of the equator (Andreae et al. 1981). While DMS and CS_2 have only very short life-times in the atmosphere, probably only hours up to one day, the estimated tropospheric life-time for COS is approximately one year. The short life-time of DMS in the atmosphere leads to large variations of the ambient concentration with time and space. Bingemer (1981) measured in surface air over the middle and south Atlantic DMS-concentrations between 2 and 44 ng S (DMS)/m^3 with an average of 6 ng S (DMS)/m^3. On a recent cruise to the Azores an average concentration of 2.8 ng S(DMS)/m^3 was observed. Nguyen et al. (198o) measured in the Mediterranean sea and in the Atlantic ocean an average DMS-concentration of 3 - 6 ng S(DMS)/m^3. Obviously, there exist also anthropogeneous sources of DMS on the continents, possibly in connection with the release of sulfuric emissions from refineries. In the Rhein-Main basin DMS-concentrations of 25 to 5o ng S (DMS)/m^3 were observed during summer 1981.

On the distribution of CS_2 there exists hardly any data. Sandalls and Penkett (1977) report a concentration of 7o to 37o ppt with great variability. Bandy (1981) has recently reported average CS_2-concentrations of 15 ppt in the boundary layer above the Pacific Ocean.

The horizontal distribution of COS appears to be more homogeneous. All authors agree on a rather uniform level of 5oo ppt.

The removal mechanism for DMS from the atmosphere is not sufficiently understood. Homogeneous photochemical reaction appears to be one important sink. Cox and Sandalls (1974) studied the oxidation of DMS in the laboratory by using ppm-concentrations. They found that free radicals such as OH reacted rapidly with DMS . It is suggested that DMS is a precursor for SO_2 found over the oceans. Otherwise the SO_2-mixing ratio of about o.1 to o.3 ug SO_2/m^3 found over the tropical Atlantic ocean by Ockelmann et al. (1981) is difficult to understand.

With respect to COS, little information is available on anthropogeneous sources. Crutzen et al. (1979) mention biomass-burning, mainly due to man's activities in the tropics, as one possible source. There are also reports of COS and CS_2 being released from refinery sulfur recovery plants. Our present knowledge does not permit yet any estimation with respect to an increase of anthropogeneous emissions of COS in order to speculate on a trend towards future higher concentrations.

TRANSFORMATION MECHANISMS IN THE ATMOSPHERE

As mentioned above the residence time of SO_2 is limited to a few days. The oxidation of SO_2 leads to the formation of sulfate-aerosols. Although the mechanism of SO_2-oxidation is not yet fully understood, according to Wilson (1978), we can differentiate the following five oxidation-processes:

Table 6 Mechanisms of conversion of SO_2 to sulfate

Mechanisms	reaction	dominant factors
1. Direct Photooxidation	$SO_2 \longrightarrow H_2SO_4$	SO_2-concentration Intensity of sunlight
2. Indirect Photooxidation	$SO_2 \xrightarrow[OH,\ HO_2,\ RO_2]{} H_2SO_4$	SO_2-concentration, Oxidants, OH, NO_x, HO_2, RO_2
3. Heterogeneous oxidation in droplets	$SO_2 \xrightarrow{\text{liquid water}} H_2SO_3$ $NH_3 + H_2SO_3 \longrightarrow NH_4^+ + SO_4^=$	pH of liquid water NH_3-concentration
4. Heterogeneous catalytic oxidation in droplets	$SO_2 \xrightarrow[\text{heavy metal ions}]{\text{liquid water}} SO_4^=$	Concentration of heavy metals (Fe, Mn, V)
5. Heterogeneous catalytic oxidation in aerosols	$SO_2 \xrightarrow[\text{heavy metal ions}]{\text{liquid film}} SO_4^=$	surface area of particles heavy metal aerosols (Fe, Mn, V)

The two homogeneous oxidation mechanisms (1) and (2) have different importance. The quantum yield by sunlight is only about 3 x 10^{-4}. This reduces the maximum rate of SO_2-oxidation by direct absorption of sunlight to 10^{-2}%/hr leading to a life-time of SO_2 of many months. This mechanism is therefore insignificant. The oxidation mechanism of SO_2 with OH-radicals is now assumed to be of dominant importance. Experimental results show that the HSO_3^- -radical, produced by the reaction:

$$M + OH + SO_2 \longrightarrow HSO_3^- + M$$

is rapidly transformed into an aerosol-particle consisting mainly of sulfuric acid. The reaction with HO_2-radicals is obviously slower by 3 orders of magnitude. The following table shows the most important homogeneous reactions according to Schurath et al. (1978).

Table 7 SO_2-oxidation by homogeneous gas-reactions

reactions-partner	average concentration molecules/cm^3	oxidations-rate h/r max.	typical
OH	5 x 10^5	2 %	0.2 %
HO_2	10^8	0.15%	0.03%
RO_2	10^8	0.35%	0.04%
RO	5 x 10^5	1 %	0.01%
photochemical Reaction		0.01%	0.01%

With the estimated concentrations of OH-radicals we can expect a life-time of SO_2 on the basis of the dominance of homogeneous reactions of 3-5 days for temperate latitudes. In our climate regime heterogeneous oxidation-processes, as mentioned under 3) to 5) may be of greater relevance at least during the winter-months. Heterogeneous oxidation occurs in droplets of clouds and fog. The rate determining factor is the oxidation and not the transport to the droplet or within the droplet (Beilke 1980). The oxidation in absence of catalysts is a process which is strongly pH-dependant and leads to oxidation-rates above 1%/hr only in such cases when the pH-value of the liquid phase is above 6. Owing to the fact that the pH-value of cloud-water is generally between 3 and 5, this process is of little relevance in the atmosphere. A special type of the SO_2-oxidation in the absence of catalysts is the oxidation in the presence of NH_3. In the past much attention has been paid to this mechanism which was thought to be the most important SO_2-transformation process by some authors. The effect of NH_3 is to keep the droplet pH high when S (IV) is oxidized to sulfate. However, it must be noted, that NH_3 has to compete in the real atmosphere with many other soluble substances, for example with the acidic background aerosol. Therefore at a droplet-pH below 5 NH_3 can hardly enhance the oxidation-process (Beilke and Gravenhorst 1978). Catalytic oxidation (4) occurs mainly in the presence of transition metals of the 4th period, especially manganese and iron.

On the basis of the experiments of Barrie and Georgii (1976) the following SO_2-decay-rates were found for cloud- or fog-water:

Table 8

SO_2-decay rates for catalytic oxidation

liquidwater content	pH	temperature °C	Mn^{++} molar	Fe^{++} molar	SO_2-decay-rate
o.1 g/m³	3.5	8	$1o^{-5}$	$1o^{-5}$	1 %/hr
	3.5	25	$1o^{-5}$	$1o^{-5}$	1 %/hr
	3.5	8	$1o^{-6}$	$1o^{-5}$	o.25%/hr
	3.5	25	$1o^{-6}$	$1o^{-5}$	o.75%/hr
	4.o	25	$1o^{-6}$	$1o^{-5}$	3 %/hr

On the basis of these experiments the conclusion can be drawn that sulfate formation due to catalytic SO_2-oxidation can be significant in clouds and fogs of polluted regions.

Finally, we want to mention the investigations by Penkett et al. (1979) on the SO_2-oxidation in the presence of ozone and in the presence of H_2O_2.

To quantify the latter process more atmospheric measurements of H_2O_2 are required. We can summarize that heterogeneous oxidation in the presence of heavy metals as catalysts may lead to a similar SO_2-decay rate as homogeneous oxidation. However, the heterogeneous process depends largely on the liquid water content of clouds and fogs, of the pH-value and the temperature of the droplets and the concentration of heavy metal ions in the liquid phase. With respect to the SO_2-oxidation after absorption on aerosol-particles, the experiments by Runca-Köberich and Georgii (1978) have shown that the rate of sulfate-formation depends on the relative humidity and the chemical behaviour of the aerosols as catalysts. If the particles are covered by a liquid film at higher humidities, they act as droplets do.

DRY AND WET DEPOSITION OF SULFUR-COMPOUNDS

The residence-time and the travel-distance of sulfur-compounds are essentially determined by the removal from the atmosphere. Four major mechanisms are to be mentioned: dry and wet removal of SO_2 and dry and wet removal of sulfate-aerosols. Dry deposition is mostly described by the deposition velocity. Measured deposition velocities for SO_2 fluctuate between o.2 and 2 cm/sec. An average value suggested by Garland (1978) is o.8 cm/sec. Our knowledge of the influence of thermal stability or surface conditions on the dry deposition of trace gases is still insufficient. The dry deposition of sulfate-aerosols is size-dependant. Experimental investigations are very rare. Direct deposition measurements by C.Koch (1979) show a dry deposition of SO_2 of 3.15gS/m^2year and a dry sulfate deposition of o.32 gS/m^2year for the polluted environment of Frankfurt/Main. These fractions are in agreement with estimates by Meszaros et al. (1978) who supposed a fraction of dry sulfate-deposition of 1o% of the total dry sulfur deposition.

Wet deposition patterns are mainly determined by precipitation-chemistry and precipitation-rate. On the other hand the concentration of various ions is often highest with light precipitation. A full study of the relationship between ion-concentration and precipitation rates has not yet been made. The annual wet sulfate-deposition for Germany is about 2 g S/m^2. Estimations by C.Perseke (1978) for the year 1974 revealed maximum values of wet deposition for regions with high precipitation - rate like Upper Bavaria with

2-3 g S/m^2year or for regions with high level of pollution like the Rhein-Ruhr area with 4 g S/m^2year.

Direct measurements of dry and wet deposition of sulfur-compounds were performed in the frame of a deposition-network in the Federal Republic of Germany. The following comparison of dry and wet sulfur-deposition for the winter halfyear 1979/8o shows that the wet sulfate-deposition is the more efficient removal-process (Georgii and Perseke 198o).

Table 9 Comparison of wet and dry sulfur-deposition during winter halfyear 1979/8o g S/m^2

station	wet deposition	dry deposition	wet/dry
Schleswig	o.66	o.23	2.8
Hamburg	o.5	o.18	2.7
Braunschweig	o.42	o.22	1.9
Essen	o.84	o.4	2.1
Jülich	o.46	o.27	1.7
Deuselbach	o.34	o.14	2.5
Frankfurt/M.	o.6	o.21	2.8
Hohenpeißenberg	o.43	o.18	2.3

On the whole it appears from recent investigations, that dry deposition of sulfur is an important sink in heavily polluted areas while the importance of wet deposition of sulfur increases in less polluted areas and in regions with high rainfall-rate. An evaluation for the year 1979 to 198o for the Federal Republic of Germany reveals that the wet deposition of acid pollutants is highest during the summer season due to high precipitation rates.

CLIMATIC EFFECT OF SULFUR COMPOUNDS

With respect to relevance to climate we can neglect the gaseous sulfur compounds and have only to consider the climatic impact of sulfate-aerosols. Although very little quantitative data are available, we can expect that in Europe, in the Eastern part of North America and in the plume of the emissions of these continents, the impact of anthropogeneous aerosols is noticeable as measurements of the excess-sulfate over the Atlantic have shown (Gravenhorst 1979). Sulfate-particles in non-cloudy air cause scattering and absorption of outgoing radiation. As a result, aerosol particles are involved in the radiative climate of parts of the globe. This involvement depends on number, size distribution and refractive index at various wave-lengths as has been outlined by Bolin and Charlson (1976). According to these authors, non-cloud aerosol actually modifies incoming visible radiation, although the details and the extent of the effect on the radiative climate are still not well known. Sulfate aerosols are distributed over a scale of the order 1000 to 3000 km and are therefore not limited to urban areas (Georgii 1979). This suggests an influence of human activity on the short radiative climate of synoptic scale regions. Nephelometric measurements in the U.S. and in Europe, particularly in Sweden, indicate a relation between the sulfate content in the atmosphere and the extinction coefficient (Charlson et al.1977). A strong correlation is found between light-scattering and the concentration of sulfate-aerosols in the New York area. Turbidity data collected in US rural areas show that regions of high turbidity coincide with regions of high sulfate concentration. The calculations by Bolin and Charlson indicate that the change of scattered solar radiation distributed over the whole northern hemisphere may be equivalent to a small drop of the average temperature by possibly 0.05°C. It is understood that in polluted areas the effect of scattering due to sulfate aerosols may be greater but the circulation of the atmosphere reduces this effect. This effect takes only in consideration sulfates concentrated in the lower troposphere. It may, however, very well be that the aerosol in the upper troposphere and in the stratosphere is more influental.

We want to close this chapter by stressing that COS seems to be a dominant source of the stratospheric aerosol and that approximately 50% of COS is produced from anthropogeneous activities, mostly related to fuel processing and consumption (Turco et al. 1980). With an increasing use of coal, COS-release may rise in future leading to an additional contribution to the stratospheric aerosol-formation. The above - mentioned authors estimate that a tenfold increase of the COS-mixing ratio may lead to a factor of five increase in aerosol-mass mixing ratio. This may produce an effect comparable to that of a moderate semi-permanent volcanic eruption, lasting as long as the

COS-emissions persist.

The research discussed in this paper has been partly sponsored by Deutsche Forschungsgemeinschaft through Sonderforschungsbereich 73 "Atmospheric trace Substances". The support is gratefully acknowledged.

REFERENCES

1. Barnard, W.R., Andreae, M.O., Watkins, W.E., Bingemer, H., Georgii, H.W. : 1981, The flux of dimethylsulfide from the oceans to the atmosphere, To be published in J. Geophys.Res.
2. Barrie, L. and Georgii, H.W. : 1976, An experimental investigation of the absorption of sulfur dioxide by water drops containing heavy metal ions, Atm. Environment 1o, 743-749.
3. Beilke, S.: 198o, Luftchemisches Verhalten von SO_2, Ergebnisse der VDI-Arbeitsgruppe "Luftchemie" 12-25.
4. Beilke, S. and Gravenhorst, G.: 1978, Heterogeneous SO_2-oxidation in the droplet phase, Atm. Environment, 12, 231-241.
5. Berresheim, H.: 198o, Der Beitrag vulkanischer Exhalationen zum globalen atmosphärischen Schwefelhaushalt, Diplomarbeit, Inst.f.Meteorologie und Geophysik, Univ.Frankfurt/M.
6. Bingemer, H.: 1981, Personal communication.
7. Bolin, B. and Charlson, R.: 1976, On the role of the tropospheric sulfur cycle in the short-wave radiative climate of the earth, Ambio 5,47-54.
8. Büchen, M. and Georgii, H.W.: 1971, Ein Beitrag zum atmosphärischen Schwefelhaushalt über dem Atlantik, "Meteor"-Forsch.-Ergebnisse B 7,71-77.
9. Charlson, R.J., Vanderpol, A.H., Waggoner, A.P., Covert, D.S., Baker, M.B.: 1977, The dominance of tropospheric sulfate in modifying solar radiation, Proc.Symposium on Radiation in the Atmosphere, Science Press. 32-37.
1o. Cox, R.A. and Sandalls, F.S.: 1974, The photo-oxidation of hydrogen sulfide and dimethylsulfide in air, Atm.Environment, 8, 1269-1281.
11. Crutzen, P.J.: 1981, On the vertical distribution of gaseous sulfur compounds, Unpublished Manuscript.

12. Cullis, C.F. and Hirschler, M.M.: 1979, Emissions of Sulphur into the Atmosphere, Proc, Int. Symposium Sulphur Emissions and the Environment, 1-23.
13. Garland, J.A.: 1978, Dry and wet removal of sulfur from the Atmosphere, Atm. Environment 12,349-363.
14. Georgii, H.W.: 1979, Large scale distribution of gaseous and particulate sulfur compounds and its impact on climate, in:Bach,Pankrath, Kellogg Man`s Impact on Climate, Elsevier Scientific Publishing Comp. Amsterdam,181-193.
15. Georgii, H.W.: 1978, Large scale spatial and temporal distribution of sulfur compounds, Atm. Environment, 12,681-691
16. Georgii, H.W., Jost, D.: 1964, Untersuchung über die Verteilung von Spurengasen in der freien Atmosphäre, Pure and Appl. Geophys. 59, 217-224.
17. Georgii H.W. and Meixner, F.: 198o, Measurement of the tropospheric and stratospheric SO_2-distribution, J.Geophys.Res.85, 7433-7438.
18. Georgii, H.W. and Gravenhorst, G.: 1977, The ocean as source or sink of reactive trace gases, Pure and Appl. Geophys.115, 5o3-511.
19. Georgii, H.W. and Perseke, C.: 198o, Some results of wet and dry deposition of sulfur compounds, Proc.European Symposium Physico-chemical behaviour of atm.pollutants, Ispra 41o-418.
2o. Georgii, H.W. and Vitze, W.: 1971, Global and regional distribution of sulfur components in the atmosphere, idöjaras 294-299.
21. Granat, L., Hallberg, R.O. and Radke, H.: 1976, The global sulphur cycle, SCOPE Report 7, Ecol. Bull (Stockholm), 89-135.
22. Grünert, A., Ballschmitter, K.H. and Tölg, G.: 1968, Fluoreszenzanalytische Bestimmung von Sulfidionen im Nanogrammbereich, Talanta 15, 451-457.
23. Harrison, H. and Larson, T.: 1974, The oxidation of SO_2 in the stratosphere, J.Geophys.Res.79, 3o95-3o97.
24. Harker, A.B.: 1975, The formation of sulfate in the stratosphere through the gas-phase oxidation of SO_2 , J.Geophys.Res.8o, 3399-34o1.
25. Jaeschke, W., Claude, H., Hermann, J.: 198o, Sources and Sinks of Atmospheric H_2S, J.Geophys.Res.85, 5639-5644.
26. Jaeschke, W. and Haunold, W.: 1977, New methods and first results of measuring atmospheric H_2S and SO_2 in the ppb-range, WMO Special Env. Report No.1o, 193-198.

27. Jost, D.: 1974, Aerological Studies on the atmospheric sulfur-budget, Tellus 26, 2o6-213.
28. Junge, C.: 1974, Sulfur budget of the stratospheric aerosol layer, Proc.IAMAP-Conference Melbourne Vol.1, 85-97.
29. Junge, C., Chagnon, C.W., Manson, J.E.: 1961, Stratospheric aerosols, J.Met.18, 81-1o8.
3o. Koch, C.: 1979, Experimentelle Untersuchung der trockenen und nassen Deposition der atmosphärischen Schwefelverbindungen, Diplomarbeit Inst. f.Met. und Geophys. Univ. Frankfurt/M.
31. Meszaros,E., Varhelyi, G. and Haszpra, L.:1978, On the atmospheric sulfur budget over Europe, Atm.Environment 12, 2273-2277.
32. Moortgart, G.S. and Junge, C.: 1977, The role of SO_2-oxidation for the background stratospheric sulfate-layer in the light of new reaction data, Pure and Appl. Geophys.115, 769-774.
33. Nguyen,B.C., Gaundry, A., Nadaud,P., Lambert, G.: 1981, Dimethylsulfide production at the air/sea interface, Paper presented at the IAMAP-Assembly Hamburg.
34. Ockelmann, G., Meixner, F. and Georgii, H.W.: 1981, SO_2-measurements over the Artic and Atlantic Ocean, Paper presented at IAMAP-general Assembly Hamburg.
35. Penkett, S.A.: 1979, Chemical changes in the Air, Proc.Int.Symposium Sulfur Emissions and the Environment 1o9-122.
36. Perseke, C., Beilke, S. and Georgii, H.W.: 198o, Die Gesamtschwefeldeposition in der Bundesrepublik Deutschland auf der Grundlage von Meßdaten des Jahres 1974, Berichte des Inst.f. Meteorologie und Geophysik der Univ.Frankfurt/M. Nr.4o.
37. Runca-Köberich, D. and Georgii, H.W.: 1978, Surface Reactions of SO_2 on Aerosol Particles, J.Aerosol Sci. 9, 238-24o.
38. Salmon,L., Atkins, D.H.F., Fisher, E.M.R., Healy, C. and Law, D.V.: 1978, Retrospective Trend Analysis of the Content of U.K.Particulate Material 1957-1974, Science of the total Environment 9, 161-2oo.
39. Schurath, U., Seitz, H., Löbel, J.: 1978, Oxidation von SO_2 in der Gasphase, VDI-Berichte Nr.314, 33-4o.
4o. Sandalls, F.J. and Penkett, S.A.: 1977, Measurements of carbonylsulfide and carbondisulfide in the atmosphere, Atm.Environment, 11, 197-199.

41. Stauff, J. and Jaeschke, W.: 1975, A chemiluminescence technique for measuring atmospheric trace concentrations of sulfur dioxide, Atm.Environment 9, 1o38-1o39.
42. Turco, R.P., Whitten, R.C., Torn, O.B., Pollack,J.B., Hamill, P.: 198o, COS, stratospheric aerosols and climate, Nature 283, 283-286.
43. Turco, R.P., Hamill, P. Torn,O.B., Whitten, R.C., Kiang, C.S.: 1979, A one-dimensional model describing aerosol-formation and evolution in the stratosphere I, J.Atm.Sci. 36, 699-717.
44. Varhelyi, G. and Gravenhorst, G.: 1981, An attempt to estimate biogenic sulfur emission into the atmosphere, Idöjaras 85, 126-134.
45. Wilson, W.E.: 1978, Sulfates in the atmosphere, Atm.Environment 12, 537-547.

SULFUR EMISSIONS FROM VOLCANOES

Harald Berresheim and Wolfgang Jaeschke

Center of Environmental Protection
University of Frankfurt, FRG

ABSTRACT

Measurements of the mass fluxes of volcanic SO_2, H_2S and SO_4^{2-} into the atmosphere, which have been performed at Mt. Etna (Sicily), are described in the first part of this paper. It was found that the emission rates vary between 142-1600 t SO_2/d, 1-10 t H_2S/d and 182-232 t SO_4^{2-}/d with respect to different non-eruptive activity phases of this volcano.
In the second part of this paper a new assessment of the global volcanic emission rates of the above sulfur compounds is presented. It is based on a quantitative classification of volcanic activity which is applied to the measurements at Mt.Etna and Mt. Vulcano (Lipari Islands) as well as to other measuring data available from the literature concerning the sulfur source strength of different volcanoes during different phases of activity. The total global emission of volcanic SO_2 into the atmosphere is estimated to 1.52×10^7 t/a. Only a small fraction of this amount, namely 1.0×10^6 t/a, is emitted during eruptions, while the major part, namely 1.42×10^7 t/a, is contributed by non-eruptive activity. The average global emission of volcanic H_2S and SO_4^{2-} is found to be around 10^6 t/a and 10^7 t/a, respectively. From these estimates it is concluded that volcanoes contribute to about 5-7 % of the total global sulfur emission into the atmosphere.

INTRODUCTION

Both the emissions from anthropogenic and natural sources contribute to the presence of sulfur pollutants in the atmospheric

H. W. Georgii and W. Jaeschke (eds.), Chemistry of the Unpolluted and Polluted Troposphere, 325–337.

environment. The total global mass fluxes of these pollutants into the atmosphere can satisfactorily be estimated only if the individual contributions of each of these sources are determined by direct measurements. Furthermore, such measurements should provide a better evaluation of the role that anthropogenic sources play in the atmospheric sulfur budget compared to the contribution of the natural sulfur sources.

One of these natural sources for atmospheric sulfur compounds is volcanism, which emits both reduced and oxidized sulfur species. Among these, H_2S, SO_2 and SO_4^{2-} are the predominant components, which represent the valence states (-2), (+4) and (+6), respectively.

Since there is a considerable lack of measuring data on the sulfur source strength of volcanoes, especially regarding H_2S and SO_4^{2-}, the present knowledge of the global volcanic emission of these components is still very incomplete. Thus, in order to obtain more information about the volcanic sulfur emission into the atmosphere, measurements have been carried out by Jaeschke et al. [1] during 3 field trips to the Mt. Etna volcano in Sicily. In the first part of this paper a short survey of the results of these measurements is presented, which gives an idea, first, of the amount of sulfur masses that are released into the atmosphere by such a type of volcano and, second, of the relationship between this amount and the actual volcanic activity.

Based upon these results and upon other data available from the literature, an estimation of the global volcanic emission of SO_2, H_2S and SO_4^{2-} into the atmosphere is presented in the second part. In this estimation the variation of the sulfur source strength within the spectrum of volcanic activity is considered by classifying the activity into 8 categories according to the eruptive phase and into 4 other categories according to the non-eruptive phase. This procedure leads to a differentiated picture of the global volcanic input of sulfur into the atmosphere, which is then finally discussed in comparison with the contribution of other sources to the atmospheric sulfur budget.

PART I : MEASUREMENTS OF THE SO_2, H_2S AND SO_4^{2-} EMISSION RATES OF MT. ETNA, SICILY

Mt. Etna represents the greatest volcanic massif in Europe with a summit height of 3340 m NN and a ground level circumference of nearly 200 km. Furthermore, it belongs to the few permanently active volcanoes in the world.

During our field trips to Mt. Etna in Sept. 1978 and June 1979 only one of its main craters, the so-called 'Bocca Nuova', showed a strong fuming activity, while the other three central craters of the volcano were blocked. The activity was non-eruptive, and the plume spread out continuously in the open atmosphere.

The measuring program was focused on three main subjects:

1) Ground based measurements in the central crater region to determine the source concentrations of H_2S, SO_2 and SO_4^{2-}.
2) Concentration measurements of these sulfur compounds in the spreading plume which was tracked by an array of flight patterns using an aeroplane.
3) Determination of the removal rates of H_2S and SO_2 as well as the production rate of SO_4^{2-} after the emission of these gases into the atmosphere.

Table 1a briefly summarizes the chemistry of the individual measuring methods. The concentrations of the sulfur compounds were measured discontinuously. As an additional reference a continuous log was recorded during each flight for the SO_2 and CO_2 concentrations (Table 1b). The SO_2 concentration was measured with a flame photometric detector and the CO_2 concentration with an infrared absorption spectrometer [2].

An extensive depiction of the measurements and a detailed discussion of the results have been the subject of an earlier paper [1]. Therefore, considering the results for merely the sulfur source strength of Mt. Etna in this paper, I will only enter into the particulars of the above mentioned point 2) in this context, since the calculation of the individual emission rates of H_2S, SO_2 and SO_4^{2-} is, without exception, based on the data obtained during the flight measurements.

The concentrations of these sulfur compounds in the source region were determined by taking samples in the center of the plume directly at the rim of the 'Bocca Nuova' crater. Table 2 shows the results for both measuring periods. It is remarkable that the SO_2 and H_2S concentrations in June 1979 were significantly higher compared to Sept. 1978, whereas the sulfate concentration did not change significantly. A possible explanation for this phenomenon may be the thermodynamic variation in the volcanic gaseous system connected with a distinct change of the volcano's activity [3].

The flight measurements should therefore give an answer to the question, whether this assumption of a change in activity is confirmed by a change in the source strength of the individual sulfur compounds regarding the two measuring periods. A special flight program for the determination

TABLE 1a: PRINCIPLES OF SAMPLING AND ANALYSIS OF THE DISCONTINUOUSLY MEASURING METHODS

COMPOUND	ACCUMULATION	ANALYSIS
H_2S	$H_2S + AgNO_3 \rightarrow Ag_2S + HNO_3$	FLUORESCENCE QUENCHING
SO_2	$2SO_2 + Na[HgCl_4] \rightarrow Na[Hg(SO_3)_2] + 4HCl$	CHEMI-LUMINESCENCE
SO_4^{2-}	PARTICLE FILTER	ISOTOPIC DILUTION

TABLE 1b: DETECTION PRINCIPLES OF THE CONTINUOUSLY MEASURING INSTRUMENTS

COMPOUND	TYPE OF DETECTOR	REMARKS
CO_2	INFRARED ABSORPTION	BOTH PRESSURE DEPENDENT
SO_2	FLAME PHOTOMETRIC	

TABLE 2: RESULTS OF FIELD MEASUREMENTS AT THE RIM OF THE BOCCA NUOVA CRATER (Mt. ETNA)

DATE	H_2S [µg/m^3]	SO_2 [µg/m^3]	SO_4^{2-} [µg/m^3]	CO_2 [ppm(V)]	Rel. Hum. [%]
Sept. 1978	0.54 ± 0.03	(22.84 ± 2.3) × 10^3	6.07 × 10^3		14
June 1979	330 ± 210	(57.10 ± 5.7) × 10^3	4.05 × 10^3	1077 ± 91	

of the individual source strengths was developed. The program consisted of different measuring flights across the plume of Mt. Etna following a rigid flight pattern. Fig. 1a and 1b show a viewgraph of the measuring region including the Catania airport being the starting and landing point. In both figures the mean wind directions on the individual measuring days in Sept. 1978 and June 1979 are shown, respectively, as well as the flights on each of these days carried out in different distances to Mt. Etna.

A more detailed scheme of the flight pattern is given in Fig. 2. From this schematic diagram it is obvious that the concentration distributions of the sulfur compounds must be determined in three dimensions within the plume, as indicated by the cartesian coordinate system. In reality this is achieved by measuring these

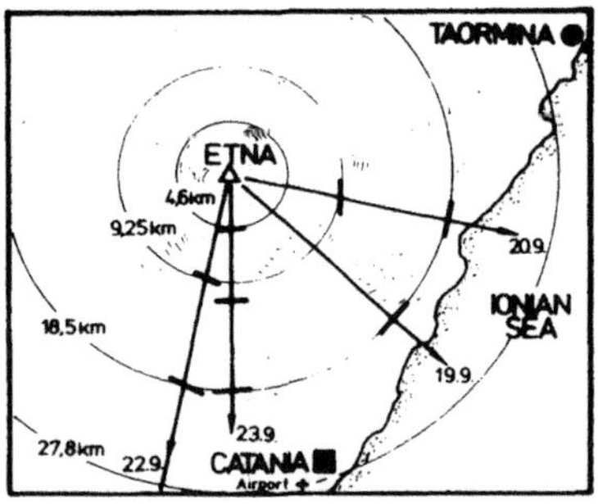

Fig. 1a

Viewgraphs of the measuring region including mean wind directions (arrows) and crosswind flights through the plume (cross- beams) for each measuring day.

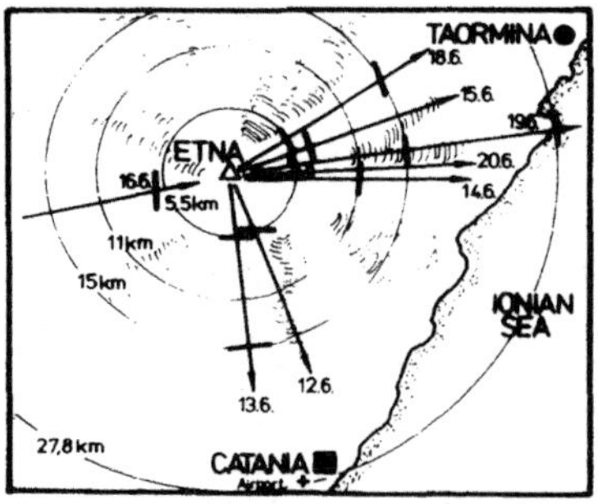

Fig. 1b

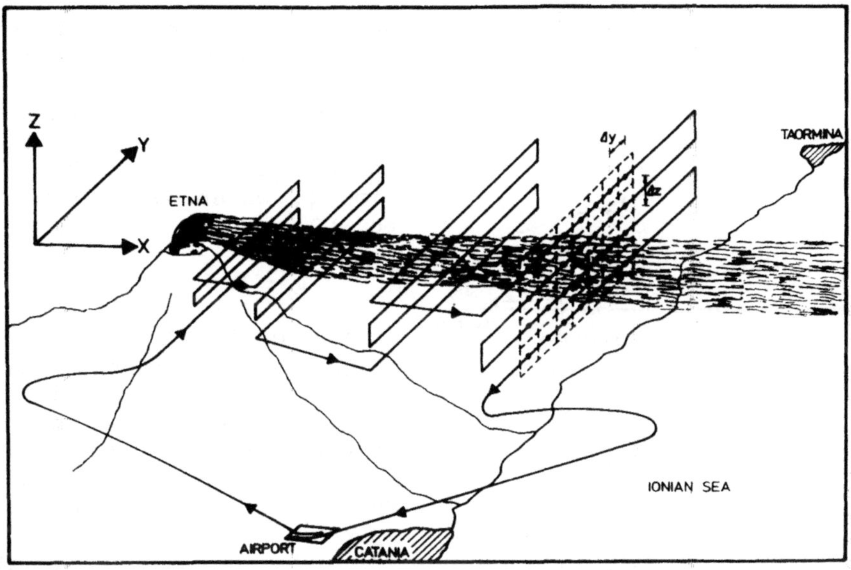

Fig. 2:
Relationship between the flight pattern in the measuring region and the associated mass transport model.

distributions consecutively within different vertical cross-sections of the plume at different preselected distances from the volcano.

In Fig. 2 each of these cross-sections is subdivided into small areas, for which mean concentrations are determined. For each cross-section a two-dimensional concentration grid is obtained with an array of data for the concentrations of the corresponding sulfur compound.

For example, if SO_2 is this sulfur compound, then the volcanic SO_2 source strength is computed from the individual SO_2 mass flux through each of these grids, i.e. cross-sections, by using eq. (1).

$$\emptyset = \sum_{i=1}^{N} \bar{C}_i \cdot \Delta y \cdot \Delta z \cdot \bar{u} \cdot \cos \bar{\theta} \qquad (1)$$

$\emptyset$ = Mass flux of a sulfur compound through a cross-section of the plume
$\bar{C}_i$ = Mean concentration in an area element $\Delta y \Delta z$
N = Total number of area elements $\Delta y \Delta z$ in a cross-section of the plume
$\bar{u}$ = Mean wind speed in the altitude of the plume axis
$\bar{\theta}$ = Mean angle between the normal of a cross-section and the mean wind direction

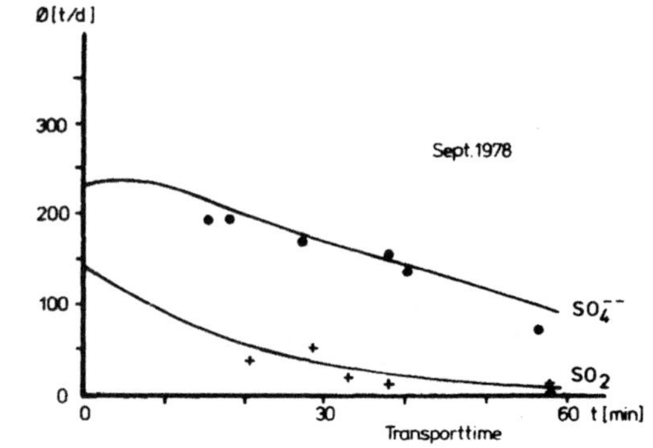

Fig. 3a (Sept. 1978)

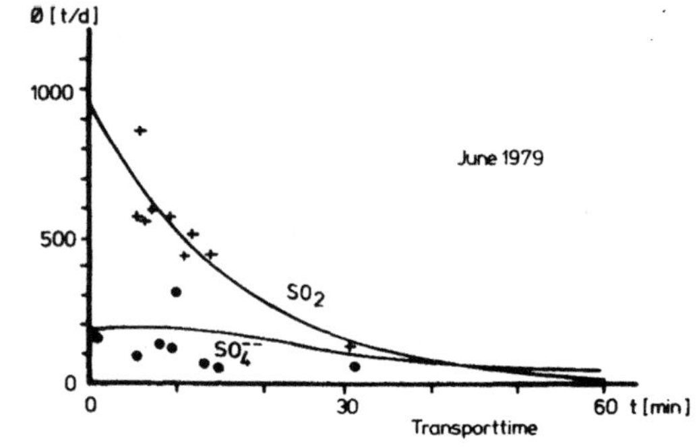

Fig. 3b (June 1979)
Mass fluxes of SO_2 and SO_4^{2-} as a function of travel time t.

Data for the mean wind speed $\bar{u}$ were taken from radiosonde stations or were measured by pilot-balloon ascents.
The mass fluxes calculated from eq. (1) correspond each to an individual transport time t, which is defined by eq. (2).

$$t = D/\bar{u} \qquad (2)$$

In case of SO_2 this means that if D is the distance between the volcano and a cross-section of the plume, then t is the transport time for SO_2 over this distance, with respect to the mean wind speed $\bar{u}$.
The data calculated for the individual mass fluxes using eq. (1), as well as for the corresponding transport times using eq. (2), are now plotted against one another in a diagram, as shown in Fig. 3a and 3b for the case of SO_2 and SO_4^{2-}, regarding the two measuring periods. By calculating the best fitting curves for these values [1] and extrapolating these curves to the transport time

TABLE 3: CALCULATED SULFUR EMISSION RATES FROM Mt. ETNA

DATE	PHASE	H_2S [t/d]	SO_2 [t/d]	SO_4^{2-} [t/d]
SEPT. 1978	POST-ERUPTIVE	1 - 3	142	232
JUNE 1979	PRE-ERUPTIVE	6 - 10	955 (max. 1600)	182

t = 0, i.e. to the distance D = 0, we obtain the desired emission rates $\phi^o = \phi(t=0)$ for the corresponding sulfur compounds. Table 3 shows the results of the measurements at Mt. Etna. It supplies a summary of the emission rates of H_2S, SO_2 and SO_4^{2-}, calculated for both measuring periods. These results confirm indeed the supposition made above, that not only the source concentrations of these sulfur compounds, but also their individual emission rates, reflect two different activity phases of the volcano regarding the two measuring periods.
Since Mt. Etna showed only a few insignificant eruptions between 1977 and 1978, its activity during Sept. 1978 is called 'post-eruptive'. The kind of activity during June 1979 is called 'pre-eruptive', because on August 5, 1979, a large fissure eruption happened below the central crater region.

PART II: ESTIMATION OF THE CONTRIBUTION OF VOLCANOES TO THE GLOBAL ATMOSPHERIC SULFUR BUDGET

An assessment of the contribution of volcanoes to the global atmospheric sulfur budget can only have satisfactory results, if it is based on direct measurements of the sulfur source strength of different volcanoes. Moreover, it plainly follows from the results presented in part I of this paper, that for such an assessment, the individual types of volcanic activity must be exactly defined and their global frequencies must be determined. Therefore, first a clear distinction is made in the following assessment between the eruptive and the non-eruptive activity, and then both activity phases are classified separately.
The broad spectrum of the eruptive activity is compiled in Table 4 using a logical order of eight phenomenological categories (A1-A8), which are often referred to in the literature [4,5,6]. The typical outward appearance of the activity categories A1-A7 is schematically shown in Fig. 4, including an approximate scale for the height of the corresponding eruption clouds.
The categories A1 and A2 represent the so-called 'Icelandic' and 'Hawaiian' eruption types, which both produce large volumes of very hot basaltic lava. The lava is emanating from long fissures or from a central crater, according to eruption type. The 'Strombolian' activity A3 is characterized by more explosive eruptions with comparatively lower temperatures and higher viscosities of the ejecta. The following categories, called 'Vulcanian' (A4), 'Vesuvian' (A5), 'Plinian' (A6), 'Pelean' (A7) and 'Krakatoan' (A8), are arranged in Table 4 according to an increase in the explosivity of the eruptions, the viscosity and volume of the ejected material and the eruption cloud height. All of these criteria are significantly different for the individual categories and thus are suitable for a classification of eruptive events. The estimated values for the eruption cloud height given in Table 4 are assumed to be typical for the eight categories.

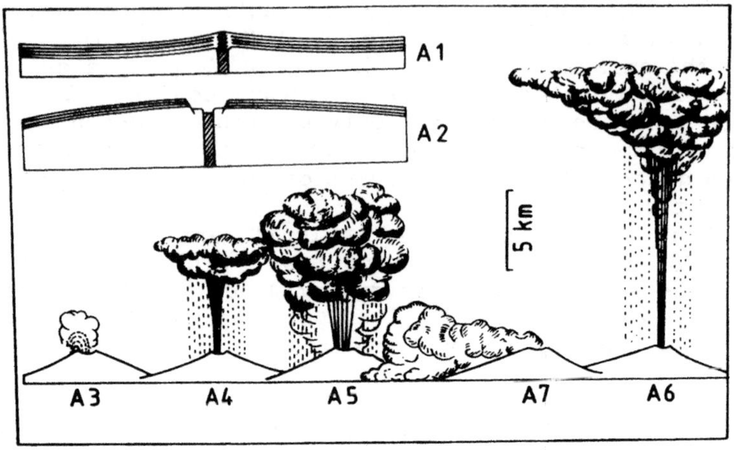

Fig. 4:
Schematic illustration of different types of eruptive activity (see text for further details).

TABLE 4: CLASSIFICATION OF THE ERUPTIVE VOLCANIC ACTIVITY

PHENOMENOLOGICAL CATEGORIES	E [%]	INTENSITY	EXPLOSIVITY	CLOUD HEIGHT [km]
A1: ICELANDIC	0 - 5	VIII	a, ab	< 1
A2: HAWAIIAN	0 - 5	IV - VI	a	< 1
A3: STROMBOLIAN	5 - 20	I	ab	max. 2
A4: VULCANIAN	20 - 40	II - IV	ab, b	2 - 7
A5: VESUVIAN	40 - 60	V	ab, b	5 - 10
A6: PLINIAN	60 - 80	VI	b, bc	10 - 20
A7: PELEAN	80 - 99	VII	b, bc	20 - 40
A8: KRAKATOAN	99 - 100	VIII - IX	b, bc	> 40

DEFINITIONS

E (EXPLOSIVITY INDEX): $\frac{\text{MASS OF SOLID EJECTA}}{\text{TOTAL MASS OF SOLID AND LIQUID EJECTA}}$

INTENSITY: CLASSIFIED BY THE TOTAL VOLUME OF THE EJECTED MATERIAL (I < 10^{-5} km^3 - IX > 10^2 km^3)

EXPLOSIVITY: CLASSIFIED BY THE KIND OF EJECTED MATERIAL
(a = DOMINANCE OF LAVA;
b = DOMINANCE OF FRESH SOLID MATERIAL;
c = DOMINANCE OF OLD ROCK MATERIAL)

The explosivity and the intensity of the different eruption types can also be expressed quantitatively. The intensity is classified by the volume of the ejecta, whereas the explosivity is characterized by the dominance of either the solid or the liquid ejecta.
Hédervári [7] has given a quantitative classification of the intensity as well as a qualitative scheme to characterize the explosivity by the kind of the ejected material. A quantitative expression for the explosivity of an eruption can be obtained by Rittmann's explosivity index E [8]. The classifications given by both authors do not take into account the gaseous output during an eruption. Three classifications are included in Table 4 by assigning the data to the individual phenomenological categories.
The classification of the non-eruptive activity used in this paper is given in Table 5. It is subdivided into four categories (B1-B4), which are called the 'pre-eruptive', 'intra-eruptive', 'post-eruptive' and 'extra-eruptive' phase, similar to the definitions given by Stith et al. [9]. In addition to a phenomenological description of the non-eruptive activity, this classification is mainly related to the characteristic time span of the individual activity categories. Therefore, measurements of the volcanic sulfur source strength during periods of non-eruptive activity can be associated with the above classification by considering the volcano's activity before, during and after these measurements using the time scale given in Table 5. This has been shown in part I of this paper using the Mt. Etna measurements as an example (see Table 3). The total spectrum of volcanic activity can now be roughly described by the above classifications of the eruptive and the non-eruptive phase.
In a further step to an assessment of the global volcanic sulfur source strength these classifications are associated with measuring data available from other publications concerning the sulfur emission rate of different volcanoes during different

TABLE 5: CLASSIFICATION OF THE NON-ERUPTIVE VOLCANIC ACTIVITY

CATEGORIES	TIME SCALE	OUTWARD APPEARANCE
B1: PRE-ERUPTIVE	DAYS - WEEKS	INTENSIFICATION OF ACTIVITY BEFORE AN ERUPTION, e.g. BY RHYTHMIC FUMING ACTIVITY.
B2: INTRA-ERUPTIVE	HOURS - DAYS	BETWEEN PAROXYSMAL ERUPTIONS
B3: POST-ERUPTIVE	WEEKS-YEARS	PERMANENT FUMING OR FUMAROLIC ACTIVITY.
B4: EXTRA-ERUPTIVE	YEARS - CENTURIES	EXCLUSIVELY FUMAROLIC AND SOL-FATARIC ACTIVITY.

TABLE 6: SO_2 EMISSION RATES OF DIFFERENT VOLCANOES AND CATEGORIZATION OF THE CORRESPONDING VOLCANIC ACTIVITY PHASES

VOLCANO	SO_2[t/d]	CATEGORY	VOLCANO	SO_2[t/d]	CATEGORY
Etna (10)	3740	A3	Fuego (17)	423	B2/3
Etna (11)	3325	B1	Fuego (15)	300/1500	A3
Etna (11)	1130	B3	Kilauea (18)	1280**	A2/3
Etna (12)	1000	B2	Kilauea (18)	280	?
Etna (1)	142	B3	Mauna Ulu (19)	30	B3/4
Etna (1)	955/1600	B1	Sulfur Banks (19)	7	B4
Telica (13)	20	B3	St. Augustine (9)	86400*	A5
Momotombo (13)	50	B3	St. Augustine (9)	432*	B3
San Christobal (13)	360	B3	St. Augustine (9)	8640	A4
Masaya (13)	180	B1	St. Augustine (9)	86	B3
Cerro Negro (14)	120+	A3	Mt. Martin (9)	3	B4
Cerro Negro (14)	2000+	A4	St. Helens (20)	1000/1900	B1
Pacaya (13)	260	A3	St. Helens (21)	900	B2
Pacaya (15)	300	B3	St. Helens (21)	600/3400	B3
Santiaguito (13)	420	A3	Mihara (22)	345	B3
Santiaguito (13)	300/1500	A3	Asama (23)	142	B3
Fuego (16)	10000+	A4	Asama (23)	780	B3
Fuego (13)	40	B3	Karimski (24)	173	?
Fuego (17)	55000*	A5			

*estimated value; **calculated from the SO_2 content of lava;
+calculated from the SO_4^{2-} content of ash particles.

phases of activity. This is shown in the case of SO_2 in Table 6. The actual activities of these volcanoes during the period of the measurements have been described in detail. Thus, using the above classifications, the individual activity categories can be rather easily identified.

Based upon both the available measuring data summarized in Table 6 and indirect estimations of the SO_2 emission during volcanic eruptions of the A6/A7 type, e.g. the Agung eruption in 1963 [25], the average emission rates have been calculated for each of the individual categories. The average values of the SO_2 emission rates calculated for the eruptive categories A1-A8 are shown in Table 7a. In addition to these values, the global frequencies of the eight categories are included in this table. The calculation of the individual frequencies is based upon the annual reports published by the IUGG since 1961 [26], which compile all eruptive events that have been observed during the course of a year. It follows from these reports that on the average about 55 volcanoes have erupted at least once annually in the period 1961-1979. With the above classifications each of these eruptive events has been categorized, and in doing so the average yearly frequency of each category during the above period has been determined. The global SO_2 source strength can now be calculated by summing up the individual products between the yearly frequencies of the categories A1-A8 and the corresponding mean SO_2 emission rates. A value of 1.0×10^6 t SO_2/a is obtained.

TABLE 7: ASSOCIATION OF ACTIVITY CLASSES WITH CHARACTERISTIC SO_2 EMISSION RATES

A) ERUPTIVE ACTIVITY

CATEGORY	SO_2 [t/d]	FREQUENCY PER YEAR
A1: ICELANDIC	10^3	-
A2: HAWAIIAN	10^3	10.46
A3: STROMBOLIAN	300 - 10^3	760.00
A4: VULCANIAN	10^4	9.40
A5: VESUVIAN	7 x 10^4	4.13
A6: PLINIAN	1 - 2 x 10^5	0.53
A7: PELEAN	1 - 2 x 10^5	0.12
A8: KRAKATOAN	?	-
AVERAGE NUMBER OF ERUPTING VOLCANOES: 55 / YEAR		

B) NON-ERUPTIVE ACTIVITY

CATEGORY	SO_2[t/d]	NUMBER OF VOLCANOES
B1: PRE-ERUPTIVE	500 - 10^3	
B2: INTRA-ERUPTIVE	500 - 10^3	
B3: POST-ERUPTIVE	50 - 500	365
B4: EXTRA-ERUPTIVE	0 - 10	100

For the non-eruptive activity (Table 7b) it is assumed that about 365 volcanoes show a post-eruptive phase annually, while 100 volcanoes show an extra-eruptive phase. These assumptions are based on a list of active volcanoes given by MacDonald [5]. The SO_2 emission by the activity categories B1 (pre-eruptive) and B2 (intra-eruptive) is negligible with respect to the total global emission of SO_2 by volcanoes, since the time span assumed for these activities is relatively short (see Table 5).
An average SO_2 source strength of 275 t/d for the 365 volcanoes with post-eruptive activity was calculated from the data in Table 7b. However, using the data from MacDonald it is assumed that this value is representative for only 25 % of these volcanoes, whereas 75 % of them emit 50 t SO_2/d at the most. This assumption yields a global SO_2 source strength of 1.4 x 10^7 t SO_2/a with respect to the B3 category. For the extra- eruptive B4 category a value of 2.0 x 10^5 t SO_2/a is calculated using the corresponding data in Table 7b.
There are only few measuring data available concerning the H_2S source strength of volcanoes [1,9]. The following estimation of the global volcanic H_2S emission rate is based on our measurements at Mt. Etna (see part I) and at Mt. Vulcano [1], the latter one showing extra-eruptive activity since historic times. During the measurements at Mt. Etna in Sept. 1978 and June 1979 an average source strength of 5 t H_2S/d was determined. This value can be considered as representative for the afore mentioned 365 volcanoes showing post-eruptive activity. Thus, the global H_2S production by this kind of activity is about 6.7 x 10^5 t H_2S/a.
For Mt. Vulcano an average source strength of 10 t H_2S/d was measured which results in a global estimate of 3.7 x 10^5 t H_2S/a for the 100 volcanoes showing extra-eruptive activity (see Table 7b). Based upon thermodynamic calculatons [3] it is assumed that only the H_2S emission by the B3 and B4 category is relevant for the atmospheric sulfur budget.
For the estimation of the global SO_4^{2-} emission of volcanoes use is made again of our own measurements at Mt. Etna (part I of this paper), since no other literature sources are available which

TABLE 8: ESTIMATION OF THE GLOBAL VOLCANIC SULFUR SOURCE STRENGTH

SO_2:	1.0×10^6	t/a	(A1 - A8)
	1.4×10^7	t/a	(B3)
	2.0×10^5	t/a	(B4)
	1.52×10^7	t/a	(total)
	(ANTHROPOGENIC: 15×10^7 t SO_2/a)		
H_2S:	ca. 10^6 t/a		
SO_4^{2-}:	ca. 10^7 t/a		

report data about the direct volcanic emission of sulfate into the atmosphere. Since our measurements were carried out exclusively during non-eruptive activity phases of Mt. Etna, the yearly SO_4^{2-}-production of 7.3×10^4 t/a calculated for this volcano [1] must be regarded as a minimum value. Assuming this value on the one hand for the above 55 volcanoes showing eruptive activity and on the other hand for the 365 volcanoes with post-eruptive B3 activity, a minimum and a maximum estimation is obtained, which yields an average value of 10^7 t SO_4^{2-}/a.

Table 8 summarizes the results of part II of this paper. It is noticeable that the SO_2 contribution to the atmosphere by the non-eruptive activity is about 14 times higher than that of the eruptive activity. This contribution has been neglected in earlier assessments because of a lack of measuring data [27]. The total SO_2 emission by volcanoes is estimated in this paper to the amount of 1.5×10^7 t/a, which corresponds to about 10 % of the anthropogenic emission.

The global volcanic H_2S emission is calculated to about 10^6 t/a, which is a rather low value compared to the contribution by other natural sources [28].

Another remarkable result of this assessment is that the volcanic SO_4^{2-} source strength is of the same order of magnitude as the value calculated for the SO_2 emission rate.

Finally, one can calculate from the data given in Table 8 a global sulfur emission of about 1.18×10^7 t S/a, which corresponds to 5 - 7 % of the total global sulfur emission into the atmosphere [28,29,30]. From this point of view the volcanic sulfur source strength may not be neglected in the calculation of the atmospheric sulfur budget.

REFERENCES

(1) Jaeschke, W., Berresheim, H., and Georgii, H.-W.: 1982, (submitted to J. Geophys. Res.).

(2) Schmitt, R.: 1976, Dissertation, Universitaet Frankfurt/M.

(3) Heald, E.F., Naughton, J.J., and Barnes, I.L.: 1963, J. Geophys. Res. 68, pp. 545-557.

(4) Ollier, C.D.: 1969, Volcanoes. Cambridge, Massachusetts, London: The MIT Press.

(5) MacDonald, G.A.: 1972, Volcanoes. Prentice-Hall, Englewood

Cliffs, N.J.
(6) Francis, P.: 1976, Volcanoes. Harmondsworth, Middlesex, Penguin Books.
(7) Hédervári, P.: 1963, Bull. Volc. 25, pp. 373-385.
(8) Rittmann, A.: 1960, Vulkane und ihre Taetigkeit. F. Enke Verlag, Stuttgart.
(9) Stith, J.L., Hobbs, P.V., and Radke, L.F.: 1978, J. Geophys. Res. 83, pp. 4009-4017.
(10) Haulet, R., Zettwoog, P., and Sabroux, J.C.: 1977, Nature 268, pp. 715-717.
(11) Zettwoog, P. and Haulet, R.: 1978, Atm. Env. 12, pp. 795-796.
(12) Malinconico, L.L.: 1979, Nature 278, pp. 43-45.
(13) Stoiber, R.E. and Jepsen, A.: 1973, Science 182, pp. 577-578.
(14) Taylor, P.S. and Stoiber, R.E.: 1973, Geol. Soc. Am. Bull. 84, pp. 1031-1042.
(15) Stoiber, R.E. and Bratton, G.: 1978, EOS 59, p. 1222.
(16) Rose, W.I., Bonis, S., Stoiber, R.E., Keller, M., and Bickford, T.: 1973, Bull. Volc. 37, pp. 338-364.
(17) Crafford, T.C.: 1975, Bull. Volc. 39, pp. 536-556.
(18) Naughton, J.J., Lewis, V., Thomas, D., and Finlayson, J.B.: J. Geophys. Res. 80, pp. 2963-2966.
(19) Stoiber, R.E. and Malone, G.B.: 1975, EOS 56, p. 461.
(20) Casadevall, T.J., Johnston, D., Harris, D.M., Stoiber, R.E., Williams, S.N., and Malinconico, L.L.: 1980, EOS 61, p. 1138.
(21) Rose, W.I., Harris, D.M., Bornhorst, T., and Casadevall, T.S.: 1980, EOS 61, p. 1139.
(22) Okita, T.: 1971, ASCO Report 8/7.
(23) Okita, T. and Shimozuru, D.: 1975, Bull. Volc. Soc. Japan 19-3, pp. 153-157.
(24) Gusseff, N.A. and Shilin, B.V.: 1973, cited in: [17].
(25) Cadle, R.D.: 1975, J. Geophys. Res. 80, pp. 1650-1652
(26) IUGG: 1961-1979, Bulletin of Volcanic Eruptions, No. 1- 19, Volcanological Society of Japan, Tokyo.
(27) Kellogg, W.W., Cadle, R.D., Allen, E.R., Lazrus, A.L., and Martell, E.A.: 1972, Science 175, pp. 587-596.
(28) Robinson, E., and Robbins, R.C.: 1975, The changing global environment. S.F. Singer (Ed.), Reidel, Dordrecht, Netherlands, pp. 111-123.
(29) Friend, J.P.: 1973, in: Rasool (Ed.), Chemistry of the lower atmosphere, Plenum Press, N.Y., pp. 177-201.
(30) Granat, L., Rhode, H., and Hallberg, R.O.: 1976, Ecol. Bull. 22, pp. 89-134

PART IV

HETEROGENEOUS SYSTEMS IN THE TROPOSPHERE

PHYSICAL ASPECTS OF THE ATMOSPHERIC AEROSOL

R. Jaenicke

Institut für Meteorologie, University of Mainz,
D-6500 Mainz, Germany F.R.

ABSTRACT

The atmospheric aerosol is a trace substance of the troposhere as well as the trace gases. Therefore a number of similiarities should be expected. As an example a common model on residence time is developed and discussed in its results. The residence time connects the content of a reservoir (size distribution for aerosols) with its source strength or removal rate. This is presented for the aerosol. In addition other physical properties of the tropospheric aerosol like optical, electrical and water uptake properties are discussed.

H. W. Georgii and W. Jaeschke (eds.), Chemistry of the Unpolluted and Polluted Troposphere, 341–373.

1. INTRODUCTION

Usually trace gases and aerosols are regarded as opposite. There are reasons given for that assumption:

(*) usually trace gases are invisible but aerosols influence the visibility
(*) aerosols play an unrenouncable role in the forming of cloud and precipitation elements
(*) aerosols are rather inactive chemically while trace gases are reacting with each other in various ways
(*) aerosol particles are seizable and handy.

However, trace gases and aerosols have much in common and many connection exists between them:

(*) Approximately half of the climatologically important aerosol mass is produced by gas-to-particle conversion as a compilation (Jaenicke, 1980 a) of various estimates show. This way, aerosol formation acts as sink for trace gases.
(*) The total aerosol mass is comparable to the mass of some trace gases.
(*) Aerosol and trace gases have geochemical cycles. They are produced at the earth surface or within the atmosphere, transported, diluted and mixed during the residence time, and finally removed from the atmosphere.
(*) Some trace gases and aerosols might influence the world climate (Kellog, 1978; Dansgaard, 1981).

Fig. 1 tries to emphasis the mutualities of trace gases and aerosols in still another way. Usually the aerosol is understood as the mixture of atmospheric gases and suspended particles. The amount of aerosol particles suspended in air is expressed as a property (number, suface, volume, mass, and others) per unit volume. If we express - for the purpose of this demonstration only - the concentration of gases as number of molecules per unit volume, we obtain size distributions for atmospheric gases and aerosols. As Fig. 1 shows, we have presently several distinct "aerosols" in the atmosphere:

1. The atmospheric gases, only for this purpose, regarded as an aerosol.
2. The small ions as clusters of atmospheric molecules with a weak binding from a center electrical charge (Chalmers, 1967).
3. The "atmospheric aerosol" in its present understanding with a typical size distribution of a continental aerosol (3a) and maritime aerosol (3b) (Jaenicke, 1978a).
4. The clouds, mainly consisting of water droplets (Mason, 1975)
5. The precipitation, in Fig. 1 presented as rain (Mason, 1975).

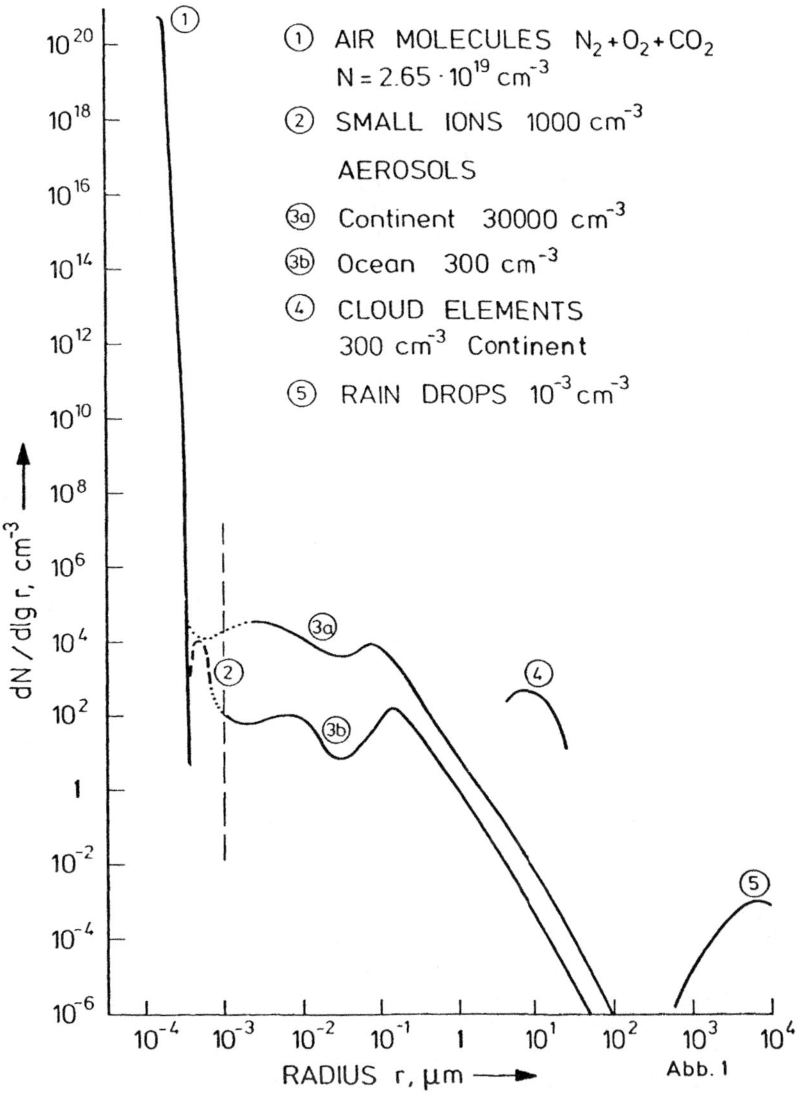

Fig. 1. Size distribution of "aerosols". It is the aim of this figure to emphasize the continuum in atmospheric aerosols. For this purpose the following "aerosols" are presented: atmospheric gases, small ions, aerosols, cloud elements, rain drops. The vertical dashed line is the lower size limit of aerosols per definition.

The "aerosols" cover a radius range of 8 orders of magnitude and 30 orders of magnitude in concentration.

Usually clouds and precipitation are not included as atmospheric aerosols, because they consist mainly of water. At least cloud elements have droplet sizes comparable to that of aerosols. The distinction between aerosols and cloud elements is not well defined, but certainly depends on the liquid water content. At least for rain droplets, the short residence time in the atmosphere was another reason to exclude it from aerosols. As we will see, the residence time cannot be used to define an aerosol. Small ions are treated separately, but it is now well established (Jaenicke, 1978b) that the residence time of small ions is described with the same equation which is valid for aerosols.

The present situation - outlined in Fig. 1 - resembles the situation in 1952 for aerosols. Up to 1952 Aitken particles, Dust particles, Small medium ions, Large medium ions, Langevin ions, Ultra large ions were treated as independent populations in the atmosphere. In his classical paper, Junge (1952) for simplicity converted the line spectra of those populations to a continuous distribution and thus formed our present understanding of atmospheric aerosols. In all modesty, an extension of the continuous aerosol size distribution is tried in Fig. 1 to include the small ions and open the connection to atmospheric trace gases.

After placing the atmopheric aerosol in proper relation to the atmospheric gases and the atmospheric liquid water elements, this paper will discuss physical properties of tropospheric aerosols.

2. RESIDENCE TIME FOR TRACE GASES AND AEROSOLS

The above attempt to connect the trace gases and aerosols was triggered from a study about residence time of trace constituents in the atmopshere. Junge (1963a) in his chapter about trace gases complained about the lack of common features among atmospheric trace gases. However, he admits: "One general feature is the relationship of the variability of a gas to its life or residence time in the atmosphere". In a later paper Junge (1974) reported about an inverse relationship between residence time and spatial and time variability of a trace gas. Species with a short residence time like water exhibit a large variability, while those with long residence times like helium show practically no variations in time and space. As we will see, the same is true for aerosols. Aerosol particles with a large variation in time and space like the Aitken particles and the giant particles have the shortest residence time, while the large particles with "little" variability show the longest residence time. This common feature of aerosols and trace gases should result in a uniform model of the residence time.

With a first order removal process, the concentration of a gas or aerosol is described by the well known equation

$$N(t) = N_0 \exp(-R\,t) = N_0 \exp(-t/T)$$

with N_0 = concentration in the source or at time t = 0
$N(t)$ = concentration at time t
R = removal rate (s^{-1})
T = residence time (s)

Fig. 2 shows the ratio N/N_0 for two cases (T = 5 days, 15 days) as a function of time. We now use the model that each measurement of any species is a measurement on the time axis of Fig. 2. This means that each measurement inspects an air parcel which has been given a certain time to age while traveling from the source to the location of the measurement. We further assume that averaging those measurements over time and space is averaging over the time axis for Fig. 2. It immediately becomes obvious, that a time t_{max} has to be introduced as upper limit of the averaging period.

In such a model we can derive a functional relationship between the residence time T and the standard deviation s or relative standard deviation (deviation coefficient) $v = s/\bar{N}$.

Averaging over the time axis in Fig. 2 between t = 0 and $t = t_{max}$ means, that we have a rectangular density function f(t) of time - observations at any time are of equal probability. This function f(t) has to be transformed into a density function g(N) of the observations from which we can calculate mean and standard deviation.

We need in addition

$$t = -T \cdot \ln(N/N_0)$$

$$N_{min} = N_0 \exp(-t_{max}/T)$$

$$f(t) = \frac{1}{t_{max}} \qquad 0 \leq t \leq t_{max}$$

The transformation gives (any standard probability testbook like Fisz (1966))

$$g(N) = f(t) \cdot \left|\frac{dt}{dN}\right| = f\left(-T \cdot \ln\frac{N}{N_0}\right) \cdot \left|\frac{d}{dN}\left(-T \cdot \ln\frac{N}{N_0}\right)\right|$$

$$g(N) = \frac{1}{t_{max}} \cdot \left|-\frac{T}{N}\right| = \frac{1}{N \cdot \ln\frac{N_0}{N_{min}}}$$

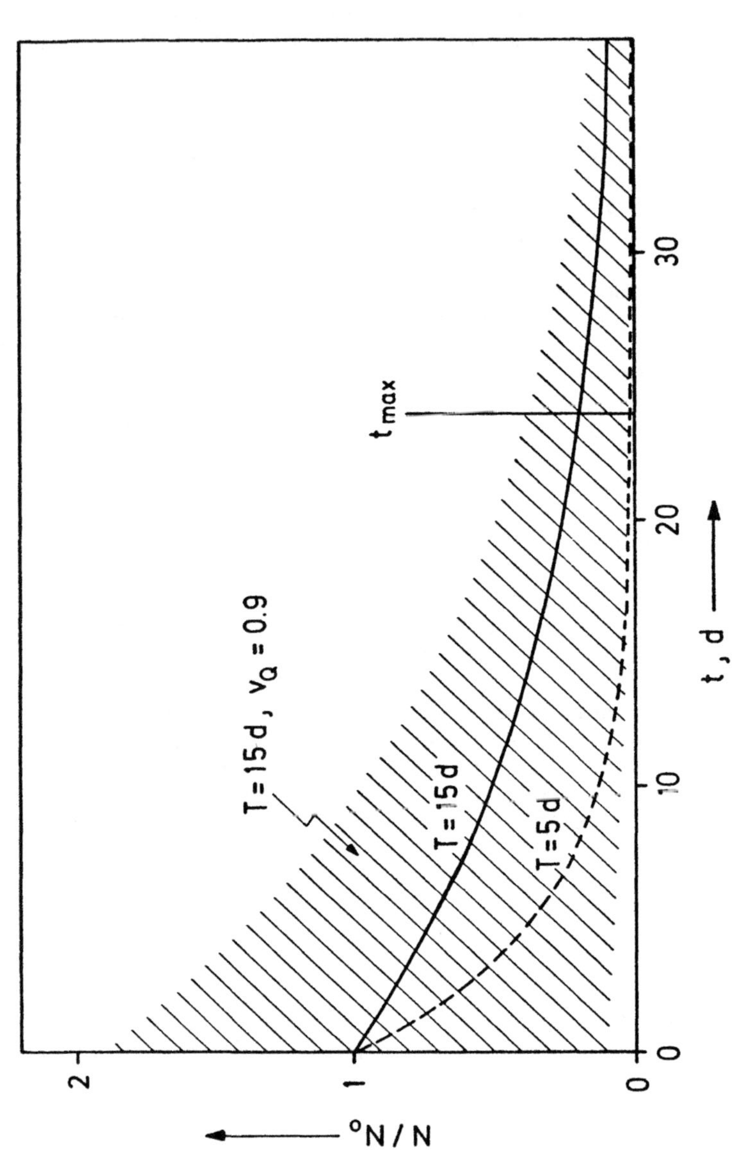

Fig. 2. Ageing of aerosols. The relative concentration N/N_0 of trace substances is shown as function of time t in days following the equation $\frac{N}{N_0} = e^{-\frac{t}{T}}$ with T = residence time. Given are two examples (T = 5 d, 15 d). The hatched area represents the case of residence time T = 15 d with a relative standard deviation v = 0.9 of the source. t_{max} is discussed in the text.

Integration over g(N)

$$\int_{-\infty}^{\infty} g(N)\, dN = \frac{1}{\ln \frac{N_0}{N_{min}}} \int_{N_{min}}^{N_0} \frac{dN}{N} = 1$$

reveals that g(N) indeed is a probability density function.
v is defined as

$$v = \sqrt{\frac{m_2 - m_1^2}{m_1^2}} = \sqrt{\frac{m_2}{m_1^2} - 1}$$

with

$$m_1 = \int_{-\infty}^{\infty} N \cdot g(N)\, dN = \int_{-\infty}^{\infty} N \cdot \frac{1}{N \cdot \ln \frac{N_0}{N_{min}}}\, dN =$$

$$= \frac{1}{\ln \frac{N_0}{N_{min}}} \cdot N \Big|_{N_{min}}^{N_0} = \frac{N_0 - N_{min}}{\ln \frac{N_0}{N_{min}}}$$

$$m_2 = \int_{-\infty}^{\infty} N^2 g(N)\, dN = \int_{-\infty}^{\infty} N^2 \frac{1}{N \cdot \ln \frac{N_0}{N_{min}}}\, dN =$$

$$= \frac{1}{\ln \frac{N_0}{N_{min}}} \cdot \frac{N^2}{2} \Big|_{N_{min}}^{N_0} = \frac{N_0^2 - N_{min}^2}{2 \cdot \ln \frac{N_0}{N_{min}}}$$

Introducing again t_{max} we obtain

$$v = \sqrt{\frac{1 - e^{-\frac{2t_{max}}{T}}}{2\left(1 - e^{-\frac{t_{max}}{T}}\right)^2} \cdot \frac{t_{max}}{T} - 1}$$

(A further derivation of v as function of T is given in Appendix 1).

For a given t_{max}, the residence time T can be calculatd for any relative standard deviation v. It is also obvious, that v can be calculated from the ratio N_0/N_{min}, the ratio of the maximal and minimal value in the set of observations. This is a very useful property, because for many species v is known only vaguely, but N_0/N_{min} can be given with better accuracy.

Fig. 3 shows $v = s/\bar{N}$ as function of the residence time for various values of t_{max}. For larger values of T, the curves show a relationship of $v \sim T^{-1}$, while for smaller T, the relationship $v \sim T^{-1/2}$ is obtained. Both results have been published in two earlier papers (Junge, 1974; Gibbs et al, 1973).

If t_{max} = 30 d, the agreement with Junge's (1974) model calculations is surprising. In his model, Junge (1974) assumed a latitude independent west wind circulation, which passes alternatively over continents and oceans of 6000 km length. The continents only were assumed as a uniform source for trace gases. Of interest is the vertical wind profile assumed with 2.2 ms^{-1} in an altitude of 0.1 km. To pass the assumed source free ocean with that speed, 31.6 days are required, quite in agreement with the above t_{max} = 30 d. Obviously it is possible describe the elaborate model of Junge (1974) with the above simple equation.

Fig. 3 contains independent estimates of v and T for various trace gases. The data are taken from Junge (1974) or updated if possible (Slemr et al (1981), Weiss (1981), Rohde et al (1980), Meixner (1981), Ockelmann *), Crutzen and Hahn *)). H_2O has been recalculated, because it obviously was overestimated by Junge (1974). Junges values of a v_{H2O} = 50 means that differences of a factor of 10^{50} in the H_2O concentration have been observed, a difference not possible to be resolved in present day measurements. It is one of the advantages of our simple model, that v can be calculated from the value N_0/N_{min}. For the troposphere, we have selected the N_0 or N_{max} value of water to occur at 100% relative humidity and 50°C resulting in 129 mbar. The value N_{min} was assumed as 1.4 10^{-6} mbar at 10% relative humidity and -100°C. This value of N_0/N_{min} = 9.21· 10^7 results in a v_{H2O} = 2.86.

All data cluster around our model curve with t_{max} = 0.5 yr. For T> 0.3 yr, this is in agreement with Junges (1974) data average. At that time, Junge (1974) could not resolve the discrepancy between his model and the actual data he collected, however, he attributed it to certain simplifications in his model. One was the lack of latitudional

----

*) personal communication by G. Ockelmann of Frankfurt and P. Crutzen and J. Hahn of Mainz, Germany.

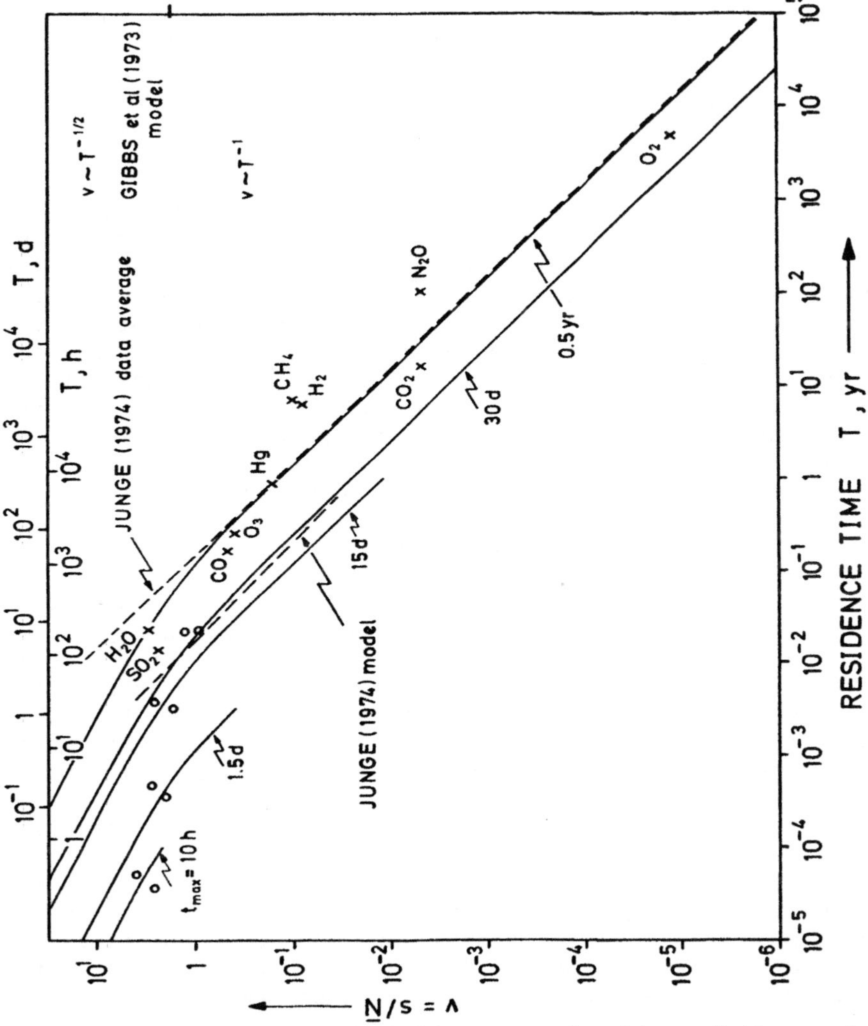

Fig. 3. Relative standard deviation v as function of the residence time T following the model developed in the paper. Trace gas data (x) have been taken from Junge (1974), Slemr et al (1981), Weiss (1981), Rohde et al (1980), Meixner (1981), Ockelmann (pers. comm.), Crutzen and Hahn (pers. comm.). H_2O has been calculated separately. Aerosols data (o) are explained in the text and valid for various particle sizes. The region $v \sim T^{-1/2}$ (comparable to Gibbs et al (1973)) is separated from $v \sim T^{-1}$ (comparable to Junge (1974)). Our model is shown for various values of t_{max} depending on the best fit.

variations of any parameter. So the interhemispherical exchange was not considered, which is in the order of the above t_{max} = 0.5 yr. Our model suggest that the variation of data in time and space can be explained only if air parcel are permitted to age for 0.5 yr in the troposphere. Presently it remains open, if this indicates a rather unmixed state of the troposhere, a problem addressed by Bolin et al (1973).

The SO_2 and H_2O values in Fig. 3 are in a v domain, where our model changes from a T^{-1} to a $T^{-1/2}$ dependency as proposed by Gibbs et al (1974). Gibbs et al (1974) assume the standard deviation of the measurements to be caused by the variation of the source strength and the source distribution only. This is different from Junge (1974) and the model of this paper, using the assumption that the standard deviation of the measurements is caused by the ageing of the trace constituent during transport. To discuss this in detail, Fig. 2 includes a variation of the source of s = 0.9 and its influence in the time development of the trace gas concentration. If T is short compared to t_{max} (example T = 5d), most measurements are carried out in that part with the concentration level fading away completely.

Our model suggest that the domain of the relationship $v \sim T^{-1/2}$ is for values $v > 1.78$, being equivalent to $T < 0.16 \cdot t_{max}$. For t_{max} = 30 d, this is indicated in Fig. 2 in drawing the ageing curve for T = 5 d in addition to T = 15 d. It can be seen that the curve T = 5 d covers the whole range of N/N_0 of the case T = 15 d with source standard deviation of s_1 = 0.9. For most trace gases used in Fig 3, $T \geq t_{max}$ and $v \sim T^{-1}$. In this case, the standard deviation of the measurements is influenced from the transport time to a greater extend than from source variations.

We now will try estimating a value of t_{max} for the troposphere. Obviously t_{max} is determined from meteorological parameters and the general mixing of the troposphere and is independent from the individual trace gas species. For normal probability density distributions exist the following rule of summation

$$s^2_{x+y} = s^2_x + s^2_y$$

the square of the standard deviation of a sum of two independent variables equals the sum of the square of the individual standard deviations. In the atmoshere, the source strength and/or source distribution (as x) might vary according to a normal distribution, however, the variations due to transport (y) do not follow a normal distribution. Thus the above equation is not valid, but most probably

$$s_0^2 > s_S^2 + s_T^2$$

with s_O = standard deviation of observations
s_S = standard deviation of source strength in the observations
s_T = standard deviation due to transport

Because $s_S \neq 0$ we must have $s_O > s_T$. Then consequentely, a t_{max} = 0.5 yr cannot be assumed for the troposphere, because in Fig. 3, the data cluster around this curve. We must look rather for a t_{max} with all observed v above it. A t_{max} = 30 d is too short, because of the agreement with Junges (1974) model, excluding latitudial variations. The gases with the smallest relative standard deviation v as compared to our model (SO_2, O_3, CO_2, O_2) suggest an apparent t_{max} = 3 month. Because that includes varia-tions of the source, we tend to assume a t_{max} = 2 month as valid for transport in the troposphere.

It remains for future research to proof the validity of the above model and the meteorological significance of t_{max} in general and t_{max} = 2 month as a quantity.

The above discussion was carried out for trace constituents in a reservoir in general, but quantitative data of troposheric gases have been used only so far. How does the aerosol fit into that picture? For the aerosol, data points have been added to Fig. 3. These data points have been derived from the model size distributions (N_{max}/N_{min}) which will be discussed later (Fig. 5) as well as the residence time (Fig. 4). Because the residence time depends on the particle sizes, the data points in Fig. 3 represent different particle sizes.

It is not possible to describe the aerosol data points in Fig. 3 with a single model curve, instead curves with t_{max} = 10 h/ 1.5 d/15 d/30 d are included. This was to be expected, because if only the curve would describe the aerosol, one single parameter (t_{max} in this case) would be sufficient to describe the aerosol in general. It can be shown with ease that it is not possible to describe the aerosol with one single parameter.

Most data points of the aerosol are in the $T^{-1/2}$ domain of Fig. 3 indicating that souce variations are of greater importance than ageing due to transport. This observation suggests the conclusion, that aerosol particles with short residence times are not in equilibrium between source and sink strength. Only particles with longer residence times - those in the optical important size range and also effective in cloud forming processes- seem to be in equilibrium with sources and sinks. Furthermore this is supported by the observation of a rather broad size distribution in tropospheric aerosols. Aerosols having time enough to age - as the stratospheric aerosols - exhibit a rather narrow size distribution.

In the present state of discussion of the model presented in this paper, one should not overstress the quantities given. Thus we will discuss only the case with t_{max} = 10 h. The data points are from particles with 10^{-3} µm in radius and 10^{2} µm respectively. The tiny particles are produced by gas-to-particle conversion and the source can be regarded as a volume source. The time (t_{max}) left at the most between production and measurement always is very short. This is not so easy to see for 10^{2} µm particles. If these particles are produced at continental surfaces, they are present in the aerosol and measured even if t_{max} is short. The presence of such large, non seasalt particles over the ocean (Jaenicke et al, 1971) can only be explained, if these particles are produced by evaporated cloud and precipitation elements (t_{max} is short).
It remains rather speculative, what we might learn from the various t_{max} in Fig. 3 about the source distributions and the source character (volume-, area-, or point sources) of the aerosol. Presently of importance is the fact, that aerosols and atmospheric gases are presented by a single model in the relationship of relative standard deviation of the measurements and residence time.

As seen in Fig. 3, the residence time of aerosols is given with several values. This is in opposition to earlier estimates, which understood the aerosol with one single residence time only. Various estimates are summarized in Pruppacher et al (1978). If these data are examined critically, it turns out that they are valid for various size ranges of the aerosol. This is caused, because various properties of the aerosol, like particle concentrations, coagulation behaviour, cloud forming properties, radioactivity, settling velocity are used for these estimates. Jaenicke (1978b) used the earlier data to derive an empirical residence time function

$$\frac{1}{T} = \frac{1}{C}\left(\frac{r}{R}\right)^{2} + \frac{1}{C}\left(\frac{r}{R}\right)^{-2} + \frac{1}{T_{wet}}$$

with r = particle radius in µm
R = 0.3 µm, standard radius
C = $1.28\cdot10^{8}$ s, constant
T_{wet} = $6.91\cdot10^{5}$ s

Fig. 4 shows the various estimates and the empirical function. Obviously T_{wet} is that part of the residence time of aerosols, which is determined from wet removal. Consequentely, T_{wet} later (Jaenicke, 1980a) was connected to the water content of the troposphere at various altitudes. These additional residence times curves for the middle troposphere and the tropopause are added in Fig. 4. The estimates (Pruppacher et al, 1978) for higher altitudes in the troposphere do agree with Fig. 4, but indepth experimental confirmation is still lacking.

Following Fig. 4, particles around 0.3 µm in radius have the longest residence times. Particles with radii larger 10 µm remain in the atmosphere only up to 1 day. The settling velocity of these particles is rather large. As we will see later, these particles are constantly present and that requires a constant production. Also particles smaller than 0.1 µm remain in the air only up to 1 day as individual particles. They are lost as individual particles through coagulation, the recombination with larger particles because of Brownian motion. At this point, one has to distinguish clearly between the residence time a particle exists as individuum and the residence time of the chemical compounds or the mass of particles. Because of particle loss due to coagulation, the particle mass is moved from smaller particle sizes to larger ones. This has pronounced impacts upon the effects of aerosols, as is demonstrated in one example. Particles smaller 0.1 µm in radius usually do not act as condensation nuclei for cloud elements (Pruppacher et al, 1978). Through coagulation these particles might grow into the cloud active particle size range. The residence time of aerosols should be longer in higher altitudes because of the lack of water. However, this applies only for particles in the range 0.1 µm to 1.0 µm. That should have the consequence of a narrow size distribution. Direct proof for that is lacking, because only parts of the size distribution are examined at higher tropospheric altitudes (Blifford et al, 1969). The stratospheric aerosol in the sulphate layer around 20 km altitude exhibits a narrow size distribution with a maximum around 0.3 µm in radius (Rosen et al, 1978). The stratosphere does not experience wet removal.

3. THE DISTRIBUTION OF AEROSOLS

The word "distribution" has several meanings with regard to the atmospheric aerosol. That includes
(*) size distribution in its number, surface, and volume order
(*) vertical distribution in the troposphere
(*) horizontal distribution over the globe.
Horizontal distribution is closely connected to the transport of aerosols and will be discussed in a later chapter.

In discussing the atmospheric aerosol, usually the size distribution is presented first. The size distribution is one of the most prominent properties of the aerosol. Only with the knowledge of this distribution, most effects of the aerosol can be understood. The aerosol is polydisperse, meaning particles of various sizes are present simultaneously. It was one of the highlights in aerosol research as Junge (1952) introduced a continuous size distribution. At that time it was a major result that the size distribution followed a r^{-3} relationship. This result suddenly

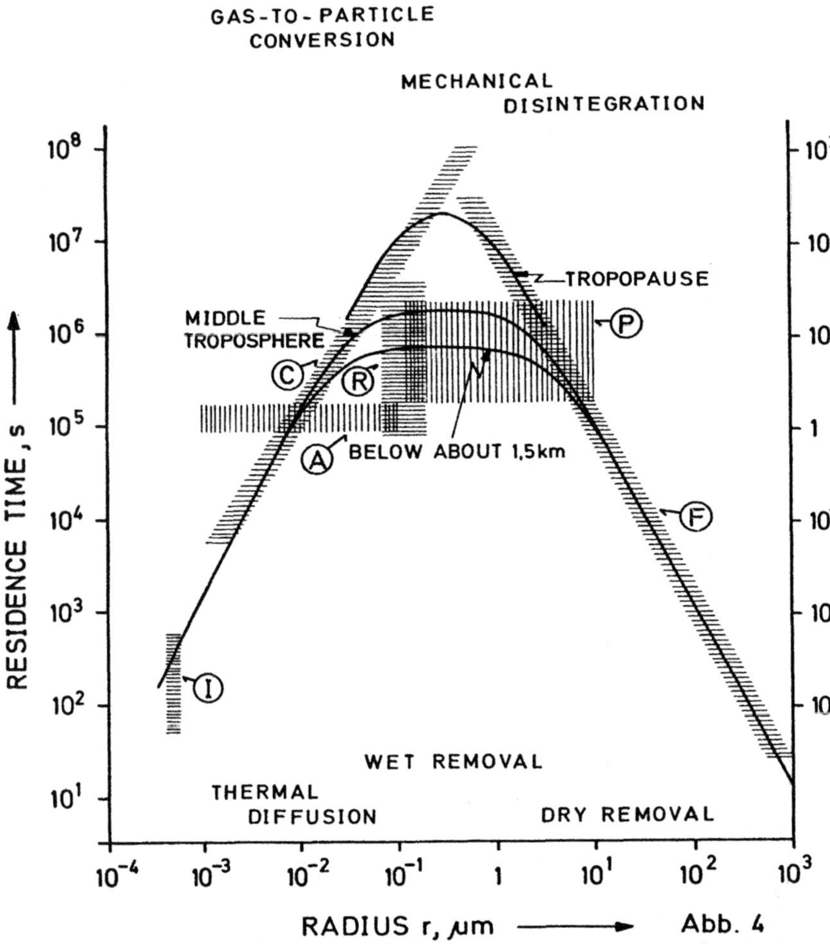

Fig. 4. Residence time of tropospheric aerosols as function of particle size. The hatched areas represent published estimates: I - small ions, A - Aitken particles, C - coagulation (thermal diffusion), R - radioactivity, P - precipitation, F - sedimentation according to Jaenicke (1978b). The curves represent various regions in the troposphere according to Jaenicke (1980a). They are not in disagreement with Pruppacher et al (1978). The major production mechanisms and removal processes are indicated in the most important particle size ranges.

explained a number of mainly optical observations (Angström exponent) and thus it even was thought to be the "r^{-3}-law". The historical development is well described in Junge (1963 a). In recent years a large body of measurements from various parts of the world was added. To a certain extend Junges picture was confirmed, but modifications and further details were added. Fig. 5 present the size distributions as models in the familiar way dN/dlgr as function of the radius r. dN/dlgr represents the concentration of particles (cm^{-3}) in the increment dlgr. Strictly speaking dN/dlgr is a density, the integral of it gives the total particle concentration.

In particle sizes, the aerosol is subdivided in various ranges, depending on the "school". Junge uses the term Aitken, Large, and Giant particles as indicated in Fig. 5. These terms are related to the particles physical size or to the method of observation (Aitken particles). Whitby (e.g. 1980) uses the terms Nucleation mode, Accumulation mode, Coarse particle mode with the limits 0.1 µm and 1 µm in diameter. This nomenclature implies production terms and the observation of various modes in the size distribution, mainly as volume distributions.

Fig. 5 shows the basic aerosol of the natural troposphere (background - including maritime -, remote continental, and polar (recent data from Jaenicke et al, 1981)) together with the desert dust storm aerosol (Schütz et al, 1974; d'Almeida et al, 1981) and anthropogenic influenced aerosols (city, rural). The stratospheric aerosol is given for comparison. While the radius range 10^{-3} µm to more than 10^{2} µm is covered, the concentration varies between 10^{6} cm^{-3} and 10^{-8} cm^{-3}. The size range 10^{-1} µm to 10 µm shows a rather uniform concentration decrease. The feature most important in connection with the discussion of the residence time is the large variation in the Aitken and Giant particle range, which is of more than 6 orders of magnitude, while around 0.3 µm only 2 orders of magnitude are observed. This is an immediate consequence of the residence time as discussed earlier. The existence of particles around 0.001 µm indicates continuous gas-to-particle production. The sharp cut-off around that radius is basically not satisfactory, but presents the knowledge of today. Fig. 1 tried a connection to the gas phase in a rather speculative form.

The models given in Fig. 5 try to summarize a large number of individual measurements as published by several authors. Examples are given in Jaenicke (1978a). However, in natural aerosols only few complete size distributions have been measured.

It was Whitby et al (1970) who emphazised the importance of the presentation of volume size distributions. Fig. 6 shows the tropospheric models of Fig. 5 as volume distribution dV/dlgr, $cm^{3}cm^{-3}$.

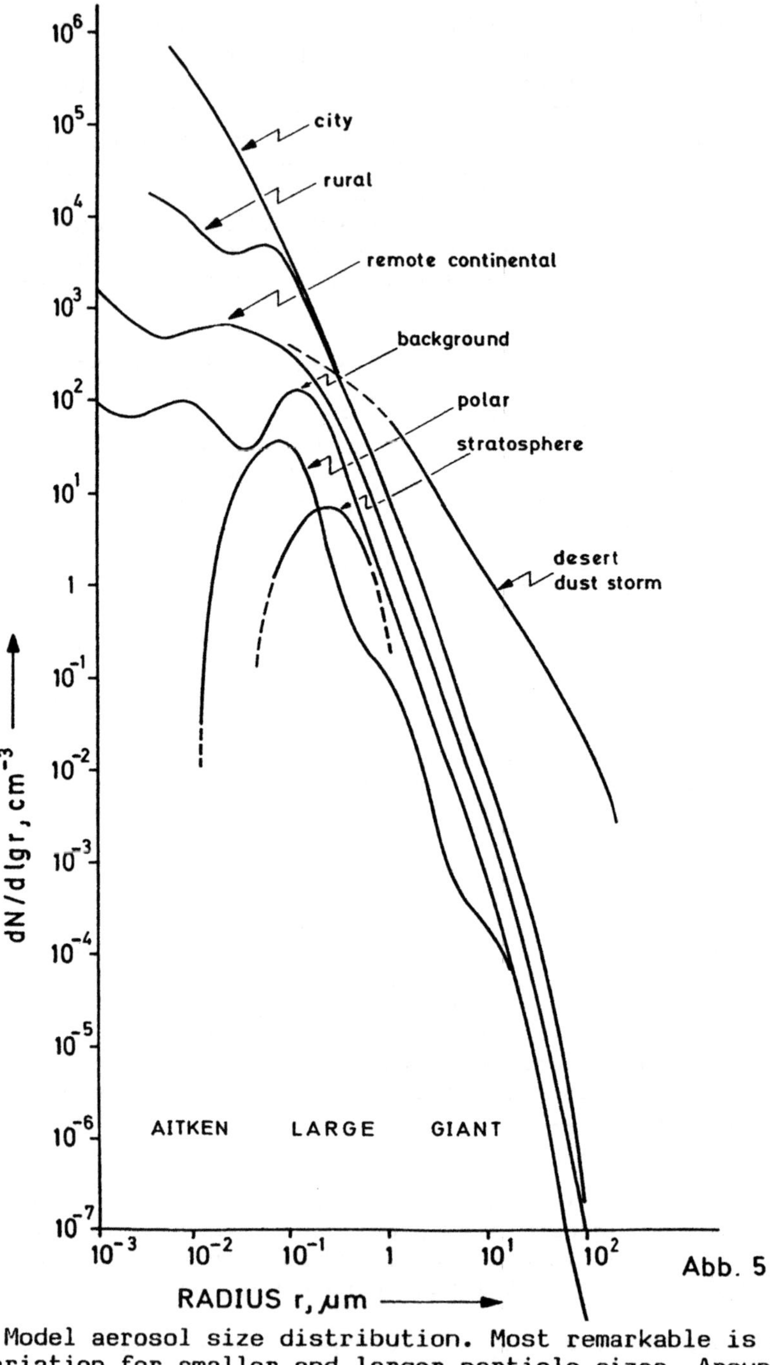

Fig. 5. Model aerosol size distribution. Most remarkable is the large variation for smaller and larger particle sizes. Around 0.3 µm in radius the variation is smallest.

This shows clearly that the volume and consequentely the mass of Aitken particles even in the city model is only a fraction of the total mass. Except for the desert dust storm with the maximum mass shiftet to the Giant particle size end, the main mass is in the range 0.1 µm to 10 µm. This is the range for which Junge (1952) first postulated the r^{-3} relationship for the number size distribution which converts to a constant volume distribution. Finally, I would like to stress the fact, that our survey of typical tropospheric size distributions did not show very pronounced the distinct modes or individual maxima in the distributions of volume, surface or number as reported by Willecke et al (1975). In the volume distributions, their measurements show a minimum around 0.1 µm diameter or 0.05 µm radius. Most of the measurements have been carried out under anthropogenic influence which might be the reason for that observation.

One word should be given to the maritime aerosol, of wich Podzimek (1980) published a survey recently. In the number distribution the maritime and background aerosol hardly can be distinguished, because particles produced from the sea make up only some 20 cm^{-3} and are missing in polar air. In the surface and volume distribution their influence is limited to particle sizes larger than 0.1 µm in radius (Jaenicke, 1980a).

All distributions discussed above are surface bound and do not reflect the situation in the free troposphere. It was mentioned above that a narrow size distribution should be expected there. It is only recently that some measurements became available (Patterson et al, 1980). They stress the fact that with increasing altitude the size distribution shows a decrease in particle counts for all sizes (0.1µm to 3 µm have been measured) relative to the boundary-layer size distribution, with a relatively larger decrease for the largest and smallest particles. The smallest change occured around 0.3 µm in radius. This fits into the above picture of the residence time and the variability. It indicates that the size distribution becomes narrow with increasing altitude and approaches that of a monodisperse aerosol.

The size distributions of Patterson et al (1980) show - without exception - a discontinuity around 0.3 µm in radius - which could be the biased result of the optical particle counter used (Knollenberg, 1976). Knollenberg (1976) discussed the effect of ambiguity in those optical counters. These measurements therefore will not be discussed further.

If a size distribution becomes narrow, it has two effects:
(*) volume-, surface-, and number distribution have the maximum at the same radius,

(*) the aerosol can be characterized with single parameters like number concentration, mass, or optical properties. This is not feasable for the broad aerosol size distribution in the tropospheric boundary layer.

Instead of discussing the aerosol size distribution as function of altitude, the particle concentration now will be discussed. Over the continents, a rather rapid decrease of the total concentration occurs up to altitudes of 5 km (Schmidt et al, 1974). This means a change from roughly 15000 cm^{-3} to 300 cm^{-3}. In higher altitudes, the concentration remains rather constant with the same value of 300 cm^{-3} over the continents and the oceans. This indicates that over the oceans with a surface value of 300 cm^{-3} the vertical profile is constant.

It was this observation that Junge (1963b) led to introduce three global aerosol bodies

(*) continental aerosol, (*) maritime aerosol, (*) background aerosol.

The background aerosol was expected to be an aged continental aerosol filling roughly 80% of the troposphere.

It is in agreement with that picture that over the ocean - as a strong source for sea salt particles - the large particles concentration (or the aerosol mass) decrease rapidly up to 3 km altitude (Dinger et al, 1970; Patterson et al, 1980). A similiar rapid decrease over the continents is observed up to 5 km altitude (Patterson et al, 1980). This rapid decrease results from the very effective wet removal through precipitation for large particles.

However, it still remains open, if the background aerosol is an aged continental aerosol. Recent measurements in surface background aerosols (as summarized in Jaenicke (1978a) and recently published from the Antarctic (Hogan et al, 1979) and the Arctic by Jaenicke et al (1982)) indicate a continuous presence of gas-to-particle conversion what should not be expected in aged aerosols. Only under rare circumstances (Haaf et al (1980), Jaenicke et al (1981)) no gas-to-particle conversion was observed, but only with particle concentrations below 50 cm^{-3} down to 5 cm^{-3}.

4. PRODUCTION AND REMOVAL OF AEROSOLS

Fig. 4 contains informations about the production and removal of aerosols. Basically there are two production mechanisms
(*) gas-to-particle conversion, (*) bulk-to-particle conversion.

Gas-to-particle conversion includes the production of particulate material, whether liquid or solid, from the gas phase. This might occur as the formation of new particles (homogeneous condensation) or the condensation on existing particles (heterogeneous condensation). Whitby (1978) estimated that 95% of the particle mass production condenses on existing particles. For an urban plume, Whitby (1980) permits the conclusion that 20% of the mass transformed by gas-to-particle conversion is used to produce new particles while 80% condense on particles. Gases included in such processes are sulfur-, nitrogen compounds and hydrocarbons. These particles tend to be smaller than 1 µm.

Bulk-to-particle conversion includes such processes like weathering, raising dust, sea salt spray, biological litter, volcanic fallout.
This mechanical disintegration tends to produce larger particles, usually larger than 0.1 µm.

This classification does not include the evaporation of cloud droplets and precipitation elements which leave behind residues of water-soluble or -insoluble material. It is our suspicion (s.a.) that giant aerosol agglomerates are formed this way (Rosinski et al, 1970).

The existence of such giant particles in the aerosol with extremely short residence times requires a clearcut decision, what particles should be included, if production rates are estimated. Jaenicke (1980b) has shown that easily discrepancies of 0.1 to 4000 Tg/yr of global aerosol production rates occur depending if particles of up to 1 µm or 1000 µm are included. It seems to be appropriate, to include in production estimates particles with residence times larger than 5 days, because only these particles can be transported intercontinentally and thus could be of global climatic interest (Jaenicke, 1980a). Tab. 1 gives those estimates which can be scaled down to the important size range ($\sim 3 \cdot 10^{-2}$ µm - ~ 3 µm) from the original literature (Smic, 1971; Jaenicke, 1978c; Schütz, 1980). The table shows clearly that for the natural aerosol gas-to-particle conversion is of equal importance as bulk-to-particle conversion. For the anthropogenic contribution this is different. Anthropogenic aerosol particles are produced 6 - 30 times more often through gas-to-particle conversion than as direct particles.

As Fig. 4 indicates, particles are removed fom their respective size range in the atmosphere by coagulation, wet removal, sedimentation and impaction on ground obstacles. Coagulation usually is not considered as a removal process, because it is part of the dynamics of the aerosol. Coagulating particles shift mass from smaller particle sizes to larger ones. With coagulation, the

Natural sources	
Seasalt	180
Mineral dust	60 - 360
Volcanoes	15 - 90
Forest fire	3 - 150
Biological material	80
Subtotal	338 - 860
Converted sulfates	130 - 200
Converted nitrates	140 - 700
Converted HC	75 - 200
Subtotal	345 - 1100
Total natural	683 - 1960
Anthropogenic sources	
Particles	6 - 54
Converted sulfates	130 - 200
Converted nitrates	30 - 35
Converted HC	15 - 90
Subtotal	175 - 325
Total anthropogenic	181 - 379
Total anthropogenic / Total natural	(181 - 379) / (683 - 1960) = 27 - 19%

Table 1. Global production of aerosols from various sources in Tg/yr or million of tons/yr. The values mainly have been taken from Smic (1971) because they could be scaled down to the climatological important size range ($r < 3$ µm). For further details see Jaenicke (1980a).

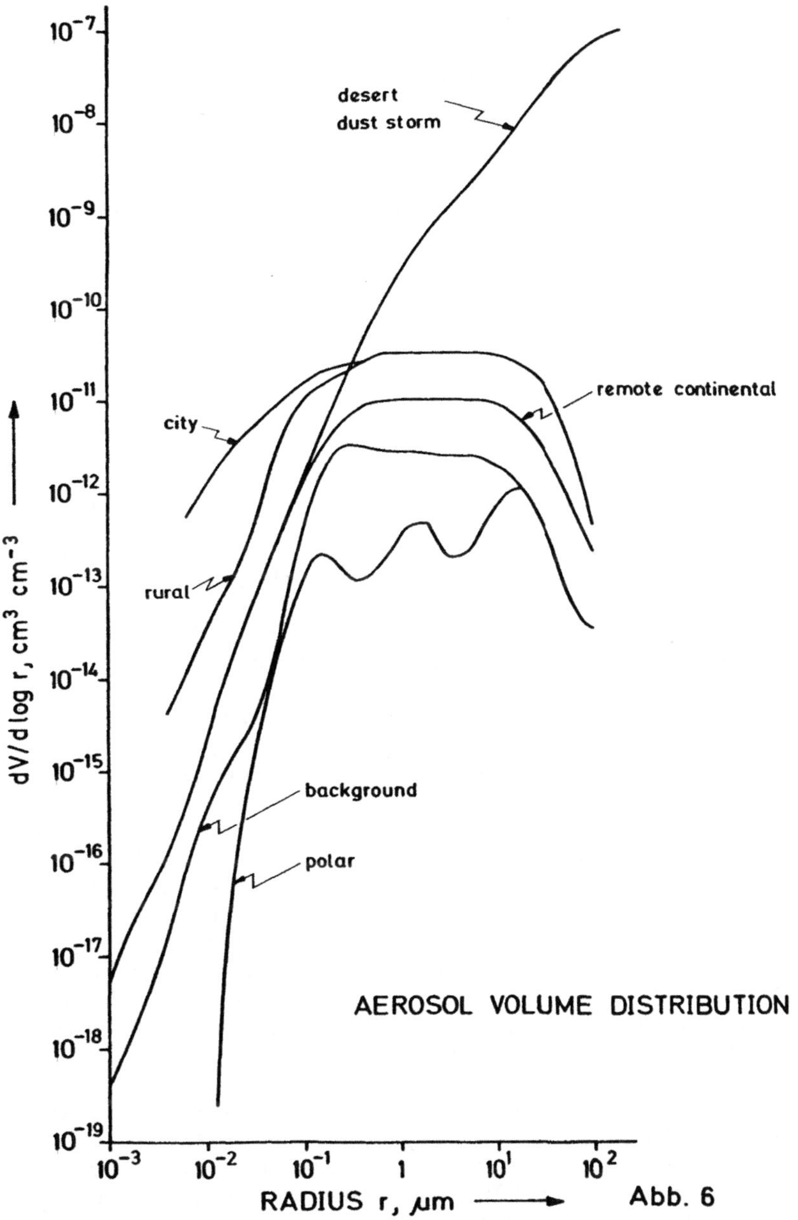

Fig. 6. Same as Fig. 5, but as volume distributions. Between 0.1 µm and 10 µm most distributions exhibit a plateau.

total mass of the aerosol remains unchanged. However, coagulation can be treated as removal if we consider the effects of the aerosol. It removes particles from a given size range and thus can be treated in accord with the term residence time used above.

This residence time developed above was determined in the sedimentation branch using a scale height H = 1.5 km of the aerosol. With such a scale height H, a deposition- (or emission-) velocity V_d can be determined

$$V_d = \frac{H}{T}$$

with T = residence time
or

$$V_d = H \cdot \left[\frac{1}{C'} \left(\frac{\tau}{R} \right)^{2} + \frac{1}{C'} \left(\frac{\tau}{R} \right)^{-2} + \frac{1}{T_{wet}} \right]$$

if we use the above equation for the residence time. With $T_{wet} = \infty$, meaning no wet removal, one obtains a "dry deposition" velocity. With H = 1.5 km this was done in Fig. 7, giving large deposition velocities for giant particles and for Aitken particles. The deposition velocity for particles in the range 0.1 µm to 1 µm is smallest. Twomey (1977) published a similiar curve remarking that the gap in effective removal between the radii 0.01 µm and 1 µm is noteworthy. Experimental data for dry deposition from Möller et al (1970) and McMahon et al (1979) are included in our figure. The agreement with the given function is quite good. The agreement could even be better, if we modify the residence time curve by assuming a larger particle bulk density - to fit the sedimentation branch - and smaller particle concentrations -to fit the diffusion branch. The theoretical deposition velocity curve would shift to smaller values if we assume a scale height of H = 1000 m, which seem to be present in Patterson et al (1980) data. In addition, excellent agreement of the theoretical curve in Fig. 7 and theoretical considerations of Twomey (1977) can be stated.

Regardless of this "agreement" between experimental data and the theoretical deposition velocity curve, the deposition velocity in the Aitken range must be discussed critically. The experimental data by Möller et al (1970) have been obtained using monodisperse aerosols being deposited on an external surface. Our theoretical calculation has been done for a polydisperse aerosol without any external deposition surface except the surface of the particles. In this model removal of Aitken particles was assumed by thermal coagulation only. If the agreement of Fig. 7 exists, it can be interpreted in the following way. The increase in deposition

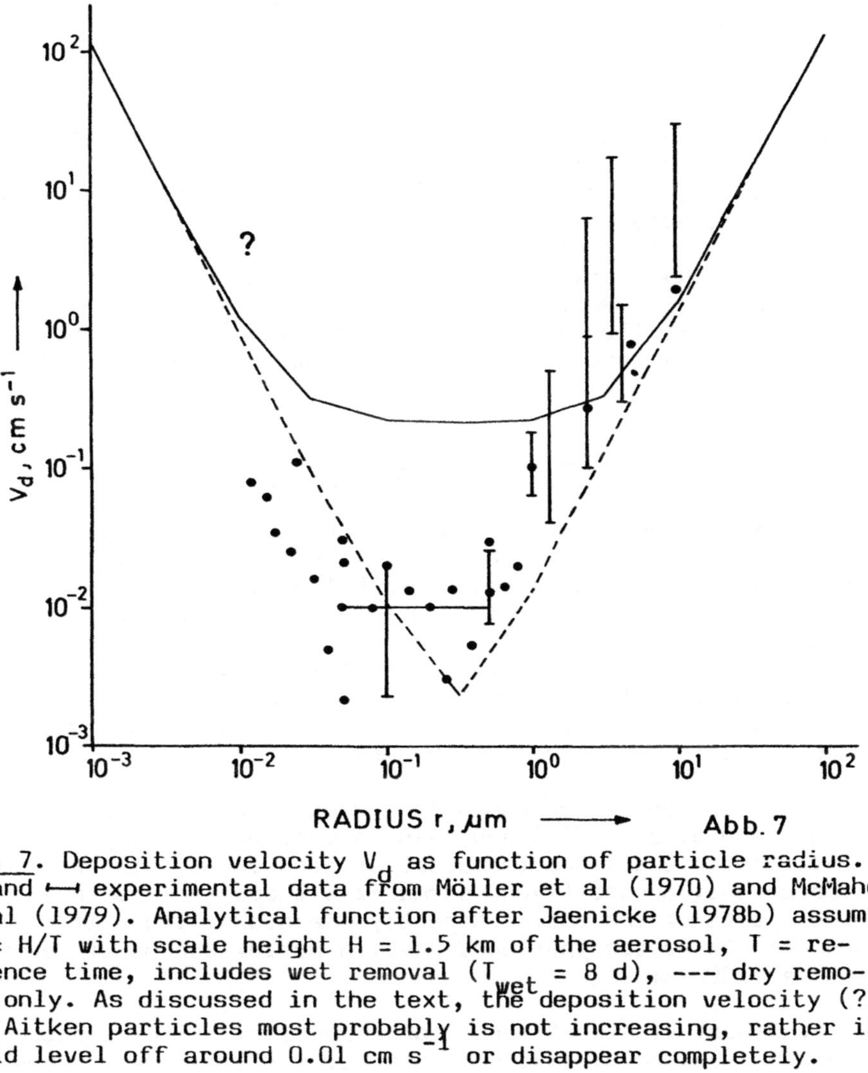

Fig. 7. Deposition velocity V_d as function of particle radius. ...and ⟼ experimental data from Möller et al (1970) and McMahon et al (1979). Analytical function after Jaenicke (1978b) assuming $V_d = H/T$ with scale height H = 1.5 km of the aerosol, T = residence time, includes wet removal (T_{wet} = 8 d), --- dry removal only. As discussed in the text, the deposition velocity (?) for Aitken particles most probably is not increasing, rather it could level off around 0.01 cm s^{-1} or disappear completely.

velocity to an external surface in the Aitken range was measured only because monodisperse aerosols were used in the experiment. In a polydisperse aerosol, as we have it in the atmosphere, the deposition velocity towards an external flat surface most probably is reduced because coagulation among particles of different size is larger than among particles of equal size (Walther, 1973). It is our conclusion that deposition velocities for Aitken particles toward the earth surface should be smaller than previously indicated.

If we permit wet removal (T_{wet} = 8 days), the second curve in Fig. 7 is obtained. We now can estimate the importance of dry removal as compared to wet removal. In the 0.1 µm particle radius range the deposition velocity including wet removal is 30 times faster than for dry removal. Thus wet removal is 30 times more effective as dry removal in this size range. Only for giant particles the influence of wet removal becomes smaller and particles larger than 10 µm are removed from the atmosphere exclusively by dry removal. This is in good agreement with recent estimates by Whelpdale (1982).

5. HORIZONTAL DISTRIBUTION AND TRANSPORT OF AEROSOLS

The residence time of the aerosol is the limiting factor if we discuss the question of transport of the aerosol. After the aerosol is produced, it is moved with the general circulation of the atmoshere. Particles with short residence time will not spread far and remain in the local area. Other particles might travel on a regional scale and some of them even intercontinentally. Because of the limited residence time however it is unlikely that aerosols are transported on a global scale which includes the interhemispheric exchange.

Especially for desert aerosols long range transport has been reported (Rahn et al, 1979) and even is monitored: Prospero et al (1979) reports about the transport of Saharan dust to the Caribbean Sea. Schütz (1980) summarizes these transports in various parts of the world. Recently (Jaenicke, 1981) it could be shown for the Arctic that condensation nuclei can be traced for more than 1000 km.

This transport creates "transition aerosols", mixtures between the various aerosols, like continental, maritime, background, polar or others. Impressive examples have been collected by Hogan et al (1971). In this way, the aerosol contributes to the formation of sediments in the oceans (Goldberg et al, 1963). With such an effective transport, it is difficult to draw a picture of the horizontal distribution besides what has been told about continental and maritime aerosols. It seems to be a consequence that none of the more recent books on atmospheric aerosols treat this topic. For the oceans, Podzimek (1980) has collected data and developed a horizontal distribution of the giant sea salt particle concentration (mass $3\ 10^{-12}$ g to 10^{-8} g). This horizontal distribution resembles a map of wind velocities with maximal values of 1.6 cm^{-3} in the region of the westerlies. The doldrums are characterized with 0.6 cm^{-3}. To our knowledge, no other distribution has been compiled so far.

6. RADIATIVE, HYDROLOCIGAL, AND ELECTRICAL PROPERTIES OF AEROSOLS

For the relation between the air and the particles therein the radiative properties of the atmospheric aerosol are of interest. The optical bulk properties are described with the complex index of refraction. The imaginary portion of this index describes the absorption properties of the aerosol. Eiden et al (1975) have compiled all data available. Within the wavelength range of the sun radiation, the real portion of this index, describing the scattering properties, is rather uniform around $n = 1.53$. This could be expected, because a large number of substances in the aerosol have similiar values. Minerals like rock salt ($n = 1.544$), calcspar ($n = 1.658$), quartz ($n = 1.544$) show rather uniform data, but also other substances being present in the aerosol like ammonium sulfate ($n = 1.53$) and sulfuric acid ($n = 1.44$). The importance of ammonium sulfate for the atmospheric aerosol was demonstrated by Charlson et al (1978). They observed a close correlation of the sulfate content and optical properties. Winkler (1981) showed the presence of free acids in the aerosol. Not as uniform is the absorption coefficient, variing in the range $n \cdot \kappa = 0.004 - 0.07$. A value of 0.03 seems to be a good average. For the terrestrial radiation, the radiation emitted from the earth's surface, the optical properties of the aerosol show a selective character. In this range the real part as well as the imaginary part of the complex index of refraction have rather large anomalies around the wavelength $\lambda = 9\ \mu m$ in the socalled window range. n in-creases from 1.2 to 2.0 (Volz, 1972) and $n \cdot \kappa$ from 0.01 to 0.3 (Fischer, 1975). Grassl (1973) estimates the radiative effects of aerosols to be of the same order as those of atmospheric water vapor. The data of desert aerosols are comparable to the values given above, as Carlson et al (1980) published. In the absence of clouds, the planetary albedo for the solar radiation mostly depends on the imaginary part of the complex index of refraction (Grassl, 1979). This planetary albedo increase with increasing particle concentration. Only over high surface albedo areas, the planetary albedo might be reduced in this case. With an increased planetary albedo, the surface of the earth receives less radiation and the heating is reduced. The aerosol instead is heated in the order of 1 K per day, depending on the absorption coefficient. This leads to an increased emission of terrestrial radiation from the aerosol. Because of the uncertainties in aerosol properties and spatial distribution, no estimates on the total effect of the aerosol are given yet.

Besides its direct effect upon radiation, the aerosol acts indirectly through the formation of clouds. In forming clouds, a fraction of the aerosol acts as cloud condensation nuclei. This mainly depends on the chemical nature of the aerosol particles. In general, atmospheric aerosols and atmospheric water are connected in two ways. At relative humidities below 100%, aerosol particles grow in absorbing water.

Above 100% relative humidity the formation of clouds is influenced. The water uptake below 100% relative humidity changes the scattering properties of the aerosol. Hänel et al (1978) has calculated that for 70% relative humditiy 3 times more liquid water is attached on an anthropogenic influenced aerosol than on a remote continental aerosol.

In clouds the cloud condensation nuclei and thus the aerosol acts in determining the optical depth, the shape of the scattering function and the absorption of radiation in clouds (Grassl, 1979). An increase in cloud condensation nuclei concentration usually results in an increased optical depth of the clouds and flattening of the scattering function. An increased optical thickness first increases the cloud albedo but with further increase of the optical thickness the cloud albedo decreases.

While the influence of aerosols in the radiation budget of the atmosphere whether directly or indirectly through clouds easily can be shown, the general significance of the electrical charge on aerosols has not been demonstrated. As Twomey (1977) summarizes, the aerosol has a profound effect on the electrical conductivity of the air and on other related atmospheric electrical parameters, but the charges which are likely to reside on aerosol particles and cloud droplets are not such as to effect appreciably the coagulation rate or other mechanical properties of these aerosols. We therefore will not discuss the electrical properties of the aerosol but open one point for discussion.

Usually it is assumed that in the aerosol an equilibrium exists between the formation of electrical charges on aerosol particles through the coagulation of small ions with aerosol particles and the destruction through recombination with charges of the opposite sign. It seems to be noteworthy that the residence time of small ions obviously follows the same function as the Aitken particles and thus the residence time can be described with the same equation as Fig. 4 shows. Usually the number of electrical charges on particles is calculated using the Boltzmann hypothesis of the distribution of energetically different states in a system under equilibrium (Twomey, 1977). This assumption worked satisfactorily in the past, but there is recent evidence that the distribution of single charges on Aitken particles is not described properly. This became evident, after measurements of the aerosol size distribution were carried out with a sufficient sensitive electrical mobility analyzer (Haaf, 1980). Obviously a Boltzmann distribution underestimates the number of single charged Aitken particles considerably. But if that number is larger in atmospheric aerosols, again coagulation of aerosols and consequentely the residence time and the dynamics of aerosols might be effected.

7. CONCLUSIONS AND SUMMARY

Aerosols and trace gases both are trace substances of the atmosphere. While aerosols consist of the solid and liquid phase, trace gases belong to the third phase. In this paper, emphazise was given to common properties. So it was tried to link the aerosol size distribution with its open end at small particle sizes to the atmospheric gases. Furthermore, a common model of the residence time was developed and compared to data.

With such a residence time model other properties of the aerosol become understandable. Especially this residence time model is a step further to the understanding of the aerosol as a continuum phase.

APPENDIX 1

Derivation of v as function of T.
Given is the function N(t) with $N_0 = 1$

$$N(t) = e^{-\frac{t}{T}}$$

If $N_0 \neq 1$, N(t) has the meaning of N/N_0. Over the time t = 0 to $t = t_{max}$ samples $N(t_i)$ are taken evenly distributed over the time period. We use the equation

$$s^2 = \frac{1}{n} \sum_{i=1}^{n} N^2(t_i) - \bar{N}^2 = \overline{N^2} - \bar{N}^2$$

with
- s = standard deviation
- $N(t_i)$ = samples at the time t_i
- n = number of samples.

With $n \to \infty$ we obtain

$$\bar{N} = \frac{1}{t_{max}} \int_0^{t_{max}} N(t)\,dt$$

$$\overline{N^2} = \frac{1}{t_{max}} \int_0^{t_{max}} N^2(t)\,dt$$

In detail

$$\bar{N} = \frac{1}{t_{max}} \int_0^{t_{max}} e^{-\frac{t}{T}}\,dt = -\frac{T}{t_{max}} e^{-\frac{t}{T}} \Big|_0^{t_{max}}$$

$$= \frac{T}{t_{max}} \left[1 - e^{-\frac{t_{max}}{T}}\right]$$

$$\overline{N^2} = \frac{1}{t_{max}} \int_0^{t_{max}} e^{-\frac{2t}{T}}\,dt = -\frac{T}{2t_{max}} e^{-\frac{2t}{T}} \Big|_0^{t_{max}}$$

$$= \frac{T}{2t_{max}} \left[1 - e^{-\frac{2t_{max}}{T}}\right]$$

It follows

$$s^2 = \frac{T}{2t_{max}}\left[1-e^{-\frac{2t_{max}}{T}}\right] - \frac{T^2}{t_{max}^2}\left[1-e^{-\frac{t_{max}}{T}}\right]^2$$

$$v^2 = \frac{s^2}{\bar{N}^2} = \frac{t_{max}}{2T}\,\frac{\left(1-e^{-\frac{2t_{max}}{T}}\right)}{\left(1-e^{-\frac{t_{max}}{T}}\right)^2} - 1 .$$

LITERATURE

d'Almeida, G.A., R. Jaenicke (1981), The Size Distribution of Mineral Dust. J. Aerosol Sci. 12, 160-162.

Blifford, I.H., L.D. Ringer, (1969), The Size and Number Distribution of Aerosols in the Continental Troposphere. J. Atmosph. Sci. 26, 716-726.

Bolin B., H. Rohde (1973), A Note on the Concept of Age Distribution and Transient Time in Natural Reservoirs. Tellus 25, 58-62.

Carlson, T.N., S.G. Benjamin (1980), Radiative Heating Rates for Saharan Dust. J. Atmosph. Sci. 37, 193-213.

Chalmers, J.A. (1967), Atmospheric Electricity. Pergamon Press, Oxford.

Charlson, R.J., D.S. Covert, T.V. Larson, A.P. Waggoner (1978), Chemical Properties of Tropospheric Sulfur Aerosols. Atmosph. Environm. 12, 39-53.

Dansgaard, W. (1981), Palaeo-Climatic Sudies in Ice Cores. In: Climatic Variations and Variability: Facts and Theories, A. Berger, ed., Reidel Dordrecht, 193-206.

Dinger, J.E., H.B. Howell, T.W. Woijeciechowski (1970), On the Source and Composition of Cloud Nuclei in a Subsident Air Mass Over the North Atlantic. J. Atmosph. Sci. 27, 791-797.

Eiden, R., G. Eschelbach (1975), Das atmosphärische Aerosol und seine Bedeutung für den Energiehaushalt der Atmosphäre. Z. f. Geophysik 39, 189-228.

Fischer ,K. (1975), Mass Absorption Index of Various Types of Natural Aerosol Particles in the Infrared. Appl. Optics 14, 2851-2856.

Fisz, M. (1976), Wahrscheinlichkeitsrechnung und mathematische Statistik. VEB Deutscher Verlag der Wissenschaften, Berlin.

Gibbs, A.G., W.G.N. Slinn (1973), Fluctuations on Trace Gas Concentrations in the Troposhpere. J. Geophys. Res. 78, 574-576.

Goldberg, E.D., M. Koide, J.J. Griffin, N.N.A. Peterson (1963), A Geochronological and Sedimentary Profile Across the North Atlantic Ocean. In: Isotropic and Cosmic Chemistry. C.H. Craig, S.M. Miller, G.J. Wasserburg eds., 211-219, North-Holland Amsterdam.

Grassl, H. (1973), Aerosol Influence on Radiative Cooling. Tellus 25, 386-395.

Grassl, H. (1978), Possible Changes of Planetary Albedo Due to Aerosol Particles. In: Proc. Man's Impact on Climate. W. Bach, J. Pankrath, W. Kellog eds., 229-241, Elsevier, Amsterdam.

Haaf, W. (1980), Accurate Measurements of Aerosol Size Distribution -II. Construction of a New Plate Condenser Electric Mobility Analyzer and First Results. J. Aerosol Sci. 11, 201-212.

Haaf, W., R. Jaenicke (1980), Results of Improved Size Distribution Measurements in the Aitken Range of Atmospheric Aerosols. J. Aerosol Sci. 11, 321-330.

Hänel, G., K. Bullrich (1978), Physico-Chemical Property Models of Tropospheric Aerosol Particles. Beitr. Phys. Atmosph. 51, 129-128.

Hogan, W.W., M.H. Degani (1971), Comprehensive Study of Maritime Aerosols. Final Report EPA-CPA 70-64.

Hogan, A.W., S.B. Arnard, J. Bortiniak (1979), Physical Properties of the Aerosol at the South Pole. Geoph. Res. Letters 6, 845-848.

Jaenicke, R. (1978a), Aitken Particle Size Distribution in the Atlantic North East Trade Winds. Meteor Forsch.-Erg. B13, 1-9.

Jaenicke, R. (1978b), Über die Dynamik atmosphärischer Aitkenteilchen. Ber. Bunsenges. Phys. Chem. 82, 1198-1202.

Jaenicke, R. (1978c), The Role of Organic Material in Atmospheric Aerosols. Pageoph 116, 283-292.

Jaenicke, R. (1980a), Atmospheric Aerosol and Climate. J. Aerosol Sci. 11, 577-588.

Jaenicke, R. (1980b), Natural Aerosols. Ann. N.Y. Acad. Sci. 338, 317-322.

Jaenicke, R. (1981), "Schmutzige" Luft über den Polen. Umschau 81, 525-528.

Jaenicke, R., C. Junge. H.J. Kanter (1971), Messungen der Aero solgrößenverteilung über dem Atlantik. Meteor Forsch.-Erg. B7, 1-54.

Jaenicke, R., L. Schütz (1982), Artic Aerosols in Surface Air. Idöjárás (J. Hung. Meteor. Service) to be published.

Junge, C. (1952), Gesetzmäßigkeiten in der Größenverteilung atmosphärischer Aerosole über dem Kontinent. Ber. d. Deutschen Wetterdienstes US-Zone 35, 261-277.

Junge, C. (1963a), Air Chemistry and Radioactivity. Academic Press, New York.

Junge, C. (1963b), Large Scale Distribution of Condensation Nuclei in the Troposphere. J. Rech. Atmosph. 1, 185-189.

Junge, C. (1974), Residence Time and Variability of Tropospheric Trace Gases. Tellus 26, 477-488.

Kellog, W.W. (1978), Global Influences of Mankind on the Climate.
In: Climatic Change. J. Gribbin ed., 205-277, Cambride University Press, Cambridge.

Knollenberg, R.G., R. Luehr (1976), Open Cavity Laser "Acitve" Scattering Particle Spectrometry from 0.05 to 5 Microns. In: Fine Particles. B.Y.H. Liu ed., 669-696, Academic Press.

Mason, B.J. (1975), Clouds, Rain and Rainmaking. Cambridge University Press, Cambridge.

McMahon, T.A., P.J. Denisn (1979), Empirical Atmospheric Deposition Parameters - A Survey. Atmosph. Environm. 13, 571-585.

Meixner, F. (1981), Die vertikale Verteilung des atmosphärischen Schwefeldioxids im Tropopausenbereich. Berichte des Instituts für Meteorologie und Geophysik der Universität Frankfurt/ Main.

Möller, U., G. Schumann (1970), Mechanisms of Transport from the Atmosphere to the Earth's Surface. J. Geoph. Res. 75, 3013-3019.

Patterson, E.M., C.S. Kiang, A.C. Delany, A.F. Wartenburg. A.C.D. Leslie, B.J. Huebert (1980), Global Measurements of Aerosols in Remote Continental and Marine Regions: Concentrations, Size Distributions, and Optical Properties. J. Geoph. Res. 85, 7361-7376.

Podzimek, J. (1980), Advances in Marine Aerosol Research. J. Rech. Atmosph. 14, 35-61.

Prospero, J.M., D.L. Savoie, T.N. Carlson, R.T. Nees (1979), Monitoring Saharan Aerosol Transport by Means of Turbidity Measurements. In: Saharan Dust: Mobilisation, Transport, Deposition. Scope 14, C. Morales, ed., 171-186, Wiley

Pruppacher, H.R., J.D. Klett (1978), Microphysics of Clouds and Precipitation. Reidel Dordrecht.

Rahn, K.A., R.D. Borys, G.E. Shaw, L. Schütz, R. Jaenicke (1979), Long-range Impact of Desert Aerosol on Atmospheric Chemistry: Two Examples. In: Saharan Dust: Mobilisation, Transport, Deposition. Scope 14, C. Morales ed., 243-266, Wiley Chichester.

Rohde, H., I. Isaksen (1980), Global Distribution of Sulfur Compounds in the Troposphere Estimated in a Height/Latitude Transport Model. J. Geophys. Res. 85, 7401-7409.

Rosen, J.M., D.J. Hofmann, S.P. Singh (1978), A Steady-State Stratospheric Aerosol Model. J. Atmosph. Sci. 35, 1304-1313.

Rosinski, J., C.T. Nagamoto, G. Langer, F.P. Parungo (1970), Cirrus Clouds as Collectors of Aerosol Particles. J. Geoph. Res. 75, 2961-2973.

Schmidt, M., P. Fabian, H. Tiefenau (1974), Simultaneous Vertical Profiles of Condensation Nuclei and Ozone in the Lower Troposphere. Pageoph 112, 887-899.

Schütz, L. (1980), Long Range Transport of Desert Dust With Special Emphasis on the Sahara. Ann. N.Y. Acad. Sci. 338, 515-532.

Schütz, L., R. Jaenicke (1974), Particle Number and Mass Distribution Above 10^{-4} cm Radius in Sand and Aerosol of the Sahara Desert. J. Appl. Meteor. 13, 863-870.

Slemr, R., W. Seiler, G. Schuster (1981), Latitudinal Distribution of Mercury Over the Atlantic Ocean. J. Geophys. Res. 86, 1159-1166.

Smic (1971), Inadvertent Climate Modification. Report of the Study of Man's Impact on Climate. MIT-Press, Cambridge.

Twomey, S. (1977), Atmospheric Aerosols. Elsevier, Amsterdam.

Volz, F. (1972), Infrared Refractive Index of Atmospheric Aerosol Substance. Appl. Optics 11, 755-759.

Walther, H. (1973), Coagulation and Size Distribution of Condensation Aerosols. J. Aerosol Sci. 4, 1-15.

Weiss, R.F. (1981), The Temporal and Spatial Distribution of Tropospheric Nitrons Oxide. J. Geoph. Res., to be published.

Whelpdale, D.M. (1982), Dry and Wet Deposition. This volume.

Whitby, K.T. (1978), The Physical Characteristics of Sulfur Aerosols. Atmosph. Environm. 12, 135-159.

Whitby, K.T. (1980), Aerosol Formation in Urban Plumes. Ann. N.Y. Acad. Sci. 338, 258-275.

Whitby, K.T., R.B. Husar (1970), The Particle Size Spectra of Atmospheric Aerosols. Proc. 2. Intern. Workshop Condensation and Ice Nuclei, Ft. Collins, Colorado.

Willeke, K., K.T. Whitby (1975), Atmospheric Aerosols: Size Distribution Interpretation. J. Air Pollut. Control Assoc. 25, 529-534.

Winkler, P. (1981), Acidity of Aerosol Particles and Precipitation in the North Polar Region Over the Atlantic. Internat. Assoc. Meteor. Atmosph. Physics, 3. Scientific Assembly Hamburg/Germany, 17 - 28 August 1981.

WET AND DRY DEPOSITION[1]

D.M. Whelpdale

Atmospheric Environment Service
Downsview, Ontario, Canada

This contribution reviews the subject of wet and dry deposition. The various precipitation scavenging processes, and factors which influence them are described. A brief quantitative description of scavenging rate coefficients and washout ratios is given. For dry deposition, the mechanisms are described and the meteorological, surface and constituent properties which affect deposition are identified. A theoretical description of dry deposition of gases is given. Finally, measurement methods are briefly reviewed, and typical values of deposition velocities given.

INTRODUCTION

The final portion of the atmospheric cycle of a substance consists of several removal processes which serve to cleanse the atmosphere of accumulated material and deliver it to the surface of the earth. Removal processes may be classified into three types: wet deposition, dry deposition and chemical transformations. Wet removal processes are associated with precipitation, whereas, dry removal processes may go on all the time. Chemical transformations can be thought of as removal processes, in that many species can be transformed into others in the atmosphere (e.g., SO_2 to sulphate), effectively removing the first (SO_2) even though the element itself (S) is not removed.

The effectiveness of a process or combination of processes may depend upon several factors:

i) meteorological conditions -- e.g., turbulence intensity,

H. W. Georgii and W. Jaeschke (eds.), Chemistry of the Unpolluted and Polluted Troposphere, 375–391.

precipitation rate and duration, amount of sunlight;

ii) properties of the substance undergoing removal -- e.g., physical state, solubility, chemical reactivity;

iii) characteristics of the earth's surface -- e.g., wet or dry, nature of vegetative cover, soil pH.

In general, the routine measurement of the amount of material that is brought to the surface is easier and more advanced in the case of wet deposition than in the case of dry deposition.

WET DEPOSITION

Precipitation Scavenging Processes

The term precipitation scavenging is commonly used to mean the removal of any material from the atmosphere to the earth's surface by various forms of precipitation. The term wet deposition is used synonymously with wet removal and precipitation scavenging to describe the processes; however, it is more commonly used as a physical quantity, the flux ($ML^{-2}T^{-1}$) of material reaching the surface by the precipitation scavenging processes.

Both natural and pollutant trace constituents in the atmosphere are scavenged by a variety of hydrometeors: cloud droplets, raindrops, snowflakes, fog droplets, etc. The processes by which incorporation of material into precipitation occurs and the effectiveness of the processes depend upon the types of hydrometeors present, their physical and chemical characteristics, and the physical and chemical properties of the scavenged material.

The scavenging of gases by cloud droplets and raindrops is similar in many respects. Gas molecules are transported to the drop surface from the surrounding air by convection and diffusion, across the air-water interface, and then into the interior of the drop by turbulent mixing and molecular diffusion. Inside the drop they may exist in a dissolved state and they may undergo chemical reactions. In contrast to particle scavenging, gas scavenging may be a reversible process (1), in that molecules may migrate into or out of the drop. The direction and magnitude of the gas flux between air and drop depend on the relative concentrations of gas in the surrounding air and dissolved in the drop, and on the mass transfer coefficient between the two phases. Precipitation scavenging is relatively unimportant for gases of low solubility, although certain hydrocarbons may form surface films on the drops which interfere with the transfer processes and enhance scavenging (2).

Under equilibrium conditions, the concentration of dissolved (i.e., unreacted, or physically dissolved) gas is related to the gas phase concentration by a solubility relationship known as Henry's Law:

$$c_g = Hc_\ell \tag{1}$$

where c_g is the gas-phase concentration (mol ℓ^{-1} of air), c_ℓ is the liquid-phase concentration (mol ℓ^{-1} of liquid), and H is the Henry's Law constant (3). The behaviour of gases that dissociate or react in water may deviate considerably from that predicted by the above solubility relationship.

Common gaseous species which undergo complex physical and chemical transformations in the liquid phase include CO_2, SO_2 and NH_3. Carbon dioxide, for example, dissolves in water to form physically dissolved $CO_2.H_2O$, which in turn dissociates in stages to bicarbonate HCO_3^-, and carbonate CO_3^{2-}. In pure rain-water in equilibrium with an ambient CO_2 concentration of about 335 ppm(v), a solution of pH between 5.6 and 5.7 would result.

The case of SO_2 is even more complex. SO_2 dissolves in water to form physically dissolved sulphur dioxide, $SO_2.H_2O$. Reversible dissociation then occurs to the forms bisulphite HSO_3^-, and sulphite SO_3^{2-}. If other reactants or catalysts are present the SO_3^{2-} may undergo irreversible oxidation to sulphate. The latter reaction depends strongly on the concentrations of other substances present in the drop (e.g., ozone, hydrogen peroxide, ammonium ion, and metal catalysts) and on the drop pH (4,5).

Most in-cloud scavenging occurs under conditions that Hales (6) refers to as "equilibrium scavenging" conditions. In such situations the gas-phase concentrations vary slowly in comparison to the time scale of the scavenging processes, and the droplet sizes are small. Liquid-phase concentrations may then be assumed to be given by the solubility relationship above.

Below clouds, however, the reversible nature of gas scavenging means that the vertical distribution of the trace consituent is important. Rain may scavenge material from an atmospheric layer with high gas-phase concentrations (e.g., a plume) and subsequently release the material in the cleaner air below. In this way, below-cloud scavenging may serve to redistribute trace constituents in the vertical.

Several mechanisms contribute to the scavenging of particles by precipitation. As noted earlier, the importance of a given mechanism depends on the characteristics of the aerosol particles. In the atmosphere particles exist in a very broad range of sizes --

from less than 0.01 μm to over 100 μm radius. Elemental composition, soluble fraction, and hygroscopicity[2] also show large variations. For example, particle composition may vary from a completely insoluble piece of crustal material to a liquid droplet of dissolved seasalt particle. In general, the continental aerosol is a mixture of soluble and insoluble fractions, whereas the marine aerosol has a dominant sea-salt (soluble) component. Particle size and hygroscopicity are among the most important properties in determining the effectiveness of the various particle scavenging mechanisms.

During the formation and growth stages of clouds the larger aerosol particles (large and giant nuclei) serve as cloud condensation nuclei. This is the most efficient in-cloud particle scavenging process, and for particles of radius greater than a few tenths μm, the efficiency is probably close to one hundred per cent. The smaller aerosol particles, those in the Aitken nuclei range below approximately 0.1 μm, may also be incorporated into cloud droplets by Brownian motion, by electrical interactions, and by phoretic effects during evaporation and condensation (7).

Below clouds the predominant scavenging mechanism is gravitational collision. Falling drops sweep through a volume of air and may collide with and collect any particles in their path. In fact, however, for raindrops of radius between approximately 50 and 2000 μm, only particles of radius greater than about 1 μm are effectively captured (8). The smaller particles tend to follow the flow of air around the falling drop. In addition to particle and drop size, the rainfall rate and rainfall duration are important to the overall effectiveness of this scavenging mechanism.

An important aspect of particle scavenging is the variation of aerosol particle size with relative humidity. This depends on the relative amounts of soluble and insoluble material in the particles, on their hygroscopicity and on the relative humidity of the air itself. A shift to larger sizes in the particle-size spectrum below cloud as a result of increasing relative humidity will result in an increased probability of their scavenging. On the other hand, complete evaporation of falling hydrometeors in the relatively drier air below cloud base may actually prevent scavenged material in a hydrometeor from reaching the surface, and result instead in a redistribution of the material in the vertical.

In many parts of the world a significant fraction of the total precipitation is in the form of snow. Unfortunately, relatively little is known quantitatively of snow scavenging processes (9). Snow crystals, with their large surface area to mass ratio, may scavenge particles and supercooled droplets efficiently by interception, but the role of diffusion, phoretic and electrostatic effects is virtually unknown.

Forms of precipitation other than rain and snow are able to scavenge material from the atmosphere and contribute to deposition. Dew and other surface "wetness" (such as intercepted rain), that may be broadly termed precipitation, can alter surface characteristics and enhance the uptake of atmospheric constituents. Brimblecombe (10) has recently reviewed this scavenging phenomenon for SO_2 and concluded that, despite the frequent occurrence of dew in the U.K., this mechanism is relatively unimportant in comparison with other scavenging processes, and is able to account for deposition of less than a few tenths of one per cent of U.K. emissions.

Contaminants are incorporated into fog droplets by the same processes that occur in clouds. Deposition then occurs by the impaction of the fog droplets on obstacles at the surface. This mechanism can result in a significant deposition rate in polluted atmospheres where liquid-phase concentrations are high. It can be very important in areas which experience persistent cloud and high winds such as coastal areas and high-elevation ecosystems (11,12). In this latter situation, the continual interception of cloud elements by vegetation may result in deposition rates of both water and dissolved substances which are considerably greater than the usual wet deposition inputs.

Quantitative Description of Scavenging

Precipitation scavenging processes and pathways are described quantitatively in terms of precipitation scavenging (or washout) rate coefficients and washout (or scavenging) ratios.

In the case of the wet removal of particles, the rate of loss of particles L from a unit volume of air, irreversibly taken up by precipitation, is proportional to their air concentration c_g:

$$L = \Lambda\ c_g \tag{2}$$

where Λ is the scavenging coefficient. This parameter can be derived from the analysis of particle-hydrometeor collisions, and it may be formulated in a space-averaged form, as commonly determined from experimental work. For example, the mass-average scavenging rate Λ integrated over the raindrop and aerosol size distributions is given by:

$$\Lambda = \Pi \int_0^\infty \int_0^\infty v_t(r)\ r^2 E(a,r)\ N(r)\ f(a)\ dr\ da \tag{3}$$

where r, $v_t(r)$ and N(r) are the radius, terminal velocity and number-density functions respectively for the raindrops, f(a) is the probability-density function for the mass distribution of aerosol particles, and E(a,r) is the capture efficiency for drop-aerosol collisions. Although such formulations are valid for both in- and below-cloud scavenging, they do not take explicit account of the nucleation process, which may be the most effective scavenging process for a number of particle species (e.g., sulphate). Inclusion of the nucleation step adds further severe complexities (6).

Slinn et al. (3) provide semi-empirical expressions for hydrometeor-particle capture efficiencies to facilitate wet removal computations. For the removal of aerosol particles of radius a by rain,

$$\Lambda_r(a) = c \frac{\dot{P}}{R_m} E(a, R_m), \tag{4}$$

and by snow,

$$\Lambda_s(a) = g \frac{\rho_w}{\rho_a} \frac{\dot{P}}{v^2_{tm}} E'(a, \lambda). \tag{5}$$

In these equations, c is a numerical factor ~ 0.5; $\dot{P}$ is the precipitation rate; R_m is the volume-mean drop radius; g is the gravitational acceleration; ρ_w and ρ_a are water and air densities, respectively; v_{tm} is an average terminal velocity for snowflakes; and E and E' are average collection efficiencies, evaluated using mean drop size and characteristic length λ data given by Slinn et al. From equations (4) and (5) the wet flux of aerosol particles to the surface D_w can be calculated by integrating over the hydrometeor's path:

$$D_w = \int_0^\infty \Lambda\, c_g dz. \tag{6}$$

Slinn (9) has compared available scavenging rate measurements with predictions and McMahon and Denison (13) compiled listings of scavenging rate coefficients (as well as many other parameters) from the literature. Values are typically in the range 10^{-4} 10^{-6} s^{-1}.

The washout (or scavenging) ratio W is defined as the ratio of a species' concentration in surface-level precipitation c_ℓ (mol ℓ^{-1} of liquid) to its concentration in surface-level air c_g (mol ℓ^{-1} of air):

$$W = \frac{c_\ell}{c_g} . \tag{7}$$

Washout ratios for aerosol particles typically are in the range 10^5-10^6. (W values near 10^3 are often found in the literature as a result of concentrations being expressed as mass ratios (13). With the use of measured washout ratios wet fluxes of aerosol particles can be estimated:

$$D_W = W\dot{P}\ c_g . \tag{8}$$

Washout ratios may be expected to show significant geographical and temporal variability. They are usually derived from near-surface measurements of two quantities, one of which c_ℓ, is effectively an integrated measure of cloud-to-surface scavenging, and the other of which, c_g, is a point measurement, not necessarily representative of in-cloud conditions. In addition, the efficiency of the scavenging processes depends strongly on the mechanisms of precipitation formation and on the precipitation rate. Aerosol particles which have gaseous precursors may have enhanced washout ratios compared with inert species of similar size distribution.

The quantitative description of precipitation scavenging of gases must take account of the reversible nature of the processes - that is, the possibility of desorption. In most situations, except near sources where large gas-phase concentration gradients may occur, the surface-level concentration of an unreactive gas in raindrops can be assumed equal to the equilibrium value given by Henry's law. In this case the washout ratio becomes the reciprocal of the Henry's law constant:

$$W = \frac{c_\ell}{c_g} = \frac{1}{H} , \tag{9}$$

and the wet flux is given by equation (8).

In the case of reactive gases it is necessary to account for possible chemical reactions in the hydrometeors in the presence of other constituents, which effectively lower aqueous phase concentrations of the dissolving species. Enhanced washout ratios result. Scavenging of reactive gases may be treated in a fashion similar to equations (8) and (9) above by defining an enhanced solubility coefficient (1) which takes account of the increased uptake of the gas as a result of liquid-phase reactions.

DRY DEPOSITION

Mechanisms

Dry deposition of atmospheric gases and particles occurs by direct transfer to and retention by various surfaces of the earth - water, soil, vegetation, snow, buildings, etc. In contrast to wet deposition, dry deposition goes on all the time. Most trace atmospheric constituents produced by man remain in the lowest thousand or so meters above the surface and dry removal requires their continuous transport down to the surface.

For gases and small particles of diamter less than about 10 μm this transfer to the near-surface region is accomplished by turbulent diffusion. For larger particles the influence of gravity is significant and sedimentation becomes a progressively more important process with particle size. Very close to the surface (i.e. within less than ca 1 mm) viscous forces become important and transfer through the resultant laminar flow layer is governed by molecular instead of turbulent processes. Transfer of gas molecules occurs by molecular diffusion and that of small particles (<ca 0.1 μm) primarily by Brownian diffusion. For particles larger than ca 1 μm inertial impaction on surface elements becomes important. The transfer of particles of diameter between ca 0.1 and 1 μm is rather inefficient.

Actual capture at the surface is accomplished by a variety of processes. Gases may dissolve in the water of oceans, lakes, other wet surfaces and vegetation; they may be physically adsorbed, for example, onto soil particles or snow crystals; they may undergo direct chemical reaction with surface material; and they may enter plants through stomata. The primary mechanism for particle capture is impact with the surface and retention there, although uptake through stomata has been suggested as a possibility. Subsequent escape from surfaces, particularly for larger particles by bounce-off or resuspension, frequently occurs.

Three sets of factors influence the mechanisms identified above, and thus the efficiency of the dry deposition pathway for particular species and surfaces. Meteorological variables control atmospheric transfer down to the near-surface region. Effective transport occurs with vigorous turbulent mixing (unstable conditions) which results from higher wind speeds and radiation inputs. In general, in northern temperate latitudes, atmospheric stability is greater during night and during winter, so that turbulent transfer to the surface is correspondingly decreased. It has been suggested that high relative humidity may contribute to increased fall velocities of particles by increasing their mass and density. In addition, the occurrence of precipitation temporarily decreases dramatically the relative importance of dry deposition, but at the same time wets all surfaces to make subsequent dry uptake more efficient.

Properties of the depositing pollutant itself affect both atmospheric transfer and surface uptake. The effectiveness with which gases cross the laminar flow layer is thought to depend upon their molecular diffusivity, and the effectiveness with which they are taken up by the surface is proportional to their solubility and chemical reactivity. In the case of particles, their atmospheric behaviour depends upon aerodynamic size, shape, and hygroscopicity; their interaction with a surface depends upon their solubility, chemical properties, and charge.

The nature of the surface also has an important influence on dry deposition. The greater the surface roughness, the more effective is the turbulent exchange above the surface. Gaseous uptake at a surface depends upon such factors as chemical composition and properties (e.g., pH), physiological state (e.g., actively growing, senescent), wetness and perhaps prior loading. In the case of particles the nature of the surface - for example, wet sticky, hairy, charged - controls the degree of retention of impacted particles.

A Theoretical Treatment of Dry Deposition of Gases

Micrometeorological studies of the exchanges of heat, water vapor and momentum between the lowest several metres of the atmosphere and the surface have provided a theoretical framework for the description of dry deposition of gases (14, 15). The downward flux F of a gas can be related to the gradient of its concentration c with height z:

$$F = K(z)\,\frac{dc}{dz} \qquad (10)$$

where K(z) is the diffusion coefficient representing turbulent and molecular diffusion. Integrating from the surface to height z yields.

$$F = \frac{c(z) - c(o)}{\int_o^z \frac{dz}{K(z)}} \qquad (11)$$

where the denominator may be considered as gas-phase resistance to transfer.

A surface resistance r_s may also be defined, under assumptions that uptake at the surface is irreversible and that it is proportional to the gas concentration there:

$$r_s = \frac{c(o)}{F} . \tag{12}$$

The flux to the surface can then be expressed in terms of a gas-phase resistance and a surface resistance:

$$F = \frac{c(z)}{r_g(z) + r_s} . \tag{13}$$

Rates of deposition are usually discussed in terms of these resistances and the deposition velocity $v_d(z)$:

$$v_d(a) = \frac{F}{c(z)} = \frac{1}{r_g(z) + r_s} = \frac{1}{r(z)} . \tag{14}$$

v_d is usually determined from measurements made at a height of approximately 1 m above the surface.

The resistance to transfer in the atmosphere down to the near-surface region can be estimated theoretically under certain conditions. In this region where turbulent diffusion prevails, transport for all gases and vapors, and for heat and momentum are expected to be similar, and thus diffusivity and resistance values may be calculated. For example, under neutral stability conditions the wind speed u is a logarithmic function of height:

$$u(z) = \frac{u_*}{k} \ln \left(\frac{z - d}{z_o}\right) \tag{15}$$

where k is von Karman's constant (ca 0.4), d is the displacement height of the surface, and z_o is the surface roughness length. The resistance to momentum transport $r_M(z)$ is then (14)

$$r_M(z) = \frac{1}{ku_*} \ln\left(\frac{z - d}{z_o}\right) \tag{16}$$

or, from equation (15),

$$r_M(z) = \frac{u(z)}{u_*^2} . \tag{17}$$

Resistance to transfer for gases is assumed to be given by equation (17), which can be modified when buoyancy effects are present.

In the near-surface layer where molecular diffusion dominates, an additional resistance to transfer r_b is encountered, which is given (16) by

$$r_b = (Bu_*)^{-1}. \tag{18}$$

Thus, Garland (15) gives the total gas phase resistance as

$$r_g(z) = \frac{u(z)}{u_*^2} + \frac{R(z/L)}{u_*} + \frac{1}{Bu_*} \tag{19}$$

where the second term on the right is a correction for buoyancy effects refered to above. R, as B, is a slowly varying function of u_*, and L is the Monin Obukhov length, which reflects the relative importance of heat and wind shear on turbulence. The first term on the right decreases with increasing surface roughness, and all three terms decrease as wind speed increases (since u_* is usually proportional to u(z) if z is not too large).

The resistance to uptake at the surface r_s cannot be predicted because of its dependence on many factors. Values of r_s are obtained by experiments and by difference between r(z), which is determined from the rate of deposition, and r_g (z).

Because the resistance to transfer is made up of an atmospheric and a surface part, even when surface uptake is very efficient (i.e., $r_s = 0$), the remaining atmospheric resistance will impose an upper limit on the deposition velocity, $r_g(z)^{-1}$.

Deposition Velocities

The effectiveness of dry deposition of gases and particles is usually measured in terms of their deposition velocity. Values of this parameter are derived either from direct measurements of pollutant concentration and flux, or from wind, temperature and humidity measurements, as noted in the previous section.

Four common measurement approaches are the direct measurement of deposition to a surface, the flux-gradient method, the eddy-correlation method, and the mass balance approach. The first method entails the direct measurement of the amount of material which has been deposited on a surface. This flux is used along with a measured air concentration above the surface to determine the deposition velocity (17). In practice, direct deposition measurements are difficult to make because of surface inhomogeneities and contamination. An exception has been the use of radioactive or other tracer materials, which have been employed successfully for both gas and particle deposition measurements

(18). A related approach to direct measurement is the use of an artificial surface or a collector (often an open bucket) to simulate the natural surface, or to collect a representative sample of the deposition, respectively. Such methods generally are not successful, both because natural surface characteristics (and thus r_s) are very difficult to reproduce, and because neither natural aerodynamic nor surface features are similar.

The flux-gradient method is a standard micrometeorological approach where measurements of gradients of concentrations and meteorological variables are used with the theory discussed to derive fluxes, and thus deposition velocity values (18). The method is most appropriate for gases, but has been attempted (rather unsatisfactorily) for particles as well.

Another micrometeorological method is the eddy-correlation approach, in which simultaneous measurements of pollutant-concentration and vertical-velocity fluctuations are used to determine vertical flux directly (19). Sophisticated instrumentation is required for eddy-correlation measurements, but the approach has been successful for momentum, heat and water vapour flux measurements (20). It is most suitable for gases, but in time could possibly be used for particles as well.

The fourth method makes use of pollutant concentration measurements to determine the rate of loss of an atmospheric constituent from a well defined volume of air. This approach has been used by Payrissat and Beilke (21) for example, and is suitable, in principle, for both gases and particles.

The quantities being measured (i.e., concentration, flux, deposited mass) are in many cases near the limit of their detection and thus high accuracy and precision are required. For the eddy-correlation method sensors with a fast time response (<ca 1 s) are essential. Because of the complex nature of dry deposition mechanisms (i.e., numerous, strongly dependent on constituent, surface and atmospheric characteristics) and because such stringent requirements exist for sampling and analysis methods and instrumentation, most dry deposition measurements to date have been done on a research basis, and not as a routine part of monitoring network operation.

Dry deposition measurements have been made for several gases and many kinds of particulate material (22, 13). Among the common pollutant gases sulphur dioxide has been studied most extensively, ozone somewhat less so, and the oxides of nitrogen relatively little. Table 1 lists typical, reported deposition velocity values for sulphur dioxide, nitrogen dioxide, ozone and for small particles. As Sehmel (22) described, experimental conditions for the many different investigations have varied

Table 1: Typical reported values of deposition velocity ($mm\ s^{-1}$).*

Species \ Surface	Vegetation	Soil	Water	Snow
Gases				
SO_2	8	8	8	1
NO_2	5	-	-	-
O_3	8	8	1	-
Small Particles				
(0.1<d<1 μm)	1**	1**	1**	1**

* "Representative" values are taken from the survey papers of Fowler (23, 24), Garland (14,15) and Sehmel (22).

** Uncertain values.

widely, and the values of v_d for a particular species may range over two or three orders of magnitude. Nevertheless, work of Garland (14,15) and Fowler (23,24) lend support to these values. Because of the wide range of atmospheric and surface conditions which a depositing substance may encounter over the surface of the earth and during the course of a year, considerable variation in deposition velocity values is to be expected.

Mean deposition velocity values for sulphur dioxide over vegetation-covered, soil, and water surfaces are quite similar. Uptake by vegetation is primarily through stomata, although other plant surfaces may be receptors also (25). Deposition to vegetation appears to vary considerably between day and night, and summer and winter. Gas-phase resistance to transfer decreases over the rougher surface of tall vegetation like trees, but surface resistance appears to increase in order to reduce water loss, with the result that total resistance and deposition velocity do not change a great deal with vegetation height. Wetted surfaces of vegetation, forests in particular, might be expected to enhance uptake considerably. Surface resistance to uptake by soils decreases as soil pH and moisture content increase. Thus, for example, SO_2 deposition to wet, calcareous soils is quite efficient. Because SO_2 is highly soluble in water, surface resistance is negligible and uptake over lakes and oceans is controlled by gas-phase resistance. Over snow surfaces stable atmospheric conditions make gas-phase transfer inefficient with a resulting lower deposition velocity. Uptake of SO_2 by surfaces of buildings and statuary appears to be similar to that of vegetation, soils, and water surfaces.

Few measurements of the deposition of nitrogen oxides (NO, NO_2, HNO_3) have been made; however, on the basis of their known chemical properties some predictions about their behaviour can be made. HNO_3, the most soluble (greater than for SO_2), could be expected to behave much like SO_2, perhaps with a smaller surface resistance and slightly larger deposition velocity. NO_2 is also highly reactive, but is apparently less readily taken up by water. Limited measurements (26) suggest a deposition velocity to vegetation of approximately 5 mm s^{-1} with strong stomatal control, while deposition velocities to other surfaces are thought to be slightly less than those for SO_2 (24). NO has a considerably lower solubility and is expected to have a significantly lower deposition velocity.

Ozone, a highly reactive gas, has a low solubility in water and thus, while its deposition velocity for vegetation-covered and soil surfaces is similar to that for SO_2, it is removed ineffectively by lake and water surfaces. Limited measurements of other reactive gases (e.g. hydrogen fluoride, iodine, chlorine, peroxyacetyl nitrate) suggest that their deposition velocity values are comparable to those for sulphur dioxide, with differences being attributable to differences in solubility and molecular diffusivity (15).

Extensive wind tunnel measurements and a few field measurements confirm the strong particle-size dependence of deposition velocity. The form of the relationship reflects the importance of different processes for different particle sizes. Below ca 0.1 μm Brownian diffusion through the near surface laminar boundary layer controls transfer; whereas, above ca 1 μm sedimentation contributes increasingly to the deposition. A minimum exists in the size range between 0.1 and 1 μm where both processes are inefficient.

Particle-size measurements show that the atmospheric particulate-sulphate mass lies primarily in this size range (27), and a significant fraction of the particulate-nitrate mass is expected to lie in this range also (28). Measurements for sulphate and for other substances in this size range (e.g. Pb, Sb, V, Cs) generally confirm the small deposition velocities, although much scatter in the values exists (22).

CONCLUDING REMARKS

Both wet and dry deposition effectively cleanse the atmosphere of trace constituents. The absolute importance of the two mechanisms depends upon the amount of a particular substance present in the atmosphere, the characteristics of the substance, ambient meteorological conditions, and the climatic and surface characteristics of a particular region.

The relative importance of wet and dry deposition depends on many of the same factors. In particular, the frequency, duration and, to a lesser extent, intensity of precipitation are of utmost importance. For example, in wet tropical areas where intense rainfall is a daily occurrence, wet deposition is rapid and relatively much more important than dry. On the other hand, in desert areas infrequent rainfall makes wet deposition insignificant.

In those mid-latitude regions of the northern hemisphere where man-made emissions dominate cycles of substances such as sulphur and nitrogen, wet and dry deposition are of comparable importance. Near source regions, where atmospheric concentrations of pollutants are higher, dry deposition is relatively more important; but wet deposition becomes progressively more important away from sources.

In the Arctic, particularly in winter, where precipitation is infrequent and mainly in the form of snow, and where stable atmospheric conditions prevail, both wet and dry removal processes are very inefficient. Trace constituents which eventually find their way into the Arctic atmosphere remain there for long periods of time.

Deposition measurements have so far been restricted primarily to the northern mid-latitudes. Further investigation is required in most other parts of the world in order that removal processes and deposition rates may be characterised is a more comprehensive way globally.

ACKNOWLEDGEMENT

The support of the Norges Teknisk-Naturvitenskapelige Forskningsrad during the preparation of this manuscript, and the opportunity provided by the Norsk Institutt for Luftforskning and the Canadian Atmospheric Environment Service are gratefully acknowledged.

1 Parts of this lecture have been adapted from Chapter 4 of the WMO Technical Note "Tropospheric Chemistry and Air Pollution", WMO-No. 581.

2 Hygroscopicity is a measure of the tendency of a substance to absorb moisture by acceleration of the condensation of water vapour.

REFERENCES

1. Hales, J.M.: 1972, Atmos. Environ. 6, pp. 635-659.

2. Garret, W.D.: 1978, Pageoph. 116, pp. 316-334.

3. Slinn, W.G.N., Hasse L., Hicks, B.B., Hogan, A.W., Lal, D., Liss, P.S., Munnich, K.O., Sehmel, G.A. and Vittori, O.: 1978, Atmos. Environ. 12, pp. 2055-2087.

4. Penkett, S.A., Jones, B.M., Brice, K.A. and Eggleton, A.E.J.: 1979, Atmos. Environ. 13, pp. 123-137.

5. Barrie, L.A. and Georgii, H.W.: 1976, Atmos. Environ. 10, pp. 743-749.

6. Hales, J.M.: 1978, Atmos. Environ. 12, pp. 389-399.

7. Hidy, G.M.: 1973, In: *Chemistry of the Lower Atmosphere* (ed. S.I. Rasool), Plenum Press, New York, pp. 121-176.

8. Junge, C.E.: 1963, *Air Chemistry and Radioactivity*. Academic Press, New York and London, 382 pp.

9. Slinn, W.G.N.: 1977, Water, Air, and Soil Pollution 7, pp. 513-543. (D. Reidel Publ. Co., Dordrecht, Holland)

10. Brimblecombe, P.: 1978, Tellus 30, pp. 151-157.

11. Reiners, W.A., Schlesinger, W.H. and Marks, R.H.: 1977, In: *Precipitation Scavenging (1974)*, Technical Information Center, USA ERDA, CONF-741003, pp. 571-589.

12. Falconer, R.E. and Falconer, P.D.: 1980, J. Geophys. Res. 85, pp. 7465-7470.

13. McMahon, T.A. and Denison, P.J.: 1979, Atmos. Environ. 13, pp. 571-585.

14. Garland, J.A.: 1978, Atmos. Environ. 12, pp. 349-362.

15. Garland, J.A.: 1979, WMO-No. 538, Paper presented at the WMO Symposium on the Long Range Transport of Pollutants and its Relation to General Circulation Including Stratospheric/Tropospheric Exchange Processes, (Sofia, 1-5 October 1979), WMO, Geneva, pp. 95-103.

16. Chamberlain, A.C.: 1966, Proc. R. Soc. A 290, pp. 236-265.

17. Dovland, H. and Eliassen, A.: 1976, Atmos. Environ. 10,

pp. 783-785.

18. Garland, J.A.: 1977, Proc. R. Soc. Lond. A 354, pp. 245-268.

19. Wesely, M.L., Eastman, J.A., Cook, D.R. and Hicks, B.B.: 1978, Boundary-Layer Meteorol. 15, 361-373.

20. Hicks, B.B.: 1970, J. Appl. Met. 9, pp. 386-388.

21. Payrissat, M. and Beilke, S.: 1975, Atmos. Environ. 9, pp. 211-217.

22. Sehmel, G.A.: 1980, Atmos. Environ. 14, pp. 983-1011.

23. Fowler, D.: 1980, In: *Effects of Acid Precipitation on Terrestrial Ecosystems,* (eds. Hutchinson, T.C. and Havas, M.), Plenum Press, New York, pp. 9-27.

24. Fowler, D.: 1980. In: *Ecological Impact of Acid Precipitation,* Proceedings of an international conference (Sandefjord, Norway, March 11-14, 1980), (eds. Drabløs, D. and Tollan, A.), SNSF, P.O. Box 61, As-NHL, Norway, pp. 22-32.

25. Fowler, D.: 1978, Atmos. Environ. 12, pp. 369-373.

26. Bengtson, C., Boström, C-A, Grennfelt, P., Skarby, L. and Troeng, E.: 1980, In: *Ecological Impact of Acid Precipitation*, Proceedings of an international conference (Sandefjord, Norway March 11-14, 1980), (eds. Drabløs, D. and Tollan, A.), SNSF, P.O. Box 61, As-NLH, Norway, pp. 154-155.

27. Whitby, K.T.: 1978, Atmos. Environ. 12, pp. 135-159.

28. Spicer, C.W.: 1977, In: *Advances in Environmental Science and Technology*, Volume 7 (eds. Pitts, J.N. and Metcalf, R.L.), Wiley-Interscience, New York, pp. 163-261.

PART V

AIR POLLUTION

ACID PRECIPITATION IN THE NORTHEASTERN UNITED STATES

Volker Mohnen,* Jerre Wilson,** and John Kadlecek*

*Atmospheric Sciences Research Center
State University of New York, Albany, New York, USA
**Science Research Lab, U. S. Military Academy
West Point, New York, USA

Wet deposition of pollution-related material (sulfates, nitrates, free hydrogen ions, etc.) was determined at selected stations within the eastern United States. The deposition events were characterized with respect to air mass origin (for example, "Midwest/Ohio Valley" versus "Great Lakes/Canadian"). From this information, a regional pattern for the deposition of dominant ion species in precipitation was developed.

The ARL-ATAD trajectory model was used to characterize individual precipitation events as to air mass origin. The data analyses suggest that the "acid rain phenomenon" covers the entire northeast. The concept of mesoscale mixing or of a "regional superbowl" emerges to account for the lack of a significant gradient in the normalized deposition pattern of sulfates, nitrates, and free hydrogen ions measured in precipitation.

In order to understand acid precipitation, one must also know the composition of cloud water. An airborne cloud and precipitation collection system is described that has been successfully deployed in many field experiments. Some typical results are presented.

I. COMPOSITION

The measurement of free hydrogen ion concentration in precipitation (pH) is not a trivial measurement, and its interpretation requires detailed information on: sampling location, chemical and physical conditions of the surrounding terrain or

H. W. Georgii and W. Jaeschke (eds.), Chemistry of the Unpolluted and Polluted Troposphere, 395–423.

surface, sampling collection procedure for precipitation, and sample handling, etc.

The diagram in Figure 1 attempts to illustrate some of the informational needs that must be included with the publication of a pH value. Most measurements are made today on a weekly basis (such as in the Multistate Atmospheric Power Production Study--MAP3S). An event is defined by the beginning and the end of precipitation. Modern collectors are normally closed and open only during precipitation through an automatic triggering device. This procedure will minimize contamination that arises from dry deposition of dust and wet debris. Precipitation samples can be collected by ground collectors or airborne platforms. Rain that fell through the last few meters of the planetary boundary layer might be of different quality than rain that was collected at a considerable height above ground. A ground collector can be located over ocean or over land, above or below a vegetation canopy, or in an open field. Precipitation can be collected from above (mountain tops, etc.) or below the planetary boundary layer. Most networks today that collect precipitation for long-term monitoring, are based on weekly or event-by-event sampling from ground stations that are located in an open field with no nearby obstacles.

The pH value obtained from such collection sites might have regional representativeness. All values analyzed and reported here originate from such networks. A very strict quality control procedure exists for both MAP3S and NADP. One should keep in mind that the pH itself is insufficient to characterize precipitation quality. It is very informative to divide the ionic constituents of precipitation into three categories or components:

Soil-derived minerals Ca^{++}, Mg^{++}, Fe^{3+}, silicate ions, etc.

Sea salt Na^{+}, Cl^{-}, $SO_4^{=}$, plus small amounts of K^{+}, Mg^{++} and Ca^{++}

Atmospheric conversion products from sulfur and nitrogen bearing molecules, i.e., $SO_4^{=}$, NO_3^{-}, in addition to NH_4^{+} and Cl^{-} (from hydrochloric acid) and, obviously, H^{+}

The dominant source of data is the Multistate Atmospheric Power Production Pollution Study (MAP3S). In this ongoing study, precipitation chemistry data are obtained on an event basis beginning in 1976 for the original MAP3S collection sites of: Whiteface Mountain/New York (WFM); Ithaca/New York (ITH); Pennsylvania State University (PEN); University of Virginia (VIR).

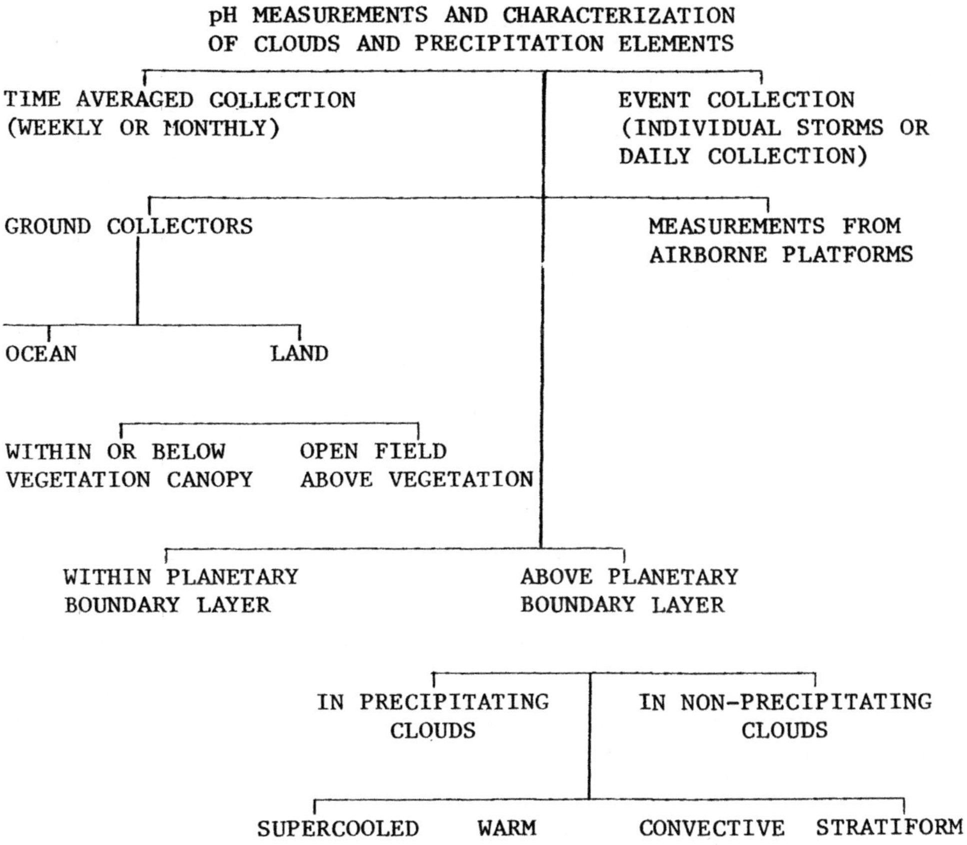

Figure 1. pH measurement classification

In 1977-78 four more stations were added to the network located at: Champaign-Urbana/Illinois (ILL); Lewis/Delaware (coastal site) (LEW); Brookhaven, Long Island/New York (coastal site) (BRO); Oxford/Ohio (OXF).

The basic precipitation collector originally used by the MAP3S network during 1977-79 was a wet deposition-only type designed by Battelle Pacific Northwest Laboratory. This collector allows a relatively large exposed collection area while minimizing splashing and turbulence. An enclosed chamber enables samples to remain refrigerated until their removal for shipment. Today all networks use a HASL type collector.

Field site operators measure pH, conductivity and sample volume before shipment to Battelle Pacific Northwest Laboratory for complete analysis.

To insure that an ionic balance was maintained between the major ions $SO_4^=$, NO_3^-, H^+ and NH_4^+ throughout the three-year period, ions or pairs of ions were analyzed for correlation or degree of association. The balance between the four dominant ions, $SO_4^=$, NO_3^-, H^+ and NH_4^+ has been tested, yielding a linear correlation coefficient of 0.95 for Whiteface Mountain. This indicates that a significant portion of the sulfate and nitrate could exist in the form of sulfuric and nitric acid and the appropriate ammonium salts. The linear correlation can further be improved by removing the soil components, indicated by Ca^{++} and Mg^{++}, from the cation summation. The correlation of 0.98 suggests an extremely good ion balance exists whenever boundary layer influences such as soil components can be reduced

This procedure can also be used to determine the degree of dependency of the pH, or H^+, upon the other three major constituents. The concentration of H^+ in precipitation is bound primarily by $SO_4^=$, to a lesser extent NO_3^-, and finally NH_4^+. This conclusion applies to all types of precipitation for all three analyses and for the northeastern United States.

The results from Whiteface Mountain are identical to those obtained from Ithaca or from any other station in the MAP3S network. This does not mean that ammonia and hydrogen ions are the only positive ions and it does not mean that sulfate and nitrate are the only negative ions. There are, depending on the station location, a host of other anions and cations present. But eventually it is the correlation between ammonia and hydrogen ion versus sulfate and nitrate ions that best classifies the precipitation in the northeastern United States. If one goes west of 85° longitude, the picture may be totally different. There may be almost no free hydrogen ions found in areas where the soil is alkaline. This has been demonstrated very clearly in some of the NADP measurements.

Knowing that the ionic balance between hydrogen ion, ammonia, sulfate and nitrate is a good, although not a complete, indicator for acid precipitation, we now turn to the problem of the ion "deposition" versus back trajectory analysis.

II. DIRECTIONAL VARIABILITY

We are now showing values primarily from the Whiteface Mountain station. This is an ASRC station in the center of the Adirondacks designated as the Rural Precipitation Chemistry Center. The 200 or so lakes that cover 10,000 acres and are allegedly without fishlife are mostly located in the high peak area (2,000 feet and above) in the vicinity of Whiteface Mountain. The other results that we are discussing here will be from the Illinois site. One can, therefore, see data from the western end of a major emission region and the eastern end with the Tennessee and Ohio Valleys located in between. This will enable us to investigate the problem of whether or not and to what degree the Tennessee Valley/Ohio Valley contribute to the wet ion deposition of pristine areas such as Whiteface Mountain in the Adirondacks.

In general, we are reporting deposition values here. Obviously, there are two parameters: one is "concentration" of pollutant-related ions in precipitation; the other is "deposition" of pollutant-related ions. Effects researchers like to see deposition numbers because it is the amount of precipitation that brings down the pollutant material that eventually will lead to effects. One could analyze every event in regard to the air mass origin. This can be accomplished by calculating back trajectories for every event for the past two-day travel of air parcels prior to their arrival at Whiteface Mountain and Illinois. In that sense, one can then see whether or not Canadian air (air that has passed during the past two days over Canada) shows a different ion deposition than air that has spent the past two days in the Midwest/Ohio Valley.

To this extent, the Air Resources Laboratories'Atmospheric Transport and Dispersion Model (ARL-ATAD) was used to develop a trajectory climatology in conjunction with the precipitation chemistry at the Whiteface Mountain and Illinois sites. Trajectories were calculated for all of 1978 and 1979 and weighted by precipitation amounts during every six-hour period to determine a dominant direction or origin of air mass for each MAP3S precipitation event. Each event was then classified in one of twelve sectors covering 30° each. Only the final two days (48 hours) of approach trajectory end points were considered. This allowed differentiation between so-called "Ohio Valley/ Midwest" and "Canadian/Great Lakes" air masses. The so-classified wet ion deposition is plotted as a function of trajectory sector in Figures 2 through 4.

The Ohio Valley/Midwest events were defined as those having their origin in the sector 165° through 285°, while the Canadian/ Great Lakes events were designated by the sector 285°-30°. The wet deposition directional patterns for all ions were found to closely follow the precipitation variability in every case. The wet deposition cumulative total of the Ohio Valley/Midwest sector was found to account for approximately 50-70% of the total wet deposition that could be classified into the 48-hour approach trajectories. Wet deposition from the Canadian/Great Lakes sector resulted largely from winter coastal storms in the form of snow or mixed precipitation. Even though ionic concentrations were generally lower for these events, the large values of precipitation sample volume resulted in almost 30% of the total wet deposition at Whiteface Mountain.

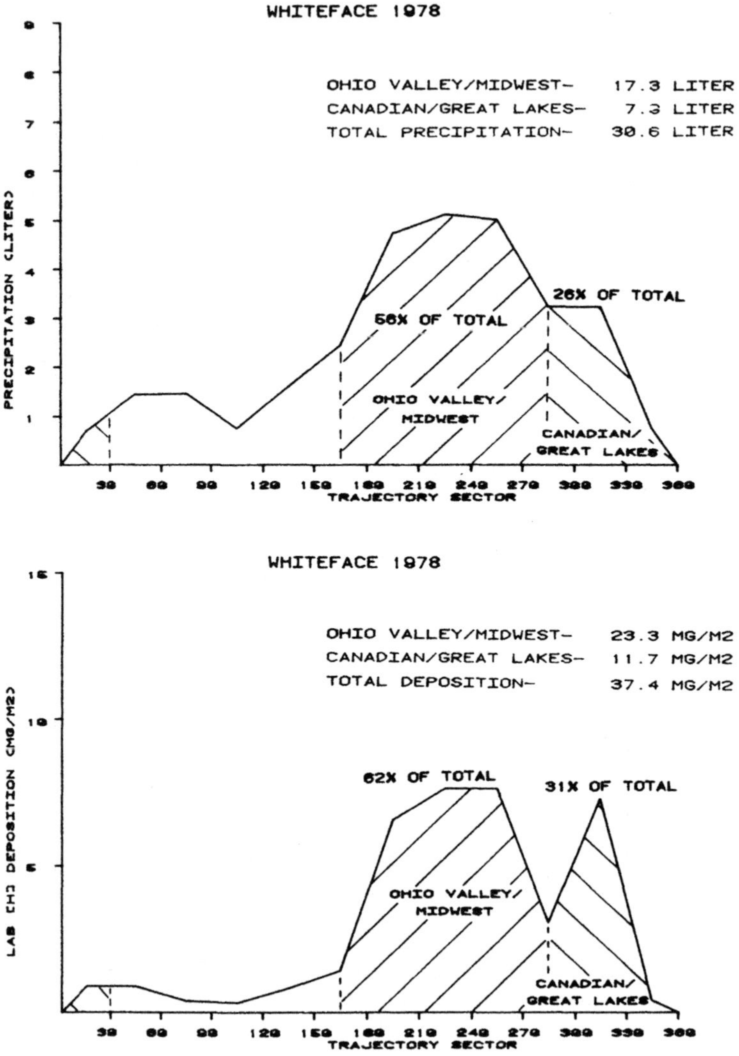

Figure 2.

a) Precipitation volume, and b) Free hydrogen ion total wet deposition per 30° trajectory for Whiteface, 1978

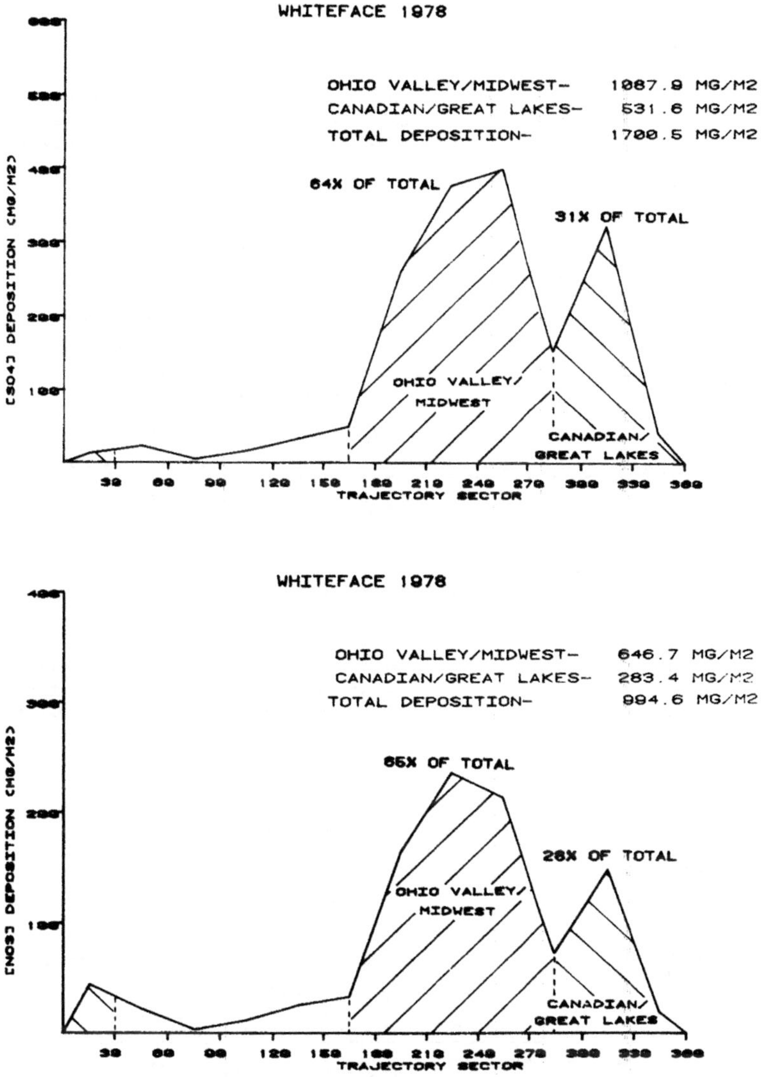

Figure 3. a) Sulfate, and b) Nitrate total wet deposition per 30° trajectory sector for Whiteface, 1978

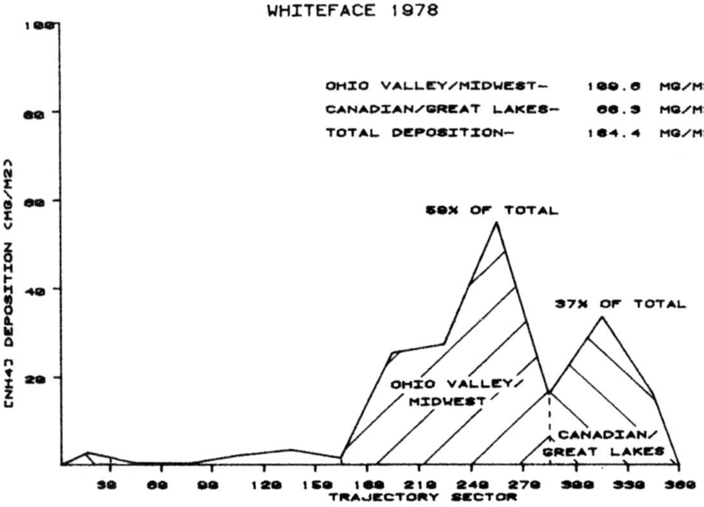

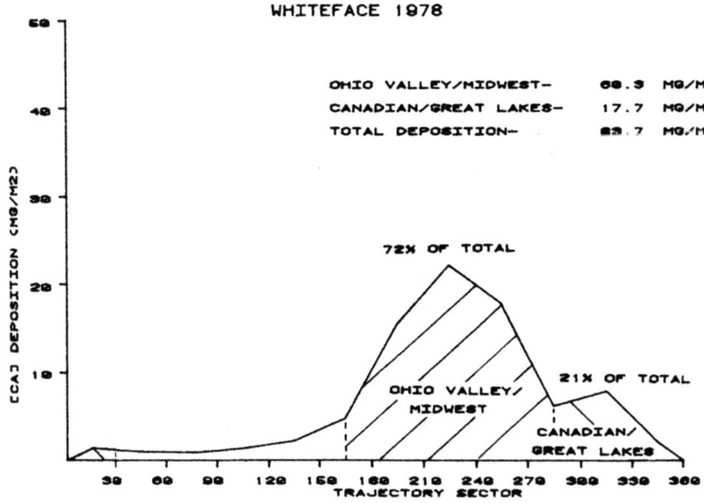

Figure 4. a) Calcium, and b) Ammonium total wet deposition per 30° trajectory sector for Whiteface, 1978

The same analysis procedure (ARL-ATAD model) was applied to the MAP3S precipitation chemistry for the Illinois site during 1978. Since large pollutant sources for SO_2, NO_x, etc. lie to the east or southeast of the Illinois station, one might expect to see a significant difference in the deposition pattern of pollution-related ions observed at Whiteface Mountain during the same year. The cumulative wet deposition total at Illinois for these events is plotted in Figures 5 and 6 revealing a very interesting picture. Approximately 70% of the total precipitation for Illinois was found to be associated with the southwest approach sector. This is consistent with synoptic considerations whereby southwesterly flow normally precedes an approaching cyclone system and subsequent deposition events. The remainder of the precipitation was found to be essentially equally distributed over the various sectors with the exception of the 0-30° segment. Air masses approaching the Illinois site from the southwest were found to deliver about 70% of the major ions (Figures 5 and 6). One can again see that wet deposition of pollution-related ions depends primarily on the precipitation amount.

One might expect to see significant differences in precipitation chemistry as air masses from different regions arrive at a receptor site. However, the analysis shows:

	Whiteface Mountain (1978)
Midwest/Ohio Valley (160°-280° sector)	56% of the annual precipitation delivering:
	62% of the annual $[H^+]$ DEPOSITION
	64% of the annual $[SO_4^=]$ deposition, and
	65% of the annual $[NO_3^-]$ deposition
Canadian/Great Lakes (280°-30° sector)	26% of the annual precipitation delivering:
	31% of the annual $[H^+]$ deposition
	31% of the annual $[SO_4^=]$ deposition, and
	28% of the annual $[NO_3^-]$ deposition
	Illinois (1978)
(160°-280° sector	71% of the annual precipitation delivering:
	78% of the annual $[H^+]$ deposition
	67% of the annual $[SO_4^=]$ deposition, and
	64% of the annual $[NO_3^-]$ deposition

It is quite apparent that the precipitation volume, more than any other single factor, determines the amount of deposition on an annual basis for the three pollution-related ions, $[H^+]$, $[SO_4^=]$ and $[NO_3^-]$. The deposition of ions from the Midwest/Ohio Valley air and Great Lakes/Canadian air does not reflect the very significant differences in emissions that are located in these

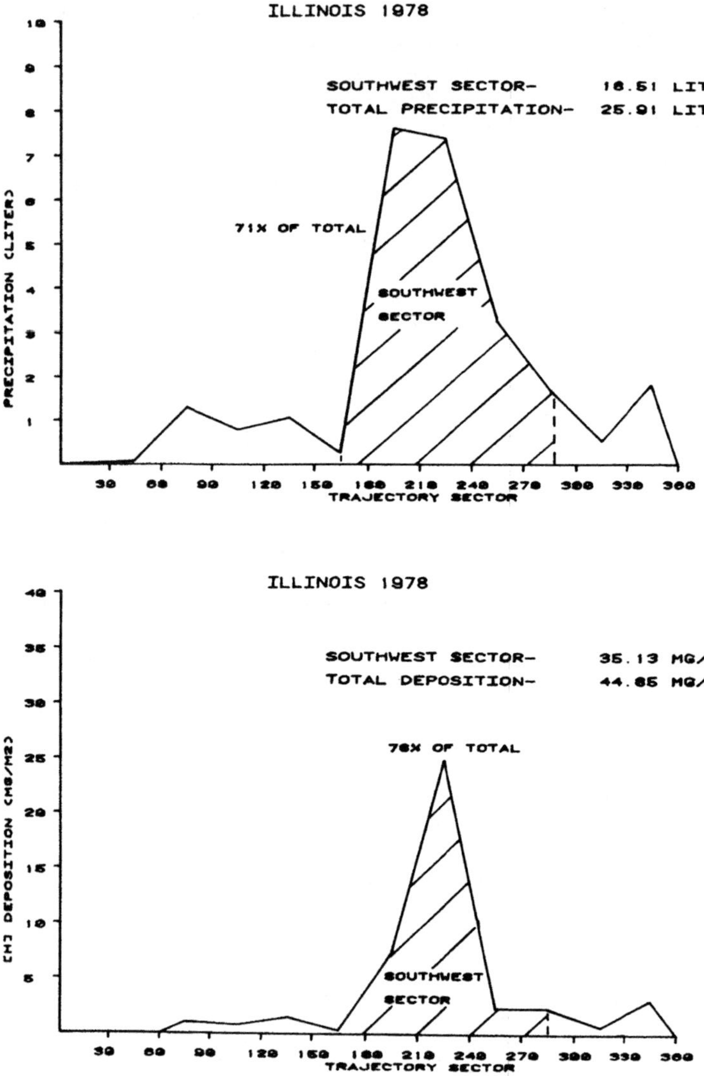

Figure 5. a) Precipitation, and b) Free hydrogen ion deposition pe
30° trajectory sector for Illinois, 1978

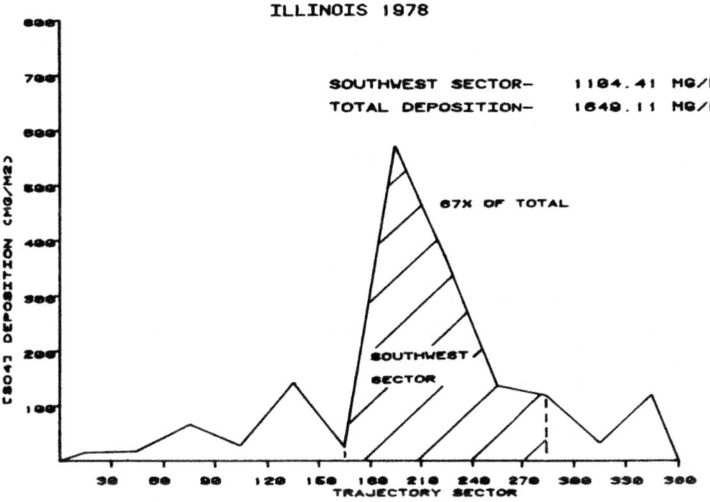

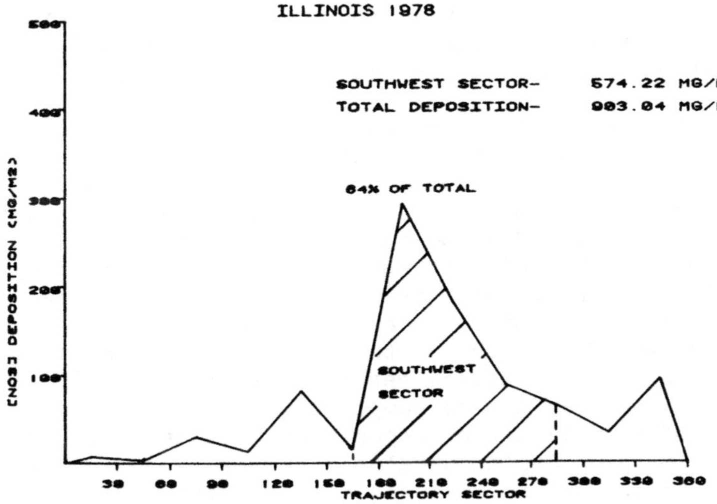

Figure 6. a) Sulfate, and b) Nitrate deposition per 30° trajectory sector for Illinois, 1978

two regions. The Illinois results further substantiate the above conclusion. There seems to exist no simple, straightforward relationship between emission source(s) for acid precipitation precursor gas(es) and receptors of "acid rain" i.e., $[H^+]$, $[SO_4^=]$ and $[NO_3^-]$ ions. The chemical transformation pathway(s) seem to be complex and insensitive to a 48-hour back trajectory analysis presented here and based upon available meteorological and chemical information. With more refined meteorological tools and more sophisticated chemical tracer techniques, one should be able eventually to establish a reliable source receptor relationship.

The concept of a "superbowl" (synoptic scale air shed) would explain to some extent the uniformity in directional variability of annual wet ion deposition in that the final products, i.e., acidic material, is being rather evenly deposited over a very large region in spite of the possibility that the precursor materials show considerable regional concentration fluctuations.

III. WET DEPOSITION

To circumvent the problem arising from different precipitation amounts for different sites, one has to normalize or standardize the wet deposition by the annual precipitation amount to obtain the normalized deposition of sulfate and nitrate ions per one centimeter of precipitation. (This is identical to a "precipitation weighted" concentration parameter.)

As mentioned earlier, several North American precipitation networks operate independently. It is, therefore, possible to correlate their results through proper analysis techniques. Two programs offering regional coverage, accessible data and reliable quality control are MAP3S and NADP. Even though MAP3S is event-oriented, while NADP collects weekly samples, their integrated ion deposition measurements are compatible when compared over time scales in excess of one year. Figure 7 shows the location of the MAP3S and NADP sites of the eastern United States. Total wet deposition was normalized per centimeter of precipitation as outlined earlier. To insure complete objectivity, spatial distributions of normalized wet deposition were analyzed using the Synagraphic Mapping System developed by the Laboratory for Computer Graphics and Spatial Analysis, Harvard University.

The resulting wet ion deposition patterns, normalized for one centimeter of precipitation, are shown in Figures 8-11. These objective analyses for the 1979 wet deposition of sulfate ($SO_4^=$), nitrate (NO_3^-), hydrogen (H^+) and ammonium ions again offer a surprising conclusion: there is no significant west-east gradient for the pollution-related ions $SO_4^=$ and NO_3^-. There exists a

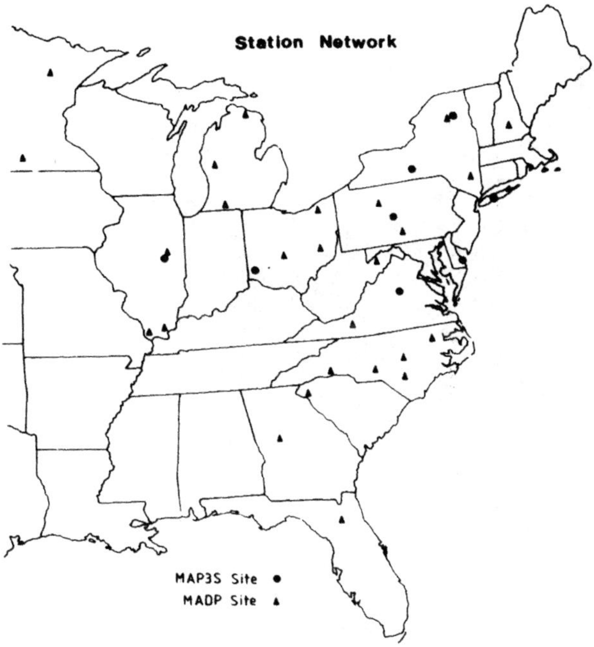

Figure 7 MAP3S and NADP station network used for 1979 spatial deposition analysis

Figure 7 (continued)

MAP3S and NADP Station Locations

MAP3S Sites

State	Site Name	Latitude	Longitude	Elevation (m)
New York	Whiteface	44.40	73.88	604
New York	Ithaca	44.38	76.72	503
Pennsylvania	Penn State	40.78	77.95	396
Virginia	Virginia	38.05	78.55	171
Illinois	Illinois	40.05	88.37	213
New York	Brookhaven	40.88	72.88	24
Delaware	Lewes	38.76	75.00	0
Ohio	Oxford	39.53	84.73	283
	NADP Sites			
Florida	Bradford Forest	29.97	82.20	44
Georgia	Georgia Station	33.18	84.40	268
Illinois	Bondville	40.05	88.37	213
Illinois	SIU	37.70	89.27	140
Illinois	Dixon Springs Ag. Ctr.	37.43	88.67	165
Michigan	U. Mich. Biological Sta.	45.57	84.68	235
Michigan	Kellogg Biological Sta.	42.48	85.38	289
Michigan	Wellston	44.22	85.85	291
Minnesota	Marcell Exp. Forest	47.50	93.47	430
Minnesota	Laberton	44.25	95.32	342
Nebraska	Mead	41.09	96.50	354
New Hampshire	Hubbard Brook	43.95	71.70	252
New York	Huntington Wildlife	44.00	74.22	500
New York	Stilwell Lake-West Point	41.35	74.03	185
North Carolina	Lewiston	36.13	77.17	25
North Carolina	Coweeta	35.02	83.45	220
North Carolina	Piedmont Research Sta.	35.67	80.57	251
North Carolina	Clinton Corps Res. Sta.	35.01	78.28	55
North Carolina	Finley	35.73	78.68	128
Ohio	Delaware	40.28	83.06	262
Ohio	Caldwell	39.78	81.52	275
Ohio	Wooster	40.77	81.93	312
Pennsylvania	Kane Exp. Forest	41.55	78.77	610
Pennsylvania	Leading Ridge	40.55	77.93	274
South Carolina	Clemson	34.67	82.83	231
Virginia	Horton's Station	37.18	80.42	322
West Virginia	Parsons	39.10	79.65	508

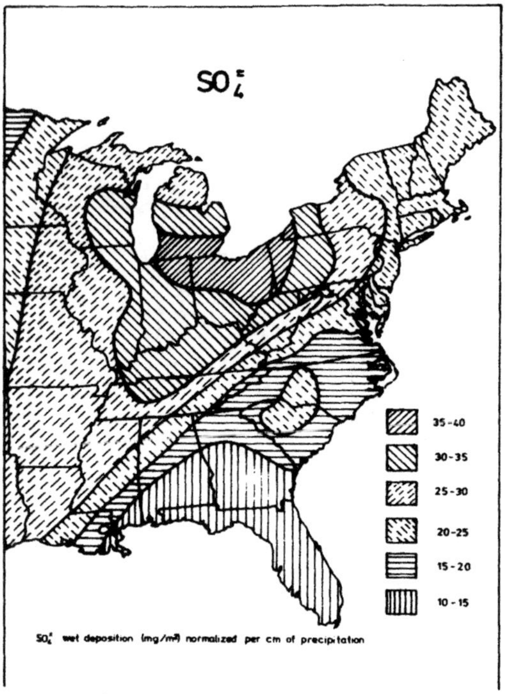

Figure 8.

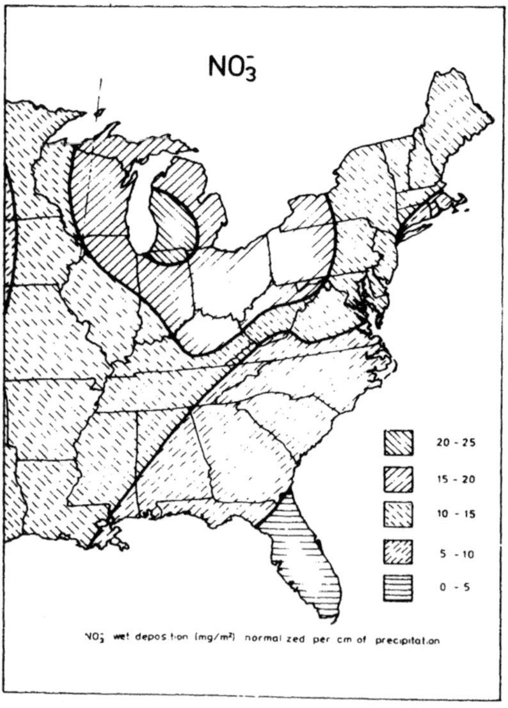

Figure 9.

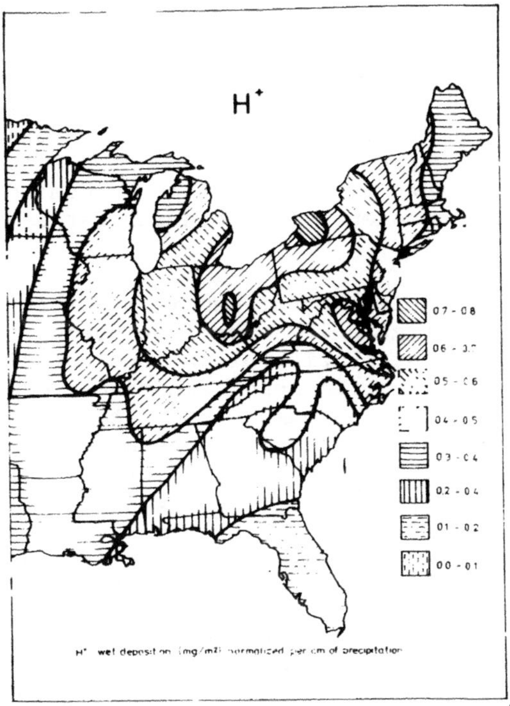

Figure 10.

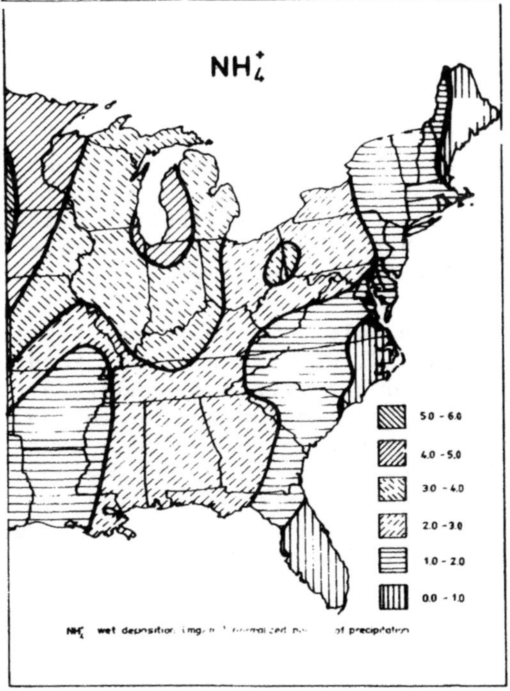

Figure 11.

rather uniform gradient in almost every direction from major emission areas that are located in the Midwest/Ohio Valley. Furthermore, no maxima can be observed in the northeastern United States in these normalized wet deposition values that would be indicative of increased wet deposition downwind of major emission areas. However, the effect of the Appalachian Mountains seems to be readily apparent, as evidenced by the strong wet deposition gradients in the northwest-southeast direction.

A slightly more complex picture emerges for the hydrogen ion deposition (H^+) in Figure 10. The free hydrogen ion deposition results from the contribution of various constituents including sulfuric acid, nitric acid, ammonium and soil-related components such as calcium and magnesium. Therefore, its spatial distribution can be expected to be more complex than the normalized wet deposition pattern of $SO_4^=$ of NO_3^- alone. This is indeed verified in Figure 11. Localized maxima in H^+ wet deposition are evident and a strong gradient emerges to the west of 85° longitude. One can speculate that soil-related material scavenged by precipitation might be responsible for the apparent change in free hydrogen ion deposition. This view is strongly supported by the normalized wet deposition emerging for the ammonium ion (see Figure 11). The normalized NH_4^+ deposition is largely a function of natural sources: increased NH_4^+ ion deposition to the west of 85° longitude correlates well with the decrease in free hydrogen ion deposition, since NH_4^+ acts as a "neutralizing agent."

Again, the concept of a regional "superbowl" can be used to describe the "obscured" source-receptor relationship (emission-wet deposition relationship) and to explain the relatively uniform west-east wet deposition distribution pattern for sulfate and nitrate ions (less than a factor of two to one between Midwest and Northeast deposition patterns).

We hope that we were able to demonstrate the complexity of acid precipitation. Indeed, we have a rather good knowledge of the emission of sulfur and nitrogen compounds. At the same time, we now seem to have good information on the wet deposition of nitrate and sulfate ions. But establishing a source-receptor relationship is still very difficult. We have not yet unravelled the complex chemistry involved in transforming SO_2 and NO_x into sulfates and nitrates. It is slowly being revealed that the processes are more complex than anything we have dealt with in air pollution so far. It involves the chemical conversion of gases such as SO_2, NO_x, hydrocarbons, etc. into material such as nitric acid and sulfuric acid that are picked up by particles, cloud droplets and hydrometeors. The rate-controlling step does not necessarily seem to be the absolute concentration of the precursor gases SO_2 and NO_x, but seems to be influenced either by meteorology, photochemical processes, the ratio of SO_2/NO_x, or other parameters such as hydrogen peroxide, soot, etc., or a combination of the above. However, the potential source-receptor relationship seems to be less obscured when individual precipitation events are considered. Preliminary results from our cloud chemistry program seem to indicate that a directional variability exists for clouds that have formed in air masses that have passed over different emission sources.

IV. CLOUD WATER COLLECTION - TECHNIQUES AND COMPOSITION

Measurements of ion concentrations in precipitation alone are not sufficient either to obtain an understanding of precipitation chemistry or to characterize changes in, and possible redistribution of, ion deposition resulting from changes in emissions. As a first step in a program to investigate the composition of cloud water, we began a measurement program of ions at our field station at Whiteface Mountain. This mountain observatory is within clouds about 50% of the time per year. The recent development of aircraft-mounted collectors at the Atmospheric Sciences Research Center can make important contributions in validating pollutant transformation and transport models. The techniques of cloud water collection used in the ASRC's summer cloud water pilot program will now be discussed and some preliminary results, intended to illustrate the value of these measurements to current research needs, will be described.

A. Ground-based cloud water collection

To permit near real time analysis on small cloud water samples (100 μl) we have developed a collection system, patterned loosely on a device used by Tabata, et al. (1), and shown in Figure 12 (Falconer and Falconer (2)). Cloud droplets collected by the vertical teflon filaments and support rods are sampled with efficiencies dependent on droplet size and wind speed (Figure 13). The accumulating water coalesces into millimeter-size droplets which then run down the filaments. The water is collected in a funnel and flows directly into the laboratory for ion chromatograph analysis. If winds exceed approximately 25 knots, the coalesced droplets, which can be blown off the filaments, are then captured by the central rods. Collection rates of a few ml/min are typical.

The site selected for the cloud water pilot project was the ASRC field station on the summit of Whiteface Mountain (1500 m) in the Adirondack Mountain region of New York State. In the past, only a few measurements of ion concentrations in clouds and fogs have been made with ground-based collectors (Figure 14). In all instances, the cloud water was collected in bulk and

Figure 12. The ASRC ground-based cloud water collector

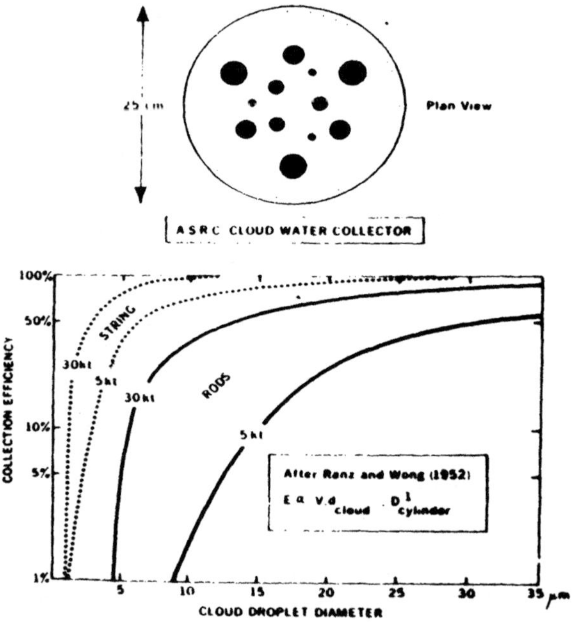

Figure 13. ASRC Cloud Water Collector design and efficiency curves

Investigator	Observation Period	Location	Sampling Method	Range of Ion Concentrations (ppm) pH	$SO_4^{=}$	NO_3^-	Cl^-	NH_4^+	Na^+
Houghton (1955)	1954	Five sites located in Northeastern U.S.	Cloud water mechanically intercepted on metal screen. Bulk samples.	4.5-7.2^{+}	0.2-125	-	0-205	-	-
Mrose (1966)	?	Four sites located In German Democratic Republic	Cloud water mechanically intercepted on pervious cloth. Bulk samples.	3.8-5.1^{+}	37-159^{+}	9-28^{+}	7-62^{+}	-	7-35^{+}
Okita (1968)	(a) July 25-26, 1963 (b) November 20, 1963	(a) Mt. Norikura (b) Mt. Tsukuba Japan	Cloud water mechanically intercepted on copper screen. Bulk samples.	(a) 3.4-4.3$^{@}$ (b) 5.6-6.5	11-60$^{@}$ 17-99	1.6-10.8$^{@}$ 0.3-2.3	2.6-8.1$^{@}$ 10-45	2.1-4.7$^{@}$ 2-17	1-3.8$^{@}$ 4-10
Lazrus et al. (1970)	1967	Pico del Oeste Puerto Rico	Cloud water mechanically intercepted on aluminum screen. Bulk samples.	4.9-5.4	3.4-9.1	-	5-70	-	-
Castillo (1979)	1976	Whiteface Mtn., New York	Cloud water mechanically intercepted on plastic strings and rods. Bulk samples	3.8±.4°	1-22	1-15	0.04-0.6	0-3.5	0.01-2
Falconer and Kadlecek	1979-1981	"	"	3.1-4.9*	1-171	0.5-54	0.02-0.6	0.07-7	0.03-0.6

$^{+}$Inter-event range of the reported mean values.
$^{@}$Intra-event range of observed values.
^{o}Range for all events mean pH ± one standard deviation.
*Inter-event range of mean
All others are inter-event range of values

Table 1. Summary of pH and dominant, inorganic ion concentrations in fog and cloud water collected at ground level.

analyzed for ion concentrations well after the collection was made. In no instances were ion deposition rates specified or was air mass origin characterized (trajectory analysis).

During the summer months of 1979 through 1981, analyses of the inorganic ions ($SO_4^=$, NO_3^-, Cl^-, F^-, H^+, Na^+, K^+) and solution conductivity were performed as part of our cloud water collection program. The pH and conductivity measurements were recorded continuously, while the concentrations of other ions were determined on a liquid ion chromatograph constructed at the Atmospheric Sciences Research Center (Bird, et al. (3)).

Two case studies have been selected to illustrate this approach. Figure 15 shows the temporal behavior of the dominant ions measured in the cloud water during August 29-30, 1979. Figure 16 shows the relative collection rates for water during the same period. The shaded peaks identify periods when the cloud water was diluted with rain blown into the collector. During periods of most intense shower activity, the concentration of all ions dropped by approximately a factor of five, suggesting that rain originating at higher elevations had substantially lower ion concentrations than clouds collected at 1.5 km. While concentrations of $SO_4^=$ and NO_3^- are typically a few tens of micromoles per liter in precipitation, they can be on the order of hundreds of micromoles per liter in clouds near cloud base.

Using trajectories generated by the Atmospheric Transport and Dispersion Model (NOAA, Air Resources Laboratory, Heffter (4)), we can determine the approximate regions which are likely to have an influence on the air masses arriving at Whiteface Mountain (Figure 17). The trajectories marked A, B and C describe the air flow history for parcels arriving near the beginning, middle and end of the event. The letters indicate six-hour intervals. As discussed extensively by Pack, et al. (5), it is not possible to assign an uncertainty to the air mass position vector (relative to Whiteface). However, the apparent uniformity of the air flows suggested by the trajectories, and supported by the 850 mb pressure level winds, indicates that air within which this cloud layer came to Whiteface from the west-southwest to southwest sector.

A second example of time dependent ion concentrations in cloud water (August 12, 1980) is shown in Figure 18. Showers occurred between the hours of 3 am and 6 am (EST) and briefly about 8:30 am, associated with the movement of a weak low system passing north of Whiteface across southern Canada. During the remainder of the collection period, only cloud water was sampled. Meteorological data tapes for August, 1980 are not yet available for mapping air mass trajectories. However, the flow of air to the summit of Whiteface Mountain can be roughly described by the

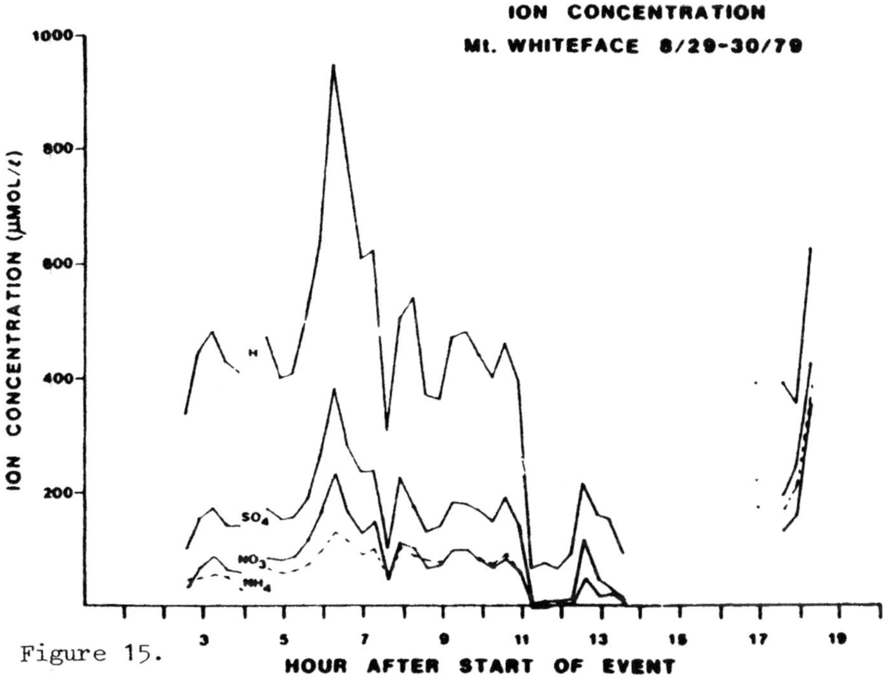

Figure 15.

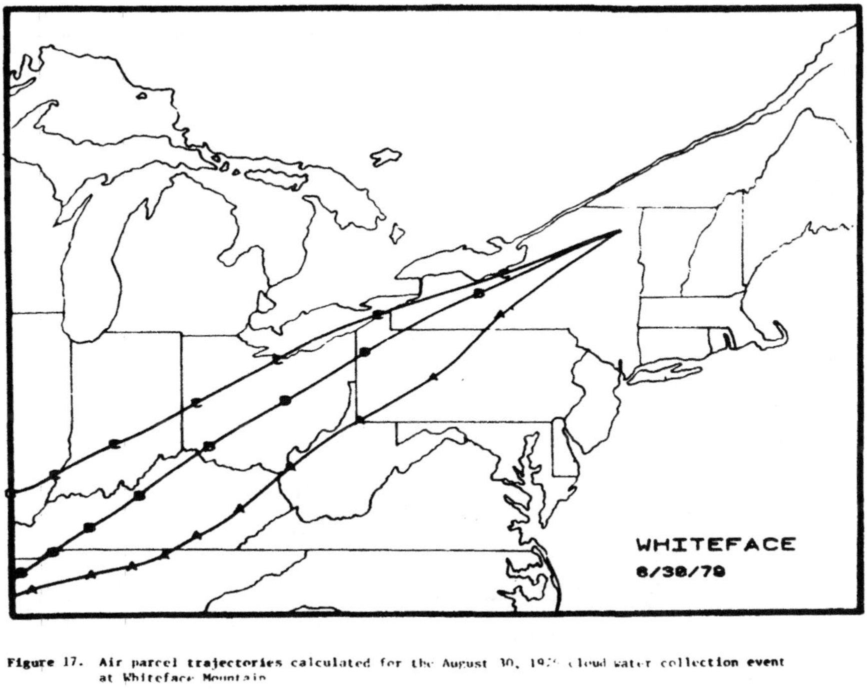

Figure 17. Air parcel trajectories calculated for the August 30, 1979 cloud water collection event at Whiteface Mountain

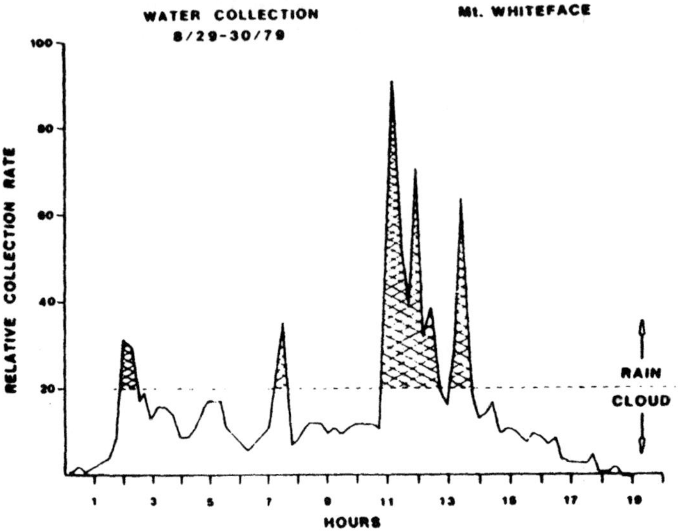

Figure 16. Water collection rate in the ASRC cloud water collector located at the summit of Whiteface Mountain, August 29-30, 1979.

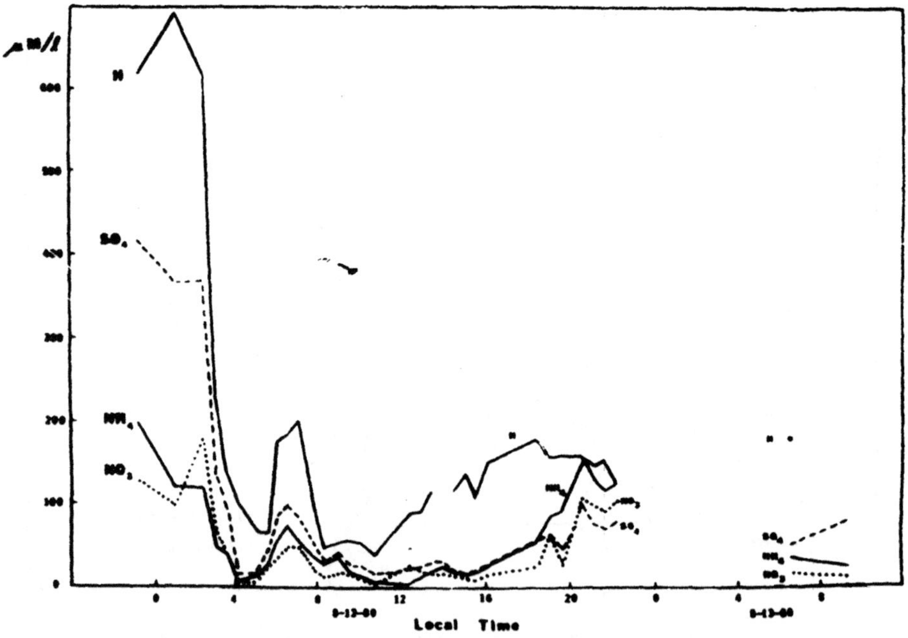

Figure 18. Time variation of the dominant ions in cloud water collected at Whiteface Mountain, August 12, 1980.

wind velocities measured at the summit and the 850 mb synoptic weather maps prepared at 7 am and 7 pm (EST) by the National Weather Service. Preceding the event, these estimated trajectories approached Whiteface from just south of the Great Lakes. By mid-day the low had moved to north of Lake Ontario with the air flow passing over the northern Great Lakes. Except for the first few hours of the event, southern and central Canada would appear to be the region most recently influencing the air mass.

The low ion concentrations observed during most of the event, notably during mid-day, are of particular interest. Normally, clouds tend to be patchy in character, creating the possibility of occasional evaporation of water from the surface of the collector. However, these low concentrations in a relatively light-cloud-density event give confidence to the much higher values observed in most of the other events studied.

Constant elevation collection can, as this data shows, provide time histories of incident cloud compositions suitable for direct composition studies. Also, classification by direction of wind flow may be more susceptible to interpretation than has been the case for precipitation (Wilson, et al. (6)), because less complicated atmospheric motions are involved for low level cloud development than for major precipitation events. To get improved transport and transformation rates, it is absolutely vital to establish a cloud chemistry data bank.

B. Airborne cloud collectors

Mounting cloud water collection devices on aircraft permits access to a wide variety of clouds not accessible to ground-mounted collectors. The principal challenge is to collect sufficient and representative water samples from nonprecipitating warm, supercooled, or glaciated clouds. The collection efficiency of any device needs to be high enough to obtain adequate sample for the desired analysis without sacrificing spatial resolution. Further, a cloud collector should avoid capturing ambient aerosol (radius $\lesssim$ 1 μm). It cannot avoid including some rain or splash, but rain-only collectors in the hands of an experienced observer can be used to identify when rain may be interfering. And finally, the collected and accumulated droplets must not be subjected to condensation or evaporation.

Prototypes using two different collection techniques--centrifugal impaction and forward stagnation--have been built by several investigators and are now being field tested. Notable designs of the first type are: 1) Kallend (7), who uses a fixed angle vane system which deflects the air flow causing the droplets to impinge on the wall of a two-inch cylinder. Any accumulated water is removed downstream through an extraction ring. Testing is currently being carried out as part of the Central Electricity Research Laboratories (CERL) European acid rain program; and 2) Scott (8) who introduced a rotating centrifuge ($\sim 10^4$ rpm) to force cloud water to the inside of an outer cone.

A design of the second type uses an air stagnation zone just in front of a slotted cylinder (Mohnen (9)). In this configuration seven cylindrical rods are arranged in two rows perpendicular to the flight direction as shown in Figure 19. Water from the remaining three backmost cylinders is collected separately should cylinder surface contamination be suspected.

The selected dimensions for flight speeds of 50 to 100 m/sec are: rod diameter = 0.476 cm, and slot width = 0.15 cm with total collection area depending upon the length of the slot as well as the number of rods. The ratio of the slot width to the rod radius cannot exceed 0.3, or the flow around the cylinder will be markedly disturbed (Langmuir and Blodgett (10)). The cloud droplets, captured on the inside walls of the slots, are protected from the passing air aerosols and trace gases and are drawn off to a removable bottle in the airplane through a capillary tube connected to each rod. In the initial flight and wind tunnel tests (using the cloud wind tunnel facility of Dr. R. Knollenberg, P.S.I., Boulder, Colorado), this device has been performing particularly well (efficiencies = 85 ± 15% with liquid water content of 0.65 g m^{-3}, speed = 60 m sec^{-1}, and mean droplet diameter of 12 µm; and no measurable evaporation of collected water occurred in the absence of a cloud--Mohnen (9). Little impaction of particles below 1 µm radius occurs, and essentially all cloud droplets above 8 µm will pass into the stagnation slot of the cylindrical rods.

The unit is held in place by a support panel mounted on the fuselage of the airplane and can be exchanged during flight within seconds from inside the airplane. If the cloud should be supercooled, the full surface of the rod exposed to the oncoming airstream will act as a collector and will ice over. In this case, the collector has to be exchanged until the sample can be removed by melting.

Shown in Figure 20 is a scaled-up version of the same collection concept to collect rain. The cylindrical rod is now sufficiently large to exclude cloud droplets and only the more massive raindrop will enter the slot, and will then flow into a bottle below. This gives the investigator the capability of separating rain from the cloud--a most valuable tool in studying atmospheric liquid phase chemistry.

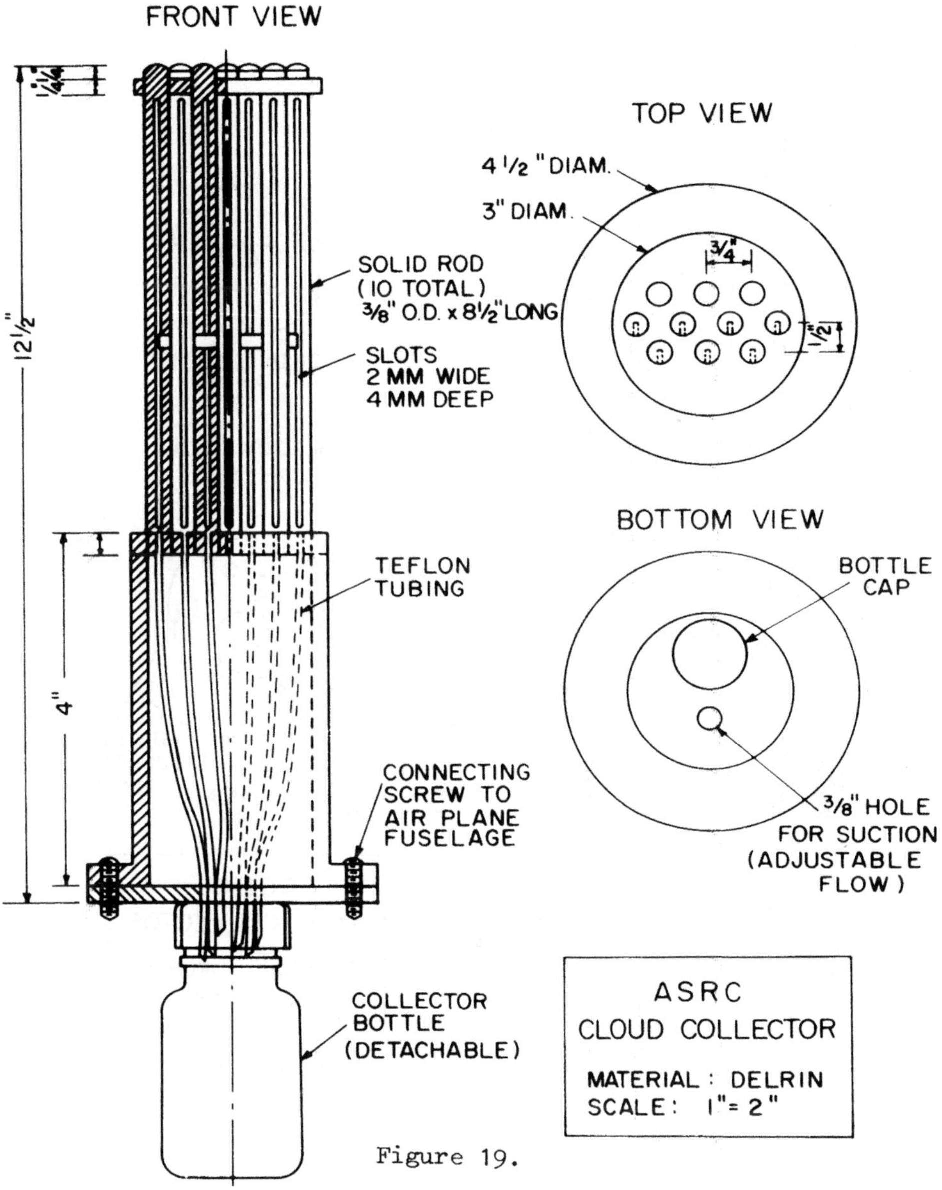
FRONT VIEW
12½"
4"
SOLID ROD
(10 TOTAL)
3/8" O.D. x 8½" LONG
SLOTS
2 MM WIDE
4 MM DEEP
TEFLON
TUBING
CONNECTING
SCREW TO
AIR PLANE
FUSELAGE
COLLECTOR
BOTTLE
(DETACHABLE)
TOP VIEW
4 1/2 "DIAM.
3" DIAM.
3/4"
1/2"
BOTTOM VIEW
BOTTLE
CAP
3/8" HOLE
FOR SUCTION
(ADJUSTABLE
FLOW)
ASRC
CLOUD COLLECTOR
MATERIAL: DELRIN
SCALE: 1"= 2"

Figure 19.

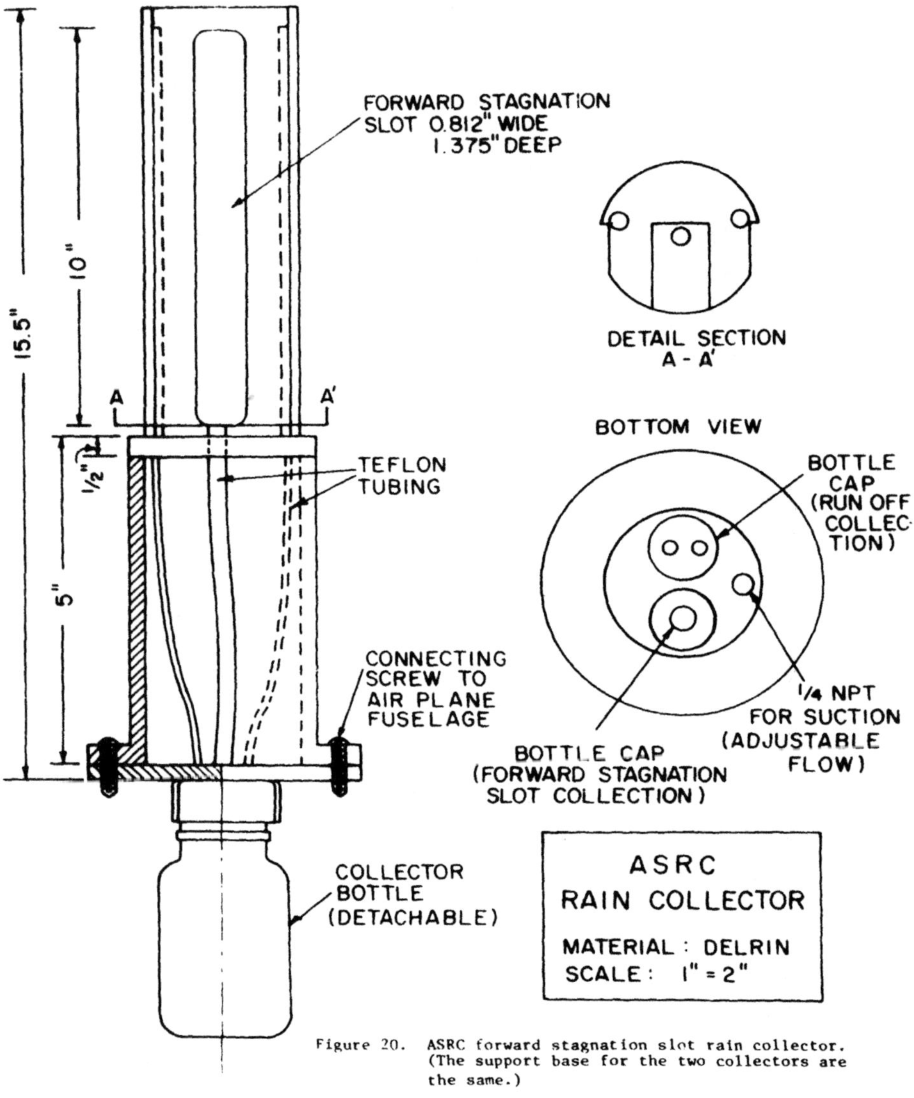

Figure 20. ASRC forward stagnation slot rain collector. (The support base for the two collectors are the same.)

The ASRC cloud collector was in-flight tested during the APEX (Acid Precipitation Experiment) August 8, 1980 flights downwind of Buffalo, New York. The concentrations obtained in cumulus clouds at 6000 ft. are shown in Figure 21, along with values obtained from clouds at Whiteface Mountain during the same event.

Even though the Buffalo cloud contained the greater anion concentrations, neutralization provided by relatively local emissions were sufficient to dominate cation chemistry. As this data shows, regional plume studies, investigations into the variability of transformation rates, and pollutant transport are now possible with improved sensitivity and reliability.

ACKNOWLEDGEMENTS

The ASRC cloud and precipitation research efforts are supported by the National Science Foundation under grant ATM-7925435, the Office of Naval Research under grant N0001476C0283, and the Department of Energy under grant DEAC0276EV02986.

DOWNWIND OF BUFFALO AT 6,000 FEET (CUMULUS)

	Ca^{++}	Mg^{++}	K^{+}	Na^{+}	NH_4^{+}	$SO_4^{=}$	NO_3^{-}	Cl^{-}	$PO_4^{\equiv}$	pH
meq/liter	0.90	0.17	0.044	0.26	0.21	0.65	0.38	0.20	0.003	5.42

WHITEFACE MOUNTAIN SUMMIT-BASED CLOUD COLLECTOR

	Ca^{++}	Mg^{++}	K^{+}	Na^{+}	NH_4^{+}	$SO_4^{=}$	NO_3^{-}	Cl^{-}	$PO_4^{\equiv}$	pH
meq/liter	0.025	0.008	0.003	0.004	0.10	0.27	0.10	0.005	-	3.67

Figure 21.

Comparison of two clouds during August 8, 1980.
Analysis performed by G. Likens, Cornell University,
with the cooperation of A. Lazrus, NCAR

REFERENCES

(1) Tabata, T., Huzioka, T., and N. Matsumura: 1953, Studies on Fogs, T. Hori (Ed.), Tanne Trading Co., Publisher, Sapporo, Japan.

(2) Falconer, R.E., and Falconer, P.D.: 1980, J. Geophys. Res., 85, pp. 7465-7470.

(3) Bird, D., Kadlecek, J.A., Mohnen, V.A., and DelSanto, J.H.: 1979, ASRC-SUNY Pub. No. 713.

(4) Heffter, J.L.: 1980, NOAA Tech. Memo ERL-81.

(5) Pack, D.H., Ferber, G.J., Heffter, J.L., Telegadus, K., Angell, J.K., Hrecker, W.H., and Machta, L: 1978, Atmos. Environ., 12, pp. 425-444.

(6) Wilson, J.W., Mohnen, V.A., and Kadlecek, J.A.: 1981, accepted for publication in Atmos. Environ.

(7) Kallend, A.S.: 1979, First annual report to EPRI, Central Electricity Research Laboratories, RD/L/R 1992.

(8) Scott, W.D.: 1978, Atmos. Environ., 12, pp. 917-924.

(9) Mohnen, V.A.: 1981, Atmos. Tech., NCAR No. 12, pp. 20-25.

(10) Langmuir, I., and Blodgett, K.: 1961, Collected Works of Irving Langmuir, Vol. 10, Pergamon Press.

(11) Castillo, R.: 1979, Ph.D. Thesis, Dept. of Atmospheric Science, State University of New York, Albany, NY.

(12) Houghton, H.: 1955, J. Meteor., 12, pp. 355-357.

(13) Kadlecek, J.A., and Mohnen, V.A.: 1975, 1st Intl. Symposium on Acid Precipitation and the Forest Ecosystem, Columbus, OH.

(14) Lazrus, A.L., Baynton, H.W., and Lodge, J.P., Jr.: 1970, Tellus, 22, pp. 106-114.

(15) Mrose, H.: 1966, Tellus, 18, pp. 266-270.

(16) Okita, T.: 1968, J. Met. Soc. Japan, 46, pp. 120-126.

THE CHEMISTRY OF THE POLLUTED TROPOSPHERE

Jack G. Calvert

Department of Chemistry
The Ohio State University, Columbus
Ohio, U.S.A.

ABSTRACT:

Several important aspects of the chemistry of the polluted troposphere are discussed in terms of the elementary reactions involved. First, the mechanism of ozone generation in the polluted atmosphere is reviewed briefly. Limitations on the use of the approximate photostationary state equation, $[O_3][NO]/[NO_2] \simeq k_1/k_3$, are discussed. Then the effects of the levels of the impurities, NO, NO_2, CO, hydrocarbons, and aldehydes on the maximum ozone development are considered through computer modelling of the chemistry of simulated atmospheres. These results are used to illustrate the development of control strategy for ozone in the polluted troposphere. The simulations are used also to predict the transformations and product distributions of nitrogen-containing products resulting from the oxidation of NO and NO_2 in the atmosphere. Finally the mechanisms of acid development in the polluted atmosphere are considered. The effect of the various pollutant levels on some of the important homogeneous and heterogeneous pathways for SO_2 oxidation are considered from the standpoint of the elementary steps involved. Interrelationships between the homogeneous and heterogeneous pathways of SO_2 oxidation are noted.

I. INTRODUCTION

Our understanding of the chemistry of the polluted troposphere has improved significantly in recent years as a result of well designed field and labaratory studies. New insights have allowed important mechanistic detail to be added to the chemical reactions

H. W. Georgii and W. Jaeschke (eds.), Chemistry of the Unpolluted and Polluted Troposphere, 425–456.

included in predictive models for the urban air shed and long range transport of polluted air masses. However, a realistic view of the extremely great complexity of the mixture of reactants in urban air and the remaining uncertainties in several aspects of the chemistry of the known impurities, continue to limit our confidence in present tropospheric chemical models. In this work the author will review briefly several important aspects of the chemistry of the polluted troposphere with two major goals in view. First, some discussion of the fundamental chemistry will be given for those scientists who are newcomers or potential workers in the area. Secondly, the anatomy of this chemistry will be sought in terms of the elementary reactions involved by utilyzing computer simulations of these complex chemical systems. We will attempt to identify the role of the various impurities and transient species in the development of ozone and identify simple strategies for its control. A brief discussion will be given of the chemistry of acid formation in the troposphere and the factors which govern these rates.

II. THE CHEMISTRY OF OZONE DEVELOPMENT

The typical urban atmosphere receives a variety of trace impurities as a result of the emissions from vehicles, power plants, industrial operations, and other human activities, as well as the natural sources such as trees, plants and soils. Commonly these are rich in oxides of nitrogen (NO, NO_2) and sulfur (SO_2), a great variety of hydrocarbons (alkenes, alkanes, and aromatics) and their oxidation products such as the aldehydes and ketones. The action of sunlight upon this complex mixture may result in the creation of secondary pollutants whose properties are often much more objectionable than those of the original brew. Thus ozone,peroxyacetylnitrate, nitric and sulfuric acids, and other reactive compounds may form. We may illustrate some of the chemistry occurring in a polluted atmosphere from the observations made during the LARPP study in Los Angeles in 1973 [1]. Instrumented helicopters sampled impurities in well defined parcels of air as they moved about the Los Angeles basin under the influence of prevailing winds.shown in Figure 1 are the concentrations of ozone and nitrogen oxides: Observed at an altitude of about 400 ft above ground level. All measurements shown here were made below the inversion layer on this day. Ozone levels near 40 ppb in the early morning hours are seen to decrease somewhat as the ozone is titrated by the insertion of NO-rich pollution from the morning traffic. The ozone concentration then rises throughout the day, reaching about 140 ppb at the 1400 hour. The NO levels of the early, high traffic hours decreases rather quickly, at a faster pace than that of the NO_2. This is seen more clearly in the data of Figure 2 in which the $[NO]/[NO_2]$ ratio is shown as a function of time for analyses made at several altitudes

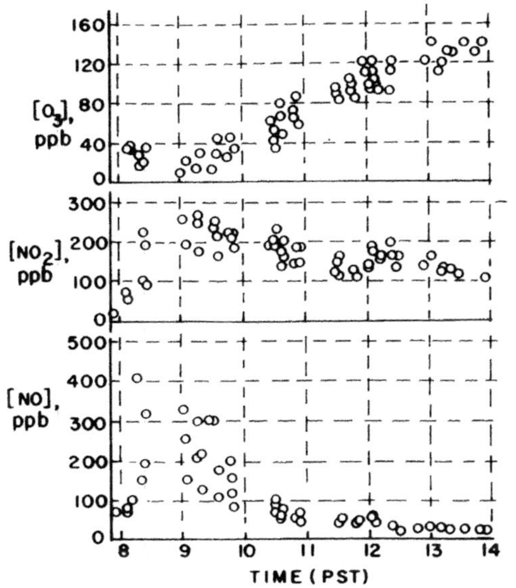

FIGURE 1. Concentrations of O_3, NO_2, and NO versus time as measured during the LARPP project in Los Angeles California in a tetroon-tagged air mass of November 5, 1973; these measurements were made at about 400 ft above ground level in LARPP operation #33.

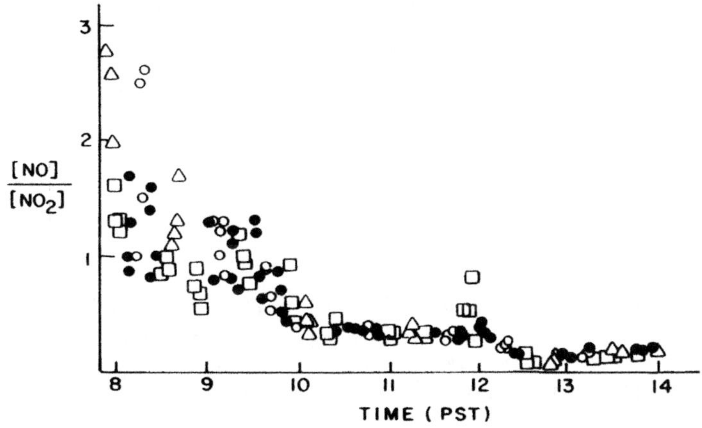

FIGURE 2. $[NO]/[NO_2]$ ratio as a function of time in the LARPP study #33; the symbols designate different elevation regions monitored (AGL): triangles, 768-838 ft; squares, 600-665 ft; darkened circles, 420-437 ft; open circles, 218-230.

within the marked air mass. Thus it seems evident that NO is converted to NO_2 as the O_3 rises. The hydrocarbons also undergo chemical change during this period. Note in Figure 3 for typical data collected at an elevation of 400 ft (AGL) in this same air mass, that the ratio of $[C_3H_6]/[C_2H_2]$ decreases with time. Acetylene is a very much less reactive compound toward O_3, $O[^3P]$,

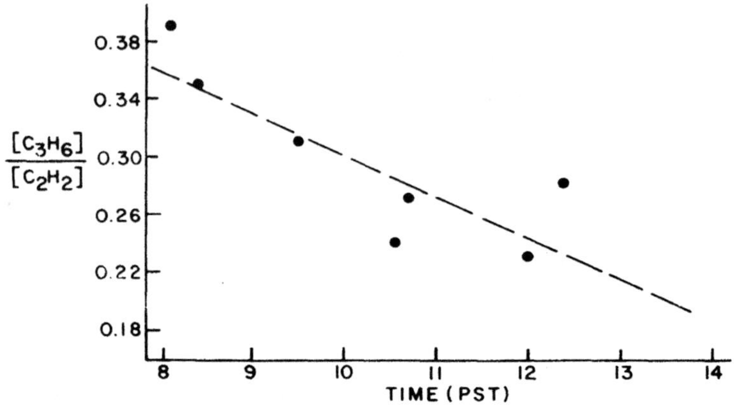

FIGURE 3. Plot of the ratio $[C_3H_6]/[C_2H_2]$ versus time for the LARPP operation #33 data obtained at 420-437 ft (AGL).

and OH radicals than is C_3H_6, and hence it can be used to normalize the data to isolate changes in concentration because of diffusion and dilution from the chemistry involved.

The chemical mechanism of ozone generation within an NO, NO_2, hydrocarbon, aldehyde-polluted air mass appears to be well understood [1,2]. The important sequence of reactions (1) to (6) can be used to rationalize O_3, NO, and NO_2 concentrations in such a highly polluted, sunlight-irradiated, urban atmosphere:

$$NO_2 + h\nu[\lambda < 3400\ A] \rightarrow O[^3P] + NO \quad (1)$$
$$O(^3P) + O_2 + M \rightarrow O_3 + M \quad (2)$$
$$O_3 + NO \rightarrow O_2 + NO_2 \quad (3)$$
$$O_3 + NO_2 \rightarrow O_2 + NO_3 \quad (4)$$
$$O_3 + ALKENE\ [RH] \rightarrow PRODUCTS \quad (5)$$
$$O_3 + h\nu[\lambda < 3100\ A] \rightarrow O[^1D, ^3P] + O_2[^1\Delta_g, ^3\Sigma_g^+] \quad (6)$$

Thus the approximate instantaneous rate of ozone generation in the absence of significant atmospheric turbulence, heterogeneity, dilution, or diffusion within the turbulent layer should be given by (7):

$$\frac{d[O_3]}{dt} = [NO_2]k_1 - [O_3]\{[NO]k_3 + [NO_2]k_4 + [RH]k_5 + k_6\} \quad (7)$$

In a highly polluted atmosphere the rate of change of O_3 with time, $d[O_3]/dt$, is often small compared to the first two terms on the right of equation (7). For these circumstances we expect the approcimate relation (8) to apply:

$$[O_3] \simeq [NO_2]k_1/\{[NO]k_3 + [NO_2]k_4 + [RH]k_5 + k_6\} \quad (8)$$

For many conditions encountered in highly polluted atmospheres, the terms involving k_4, k_5, and k_6 are small compared to that containing k_3, and the approximate relation (9) may be used:

$$[O_3][NO]/[NO_2] \simeq k_1/k_4 \quad (9)$$

It is possible to estimate k_1 from a knowledge of the wavelength dependence of $O[^3P]$ quantum yield of formation in (1), the absorption coefficients of NO_2, and the wavelength dependence of the solar irradiance within the air parcel of interest. Of course experimental values of k_3 are available for any temperature of interest. Thus the ratio of k_1/k_3 can be calculated in theory and checked against the measured values of $[O_3][NO]/[NO_2]$ within the limits imposed by theory. One such test has been made by Calvert (1a) using the LARPP data of which that in Figures 1, 2, and 3 are part. These data are shown in Figure 4 where the solid curve represents the theoretical value of k_1/k_3 for the conditions of the experiment, and the vertical lines show the spread (within one standard deviation) of the average of the $[O_3][NO]/[NO_2]$ values measured at specific times along one side of the square flight pattern around the air mass. Within the large error which is inherent in the measurements, the results appear to be consistent with theory for data collected after 0900 hour. Before this time large positive deviations form the simple theory are seen. This has been interpreted in terms of the limited response time of the measurement equipment coupled with the sampling of very nonhomogeneously mixed pockets of ozone- rich air from above with NO-rich pockets of air from below [1,3]. It should be noted that the use of the simple relaton (9) allows a rapid and convenient

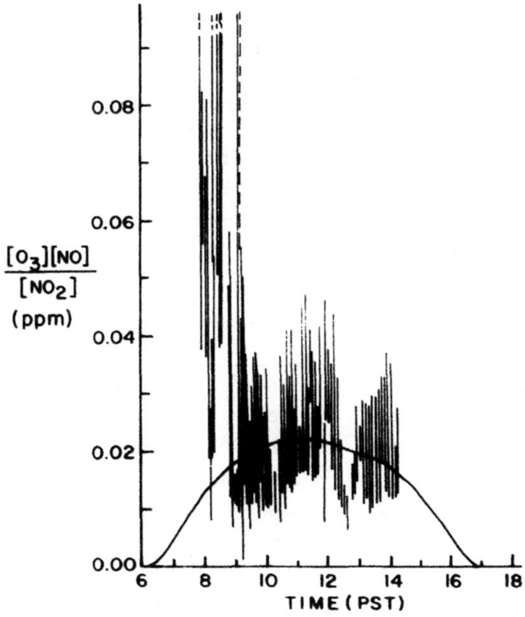

FIGURE 4. Plot of the function $[O_3][NO]/[NO_2]$ as estimated from experimental data from the LARPP operation #33; length of the vertical lines indicate the range of standard deviation of the average over given leg of flight; the solid curve is the theoretical function k_1/k_3 calculated by Calvert [1a].

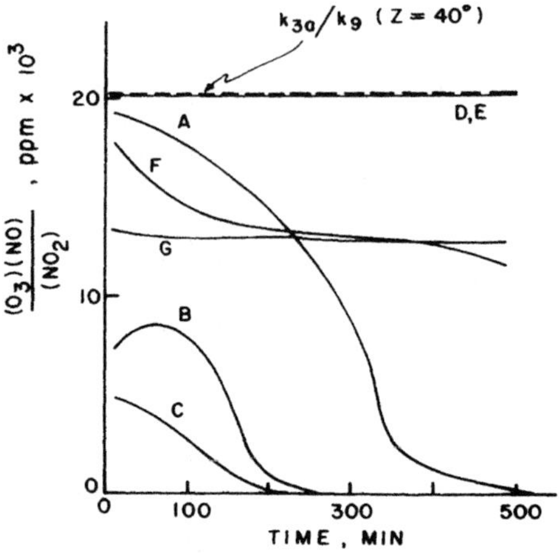

FIGURE 5. Plot of the function $[O_3][NO]/[NO_2]$ versus time for computer simulated irradiations of polluted air masses; the cases A through G shown correspond to the initial concentrations outlined in Table I.

check on the O_3 levels which can be anticipated in a sunlight-irradiated air mas which is rich in NO. However there are many circumstances which are commonly encountered in a polluted troposphere under which relation (9) is expected to fail. Computer simulations of the chemistry of synthetic polluted atmospheres for a variety of impurity levels show the size of deviations which can be expected in the use of relation (9). See Figure 5. Here the intensity of the solar radiation has been fixed at that expected for a solar zenith angle of 40°, and initial pollutant levels were varied as shown in the Table I.

The atmospheres wich are initially rich in NO and NO_2 (cases A, D, and E) with varied alkene, alkane, and aldehyde levels, all follow the approximate relation (9) quite well; see Figure 5. However for case A (the highest hydrocarbon levels) the negative deviations from relation (9) become significant at fairly short times. All other cases shown in Figure 5 have strong deviations even during the early part of the day. In case C of very low NO and NO_2 levels and high hydrocarbon levels, the deviations from (9) is most striking. The reasons for this failure can be identified in

Table I. Initial Concentrations of the Pollutants Used in Various Computer Simulations of the Chemistry of Polluted Atmospheres; $(O_3)° = 30$ ppb

Case	Initial Concentrations, ppb						$\tau_{99} > 5$ min	(Alkene)/(NO) > 160
	(NO_2)	(NO)	(CH_2O)	(CH_3CHO)	(Alkene)	(Alkane)		
A	25	75	20	10	100	400	16 min	315 min
B	2.5	7.5	20	10	100	400	0	40
C	0.25	0.75	20	10	100	400	0	<10
D	25	75	2	1	10	40	380	>480
E	25	75	0.2	0.1	1	4	>480	>480
F	2.5	7.5	2	1	10	40	0	>480
G	0.25	0.75	0.2	0.1	1	4	0	>480

these cases, and they point up the practical limitations on the use of (9). We will consider two of the important criteria for failure of (9) in these examples. First if the [Alkene]/[NO] ratio is greater than about 160 then the rate of reaction of O_3 with alkene becomes equal to or greater than about 10 % of its rate with NO in reaction (3), and it is no longer negligible. Note in the last column of Table I that this is the situation at very short times for cases C and B with highest [RH]/[NO] ratios chosen. In all other cases failure did not result from this cause. However there is another important requirement which must be met before relation (9) can hold well: the time period to generate a photostationary state of O_3, NO, and NO_2 must be short compared to the time for a significant change in any one of the species to occur. For the common case of dominance of the ozone through reaction with NO, a rough measure of the time required to reach 99 % of the photostationary state can be had from the relation: $\tau_{99} \simeq 4.6/[NO]k_3$. If τ_{99} becomes equal to or greater than a few minutes (say 5 min) then the photostationary state will not be reached or maintained during this short time period, and relation (9) will fail. At 25° this is the case for [NO] < 40 ppb. Shown in Table I are the times at which $\tau_{99} > 5$ min for the various simulations. It is seen that for all cases considered other than A, D, and E, one expects failure of the relation (9). Thus in summary we note that one must not be surprised with failure of the measured $[O_3][NO]/[NO_2]$ to equal k_1/k_3 in relatively clean air which is low in [NO]. We expect this in terms of theory.

III. THE EFFECTS OF THE IMPURITIES NO, NO_2, HYDROCARBONS, AND ALDEHYDES ON THE LEVEL OF OZONE DEVELOPMENT IN THE HIGHLY POLLUTED TROPOSPHERE

We have seen that the ozone levels present in the sunlight-irradiated, NO-rich, highly polluted atmosphere can be anticipated well from the relation (9) and the $[NO_2]/[NO]$ ratio. However note that the ozone level continues to climb slowly throughout the day, reflecting a rise in $[NO_2]/[NO]$ ratio. Obviously the conversion of NO to NO_2 occurs in the atmosphere, and it is the mechanism of this very important process which concerns us in this section. It has been well established both trough so-called "smog chamber" experiments and theoretical considerations using computer modelling of the chemistry of these systems, that hydrocarbons and their many oxidation products, e.g., the aldehydes and carbon monoxide, are instrumental in effecting the NO to NO_2 conversion. Obviously when in the photostationary state, interplay between NO, O_3,NO_2, and sunlight through reactions (1) and (3) does not result in any net NO to NO_2 conversion. Other reactions must be involved. At the very low levels of NO involved in the ambient troposphere, the thermal oxidation of NO by O_2 [$2NO + O_2 \rightarrow 2NO_2$] is much too slow to be a significant conversion mechanism. In the past twenty

Table II. The Formation of Highly Reactive Intermediate Species in Tropospheric Chemistry

O_2 + sulight (12700 A) → $O_2(^1\Delta_g)$

(7600 A) → $O_2(^1\Sigma_g^-)$

NO_2 + sunlight (λ < 4300 A) → $O(^3P)$ + NO

$O(^3P) + O_2 + M \rightarrow O_3 + M$

$O_3 + NO \rightarrow O_2 + NO_2$

$O_3 + NO_2 \rightarrow O_2 + NO_3$

$NO_3 + NO_2 \rightleftharpoons N_2O_5$

O_3 + sunlight (λ < 3100 A) → $O(^1D) + O_2$

(λ < 7000 A) → $O(^3P) + O_2$

$O(^1D) + H_2O \rightarrow 2OH$

$NO + NO_2 + H_2O$ (surface?) $\rightleftharpoons$ 2HONO

HONO + sunlight (λ < 4000 A) → OH + NO

$OH + CO \rightarrow H + CO_2$

$H + O_2 + M \rightarrow HO_2 + M$

$HO_2 + NO \rightarrow OH + NO_2$

$OH + CH_4 \rightarrow H_2O + CH_3$

$CH_3 + O_2$ (+M) → CH_3O_2 (+M)

$CH_3O_2 + NO \rightarrow CH_3O + NO_2$

$CH_3O + O_2 \rightarrow CH_2O + HO_2$

CH_2O + sunlight (λ < 3380 A) → H + HCO

$HCO + O_2 \rightarrow HO_2 + CO$

→ $HCOO_2$ (+M)?

$O_3 + C_2H_4 \rightarrow [CH_2{-}O{-}O{-}O{-}CH_2]^\dagger \rightarrow CH_2O_2 + CH_2O$

$OH + NO_2 + M \rightarrow HONO_2 + M$

$OH + NO + M \rightarrow HONO + M$

$HO_2 + NO_2 + M \rightarrow HO_2NO_2 + M$

$CH_3CHO + OH \rightarrow CH_3CO + H_2O$

$CH_3CO + O_2 \rightarrow CH_3COO_2$

$CH_3COO_2 + NO \rightarrow CH_3CO_2 + NO_2$

$CH_3COO_2 + NO_2 \rightarrow CH_3COO_2NO_2$ (PAN)

years a great deal of insight into the nature of these reactions has been gained.
In a sunlight-irradiated, polluted troposphere a great variety of highly reactive transient species are formed through both thermal and photochemical reactions involving the impurity species. Some of these interesting species are shown for illustration in Table II. These include electronically excited molecular oxygen, $O_2(^1\Delta_g)$, $O_2(^1\Sigma_g^-)$,oxygen atoms, both ground state $O(^3P)$ and electronicaly excited $O(^1D)$, nitrogen trioxide (NO_3), dinitrogen pentoxide (N_2O_5), nitrous acid (HONO), hydroxyl radicals (OH), H-atoms, hydroperoxy radicals (HO_2), alkyl radicals (e.g., CH_3, C_2H_5, etc..), alkylperoxy radicals (e.g. CH_3O_2, $C_2H_5O_2$, etc.), alkoxy radicals (e.g., CH_3O, C_2H_5O, etc.), acyl radicals (e.g., HCO, CH_3CO, etc.), acylperoxy radicals (CH_3COO_2, etc.), ozonides $CH_2 \frac{\diagup O\text{-}O\text{-}O \diagdown}{} CH_2$, Criegee intermediates (CH_2O_2, etc.), peroxyacetylnitrate or PAN ($CH_3COO_2NO_2$), and many other species. Much of the simple chemistry of these reactive fragments and compounds has been reasonable well studied today, and one can evaluate realistically their contribution to the NO to NO_2 conversion pathways. Such studies have shown clearly that the hydroperoxy (HO_2), alkylperoxy (RO_2), and the avcylperoxy radicals ($RCOO_2$) are the dominant primary driving force for NO to NO_2 conversion here, and these originate largely from the attack of the important hydroxy (OH) radical on hydrocarbon (RH), aldehyde (RCHO), and CO impurity molecules. Typical generallized reactions which occur here are the following:

$$OH + RH \rightarrow H_2O + R \qquad (10)$$
$$R + O_2 \rightarrow RO_2 \qquad (11)$$
$$RO_2 + NO \rightarrow RO + NO_2 \qquad (12)$$
$$OH + RCHO \rightarrow H_2O + RCO \qquad (13)$$
$$RCO + O_2 \rightarrow RCOO_2 \qquad (14)$$
$$RCOO_2 + NO \rightarrow RCO_2 + NO_2 \qquad (15)$$
$$RCO_2 \rightarrow R + CO_2 \qquad (16)$$
$$CH_2O + OH \rightarrow HCO + H_2O \qquad (17)$$
$$HCO + O_2 \rightarrow HO_2 + CO \qquad (18)$$
$$HO_2 + NO \rightarrow OH + NO_2 \qquad (19)$$
$$OH + CO \rightarrow H + CO_2 \qquad (20)$$
$$H + O_2 + M \rightarrow HO_2 + M \qquad (21)$$

Alkoxy radicals (RO) formed in (12) can generate efficiently the HO_2 radical; for example, the CH_3O radical reacts largely by (22) when in the lower troposphere:

$$CH_3O + O_2 \rightarrow HO_2 + CH_2O \qquad (22)$$

When one couples NO oxidation steps such as (12), (15), and (19)

with regeneration of OH in (19), and reformation of peroxy radicals in (10), (13), (17), and (20), then one has the elements of a chain reaction which can occur over again as NO is converted to NO_2 and hydrocarbon, aldehyde and CO react in these free radical processes.
The cycle of events is seen more clearly in the following diagramm:

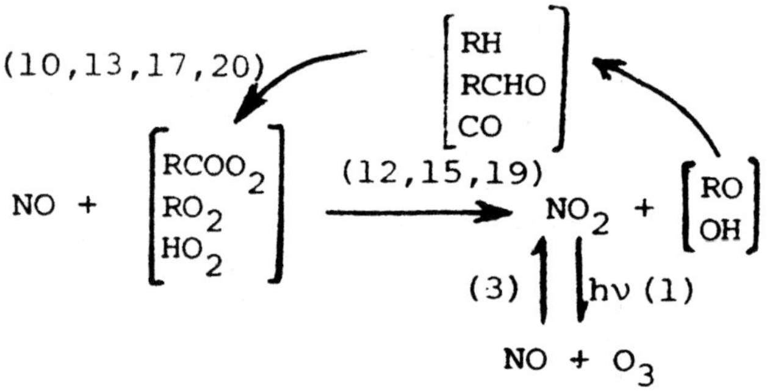

In the highly polluted troposphere the O_3 level is established largely through the rapid reactions (1) and (3), but the cycle of (10), (13), (17), (20) followed by (12), (15), and (19), followed by (10), (13), (17), and (20) etc., raises the $[NO_2]/[NO]$ ratio and accordingly raises the $[O_3]$ as required in relation (9).
We cannot claim that we understand well all of the detailed reactions that make up this NO to NO_2 conversion sequence, although the general features as outlined above can account reasonably well for laboratory and field measurements made to date. The limitations to our understanding will become somewhat clearer when one notes the extreme complexity of the reactant mixture in an urban atmosphere. Thus Laub and Smith [4] have resolved chromatographic peaks corresponding to over 850 different compounds in a sample of Shell unleaded gasolene. Not only does a small portion of these hydrocarbon enter the urban atmosphere by evaporation and incomplete combustion, but the many different products of partical oxidation of each compound form to creat a mixture of almost incomprehensible complexity when joined with the many other emissions from other human activities.
Most of these compounds undergo a series of reactions initiated by the OH-radical, and the subsequent primary reactions of the radicals formed are not only diverse but often ill-defined. For example in Table III we have illustrated some of the variety of degradation pathways for a simple hydrocarbon, n-butane. Here the primary attack of the OH-radical on butane creats either the secondary-butyl radical (about 78 % of the time) or the n- butyl radical (about 22 %). Note in Table III that every structurally allowable aldehyde and ketone are formed as well as difunctional hydroxaldehydes, etc. Ozone attack on the alkenes also leads to

Table III. Typical Hydrocarbon Reaction Pathways in the Polluted Troposphere

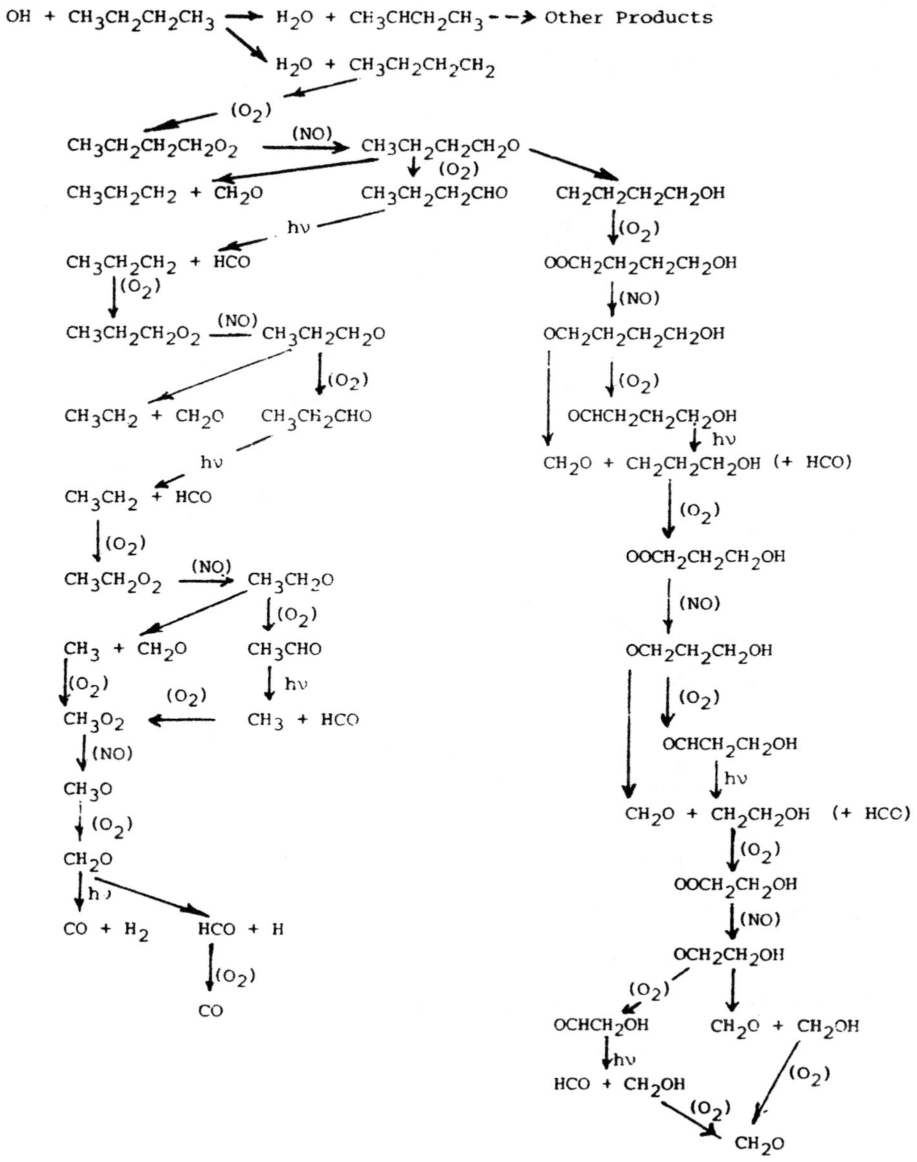

Table IV. Alkene-Ozone Reactions in the Polluted Atmosphere (ΔH in parentheses)

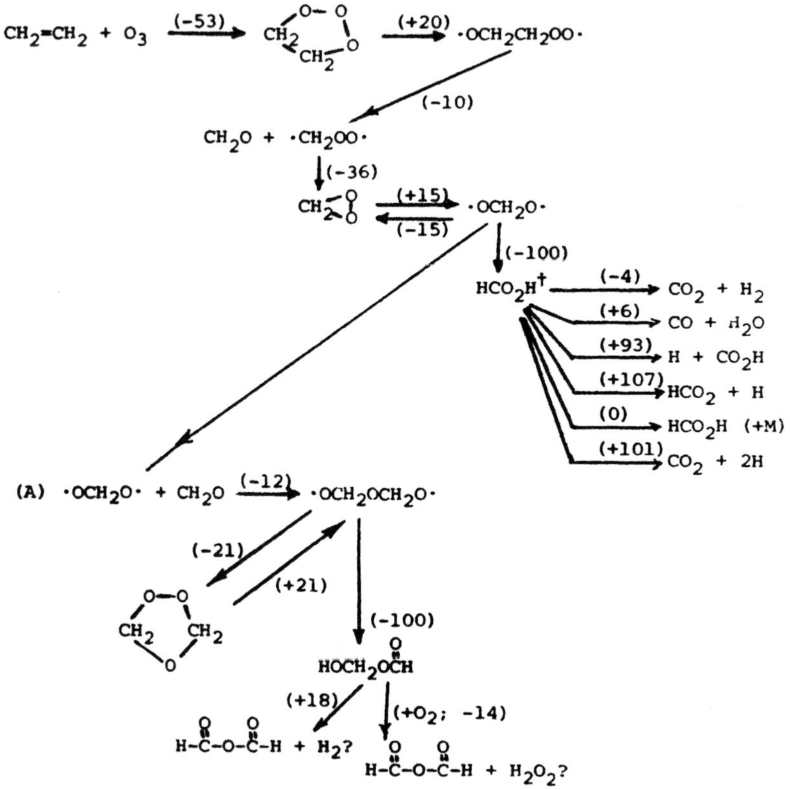

(B) $\cdot OCH_2O\cdot + SO_2 \longrightarrow (OCH_2OSO_2) \longrightarrow CH_2O + SO_3\ (H_2SO_4)$

(C) $\cdot OCH_2O\cdot + CO \longrightarrow (OCH_2OCO) \longrightarrow (HCO)_2O$

Relative Rate Constants: $k_A:k_B:k_C::100:400:0.7$[7d]

(D) $\cdot OCH_2O\cdot + NO \longrightarrow (CH_2O_2NO) \longrightarrow CH_2O + NO$

k_D is unknown today.

complex degradation pathways which give many new species in addition to the aldehydes and ketones; these include ozonides, Criegee intermediates (CH_2O_2, CH_3CHO_2, etc.), acids, etc. Some typical reactions in the sequence involving the simplest of the alkenes, ethylene (C_2H_4) are summarized in Table IV for illustration. Note that the rate constants for the reaction of NO oxidation by the Criegee intermediated (e.g.,$CH_2O_2+NO \rightarrow CH_2O+NO_2$) are unknown today, and the values which are employed in models are, et best, educated guesses.

Remember that each of the more than 840 hydrocarbons of the gasolene analysis and the axtremely number of partial oxidation products of these, coupled with the thousands of other volatile organic compounds released from the many non-combustion and other combustion sources, wich enter the atmosphere have different complex degradation pathways which lead to literally hundreds of thousands of different chemical species in a polluted troposphere. The nature of the general pattern of these reactions can be theorized reasonably well, but many reactions are not well characterized, e.g., the competitive reactive chanals for many of the alkoxy radicals is unknown. It is clear that we have little hope today of incorporating all of the detailed chemistry which occurs in a polluted troposphere into models which we develop to describe it [5]. Model development based upon generallized reaction schemes has progressed through the years, particularly in relation to ozone control strategies. At this stage of our understanding of the many complex reaction schemes there are no reasonable alternatives which we can call upon. However, we should recognize that the output from models which employ highly simplified reaction schemes may not simulate well the happenings in many real atmospheres of unusual compositions (e.g., those containing a dominance of terpenes or alkanes). It is also evident that any highly simplified model cannot be used for predictive purloses beyond those for which it has been parameterized and "validated".

A tropospheric chemical reaction scheme can be a useful bookkeeping device which enables us to dissect the complex series of coupled reactions and learn something of the general influence of the various pollutants on this chemistry. We may employ a simplified scheme here in an attempt to judge the extent of the involvement of the various pollutants in the important NO to NO_2 conversion in a heavily polluted troposphere; of course this transformation is the major factor controlling the extent of O_3 development in the polluted troposphere. The chemical mechanism employed here is similar to that used by our group in previous work, but it has been updated to include the currently accepted values of rate constants of the elementary reactions. It is based upon only two classes of hydrocarbons (alkene and alkane), three carbonyl compounds (CH_2O, CH_3CHO, ketone), and CO as reactants for the chain oxidation of NO to NO_2. The rate constants chosen for the "alkene" and "alkane" were weighted values (composition, reactivity basis) which correspond to measured composition of a typical hydrocarbon mixture seen in Los Angeles in the early morning hours. A fixed solar intensity for a zenith angle of 40°, no dilution of the air mass, and no time dependent impurity additions were assumed. The initial concentrations were picked to be typical of those observed in Los Angeles in the early morning hours during periods of high pollution and atmospheric stagnation; these are shown for most impurities in Table I, Case A. In addition to the reactants shown there the original mixture contained CO (2.0 ppm), ketone (0.005 ppm), SO_2 (0.06 ppm), O_3 (0.03 ppm), relative humidity 50 %, temperature, $25^{o}C$. The O_3 level was picked to be out of balance with the photostationary state in the typical fashion of early morning urban conditions. In Figure 6 a few product and reactant

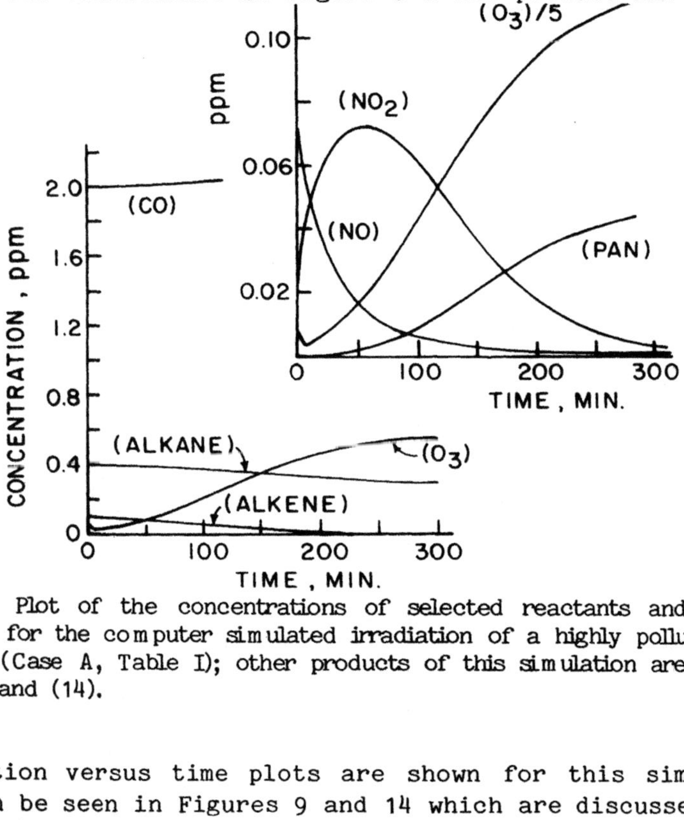

FIGURE 6. Plot of the concentrations of selected reactants and products versus time for the computer simulated irradiation of a highly polluted urban atmosphere (Case A, Table I); other products of this simulation are shown in Figures (9) and (14).

concentration versus time plots are shown for this simulation. Others can be seen in Figures 9 and 14 which are discussed later. The general features are those characteristic of the smog chemistry observed in the field. As NO is converted to NO_2 and $[O_3]$ climbs, the (alkene) and (alkane) decrease, and peroxyacetylnitrate and CO grow. It is interesting to partition these reactions by computer to judge how effective each reactant is in converting NO to NO_2 and hence in building O_3 levels. Only those fragments (RO_2, HO_2, etc.) which ultimately react with NO and were originally generated from the given reactant were counted

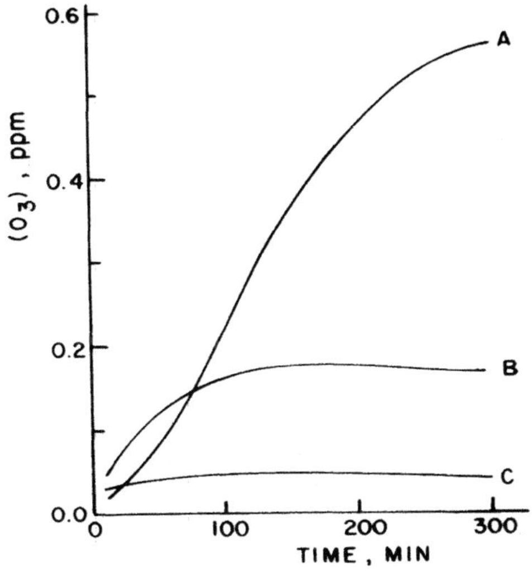

Fig. 7a

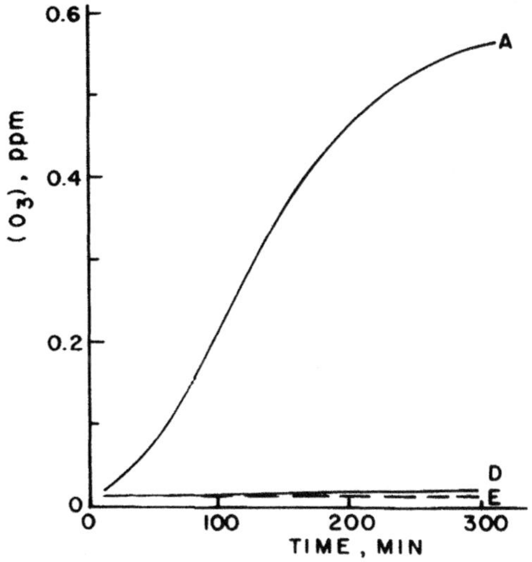

Fig. 7b

in the process. The results are summarized in Table V.

You will observe that at early times the major promoter of the NO to NO_2 conversion is the alkene. About 2 hours into the irradiation of the mixture the alkane is the most important reactant with large contributions from formaldehyde and acetaldehyde. The CO is important but less so than the other reactants at all times.

It is also instructive to observe the theoretical expectation of the initial reactant concentrations on the ozone levels produced as these factors relate directly to ozone control strategies. These results are shown in Figure 7; we can see the effect on ozone development of

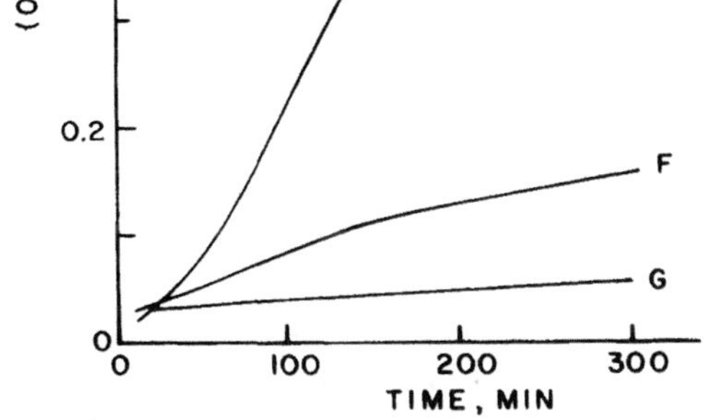

Fig. 7c

FIGURE 7. Plot of $[O_3]$ versus time for computer simulated irradiations of polluted air masses at various initial pollutant levels; these are summarized in Table I.

Table V. Percentage of NO to NO_2 Conversion Which Results from Various Reactive Pollutants in a Simulated Atmosphere (Case A, Table I)

Reactant	Time: 10 min	[illegible]0 min	120 min
Alkane	25.2%	26.3%	31.5%
Alkene	52.3	39.2	24.0
CO	6.3	6.9	8.9
CH_2O	7.4	12.0	16.9
CH_3CHO	8.8	15.5	18.6

lowering initial NO and NO_2 by a factor of 10 (Case B in Fig. 7a) and a factor of 100 (Case C, Fig. 7a) while maintaining the high hydrocarbon and aldehyde concentrations of Case A. In Figure 7b the initial hydrocarbon and aldehyde levels are lowered by these same factors, keeping NO_x high, while in Figure 7c NO_x, hydrocarbon, and aldehydes are all lowered successively by the same factors. It is immediately evident that the effects on O_3 maximum are most dramatic for the control of initial hydrocarbons and aldehydes (Cases D and E of Fig. 7b). We see as well that the control of both initial NO_x and hydrocarbons produces less effective O_3 reductions (Compare Cases F and G, Fig. 7c with D and E, Fig.7b). These results and others are summarized in Figure 8 where the $\log_{10}[O_3]_{max}$ is plotted versus the $\log_{10}[X]$, where X is the controlled variable. In order to achieve the reduction of $[O_3]_{max}$ of the high pollution case A to the 0.12 ppm, the present desired level of control agencies in the U.S.A., the hydrocarbon -aldehyde reduction factor of 0.55 is necessary; 0.12 for hydrocarbons alone; 0.042 factor for NO_x control; 0.052 factor for simultaneous reduction of NO_x, RH, and ECHO. Of course these results are not intended for

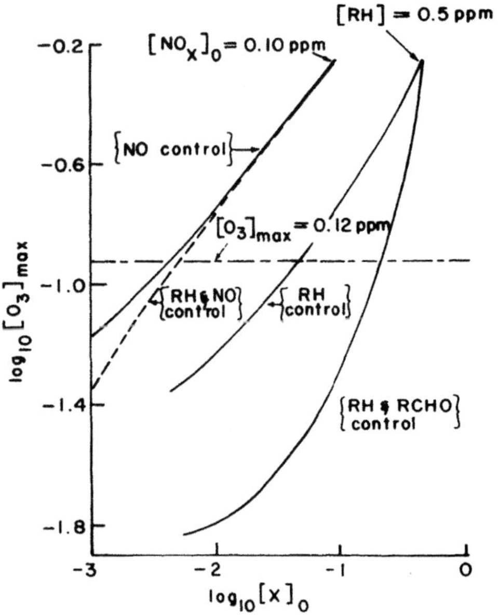

FIGURE 8. Plot of $\log^{10}[O_3]^{max}$ versus $\log^{10}(X)^o$ where X is the varied impurity in simulated polluted atmospheres; these data were derived from the simulations summarized in Table I and other similar runs.

formulation of control laws, but they simply illustrate qualitatively that the control of O_3 is not a simple procedure; one cannot by casual considerations anticipate some of the effects which become evident only through computer simulations. Present control strategy in the United States is built upon hydrocarbon reduction, and presumably, simultaneous aldehyde reduction. Extensive NO_x control has been delayed because of the lack of simple technology which can be applied to mobile and stationary combustion sources; this lack of NO_x control may fortuitously account for at least some of the limited success which has been seen in the present control strategy. The true control problem is much more complex than this simple illustration suggests. Thus in the real troposphere there are other complicating factors; for example, an air parcel may experience transport, dillution, new impurity inputs, periods of darkness, and multiday irradiations. In some cases which one can simulate in theory, the uncontrolled NO_x atmosphere which "benefits" a local community with lower $[O_3]_{max}$ values may lead to unwanted high levels of O_3 in a community downwind from the first. The health problems from large N-compound formation in th first community must also be weighed against the benefit from the lower O_3 level. Control strategies will eventually take into account the risks from each pollutant and properly minimize the total effect in the controls adopted. This cannot be done well at present.

IV. THE FATES OF THE NITROGEN OXIDES IN THE POLLUTED TROPOSPHERE

We may use our simplified chemical model to anticipate the relative importance of the nitrogeneous products which one might expect in a polluted atmosphere. These results are summarized in Figure 9. As we have seen NO to NO_2 conversion occurs rather quickly, but during the early time periods when NO is the dominant oxide present, one expects a burst of HONO formation; it builds to about 0.45 ppb in the early morning hours. Actually HONO formation has been observed to occur during the nighttime hours by some as yet undefined mechanism. Even if this formation did not occur at night, HONO would build to about 0.5 ppb in homogeneous processes (OH + NO; M $\rightarrow$ HONO + M) in the early morning hours and then decrease rapidly later in the day. Note that nitric acid is by far the major sink for NO_x expected in theory, while peroxyacetylnitrate (PAN) is also a favored product for these conditions of high hydrocarbon (hence high acetaldehyde) impurity. The interesting reactive compound peroxynitric acid peaks at about 1.5 ppb near midday. The transient NO_3 levels reach about 0.1 ppb near the end of the day when O_3 and NO_2 are high. The (N_2O_5) maximized at about 0.25 ppb at 160 min (not included in the data of Figure 9). In the real atmosphere from which we have data the levels of $HONO_2$ do not reach expected here. This is probably the result of heterogeneous removal paths for this compound (NH_4NO_3

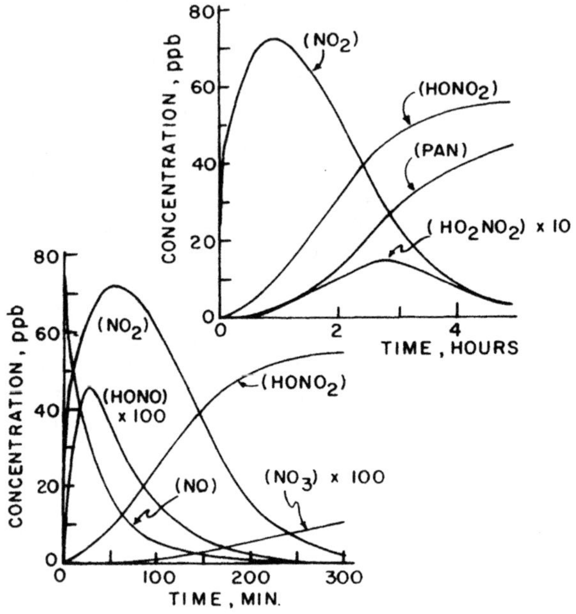

FIGURE 9. The concentration of various nitrogen-containing products of the computer simulated sunlight-irradiation of a highly polluted air mass of initial concentrations shown in Table I, Case A.

formation, incorporation into aerosols, organic nitrates, etc.) which we have not included here. However this simple model clearly suggests that we should not be surprised at the high nitrate levels which have been observed in atmospheres rich in NO_x.

V. THE CHEMISTRY OF ACID FORMATION IN THE TROPOSPHERE AND THE IMPORTANCE OF THE OH-RADICAL IN THESE PROCESSES

A. Homogeneous Chemistry.

As we have observed in our previous discussions, the OH-radical is important in initiating the chain processes which oxidize NO to NO_2 through their reactions with hydrocarbons, aldehydes, CO, etc. In addition the OH-radical reactions with NO_2 and SO_2 are believed to be the major homogeneous sources of $HONO_2$ and H_2SO_4, respectively.

$$OH + NO_2 + M \rightarrow HONO_2 + M \qquad (23)$$

$$OH + SO_2 + M \rightarrow HOSO_2 + M \ [- - - \rightarrow H_2SO_4] \qquad (24)$$

Reaction (24) is the rate determing step which leads to H_2SO_4 in the subsequent reactions [6]. Although other possibly important steps for the homogeneous oxidation of SO_2 may occur with $O[^3P]$, CH_2O_2 and other Criegee intermediates fomed from the alkene-ozone reactions [7], and CH_3O_2 [8], reaction 24 now appears to be the dominant H_2SO_4 source for the polluted troposphere. Reaction (23) is certainly the major homogeneous source of $HONO_2$ in the troposphere [9].

We may interpret the current literature [6, 10,11] to suggest the rate of (23) and (24) will be given by the approximate relations (25) and (26), respectively:

$$\frac{RATE\ OF\ (23)}{[NO_2]} \simeq [OH](4.3 \times 10^{-6})(T/298)^{-1.6},\ {}^{\circ}/_{\circ}/HR \qquad (25)$$

$$\frac{RATE\ OF\ (24)}{[SO_2]} \simeq [OH](1.6 \times 10^{-8}\ e^{1086/T}),\ {}^{\circ}/_{\circ}/HR \qquad (26)$$

Thus the OH-radical concentration controls these acid forming processes, and it is of primary interest to the atmospheric scientist to understand the sources of the OH-radical and the

Table VI. Primary OH-Radical Sources in Sunlight-Irradiated Polluted Atmospheres

1) O_3 (λ < 3200 A)	$O_3 + h\nu \rightarrow O(^1D) + O_2(^1\Delta_g, ^1\Sigma_g^-)$
	$O(^1D) + H_2O \rightarrow OH + OH$
	$O(^1D) + M \rightarrow O(^3P) + M$
2) HONO (λ < 4000 A)	$HONO + h\nu \rightarrow OH + NO$
3) CH_2O (λ < 3380 A)	$CH_2O + h\nu \rightarrow H + HCO$
	$H + O_2 + M \rightarrow HO_2 + M$
	$HCO + O_2 \rightarrow HO_2 + CO$
	$HO_2 + NO \rightarrow OH + NO_2$
4) CH_3CHO (λ < 3350 A)	$CH_3CHO + h\nu \rightarrow CH_3 + HCO$
	$\rightarrow H + CH_3CO$
	$CH_3 + O_2 \rightarrow CH_3O_2$
	$CH_3O_2 + NO \rightarrow CH_3O + NO_2$
	$CH_3O + O_2 \rightarrow CH_2O + HO_2$
	$CH_3CO + O_2 \rightarrow CH_3COO_2$
	$CH_3COO_2 + NO \rightarrow CH_3CO_2 + NO_2$
	$CH_3CO_2 \rightarrow CH_3 + CO_2$
	Followed by OH generation in the sequence involving CH_3 above and the HO_2 + NO reaction.
5) H_2O_2 (λ < 3500 A)	$H_2O_2 + h\nu \rightarrow OH + OH$

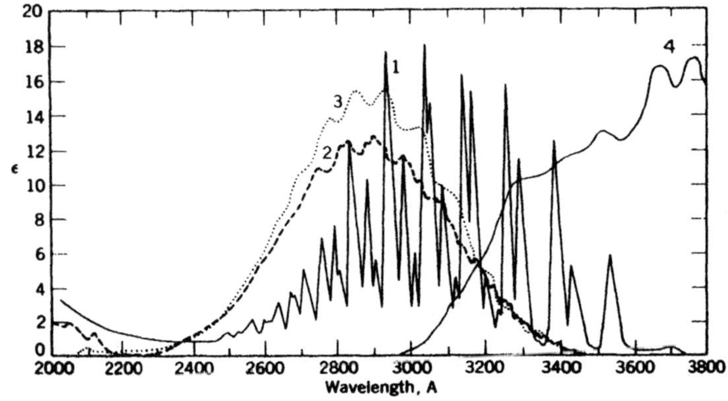

Figure 10a

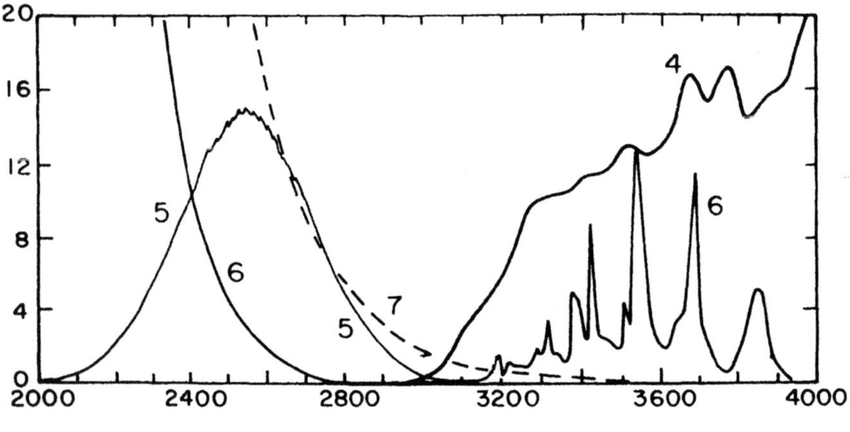

Figure 10b

FIGURE 10. Absorption coefficients versus wavelength for some common sources of the OH-radical present in the polluted atmosphere: formaldehyde, curve 1; acetaldehyde, 2; propionaldehzde, 3; relative solar irradiance at sea level [20], (Z = 40°), 4; ozone (ε/200), 5; nitrous acid (ε/10), 6; hydrogen peroxide, 7; ε has units of liters $mole^{-1}$ cm^{-1}, base 10 logarithmic function.

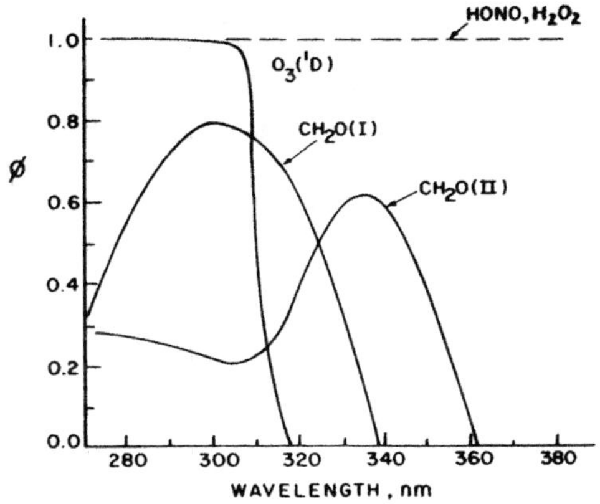

FIGURE 11. Plot of the quantum yield versus wavelength for various primary processes in the species shown; with HONO and H_2O_2, the OH radical is formed directly in the primary process; for O_3 the process forms $O[^1D]$ atoms which may react with H_2O to form OH; CH_2O(I) gives the yield of the process forming H + HCO which subsequently may from OH as shown in Table VI; CH_2O(II) is the yield of H_2 + CO formation in the intramolecular process.

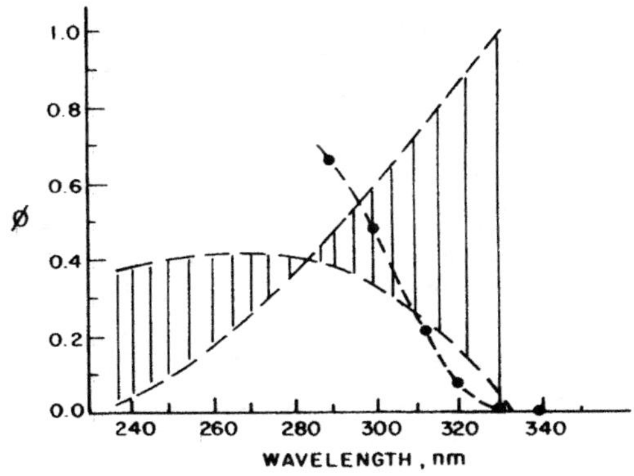

FIGURE 12. Plot of the quantum yield of the primary process, $CH_3CHO + h\nu \rightarrow CH_3 + CHO$, versus wavelength; the solid points are from Horowitz, et al. [12], and the marked area between the dashed curves is the range of values estimated by Demerjian, et al. from literature available in 1979 [20].

levels which it achieves in the atmosphere. A study of the potential sources points to the participation of the processes summarized in Table VI. O_3, HONO, CH_2O, CH_3CHO (and the higher aldehydes), and H_2O_2 all have reasonably intense electronic abosorption bands which overlap the wavelength distribution of the solar irradiance present within the lower troposphere; see Fig. 10. In each case OH-radical generation can result following primary photodissociation, and the quantum efficiencies of these reactions are reasonably well established in most cases; see Figure 11. Acetaldehyde data are limited; data from our laboratory sugesst rates of radical generation for the lower troposphere which are much lower than those currently employed in most tropospheric models; see Figure 12. If one utilyzes the seemingly well established data (excluding those for CH_3CHO at this time) we can estimate the relative efficiences of the various processes for OH-radical generation in the highlypolluted synthetic troposphere we have considered previously (Case A); see Figure 13. The length

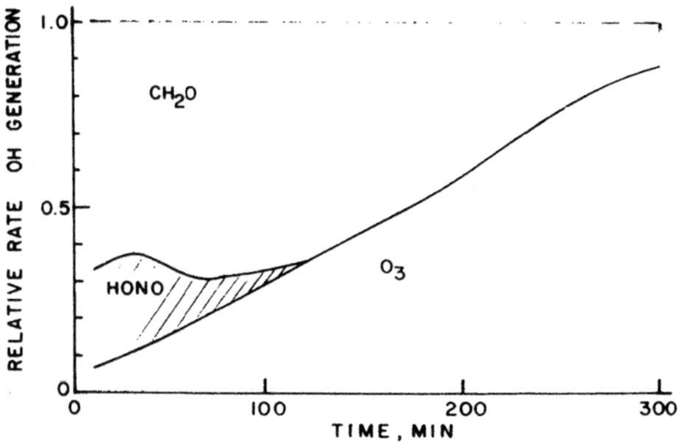

FIGURE 13. Relative rates of OH-radical generation versus time for various molecular sources in simulated sunlight-irradiation of a highly polluted atmosphere (Case A, Table I); the length of the ordinate within the area labeled for a given species is proportional to the rate of formation OH from that species at a given time; CH_3CHO and H_2O_2 should also provide significant OH, but quantitative estimates of these contributions are not possible now.

of the ordinate in a given labeled area of the plot at any given times is proportional to the rate of OH-radical generation which results from that species at that time. Note that formaldehyde is the major source early in the day for these conditions. Nitrous acid provides a fair fraction also at that time. Ozone photolysis becomes the dominant OH-radical source at late times. H_2O_2 is unimportant at short times, but may become significant at long times, provided that its removal is unimportant. We have not shown this source in the figure in that the rather low gas phase concentrations of H_2O_2 usually observed in the polluted atmosphere point to efficient removal paths which our mechanism does not include; these presumably include removal in aerosols, cloud water, etc. We can conclude from this exercise that aldehydes probably are the major sources of the OH-radical in the highly NO-polluted atmosphere where HO_2 generation is nearly always followed by OH-radical formation: $HO_2 + NO \rightarrow OH + NO_2$. This is in contrast with the dominant O_3 mechanism characteristic of the clean atmosphere at all times.

The knowledge of the [OH] in the polluted troposphere would be a large key to the estimation of rates of many atmospheric processes such as the rate of hydrocarbon removal, NO_2 transformation to $HONO_2$, SO_2 conversion to H_2SO_4, etc. Thus it is important that we examine here both the theoretical and experimental estimates of the (OH) as a function of the various pollutant reactant levels. Recent experimental estimates of the [OH] as a function of the various pollutant reactant levels. Recent experimental estamites of the [OH] within the real troposphere are summarized in Table VII. We have excluded the early estimates based upon laser induced fluorescence since the generation of OH by the laser beam itself was an unrecognized problem. Data based on the ^{14}CO- chemical method of Campbell and his co-workers [13], the laser induced fluorescence data of Wang and colleagues [14], and the absorption

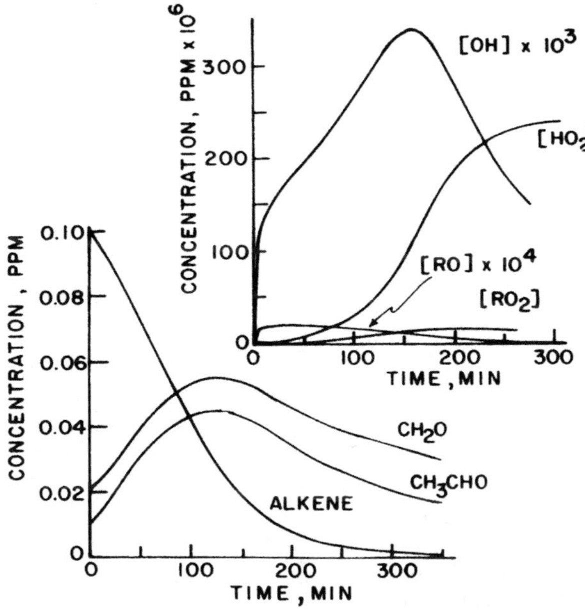

FIGURE 14. Plot of the concentrations of some reactants, transient free radicals, and products versus time in a simulated sunlight-irradiated, highly polluted troposphere (initial concentrations as for Case A, Table I); other products and reactants are shown in Figures 6 and 9 for this case.

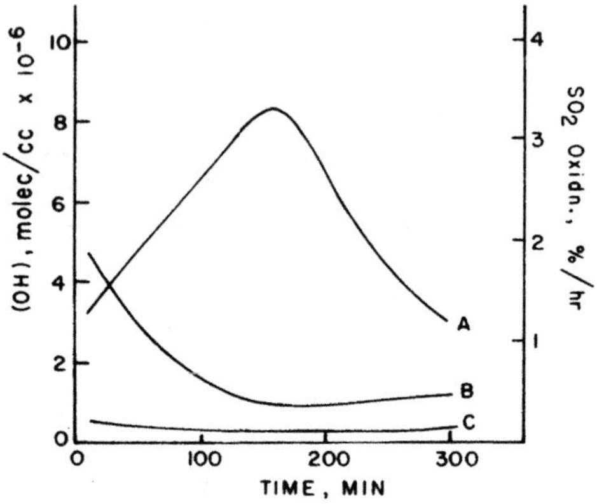

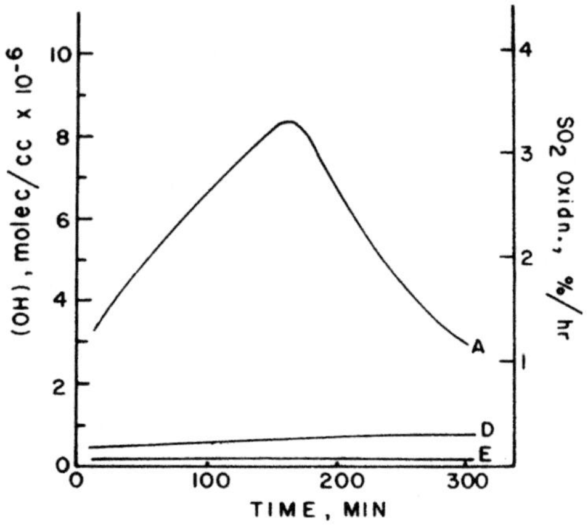

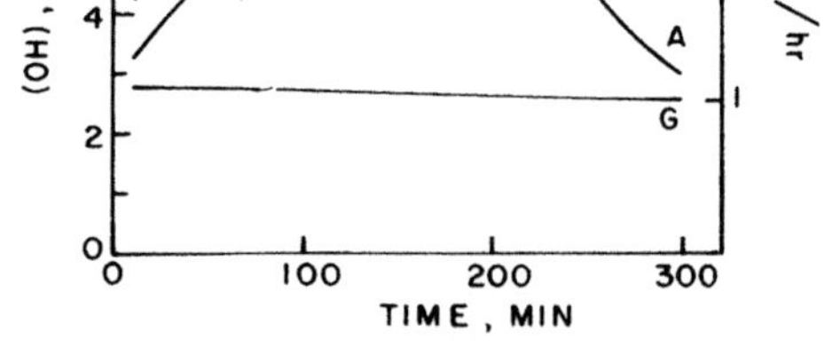

FIGURE 15. Plot of the [OH] versus time for computer simulated irradiations of polluted air masses at various initial pollutant levels; these levels for each of the cases are summarized in Table I; the right-hand ordinate shows the rate of SO_2 oxidation to H_2SO_4 (%/hr) which these levels of OH would promote in an atmosphere.

data of Ehhalt and colleagues [15] are resonably consistent among themselves and with theoretical estimates based on computer models [16]. The availability of additional relable data will be forthcoming in the years ahead. It is this type of accurate information on the fundamental species involved in tropospheric chemistry which will provide a most significant test of our theories and models of the chemistry of the polluted troposphere.
Using the simplified model employed in previous considerations in this study, we may estimate the theoretical time dependence of the important radical concentrations for the simulation of the highly polluted atmosphere, Case A. See Figure 14. As the chemical reactions progress the OH-radical concentration climbs to about 3.2×10^{-7} ppm or about 8.1×10^6 molec/cc. It is also instructive to see the effect of variation of the pollutant concentrations on the OH-radical levels and hence the potential rates of homogeneous generation of H_2SO_4 and $HONO_2$ in reactions (24) and (23), respectively. These results can be seen in Figure 15a from simulations of the various polluted mixtures, Cases A through G of Table I; the estimated rates of H_2SO_4 generation [%/hr] as estimated from equation (26) are shown on the right-hand ordinate. We see in Figure 15A that NO_x reduction is somewhat effective in lowering the OH levels at long times although at short times the ten-fold NO_x reduction results in an increase in [OH]. We note as well that rates of SO_2 oxidation as high as about 3.5 %/hr can occur from process (24) in a highly polluted troposphere.
At high hydrocarbon, low NO_x levels, rates of only a few tenths of a percent per hour are expected. In Figure 15B we note that the hydrocarbon reduction is even more effective in lowering the [OH] when NO_x is high. Perhaps the most surprising effect is seen in Figure 15C where all of the impurities (RH, RCHO, and NO_x) are decreased from their levels in highly polluted air. A factor of 10-decrease in this case results in a general increase in [OH] at most times and high rates of acid formation. Even a decrease of the impurities by a factor of 100 allows a 1 %/hr oxidation rate of SO_2.
In a similar fashion we may consider the rates of NO_2 removal to form $HONO_2$. Relations (25) and (26) show that for a given [OH], NO_2 removal in (23) is about 11-times faster than SO_2 removal at 298°K and 1 atm pressure. Thus in summary we note that the rates of homogeneous generation of H_2SO_4 and $HONO_2$ in the troposphere show a very complex relationship to the hydrocarbon and NO_x concentrations. Obviously a tropospheric model which uses only first order rate terms to describe SO_2 and NO_2 oxidations in an empirical fashion is completely unrealistic and cannot lead to meaningful results.

B. Heterogeneous Chemistry of H_2SO_4 Formation in the Troposphere.

Recent studies of Penkett (17), Hoffman (18), Martin (19), and

Table VII. Measured and Theoretical Estimates of the (OH) in the Troposphere

A. Experimental Methods

1) Chemical Method, $^{14}CO + OH \rightarrow {}^{14}CO_2 + H$ (Campbell, et al.)[13]

	(OH), molec/cc
Pullman, Wash. (46.7°N, 770 m)	$(5.8 \pm 2.4) \times 10^6$
New Zealand (44°S, 1030 m)	$(1.1 \pm 0.4) \times 10^6$
Tennessee (Rural, 36°N, 270 m)	$(1.9 \pm 0.7) \times 10^6$
Los Angeles (34°N, 270 m)	$(5.0 \pm 1.2) \times 10^6$
Arizona (Desert, 37°N, 2300 m)	$(57 \pm 23) \times 10^6$

2) Laser Induced Fluoroscence, 2820.7 A excitation, 3090 A Emission, 3020 A N_2 Raman (Wang, et al.)[14]

Niwot Ridge, CO (3048 m)	$(40 \text{ to } 6) \times 10^6$
Los Angeles, CA (11886 m)	$(2.5 \pm 2) \times 10^6$
SanBernadino, CA (10057 m)	$(20 \pm 6) \times 10^6$
SanDiego, South, CA (10667 m)	$(10 \pm 4) \times 10^6$
Denver, CO (10057 m)	$(1 \pm 3) \times 10^6$

3) Absorption Spectroscopy, 3079.95 A (Ehhalt, et al.)[15]

Jülich, Germany (51°N)	$(11 \pm 6) \times 10^6$
Usually	$(6 \pm 3) \times 10^6$

B. Theoretical Estimates of (OH) in the Urban and Rural Troposphere

Typical Recent estimates found in simulations of Calvert, Demerjian, Seinfeld, Niki, Graedel, Isaksen, etc., and their co-workers:

Daytime maximum (Summer, 40°N)	about 9×10^6
Daytime maximum (Winter, 40°N)	about 2×10^6
Nighttime values less than	about 2×10^5

(Values are sensitive to ultraviolet solar irradiance at point of interest, levels of impurities, NO, NO_2, RH's, RCHO's, etc.)

others show that SO_2 dissolved in cloud water, rain water, aqueous aerosols, etc., can be oxidized readily to H_2SO_4 through the reactants O_3 and H_2O_2 when these species are present at typical ambient levels; even O_2 can be effective, particularly when certain transition metal ions such as Mn^{++} are also present. Note that both H_2O_2 and O_3 are products of the homogeneous gas phase reactions, so it is evident that the homogeneous and heterogeneous pathways of acid generation in troposphere are not completely independent.

The major homogeneous sources of H_2O_2 in the troposphere are reactions (27) to (29) involving the HO_2 and the hydrated HO_2 radicals:

$$2HO_2 \rightarrow H_2O_2 + O_2 \qquad (27)$$

$$HO_2 + H_2O \rightleftarrows H_2O \cdot HO_2 \qquad (28)$$

$$H_2O \cdot HO_2 + HO_2 \rightarrow H_2O_2 + H_2O + O_2 \qquad (29)$$

Now in a polluted atmosphere there is a strong competition for the HO_2 radicals through reactions (19), (30), (31), and possibly others.

$$HO_2 + NO \rightarrow OH + NO_2 \tag{19}$$

$$HO_2 + CH_2O \rightarrow HO_2CH_2O \rightleftarrows O_2CH_2OH \tag{30}$$

$$HO_2 + O_2CH_2OH \rightarrow HO_2CH_2OH + O_2 \tag{31}$$

Thus the efficiency with which H_2O_2 is generated in the polluted troposphere through (27) and (29) is a complex function of the [NO] and $[CH_2O]$ as well as the hydrocarbon and other aldehyde concentrations. We may use our computer simulations to observe the theoretical expectations of the (H_2O_2) which should be generated in various simulated polluted atmospheres (Cases A-G) considered previously here. Of course if an aqueous phase is available to the H_2O_2, this compound will end up largely in that phase because of the large Henry's law constant for H_2O_2. These results are summarized in Figure 16. With the highly polluted atmosphere (Case A) we see H_2O_2 development is unimportant until the NO has been highly depleted at late times. However when we start with a mixture which is low in NO_x and high in RH and RCHO, H_2O_2 generation begins early in the day. Even with a 100-fold decrease in the NO_x, resonably high levels of H_2O_2 are found early in the sunlight-irradiated mixture; see Case C. Cases D and E which are at high NO_x lead to no measurable H_2O_2, while F and G with equally lowered levels of all pollutants present in A, small H_2O_2 amounts are expected late in the irradiation period. One may conclude from these observations that conditions of low NO_x and high hydrocarbon and aldehyde impurity levels favor H_2O_2 formation. Hence if and when the air mass encounters cloud water, rain water, high aqueous aerosol levels, etc., then a

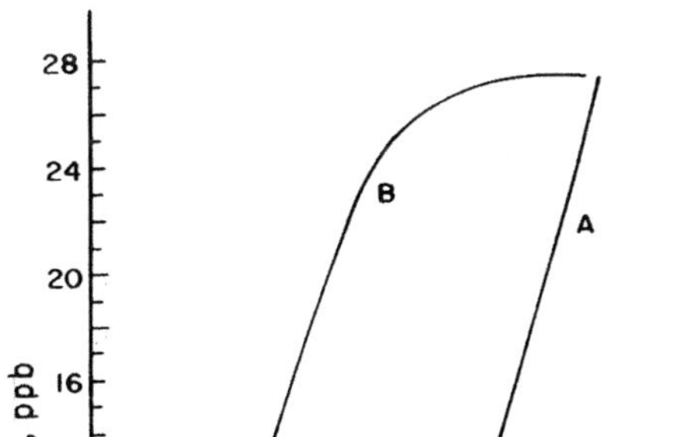

FIGURE 16. Theoretical $[H_2O_2]$ versus time for computer simulated irradiations of polluted air masses at various initial pollutant levels; these are summarized for each case in Table I; presumably these gas phase levels would be lowered in the real atmospheres through the transfer of H_2O_2 into cloud water, aqueous aerosols, etc.

heterogeneous pathway for H_2SO_4 formation if favored as well. Note in this case that the optimum impurity levels for ozone generation can be quite different then those for H_2O_2 formation. Thus the heterogeneous components of H_2SO_4 formation from O_3 and H_2O_2 will not proceed exactly in phase with fixed fractions occurring by each pathway. The rates of the homogeneous and heterogeneous pathways which lead to acid formation will depend on two quite different complex functions of the pollutant concentrations, although it is clear that the two are not entirely independent.

VI. SUMMARY

The present understanding of the chemistry of the polluted troposphere has improved signifcantly in recent years. However the extreme complexity of the mixture of reactive pollutants prevents a completely sound quantitative treatment of these systems at present. Even so there are several conclusions which can be formulated at this stage of our knowledge:

1) The existing field data and laboratory data on photochemical generation of O_3 in the polluted troposphere appear to be consistent with the generally proposed mechanism and elementary reactions (1), (2), and (3). Of course this does not rule out the dominance of ozone from stratospheric intrusions for certain "clean" tropospheric conditions.

2) The relation, $[O_3][NO]/[NO_2] \simeq k_1/k_3$, which is most useful in homogeneous, NO_x-rich atmospheres, does not hold well for poorly mixed atmospheres, for cases where [Alkene]/[NO] > 160, at low [NO] and for rapidly changing sunlight intensity (early morning or late afternoon), where the photochemical stationary state cannot be achieved or maintained quickly or ozone is removed significantly by reactions with species other than NO.

3) Many uncertainties remain in the mechanism of NO to NO_2 conversion. Thus the extent of alternate reactive channels for alkoxy radicals as a function of radical structure is not well defined in many cases; the possible Criegee intermediate involvement in this conversion remains unevaluated, etc.

4) The homogeneous oxidation of SO_2 and NO_2 (through the OH-radical) can occur at rates of up to 3.5 %/hr for SO_2 and 39 %/hr for NO_2 during the summer months in a highly polluted atmosphere at midday.

5) The heterogeneous oxidation of SO_2 in cloud and rain water may occur with reactants H_2O_2 and O_3 which are major products of the homogeneous chemical reactions. Thus the homogeneous and heterogeneous pathways of acid generation in the atmosphere are coupled in a complex fashion.

6) The unambiguous, direct measurements of the time dependence of specific key compounds in air parcels (O_3, NO, NO_2, CH_2O, CH_3CHO, $CH_3COO_2NO_2$, alkenes, alkanes, terpenes, $HONO_2$, H_2SO_4, SO_2, H_2O_2, CO, etc.) will be necessary to provide rigorous tests of current

theories and to develpo realistic control strategies based upon elemtary reactions and sound theoretical considerations.
7) The continued, improved measurement of [OH] in the troposphere, simultaneous with measurements of the species outlined in summary statement 6, should provide the most direct test of the kinetic schemes involved in the ttropospheric chemistry.
8) Obviously the development of reasonably accurate and complete tropospheric chemistry and transport models faces the strong restraints imposed by computer costs and limitations in our knowledge. However future model development must include suficient detailed elementary chemical steps to account properly for the particular chemical transformations which are of interest (O_3 generation, "acid rain" formation, PAN formation, etc.). In no case can the models based upon parameterized, first order conversion terms for SO_2 oxidation give results which are realistic for the range of pollutant levels commonly encountered in the polluted air masses.

ACKNOWLEDGEMENT

This work was supported by the United States Environmental Protection Agenxy Grant No. R-806479-03-0 for which the author is grateful.

REFERENCES

1. The basic data (both chemical and meteorological) from 35 days of operation of the LARPP study have been archived on magnetic tape and are available from the U.S. National Technical Information Service, Washington, D.C. Some detailed considerations of the chemical data from one day of the operation (Operation #33) appear in the following articles: a) Calvert, J.G., "Test of the Theory of Ozone Generation in Los Angeles Atmosphere", Environ. Sci. Technol., 10, 248 - 256 (1956); b) Calvert, J.G., "Hydrocarbon Involvement in Photochemical Smog Formation in Los Angeles Atmosphere", Environ. Sci. Technol., 10, 256 - 262 (1976).
2. For examples see the following: a) Leighton, P.A., "Photochemistry of Air Pollution", Academic Press, New York, N.Y., 1961; b) Demerjian, K.L., Kerr, J.A., and Calvert, J.G., "The Mechanism of Photochemical Smog", Chapter 1, Adv. Eviron. Sci. Technol., 4, 1 - 256 (1974); c) Niki, H., Daby, E.E., and Weinstock, B., "Mechanisms of Smog Reactions", Chapter 2, in "Photochemical Smog and Ozone Reactions", Adv. Chem. Series, 113, 16 - 57, Amer. Chem. Soc., Washington, D.C. (1972); d) Stedman, D.H. and Jackson, J.O., "The Photostationary State in Photochemical Smog", Int. J. Chem. Kinet., Symp. 1, 493 - 501

(1975).

3. a) Seinfeld, J.H., Comment, Environ. Sci. Technol., 11, 1218 - 1219 (1977); b) Feigely, C.E., Comment, Environ. Sci. Technol., 12, 843 - 845 (1978); c) Bilger, R.W., "The Effects of Admixing Fresh Emissions on the Photostationary State Relationship in Photochemical Smog", Atmos. Environ., 12, 1109 - 1118 (1978); d) Chock, D.P., and Kumar, S., "On the Photostationary State Assumption in the Atmospheric Nitric Oxide-Nitrogen Dioxide, Ozone System", Atmos. Environ., 13, 419 - 420 (1979).
4. Laub, R.J., and Smith, C.A., Department of Chemistry, The Ohio State University Columbus, Ohio, U.S.A.; recent communication to the author; a 50 m capillary column was used with FE 30 stationary phase and temperature programming (0 - 200°C).
5. For a few examples of the vast literature see references 2b and 2c and the following: a) Graedel, T.E., Farrow, L.A., and Weber, T.A., "Kinetic Studies of the Photochemistry of the Urban Troposphere", Atm. Environ., 10, 1095 - 1117 (1976); b) H¦v, O., and Isaksen, T.S.A., "Hydroxyl and Peroxy Radicals in the Polluted Troposheric Air", Geophys. Res. Lett., 6, 219 - 222 (1979).
6. Calvert, J.G., Su, J.G., Bottenheim, J.W., and Strausz, O.P., "Mechanism of the Homogeneous Oxidation of Sulfur Dioxide in the Troposphere", Atm. Environ., 12, 197 - 226 (1978).
7. a) Cox, R.A., and Penkett, S.H., "Oxidation of Atmospheric SO_2 by Products of the Ozone-Olefin Reaction", Nature, 230, 321 - 322 (1970); b) Cox, R.A., and Penkett, S.A., "Aerosol Formation from Sulfur Dioxide in the Presence of Ozone and Olefinic Hydrocarbons", J. Chem. Soc. Faraday Trans., 1, 68, 1735 - 1753 (1972); c) Su, F., Calvert, J.G., and Shaw, J.H., "A FT IR Spectroscopic Study of the Ozone-Ethene Reaction Mechanism in O_2-Rich Mixtures", J. Phys. Chem., 84, 239 - 246 (1980); d) Kan, C.S., Su, F., Calvert, J.G., and Shaw, J.H., "The Mechanism of the Ozone- Ethene Reaction in Dilute N_2/O_2 Mixtures Near One Atmosphere Pressure", J. Phys. Chem., 85, 2359 - 2363 (1981); e) Niki, H., Maker, P.D., Savage, C.M., and Breitenbach, L.P., "A FT IR Study of a Transitory Product in the Gas Phase Ozone-Ethylene Reaction", J. Phys. Chem., 85, 1024 - 1027 (1981).
8. a) Kan, C.S., McQuigg, R.D., Whitbeck, M.R., and Calvert, J.G., "Kinetic Flash Spectroscopic Study of the CH_3O_2- CH_3O_2 and CH_3O_2-SO_2 Reactions", Int. J. Chem. Kinet., 11, 921 - 933 (1979); b) Kan, C.S., Calvert, J.G., and Shaw, J.H., "Oxidation of Sulfur Dioxide by Methylperoxy Radicals, J. Phys. Chem., 85 1126 - 1132 (1981); c) Sanheusa, E., Simonaitis, E., and Heicklen, J., "The Reaction of CH_3O_2 with SO_2", Int. J. Chem. Kinet., 11, 907 - 914 (1979); d) Simonaitis, R., and Heicklen, J., "The Mechanism of SO_2 Oxidation by CH_3O_2 Radicals. Rate Coefficients for the Reaction of CH_3O_2 with SO_2 and NO", Chem. Phys. Lett., 65, 361

- 365 (1979); e) Sander, S.P., and Watson, R.T., " A Kinetic Study of the Reaction of SO_2 with CH_3O_2", Chem. Phys. Lett., 77, 473 - 475 (1981).
9. Morris, E.d., Jr., and Niki, H., "Reaction of Dinitrogen Pentoxide with Water", J. Phys. Chem., 77, 1929 - 1932 (1973).
10. a) Davis, D.D., "Absolute Rate Constants for Elementary Reactions of Atmospheric Importance: Results of the University of Maryland's Gas Kinetics Laboratory", Report No. 1, Chemistry Dept., University of Maryland, July, 1974; b) Atkinson, R., Perry, R.A., and Pitts, J.N., Fr., "Rate Constants for the Reactions of OH Radical with NO_2 (M = Ar, N_2) and SO_2 (M = Ar); J. Chem. Phys. 65, 306 - 310 (1976); c) Cox, R.A., "The Photolysis of Gaseous Nitrous Acid - A Technique for Obtaining Kinetic Data in Atmospheric Photooxidation Reactions", Int. J. Chem. Kinet., Symp., 1, 379 - 398 (1975); d) Castleman, A.W., Jr., and Tang, I.N., "Kinetics of the Association Reaction of SO_2 with the Hydroxyl Radical", J. Photochem., 6, 349 - 354 (1976/1977); e) Harris, G.W., Atkinson, R., and Pitts, J.N., Jr., "Temperature Dependence of the Reaction OH + SO_2 > HSO_3 (+M) for M = Ar and SF_6", Chem. Phys. Lett., 69, 378 - 382 (1980).
11. a) Wine, P.M., Kreutter, N.M., and Ravishankara, A.R., "Flash Photolysis-Resonance Fluorescence Kinetics Study of the Reaction OH+ NO_2 + M > HNO_3 + M", J. Phys. Chem., 83, 3191 - 3195 (1979); b) O'Brien, R.J., Green, P.J., and Doty, R.A., "Rate Constant for the Reaction NO_2 + OH + M > HNO_3 Measured Under Simulated Atmospheric Conditions Using a Novel Analysis Procedure", J. Phys. Chem., 83, 3302 - 3305 (1979); c) Anderson, L.G., "Absolute Rate Constants for the Reaction of OH with NO_2 in N_2 and He from 225 - 389 K", J. Phys. Chem., 84, 2152 - 2155 (1980).
12. Horowitz, A., Kershner, C.J., and Calvert, J.G., recent work in the author's laboratory, to be published.
13. a) Campbell, M.J., Sheppard, J.C., and Au, B.F., "Measurement of Hydroxyl Concentration in Boundary Layer Air by Monitoring CO Oxidation", Geophys. Res. Lett., 6, 175 - 178 (1979); b) Sheppard, J.C., and Campbell, M.J., "Atmospheric Hydroxyl Radical Concentration: Further Groundlevel Radiochemical Measurements", recent unpublished work from M.J. Campbell's group for which the author is grateful.
14. a) Wang, C.C., Davis, L.I., Jr., Selzer, P.M., and Munoz, R., "Improved Airborne Measurements of OH in the Atmosphere Using the Technique of Laser Induced Fluorescence", J. Geophys. Res., 86, 1181 - 1186 (1981); Davis. L.I., Jr., Hamilton, E.J., Jr., Niki, H., Wang, C.C., Weinstock, B., Bollinger, M.J., Fehsenfeld, F.C., and Liu, S.C., "Concurrent Measurements of Ambent OH and Its Controlling Factors Near Niwot Ridge, Colorado"; we are grateful for a copy of this work before publication by Dr. Wang.
15. a) Perner, D., Ehhalt, D.H., P[ty, H.W., Platt, V., R\th, E.P.,

and Volz, A., "OH Radicals in the Lower Troposphere", Geophys. Res. Lett., 3, 466 - 468 (1976); b) Ehhalt, D.H., communication to the author of recent unpublished work of Dr. Ehhalt's group.

16. For examples within the clean troposphere see: a) Levy, H., "Photochemistry of the Troposphere", Adv. Photochem., 9, 369 - 375 (1974); b) Crutzen, P.J., and Fishman, J., "Average Concentrations of OH in the Troposphere, and the Budgets of Methane, Carbon Monoxide, Molecular Hydrogen, and 1,1,1-Trichloromethane", Geophys. Res. Lett., 4, 321 - 324 (1977); c) Chang, J.S., and Wuebbles, D.J., "A Theoretical Model of Global Tropospheric OH Distributions", University of California, Livermore, Calif., 1976; for examples within the polluted troposphere see d) Hecht, T.A., and Seinfeld, J.H., "Development and Validation of Generalized Mechanism for Photochemical Smog", Environ. Sci. Technol., 6, 47 - 57 (1972); e) Clavert, J.G., and McQuigg, R.D., "The Computer Simulation of the Rates and Mechanisms of Photochemical Smog Formation", Int. J. Chem. Kinet., Symp. 1. 113 - 154 (1975); alos see references 2b, 2c, 5a, 5b.

17. Penkett, S.A., Jones, B.M.R., Brice, K.A., and Eggleton, A.E., "The Importance of Atmospheric Ozone and Hydrogen Peroxide in Oxidizing Sulfur Dioxide in Cloud and Rainwater", Atm. Environ., 13, 123 - 137 (1979).

18. Hoffmann, M.R., and Edwards, J.O., "Kinetics of Oxidation of Sulfite by Hydrogen Peroxide in Acidic Solution", J. Phys. Chem., 79, 2096 - 2098 (1975).

19. Martin, L.R., and Damschen, D.E., "Aqueous Oxidation of Sulfur Dioxide by Hydrogen Peroxide at Low pH", preprint of work under contract number 68-02-2702, U.S. Environmental Protection Agency; the author is grateful to Dr. Martin for this manuscript which is scheduled to appear in Atm. Environ., 1981.

20. Demerjian, K.L., Schere, K.L., and Peterson, J.T., "Theoretical Estimates of Actinic (Spherically Integrated) Flux and Photolytic Rate Constants of Atmospheric Species in the Lower Troposphere", Adv. Environ. Sci. Technol., 10, 369 - 459 (1980).

MEASUREMENT OF AIR POLLUTION IN TIME AND SPACE

Dr.Ir.T.Schneider

Director of the Division of Environmental Hygiene
of the National Institute of Public Health,
P.O.Box 1, Bilthoven, The Netherlands

ABSTRACT

The assessment of air quality is for a large part the collection and evaluation of data. A review is given of the specific information such as the emission sources, the ambient air quality, geographical and topographical data, meteorology etc., that is needed. A brief description is given of the methods for trend analysis, the measurement of air pollution in time. Next to the statistical distributions using the results of monitoring stations, trend analysis with biological effect determinations is demonstrated. Major emphasis is given to the air pollution measurements in space. Results of aerial determination of pollution transport are used to describe the transport within urban and industrial plumes and the transformations during transport under varying meteorological conditions over the North Sea. Total fluxes and removal rates for SO_2 and NO_2 are calculated. The overall measurement strategy needed for a good time and space description of air pollutant concentrations is discussed using recent results of the Dutch national monitoring network. Finally the relation between air pollution transport measurements and model calculations is indicated with the results of an O_3 removal study carried out over the North Sea and the bordering coastal areas.

H. W. Georgii and W. Jaeschke (eds.), Chemistry of the Unpolluted and Polluted Troposphere, 457–481.

1. INTRODUCTION

Every so often creation of ideas demands new impulses. This also is true for the field of air pollution research, in particular in relation to the measurement of air pollution. The collection of massive amounts of air pollution concentration data and the subsequent storage in tabular or grafical form in printed material or on computer tape is being questioned regarding the suitability of these information systems to answer questions raised by air pollution agencies. Not the collection of data as such but the subsequent use in research institutes and policy making agencies is one of the focal points at the moment.

In the past major emphasis was placed on urban air pollution problems. The sources within and in the vicinity of the city were the only sources of importance in determining the air pollution burden. As soon as measurement stations were located in rural areas however, results indicated that transport of pollutants did take place and that this transport occurred even on a large scale. Strong evidence has been presented of significant long range transport of man-made pollutants even into the Arctic regions. This transport occurs on a scale of thousands of kilometers especially in the winter period.

Although the use of measurement stations that are not influenced by local sources is a major improvement over the measurement stations traditionally sited near to major sources or source areas, it will be shown that investigation of the processes that occur in the atmospheric boundary layer using the results of one ore more ground level stations only, is extremely difficult and often leads to erronious conclusions.

Assessment of air quality should therefore not only comprise the determination of air pollutant concentrations at ground level or at a certain height but should also study the relationship between the measured concentrations and the transport, deposition and other removal processes that occur both in time as well as in space.

2. ASSESSMENT OF AIR QUALITY

A primary function in the assessment of air quality is the collection and evaluation of data. These data refer to present and future emission sources, existing ambient air quality, population, siting of large industrial plants, traffic, land use, topography, meteorology etc.

In some cases these data are available and only have to be compiled or transformed, in other cases special data sets have to be collected.

2.1. Geographical area and land use

The characteristics of the geographical area where the air quality should be assessed affects the selection of the control strategy to a large extend. The area therefore must be carefully examined and the existing facts documented.
Main items to be covered are:
- the identification of the area
- the boundaries and the total surface area
- the geographical and topographical characteristics such as the geographical location, the relief, the height above sea level etc.
- population, industry, traffic and their spacial distribution
- special land use areas such as farmland, recreational areas
- the energy production and consumption pattern which may affect the area population
- the surrounding areas and their characteristics as they may strongly influence the pollution found inside the assessment area

2.2. Source and emission inventories

The contents of an emission inventory and the elements of the emission inventory system are determined by the use that is made of the data. The objectives of an emission inventory system within the overall strategy should be carefully evaluated. Objectives may comprise:
- the assessment of the current emission situation in a certain area, the determination of the contribution from the various emitter categories
- the prediction of changes in the emission situation due to the addition of new sources
- the prediction of changes in the emission situation due to selected control strategies
- the prediction of changes in the ambient air quality from the above mentioned changes in the emission situation

For most of these applications information is required on the type of pollutants, the mass flow rate of the emissions, the location of the sources and the time dependence of the source strength. The objectives of the emission inventory will also determine the amount of detail required. A comprehensive inventory may include detailed information on industrial installations, their mode of operation, their exact geographic location and the time when the emissions take place. Less detailed inventories may deal only with total emissions in certain regions presented as yearly averages.

Finally the use made of the data determines the required accuracy and also dictates the frequency of updating of the emission inventory.
The existing emission inventories use the following emitter categories:
- major industrial facilities
- small industrial facilities
- domestic consumers
- transportation
- natural sources

Procedures to collect the relevant data include the use of emission factors, questionnaires, on the site source testing, and other special studies.

A further important subdivision is the collection of the data in three groups: line, point and area sources. Especially when data are used for modeling purposes, carefull planning of the appropriate details needed in each of these three categories is required.

2.3. Existing ambient air quality

To assess the existing air quality, information is needed on the present levels of harmfull or potentially harmfull pollutants. This information has to be gathered for the whole of the area involved. That is to say, measurements should not only be carried out near the emission sources, but should also cover the impact area around the emission sources and should take into consideration the "background" concentration due to regional or distant sources.

An important decision has to be made whether one should try to locate the "hot spots" within the area or whether one should determine the average for the whole region or part of the region. A scheme which only determines the average condition of the whole area needs only sampling points at large distances. To resolve the detailed pattern of "hot spots" and gradients requires a very dense network of sampling points. The same considerations apply in relation to time. Pollutant emissions are not uniform in time, the peaks and through's in the continuous measurement of pollutant concentrations are well known. The frequency of sampling that is needed to demonstrate the daily oscillation in the levels of pollutants therefore must obviously differ from the frequency of sampling needed to reveal an annual average or trend over the years.

Together with the measurement of pollutant concentrations it is often important to gather information on the effects of the pollutants as a whole. To determine these effects one can use plant indicators that are sensitive to one or more specific pollutants. Damage patterns on leaves, retardation of development and growth of the whole plant, can all give a good indication of the pollutants which cause the problems. Because pollutants usually occur in combination, synergistic effects may take place. These effects are more easily determined using direct effect measurements or observations.

2.4. Meteorological and climatological data

Concentrations of air pollutants vary in space and time partly because of the uneven distribution of sources, but also because of meteorological conditions that vary from hour to hour and from day to day. Special attention must therefore be given to the meteorological and climatological data requirements.

One of the important first steps is to make an inventory of existing meteorological stations in the survey area. In the design of a meteorological data system for air quality assessment one of the criteria to be considered is the meso-meteorology of the region. Particularly when synoptic-scale winds are light and skies are clear, the mesoscale circulations can have a major effect on air quality. The existence of mesoscale meteorological wind fields can often be determined by careful examination of topographic maps and aerial photographs and the study of published wind roses. The presence of a coastline or valley is also an indication that mesoscale winds may occur. Site inspections may be helpfull in confirming the existance of such flows. Because most mesoscale circulations show a day-night reversal, the wind roses are most informative when they have been prepared separately for day and night time hours. The following meteorological observations should be considered:
- windspeed and winddirection
- air temperature, both horizontal and vertical gradients are of importance
- relative humidity
- cloud cover and solar radiation
- precipitation
- visibility
- vertical profile data and mixing height

To determine the structure of the planetary boundary layer and the surface mixing layer one can not rely on ground based sensors only but requires: rawinsonde data to yield mixing heights and mean transport winds through the surface mixing layer, minisonde data to obtain information on the vertical profile, measurements on tall (television) towers and remote sensing devices such as lidar, sodar and acoustic sounders.

2.5. Atmospheric chemistry data

In air quality assessment there is a need for atmospheric chemistry data in relation to the following pollution problems:
- oxidant levels due to photochemical production
- acid deposition
- brown haze over cities
- pollutant removal processes
- long range transmission of air pollutants

The incorporation of data on removal processes for atmospheric pollutants is a recent development in the assessment of air quality.

Four processes contibute to the removal of pollution from thc air over a region:
1. advection to neighbourghing regions
2. chemical transformation to other species
3. absorption by the underlying surface (dry deposition)
4. precipitation scavenging (wet deposition)

Data requirements associated with these processes are:
- knowledge of chemical reactions and numerical values of their rate constants
- measurements of the concentrations of the relevant pollutants in three dimensions (aircraft surveys)
- micro-meteorological factors such as windspeed and temperature gradients near the ground surface
- wetness of the surface
- pH of the surface
- surface roughness
- physiological factors such as the degree of opening of the stomata of leaves

3. MEASUREMENT OF AIR POLLUTION IN TIME, TRENDANALYSIS

Due to the influence of changes in emission sources and emission strenght and the overriding dependence on meteorological conditions, the measured air pollution concentrations at fixed or mobile measurement stations can only be used to indicate a certain trend in the overall air pollution situation. To destillate a trend from the year to year measured data it is necessary to perform a carefull analysis of changes in emission rates and the differences in meteorological conditions and their influences on the individual measurements.

To characterise the variability of air pollution with time there are a number of possibilities to present the data. First of all one can use yearly or seasonaly averaged values in combination

province	station number	location measurement station	1976	1977	1978	1979	1980
Limburg	1.01	Afferden	3.0	2.7	2.8	3.2	3.4
	1.03	Steeg	3.1	4.4	3.8	1.9	3.8
	1.08	Born	2.9	5.3	2.9	2.6	2.9
	1.24	Oost-Maarland	5.2	5.1	4.8	3.2	4.3
Noord-Brabant	2.02	Dussen	2.5	4.1	4.7	2.5	3.4
	2.10	Heijningen	4.1	6.0	5.8	4.1	4.3
	2.17	Best/Welschap	5.1	4.9	4.8	2.3	3.4
	2.23	Ossendrecht	6.3	7.5	7.4	10.0	4.7
Zeeland	3.01	Zierikzee	8.4	9.1	8.9	7.3	5.2
	3.05	Lewedorp	5.4	6.4	8.6	9.4	4.3
	3.15	Sas van Gent	5.9	6.7	6.8	9.7	4.1
Zuid-Holland	4.02	Voorschoten	4.4	5.7	8.4	6.9	3.1
	4.09	Zevenhuizen	4.0	6.9	5.0	6.1	4.7
	4.11	Schipluiden	9.7	6.0	7.0	8.2	4.4
	4.13	Delft	10.7	8.4	5.3	6.9	3.0
	4.16	Vlaardingen	9.3	7.2	3.1	5.1	-
	4.19	Rockanje	11.6	8.9	8.9	10.0	3.8
	4.25	Spijkenisse	12.2	9.4	6.6	6.9	4.4
	4.51	Naaldwijk	12.3	9.0	7.9	6.5	4.1
	4.54	Maasland	12.2	8.8	8.6	8.0	-
	4.58	Rhoon	13.1	6.0	4.8	4.7	-
Noord-Holland	5.03	Kolhorn	11.0	10.3	7.6	12.2	7.1
	5.07	Kwadijk	7.7	9.5	6.5	8.3	5.5
	5.29	Lisserbroek	7.8	6.2	7.6	9.1	-
Utrecht	6.17	Biddinghuizen	3.0	5.5	3.1	9.9	5.6
	6.19	Bilthoven	-	-	3.6	4.9	3.8
	6.20	Cabauw	4.6	5.7	5.1	4.6	3.1
	6.26	Wageningen	2.1	4.0	3.8	3.6	2.9
Gelderland	7.03	Barneveld	3.6	3.7	1.2	5.6	3.6
	7.13	Doetinchem	3.1	3.6	2.9	4.4	3.5
	7.15	Puiflijk	2.3	5.0	4.8	5.7	3.4
	7.17	Winterswijk	2.2	3.4	2.6	7.5	3.9
Overijssel	8.01	Kuinre	3.6	4.6	5.0	12.5	5.1
	8.04	Dalfsen	3.5	5.6	2.4	6.2	5.2
	8.11	Delden	1.9	3.9	3.5	3.9	4.4
Noord-Nederland	9.04	Cornjum	4.0	7.7	6.2	11.4	6.0
	9.06	Noordwolde	4.7	6.6	4.5	3.6	4.2
	9.15	Donkerbroek	3.7	7.2	5.2	14.0	6.8
	9.21	Sellingen	3.0	7.4	6.8	4.7	6.8
	9.22	Nieuw-Ballinge	2.8	7.2	3.0	4.7	4.1
average of all measurement stations			5.8	6.4	5.3	6.6	4.4 %

Table 1. Average percentage of leaf damage by O_3 on tobacco cv. Bel W_3 at selected measurement stations of the national monitoring system in The Netherlands, 1976 - 1980.

with frequency distributions. The frequency distribution gives the relative (percentage) number of observations per concentration class. One can determine the fraction of the number of observations for which the concentration in a given year was lower than a given concentration. Assume that the 98-percentile for a measurement station over a year is 220 $\mu g/m^3$. This then means that for 98% of the total number of observations the concentration is lower than 220 $\mu g/m^3$. In most cases one uses a higher percentile (e.g. 98% of 95%) and the 50-percentile to characterise the whole distribution.

Another possibility to document the variability of the pollution concentrations with time is to study the number of exceedences of a certain value. One can show the number of stations where the daily averaged concentration is higher than the calculated 98-percentiel. Such exceedence diagrams clearly show the episodes with high pollution levels during a year.

3.1. Biological effect determinations in the Dutch national air pollution monitoring system

In the framework of the national monitoring system the Institute for Plant Research (IPO) in Wageningen performs the biological effects measurements on indicator plants at 40 measurement stations of the network. The results of these observations can be used to study the trend in the effects in relation to the measured concentrations during consequetive years. As an example the results of the damage by O_3 on the indicator plant tobacco cv. Bel W_3 during the years 1976 to 1980 are presented in Table 1. The worst damage was found in the provinces Zeeland, Zuid-Holland and Noord-Holland. On several occasions between April and November high O_3 concentrations were correlated with intensive leaf damage to tobacco. A maximum period occurred in 1979 in the end of June when damage was found over the whole of the country with up to 40% leaf damage. In general the percentages of damage are higher in the coastal area as compared with more inland locations.

4. AIR POLLUTION MEASUREMENT IN SPACE

A major influencing aspect of the whole field of ambient air pollution research is the transport of pollution that takes place. It is known from earlier experiments, among others with tower measurements and discontinuous measurements using aerial platforms, that the vertical distribution of air pollutants can be highly irregular due to the rapid changes in the meteorological conditions in the boundary layer. In recent years several series of measurements have been carried out over The Netherlands using a specially equipped airplane. Examples of the aerial determination of pollutant distribution and transport are discussed below.

In the experimental programme a twin engine Piper Navajo Chieftain has been used to determine the concentration variability and distribution of air pollutants (NO, NO_2, SO_2, O_3), both vertically and horizontally, along flight tracks over The Netherlands and the North Sea between The Netherlands, Belgium, France and the United Kingdom. The following instruments were used:

- for the measurement of the NO/NO_x concentrations a Bendix 8101 special, and a Teco 14D analyser
- the O_3 was measured with a Bendix 8002 analyser and
- the SO_2 concentrations by a Teco 43 pulsed fluorescent analyser
- air temperature was measured with a Rosemount total temperature sensor model 102BE
- the data system in the airplane consisted of a Monitor Labs 9400 datalogger, a King Radio Corporation KDC 380 computer and a Perex 6300 cartridge recorder

As an example the vertical distribution of the pollutants over land and sea is described using the results of two measurements flights.

Flight September 14, 1977

In the middle of September the synoptic weather situation was relatively stable. A high pressure system was located over the British isles with its almost stationary centre over Ireland, and an intensive low pressure system was found over Scandinavia and northern USSR. The measurement flights were carried out from 10 - 12 a.m. at 3 heights between the checkpoints H5A en H20 (see Figure 1). Under conditions with light wind air masses were transported in the south-west to westerly air stream from the North Sea over two major industrial areas in the western part of The Netherlands to the cross section covered by the airplane measurements. The results of the NO_x measurements (Figures 2 and 3) clearly indicated the major air pollution sources. The large industrial area of Rijnmond is centered around flight distance 60 - 70 km and 3 major single sources are shown at flight distances of approximately 18, 36 en 103 km. The sources indicated at 18 and 103 km are high level sources reaching into the measurement flight level of 450 m. The plume of the high level source at approximately 36 km reaches into the level of 750 m.

The O_3 results (Figure 4) demonstrate the removal of the O_3 due to the reaction with NO resulting in elevated NO_2 levels. The air mass that originally had an O_3 content over the North

Figure 1. Flight track, September 14

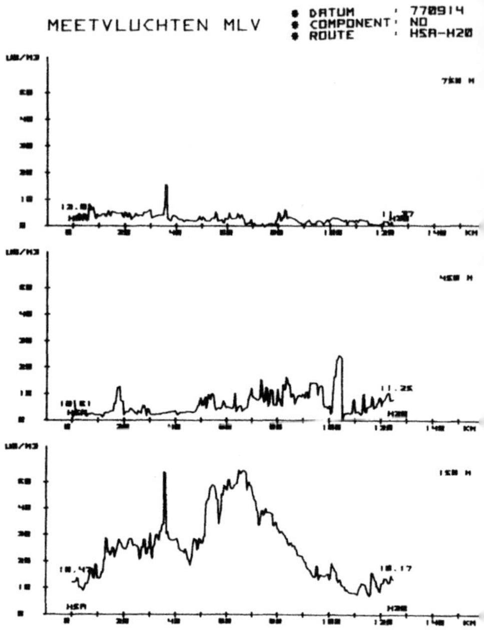

Figure 2. Flight, September 14, 1977 - NO concentration (µg/m^3)

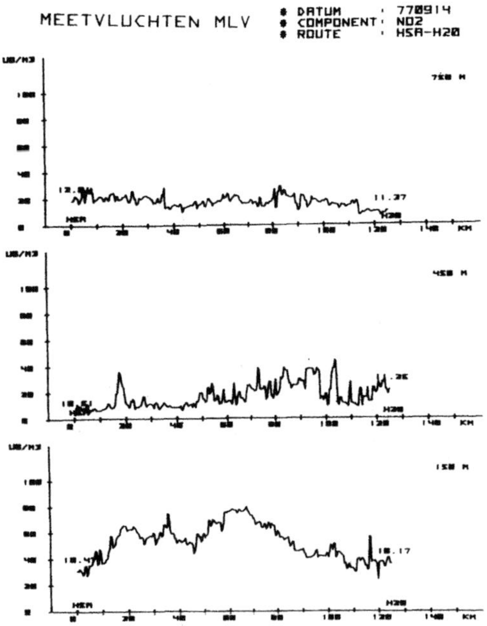

Figure 3. Flight, September 14, NO_2 concentration (µg/m^3)

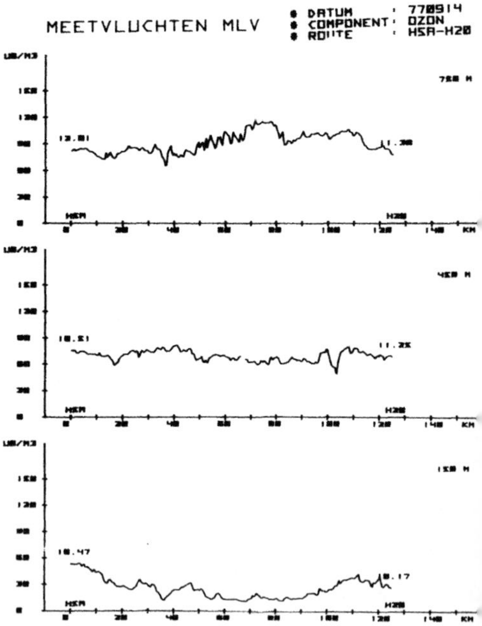

Figure 4. Flight, September 14, O_3 concentration (µg/m^3)

Sea of approximately 100 μg/m^3, looses the O_3 over land in a direct relation with the amount of NO available at the separate heights.

Flight July 26, 1979

To the north of a high pressure area located between the south of England and south-eastern Europe (Balkan area), measurement flights were carried out on the track from Knokke (Belgium) due north for 180 km. Backtrajectories (850 mb) for 24 hours for the west coast of The Netherlands are shown in Figure 5. The air masses moved over the flight track with a westerly wind of 4 - 5 m/s. The air had passed over the major source areas in the United Kingdom. Flight levels on the 26th of July were 150, 300, 600 and 1200 m.

At the higher levels the NO/NO_2 concentrations are as expected, very low (maximum 5 μg/m^3). The SO_2 concentrations varied between 25 and 50 μg/m^3 (see Figures 6 and 7 for SO_2 and NO_2). At the intermediate height of 300 m the concentration of SO_2 is 50 - 100 μg/m^3 and at the lowest level 50 - 75 μg/m^3, with a large plume producing maximum values of more than 200 μg/m^3. This plume is also evident in the NO_2 concentrations (up to 40 μg/m^3). The plume is shown in a negative sense in the O_3 concentrations at 150 m (see Figure 8). Near the Belgium coast the O_3 concentration range is 100 - 200 μg/m^3. Over the middle of the North Sea the concentration is approximately constant at 100 μg/m^3 at all levels except for the 150 m level where values up to 150 μg/m^3 are found.

Using the results of the four flight levels we can construct a total plume profile by interpolation. For interpolation a "spline" function is used. The resulting plume profile is a two-dimensional matrix with isolines. The total gasburden, the total amount of pollution in a vertical column over a surface area, is defined as the partial gasburden between the lowest and highest flight level. This gasburden in mg/m^2 follows from the plume profile matrix by integration over every vertical column of measurement values. The total flux of a pollutant over the flight track is given by multiplication of the gasburden with the windvector normal to the cross section (for the calculation the "ground wind speed" at the coast is used).

Calculated average fluxes for the period 11.30 a.m. to 2.30 p.m. on the 26th of July are:

SO_2 - 500 ton/h
NO_2 - 18 ton/h

Large variations in the calculated fluxes from day to day are found, primarily due to variations in the mean wind speed. On the 25th of July the mean wind direction was south-south west with fluxes of 250 ton/h for SO_2 and approximately 10 ton/h for NO_2.

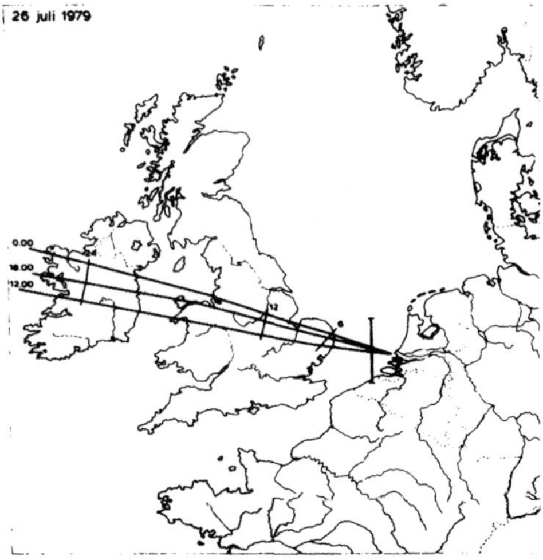

Figure 5. Flight track and backtrajectories, July 26, 1979

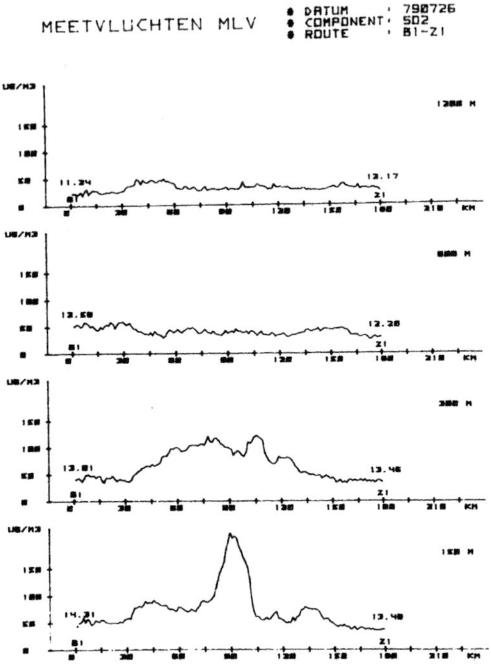

Figure 6. - SO_2 concentration ($\mu g/m^3$)

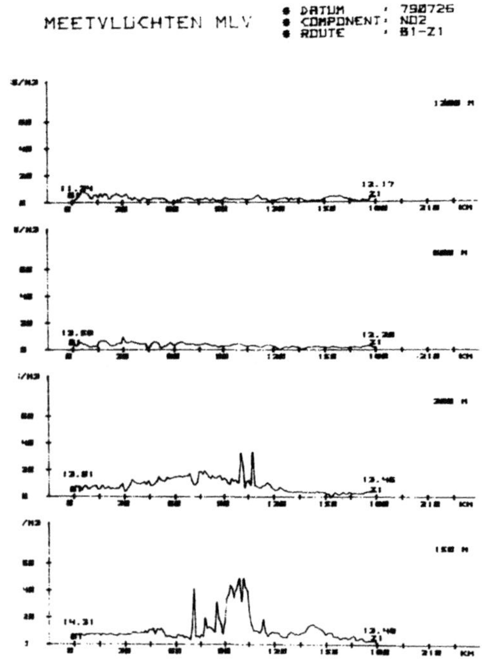

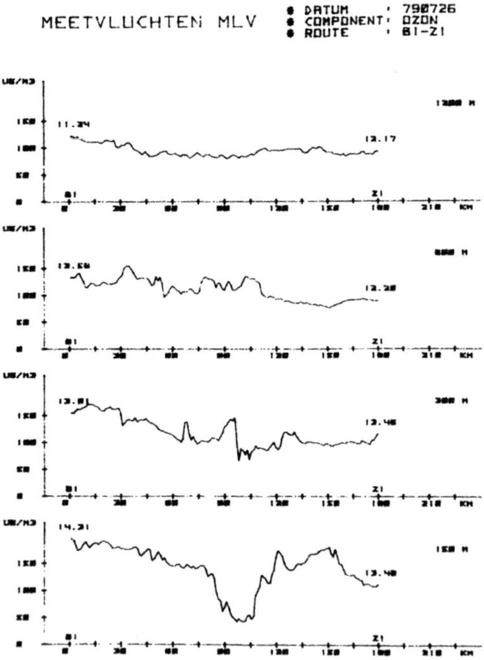

Figure 7. Flight, July 26, 1979 NO_2 concentration ($\mu g/m^3$)

Figure 8. Flight, July 26, 1979 O_3 concentration ($\mu g/m^3$)

As is shown with the results of these vertical distribution measurements it is difficult to study transport of air pollution using ground level stations only. Such a study is only possible using the results of a large monitoring network in combination with mobile measurements.

Transport studies over sea can only be carried out using an aerial platform. The results of one of the measurement flights is presented below.

Flight February 27, 1981

During a period with a high pressure system located over eastern Europe and a low pressure system above the Mediterranean, a steady easterly wind moved air masses from the continent towards the United Kingdom. On the 27th of February the transport took place with an average windspeed over the North Sea of approximately 8 m/s (winddirection was 110 - 115° at the Dutch coast).

The flight tracks are shown in Figure 9. With this configuration of tracks individual plumes of large sources could be followed over the North Sea. At the flight tracks 1 and 2 respectively RC en CB, between 9 and 10 a.m. local time an inversion layer was indicated at a height of 260 m. Over land, flight tracks 3 and 4 (BRWDT), the inversion layer was found at a height of 300 - 500 m (at 1 - 2 p.m.). Aboven this inversion layer O_3 concentrations were approximately 90 µg/m^3, air temperature 4 - 5°C with a few µg's/m^3 of NO_x and SO_2.

Over the middle of the North Sea inversion layers were later in the day (3.30 p.m.) found at 425 m and 900 m (see Figure 10). The maximum O_3 concentrations were 100 - 110 µg/m^3 at the top of the lowest inversion layer. Below this inversion layer at a height of 200 m the pollution concentrations found were; SO_2: maximum 110 µg/m^3; NO_2: 110 - 115 µg/m^3; NO: maximum 25 µg/m^3 and O_3: 50 µg/m^3.

Due to the steady mean windspeed and winddirection it was possible to study the variation in pollutant concentration with distance from the source, both for a single source (power plant stack) and a source region (Rijnmond/Rotterdam area). Results are compared for the flight tracks 09 (D-KN), 10 (KN-NS), 12 (NS-KN) and 07 (BS-NS) located at respectively 7, 60, 60 and 100 km from the sources.

In Table 2 for comparison purposes also the results of a measurement near to the source (approximately 1 km) on flight track 13 are shown.

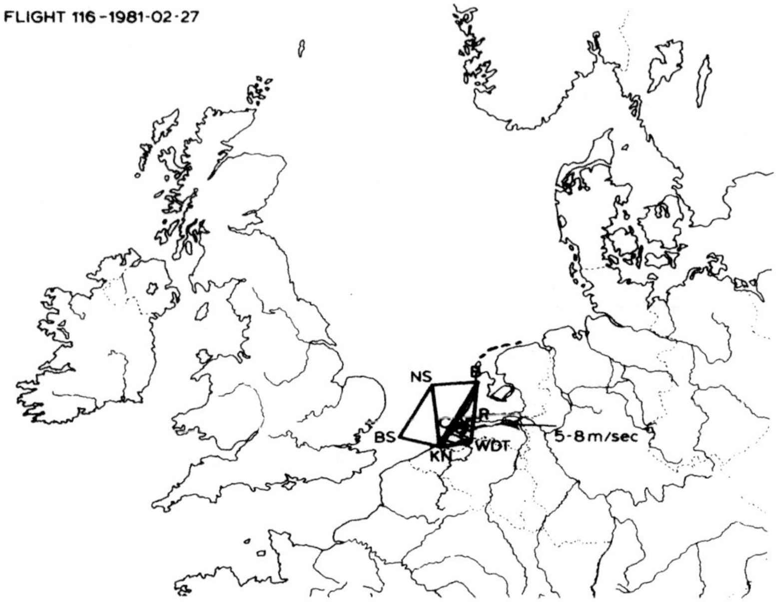

Figure 9. Flight tracks, February 27, 1981

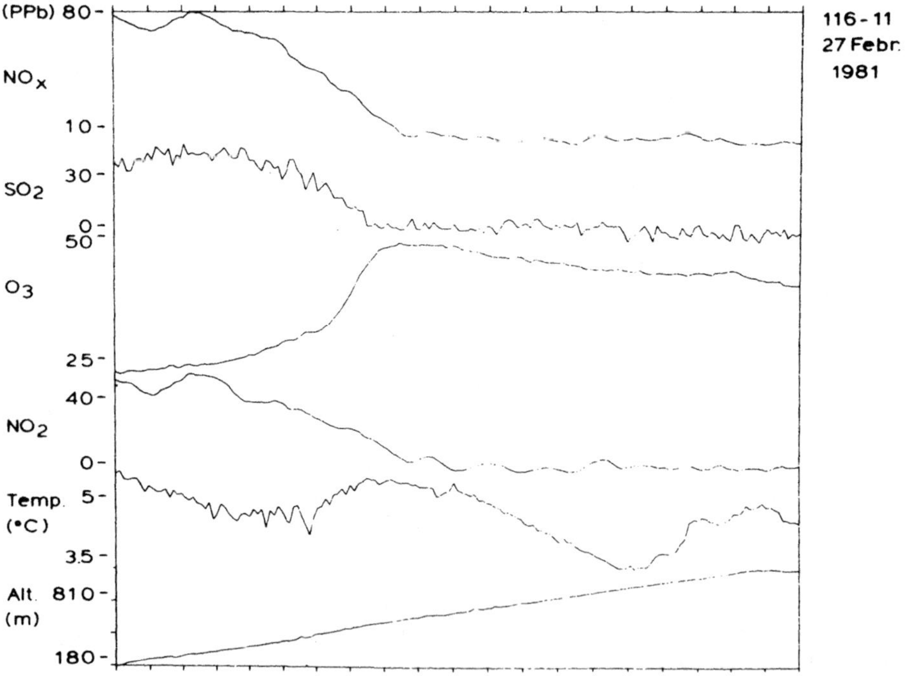

Figure 10. Flight, February 27, 1981. Pollutant concentrations measured during an ascend above the North Sea

flight track	13	09	10	12	07
distance from the source (km)	± 1	7	60	60	100
height (m)	335	330	390	350	410
NO_2	132	113	115	114	100
NO	200	35	33	22	32
SO_2	1010	430	310	306	210
O_3	25	36	44	45	40

Table 2. Air pollutant concentrations (in $\mu g/m^3$) within the plume of a power plant at different distances from the source .

Several features can be noted from these results:

- There is a marked decrease of SO_2 concentration with distance from the source. Assuming a steady mean windspeed of 8 m/s the removal rate of SO_2 is of the order of 20%/hour within the plume.
- For the nitrogen oxides on the other hand the concentration changes relatively slow; for the NO_2 concentrations within the plume the variation lies within the uncertainty in the measurements. Also the O_3 concentrations remain approximately constant, except for the flight tracks nearest to the source (09 and 13) where exess NO is available.
- The rapid removal of SO_2 from the plume of the single source is not consistent with the variation of SO_2 concentrations within the plume of a large industrial/urban source area (Rijnmond/Rotterdam area, see Table 3).
- Both in the "Rijnmond/Rotterdam" plume as well as in the region outside the main plumes (see Table 4), there is only a small reduction in the SO_2 and NO_x concentrations with growing distance from the source area. Also the O_3 concentrations show minor variations between the different tracks. (There is of course a clear reduction in O_3 in the lower layer as a whole due to dry deposition as well as to transformation where NO is sufficiently available.)
- O_3 reduction and NO_2 production is evident in the plume near the source (see Table 2, track 13). The maxumum "background" concentration over the sea of O_3 is 110 $\mu g/m^3$ at a height of approximately 1000 m.

flight track	09	10	12	07
distance from the source (km)	7	60	60	100
height (m)	335	330	390	350
NO_2	85	86	80	73
NO	24	27	19	24
SO_2	150	130	125	140
O_3	53	55	56	50

Table 3. Air pollutant concentrations (in $\mu g/m^3$) within the plume of the Rijnmond/Rotterdam area at different distances from the source area .

flight track	09	10	12	07
distance from the source (km)	7	60	60	100
height (m)	335	330	390	350
NO_2	56	56	68	56
NO	10	12	6	12
SO_2	110	92	105	100
O_3	65	67	70	60

Table 4. Average air pollutant concentrations (in $\mu g/m^3$) outside the main plumes over the North Sea.

5. OVERALL MEASUREMENT STRATEGY

On the basis of the results mentioned on the foregoing chapters one can conclude that there can be no unique conclusions drawn based on only single point measurements of pollutant components or parameters. For a better understanding and a description of thephenomena that occur a network with both fixed and mobile stations has to be used. For a more detailed investigation on the use of a monitoring network (fixed and mobile stations) in combination with a two-dimensional advection model reference should be made to: Van Egmond et al. (1978), Quantative Evaluation of Mesoscale Air Pollution Transport, Atmospheric Environment vol. 12, 2279 - 2287.

To answer most of the questions that are important in air pollution management strategies one has to deal with pollution fields. Using the data from a single station or a small number of monitoring stations one is restricted by the degree of representativeness of the individual station. With a network the possibility exists to use an interpolation scheme and describe the whole pollution field for the geographical area involved. A description of an optimum interpolation of spacial air pollution distributions is given by Van Egmond and Tissing in the proceedings of the 4th International Clean Air Congress. For individual components the following accuracies of the values are found (relative interpolation errors for hourly concentrations): SO_2 and NO_2 $\pm$ 20%; O_3 <25%. For a good interpretation inside cities and industrial areas, however, also the actual measured values should be taken into account.

6. MEASUREMENTS AND MODEL CALCULATIONS

To improve the results obtained with transport models we need to supply the modelers with better meteorological and air pollution data. In particular with specific measurements that are needed to verify model calculations or to improve details of critical modules within the transport or transformation model. For the meteorological data, wind fields from a lower level than the usual 850 mb level would be more appropriate for large-scale transport estimates. For Europe improved advection winds will become available as an increasing number of countries are expected to report upper-air data at the 925 mb level on a routine basis. Studies are being carried out where the models are not applied in the usual way to calculate concentrations and depositions, but are reversed so that emission and removal parameters are calculated from the observed chemical and meteorological data. Using three dimensional observation networks (including aircraft and other mobile platforms), it is possible to improve important parameters such as the vertical distribution of pollutants with height at different distances form source areas and the removal rates under different meteorological conditions. A practical application is given using the results of a number of flights carried out to look into the O_3 formation and removal problem.

O_3 removal

The problem of trying to deduce from ground level measurements the transport phenomena in the atmosphere is even more difficult when pollutatns are considered that can and will undergo chemical transformations on a large scale. This is also the case with secondary pollutants such as O_3. The O_3 concentrations encountered at ground level are the result of a number of processes. The major ones are: the intrusion of stratospheric O_3 into the troposphere, the O_3 production and destruction in the troposphere, and, finally the O_3 destruction at the earth surface. Although the stratosphere is clearly indicated as a potential source, and high O_3 values are occasionally found at remote stations, there is still a continuing discussion on the importance of the natural stratospheric O_3 source in comparison with the O_3 formation due to man's activities. Also other man-made sources than the well-known NO_2/hydrocarbon reactions are identified involving also the oxydation of CO. To indicate that considerable transport of O_3 over sea is taking place and that the destruction of O_3 at the earth surface can be extremely rapid in Western Europe, the results of a measurement flight on the 13th of July, 1977, are presented here.

Flight July 13, 1977

On this day the synoptic weather situation had already remained stationary during a number of days with a rather flat pressure distribution and a north-westerly wind of 3 - 6 m/s. The air masses reaching the coast-line of The Netherlands had moved rather undisturbed over the sea, additional input from pollutants due to industrial or other man-made sources had not taken place. The measurement flight on this day was carried out to the south of the Rijnmond-industrial and the Rotterdam-city area. Four flight levels were chosen from 150 - 1050 m above the flight track H21-GTR (Figure 11).

The results of the NO_x measurement (Figures 12 and 13) demonstrate the scarceness of sources to the north-west of the first 10 - 20 km of the flight track. From 40 - 70 km, the industrial area and the neighbouring town of Rotterdam are outlined in the concentration profiles. The meteorological conditions favoured a good mixing of the pollutants in the lower layers of the cross section. The large individual sources fingerprint their existence up to a level of 750 m. The NO/NO_2 concentrations found at the highest flight level are low and indicate that the pollution from the sources in this region of The Netherlands did not reach a height of 1000 m during the measurement period.

The O_3 measurement results for this morning flight are presented in Figure 14. A pronounced concentration profile is shown. At the lower flight levels O_3 concentrations were measured round 100 $\mu g/m^3$. At the intermediate height of 750 m concentrations betwee

Figure 11. Flight track, July 13,

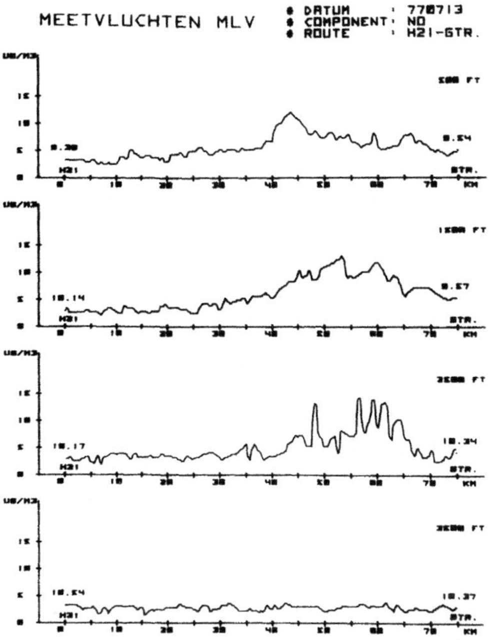

Figure 12. Flight, July 13, 1977 - NO concentration ($\mu g/m^3$)

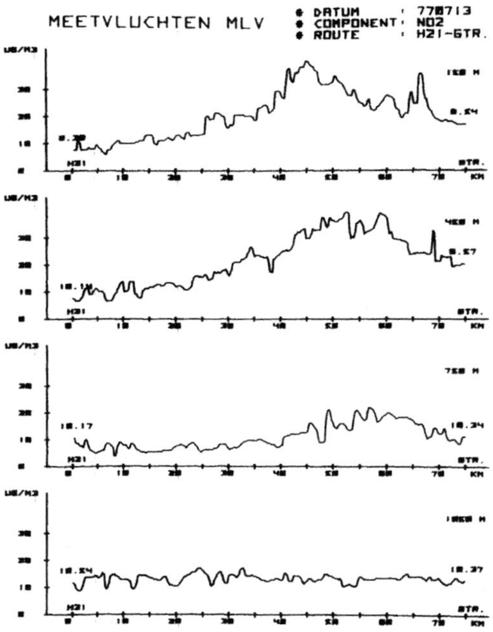

Figure 13. Flight, July 13, 1977 - NO_2 concentration ($\mu g/m^3$)

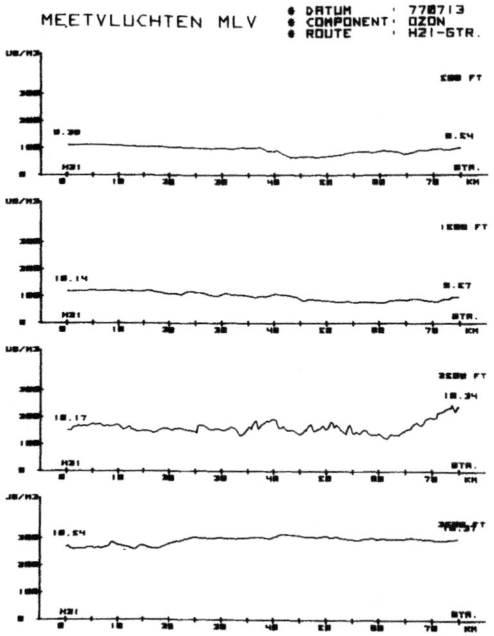

Figure 14. Flight, July 13, 1977 - O_3 concentration ($\mu g/m^3$)

150 - 200 μg/m^3 are found, while at the highest flight levels concentrations up to 300 μg/m^3 do seem to indicate the import into The Netherlands in this period of air masses with a very high concentration of O_3. This O_3 content is substantially reduced in the lower layers due to the occurence of a large number of other pollutants from industrial and urban sources that react with the O_3. The following picture emerges: The air masses transported towards The Netherlands contain O_3 in a wide range of concentrations depending on the meteorological conditions (season, level of radiation, synoptic scale weather situation etc.), and on the transport direction (over land or over sea). The air masses transported towards The Netherlands after passage over land (east or south directions) contain less O_3 than the air masses arriving over sea. Especially when the air masses are coming in from northerly or north-westerly direction over the North Sea, they can contain a large amount of O_3 as is shown by the results of the above presented flight.

This general picture is consistent with the results obtained from plant indicators in the national monitoring system. The average damage over the growing season on tobacco plant leaves due to O_3 is the highest in the whole coastal region of The Netherlands and diminishes towards the east and south-east. The other possible explanation, namely transport of O_3 precursors and subsequent O_3 formation near the receptor areas as a result of local meteorology, is inconsistent both with the pollution profiles measured over sea as well as with the relative high nocturnal O_3 concentrations occurring both over land as well as over the sea. A conclusive explanation of the high O_3 values over the North Sea cannot as yet be given. A possible explanation could be the formation of O_3 due to the offshore activities at or near the oil and natural gas drilling platforms.

This does not mean, of course, that no O_3 formation takes place over land. The photochemical production over land certainly takes place under favourable meteorological conditions in summer and autumn periods. Hourly averaged ground level concentrations of 300 μg/m^3 ore more in several areas of The Netherlands have been found. Also O_3 formation in large plumes and a semi-stagnant air mass has been observed as is demonstrated with the results of the following flights.

Flight, April 16, 1981

The general weather circulation was determined by a high pressure system located over the northern part of the North Sea. Dry air was transported from Sweden over Germany into The Netherlands and over the sea. Winddirection was east to north-east with windspeeds over the North Sea ranging from 5 - 8 m/s.

From 10.30 a.m. to 7.00 p.m. an extensive pattern of flight tracks was covered (Figure 15), including a number of ascends and descends near the coast of The Netherlands and the United Kingdom.

The profile over the North Sea (GY) indicated an inversion layer (approximately 6°C) between 525 en 575 m. Low pollutant levels were found for SO_2 and NO_2: 0 - 10 μg/m^3 above and below the inversion layer. O_3 concentrations was approximately 80 μg/m^3 below the inversion layer and 95 μg/m^3 above, with a maximum of 110 μg/m^3. The distribution of concentrations of pollutants below and aboven the inversion layer in the morning and late afternoon hours are shown with measured profiles near the south-western coast of The Netherlands (Figure 16). Clear profiles are determined in the morning, in the afternoon the distribution below the inversion layer is nearly constant with height with a sharp decrease above the inversion layer for SO_2 and NO_2. The O_3 values above the inversion layers are nearly constant over the whole area (for the coast of France (CA) the O_3 concentrations were of the same order of magnitude (minimum 75 μg/m^3, maximum 115 μg/m^3 with a mean value of 100 μg/m^3)).

Below the secondary inversion layer, however, O_3 values were slightly higher due to O_3 formation. Clear indication of such O_3 formation was also found on the flight track between Rotterdam and Calais near the coast of Dunkerque and over the Westerschelde and the south-west area from the Rijnmonddistrict. Maximum values in the urban/industrial plume near Dunkerque were 120 μg/m^3 O_3 measured at a height of 370 m. In the same plume the SO_2 concentration was 240 μg/m^3 and the NO_2 concentration 80 μg/m^3.

Flight January 30, 1981

During a number of days towards the end of January the general pressure distribution showed a high pressure system over eastern and middle Europe that was responsible for quiet weather with morning fog, relative few sunshine hours over land and a very light (1 - 3 m/s) southerly wind. The situation over the North Sea area was favourable for the build-up of high air pollutant concentrations in semi-stagnant air masses.

A flight was carried out at two levels from Rotterdam via Clacton on Sea and Dover (UK) to northern France and Belgium (Antwerp) (see Figure 17).

Measurements were made partly above and below the inversion layer located at approximately 500 m. Conditions above this layer over the sea show virtually no SO_2 or NO_x; 90 - 100 μg/m^3 O_3 with temperatures of 10 - 11°C. Also over land comparable results are found above the inversion layer as is shown in

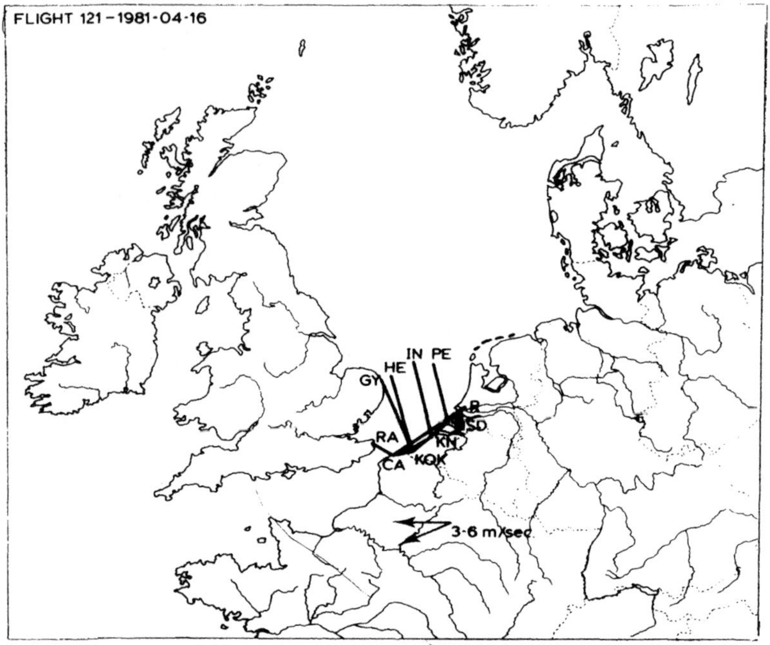

Figure 15. Flight tracks, April 16, 1981

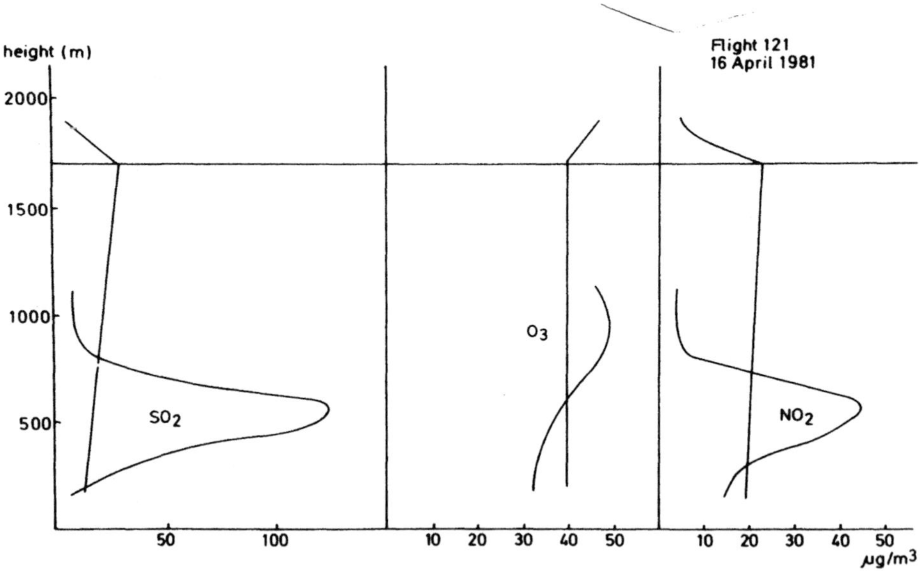

Figure 16. Morning and afternoon profiles, below and above the inversion height (inversion height a.m. at 600 m, p.m. at 1700 m)

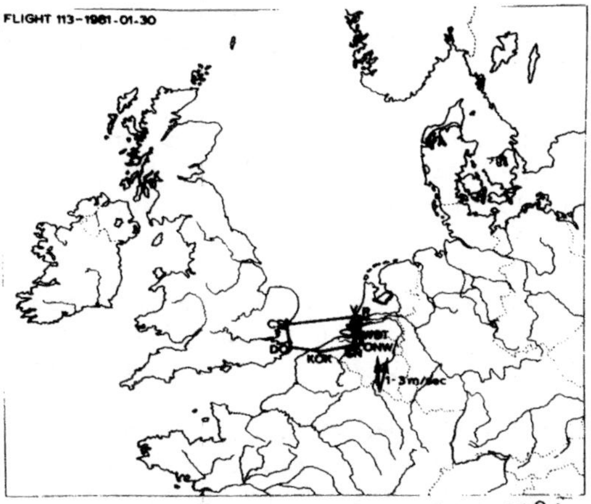

Figure 17. Flight tracks, January 30, 1981

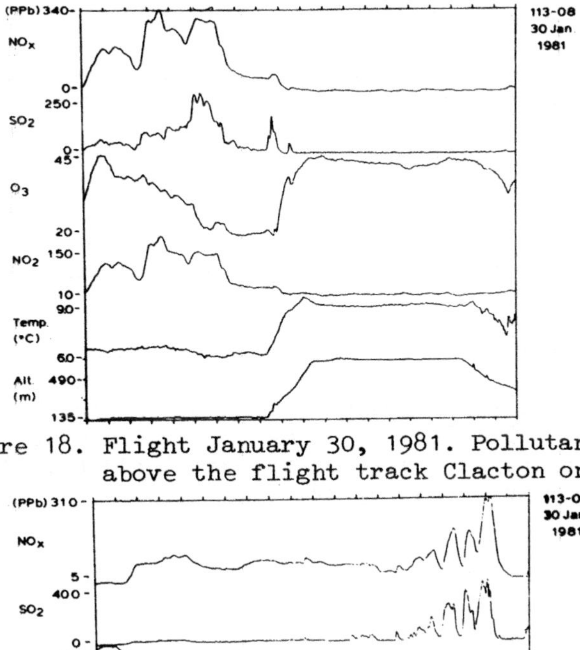

Figure 18. Flight January 30, 1981. Pollutant concentrations above the flight track Clacton on Sea - Dover (UK)

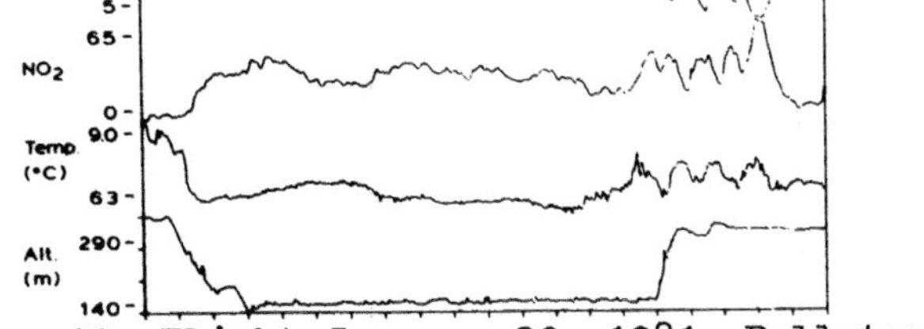

Figure 19. Flight January 30, 1981. Pollutant concentrations above the flight track Dover (UK) - KOK (Belgian coast)

Figure 18 in the second part of the flight track Clacton-Dover at a height of 660 m.

In the lower more polluted layers over the sea under the inversion layer high concentrations are found in plumes at a height of 150 - 200 m. SO_2: 700 - 1200 µg/m^3; NO_2: 200 - 400 µg/m^3.

As a result of the scavenging of excess NO (approximately 50 µg/m^3), the O_3 concentrations within the plumes decreased to values ranging from 25 to 35 µg/m^3 in the lower layers.

That the region over the North Sea with relative high pollution values is covering a large area is shown with the results of the flight track Dover-KOK. At a height of 165 m over the sea the values are: SO_2: 100 - 150 µg/m^3; NO_2: 60 - 100 µg/m^3; NO: 50 - 90 µg/m^3; O_3: 30 - 40 µg/m^3 (Figure 19).

The pollution levels do not show plumes of individual sources in contrast to the values measured near the coast of France (Dunkerque) where plumes with very high concentrations are found at a height of 325 m with maxima for SO_2: 1120 - 1660 µg/m^3; NO_2: 120 - 180 µg/m^3 and an O_3 minimum of 10 - 15 µg/m^3.

Comparing the NO_2 versus SO_2 values found over land and over the sea during the different flight tracks above and below the inversion layer, it can be concluded that over land the NO_2 concentrations in plumes are significantly higher (in northern Belgium around the Antwerp region the maximum values at a height of 325 m were 200 - 250 µg/m^3), probably due to the fact that enough NO is available for reaction with O_3 to form NO_2. O_3 reduction over land, from 80 - 90 µg/m^3 to a minimum of 10 µg/m^3, is therefore larger than over the sea (reduction from 90 - 100 µg/m^3 to a minimum of 35 µg/m^3).

7. CONCLUDING REMARKS

In conclusion we can specify the following subjects to be considered in relation to the measurement of Air Pollution in Time and Space:

- the dynamic aspect of air pollution transport
- the location of monitoring stations considering both local and remote sources
- the importance of a moving measurement system
- the validity of the measurements in point and area integrated or interpolated values
- the measurements as input for transport model calculations
- the measurements to observe the transformation in and the removal out of the transported air masses
- the measurements as tool to study the pollutant cycle or pathway from source to receptor

8. ACKNOWLEDGEMENT

The cooperation of the Environmental Research Department of the KEMA, Arnhem, is gratefully acknowledged. The efforts of the field team, J.W.Viljeer and L.van de Beld, that performed the aerial measurements in a careful as well as in an efficient way are very much appreciated.

AUTHORS

BERRESHEIM,HARALD
Z.f.Umweltschutz
Robert-Mayer-Straße 11

6000 FRANKFURT 1

CALVERT,PROF.DR.J.G.
Ohio State Univ.
140 West 18th Avenue

USA COLUMBUS,OHIO 43210

EHHALT,PROF.DR.D.H.
KFA Jülich Inst.3
Postfach 1913

5170 JÜLICH 1

FIEDLER,PROF.DR.F.
Physik Hochhaus
Kaiserstraße 12

7500 KARLSRUHE 1

GEORGII, PROF.DR.H.W.
Inst.f.Met.u.Geoph.
Feldbergstraße 47

6000 FRANKFURT 1

GIDEL,PROF.DR.LOIS T.
School of M.a.A.Sc.
4600 Rickenbacker Causew

USA MIAMI,FLORIDA 33149

HELAS, DR.GÜNTER
MPI für Chemie
Postfach 3060

6500 MAINZ

JAENICKE,PROF.DR.R.
Inst.f.Meteorologie
Postfach 3980

6500 MAINZ

JAESCHKE, DR.WOLFGANG
Ref.f.Umweltschutz
Robert-Mayer-Straße 11

6000 FRANKFURT 1

KLOCKOW,PROF.DR.DIETER
Inst.f.Chemie
Postfach 50 0500

4600 DORTMUND 50

KOHLMAIER,PROF.DR.GUNDOLF
Inst.f.phys.Chemie
Gräfstraße 39

6000 FRANKFURT 1

MOHNEN,PROF.DR.V.A.
State Univ.of N.Y.
1400 Washington Avenue

USA ALBANY,N.Y. 12222

PERNER,DR.DIETER
KFA Jülich
Postfach 1913

5170 JÜLICH 1

SCHNEIDER,DR.I.T.
RIV
Postbus 1

NL BILTHOVEN

SCHÖNWIESE,PROF.DR.
Inst.f.Meteor.
Feldbergstraße 47

6000 FRANKFURT 1

STEIN,PROF.DR.NORBERT
Uni d.Saarlandes
Fachbereich 6 / Geographie

6600 SAARBRÜCKEN

WHELPDALE,DR.D.M.
Environment Canada
4905 Dufferin Street

CDN DOWNSVIEW,ONTARIO M3H5

PARTICIPANTS

ARNOLD,DR.IGNAZ
Inst.f.Met.u.Geoph.
Feldbergstraße 47

6000 FRANKFURT 1

BALLSCHMITER,PROF.DR. K.
Inst.f.Analy.Chemie
Oberer Eselsberg

7900 ULM / DONAU

BARNES,DR.RON
Esso Res.Cent.
Abingdon

GB OXFORDSHIRE

BELTZ,NORBERT
Z.f.Umweltschutz
Robert-Mayer-Straße 11

6000 FRANKFURT 1

BENARIE,PROF.DR.M.
IRCHA
12,Rue de L'Yveline

F 91220 BRETIGNY

BINGEMER,HEINZ
Inst.f.Meteorologie
Feldbergstraße 47

6000 FRANKFURT 1

BIRKS,PROF.DR.JOHN
MPI f.Chemie
Postfach 3060

6500 MAINZ

BRÜLL,DR.

Parkweg 32

NL 6212 MAASTRICHT

BÄCHMANN,PROF.DR.KNUT
TH Darmstadt
Hochschulstraße 4

6100 DARMSTADT 1

CHATFIELD,ROBERT
MPI f.Chemie
Postfach 3060

6500 MAINZ

CRUTZEN, PROF.DR.P.J.
MPI f.Chemie
Postfach 3060

6500 MAINZ

DICKERSON,RUSSEL R.
MPI f.Chemie
Postfach 3060

6500 MAINZ

DLUGI,DR.RALPH
KFZ Karlsruhe GmbH
Postfach 3640

7500 KARLSRUHE 1

FRICKE,DR.WOLFGANG
Ges.Verb.d.d.Steink.
Friedrichstraße 1

4300 ESSEN 1

FUZZI,DR.SANDRO
Lab.Fisbat CNR
Via de'Castagnoli 1

I 40126 BOLOGNA

GRIFFITH,DR.D.W.T.
MPI f.Chemie
Postfach 3060

6500 MAINZ

GÖLZENLEUCHTER,HORST
Z.f.Umweltschutz
Robert-Mayer-Straße 11

6000 FRANKFURT 1

GÜSTEN,DR.HANS
KFZ Karlsruhe GmbH
Postfach 3640

7500 KARLSRUHE 1

HAMRUD,MATS HUBERT
Dep.of Meteorol.
University of Stockholm

S 10691 STOCKHOLM

HANS,WILFRIED
Inst.f.phys.Chemie
Wegelerstraße 12

5300 BONN 1

HERRMANN,JÜRGEN
Z.f.Umweltschutz
Robert-Mayer-Straße 11

6000 FRANKFURT 1

HUSTERT,DR.KLAUS
Inst.f.Ök.Chemie
Camerloher Straße 7a

8050 FREISING 1

JERRE,JON
Grad.Stud.
Chr.Bennechesvei 13

N OSLO 2

KANDILIS, PHEDO
Uni.Athens/Clim.Dep.
33 Hippokratus Street

GR ATHENS 144

KANTER,H.J.
Fraunhofer-Inst.
Kreuzeckbahnstraße 19

8100 GARMISCH-PARTENKIRCH

KELM,PROF.DR.HARTWIG
Inst.f.phys.Chemie
Robert-Mayer-Straße 11

6000 FRANKFURT 1

KIELLAND,GABRIEL W.
Cand.Mag.
Kringsja 10 A 316

N OSLO 8

MATTHEWS,DR.W.A.
MPI f.Aeronomie
Postfach 20

3411 KATLENBURG-LINDAU 3

MAZUREK,MONICA A.
Dep.of E.a.S.Science
405 Hilgard Avenue

USA LOS ANGELES,CA 9002

MEGIE,DR.GERARD J.
CNRS
PB 3

F 91370 VERRIERES LE BUI

MIHELCIC,DR.D.
KFA Jülich
Postfach 1913

5170 JÜLICH

MOORTGAT,DR.GEERT K.
MPI für Chemie
Saarstraße 23

6500 MAINZ

MUGLER,DR.JOHN
NASA L.R.Center
Mail Stop 401-B

USA HAMPTON ,VA 23665

NEWIGER,MADELEINE
MPI f.Meteor.
Bundestraße 55

2000 HAMBURG 13

ONO,PROF.DR.AKIRA
WRI Nagoya-Univ.
Furo-cho,Chikusa-ku

J NAGOYA 464

OTTAR,DIR.DR.BRYNJULF
NI f.Air Res.
P.O.Box 130

N 2001 LILLESTROM

PELON,DR.JAQUES
Serv.d'Anom.du CNRS
PB 3

F 91370 VERRIERES LE BUI

PIEROTTI,DAVID JOSEPH
Scripps I.o.Ocean.
A-008 Univ.of California

USA LA JOLLA, CA 92093

REICHENAUER,HERTA

An der Allee 89

6500 MAINZ

ROEDEL,PROF.DR.WALTER
Uni Heidelberg
Im Neuenheimer Feld 366

D 6900 HEIDELBERG

SCHNEIDER,WOLFGANG
MPI f.Chemie
Postfach 3060

6500 MAINZ

SCHÄFER,DR.LUTZ U.
MPI f.Chemie
Saarstraße 23

6500 MAINZ 1

SEILER, PROF.DR.WOLFGANG
MPI für Chemie
Postfach 3060

6500 MAINZ

SHADE,JOHN STUART
STEP-ONE,R+A
204 Victoria Road

GB ABERDEEN SCOTLAND

SINGH,DIR.DR.HANWANT B.
L-2023,SRI-Int.
333 Ravenswood Ave.

USA MENLO PARK,CA 94025

SJÖDIN,AKE INGEMAR
Water Research Lab.
BOX 5207

S 40224 GOTHENBURG

STORDAL,FRODE
University of Oslo
BOX 1022,Blindern

N OSLO 3

SÖDERLUND,ROLF
Dep.of Meteor.
Univ.of Stockholm

S 10691 STOCKHOLM

TANGERMANN-DLUGI,GISELA
Dipl.Phys.
Postfach 6380

7500 KARLSRUHE 1

TEUBER,ROLF
MPI f.Chemie
Saarstraße 23

6500 MAINZ

VEYRET,BERNARD
CNRS Unite de Chimie
351 Cours de la Liberati

F 33405 TALENCE CEDEX

VIGGIANO,DR.ALBERT
MPI f.Kernphysik
Postfach 103980

6900 HEIDELBERG 1

WALTER,DR.EBERHARD
Z.f.Umweltschutz
Robert-Mayer-Straße 11

6000 FRANKFURT 1

WEBER,DR.ERICH
Regierungsdirektor
Weberstraße 29

5300 BONN 1

WINCHESTER JR.,LEONARD W
MT University
Kew.Res.Center

USA HOUGHTON,MICHIGAN 49

ZAMBAKAS,PROF.DR.J.D.
Univ.of Athens
33 Hippokrates-Street

GR ATHENS 144

ZEREFOS,PROF.DR.CHRISTOS
Phys.Dep.
Univ.of Thessaloniki

GR THESSALONIKI

ZETZSCH,DR.CORNELIUS
Phys.Chemie I d.TU
Postfach 102148

4630 BOCHUM 1

ZIMMERMAN, PATRICK ROBERT
NCAR
P.O.Box 3000

USA BOULDER,COLORADO 803

INDEX

GPSR Compliance

The European Union's (EU) General Product Safety Regulation (GPSR) is a set of rules that requires consumer products to be safe and our obligations to ensure this.

If you have any concerns about our products, you can contact us on ProductSafety@springernature.com

In case Publisher is established outside the EU, the EU authorized representative is:

Springer Nature Customer Service Center GmbH
Europaplatz 3
69115 Heidelberg, Germany

Batch number: 10755248

Printed by Printforce, the Netherlands